YACOV Y. HAIMES
Risk Modeling, Assessment, and Management

MW00560321

DENNIS M. BUEDE
The Engineering Design of Systems: Models and Methods

ANDREW P. SAGE and JAMES E. ARMSTRONG, Jr.
Introduction to Systems Engineering

WILLIAM B. ROUSE
Essential Challenges of Strategic Management

YEFIM FASSER and DONALD BRETTNER
Management for Quality in High-Technology Enterprises

THOMAS B. SHERIDAN
Humans and Automation: System Design and Research Issues

ALEXANDER KOSSIAKOFF and WILLIAM N. SWEET
Systems Engineering Principles and Practice

HAROLD R. BOOHER
Handbook of Human Systems Integration

JEFFREY T. POLLOCK AND RALPH HODGSON
**Adaptive Information: Improving Business Through Semantic
Interoperability, Grid Computing, and Enterprise Integration**

ALAN L. PORTER AND SCOTT W. CUNNINGHAM
Tech Mining: Exploiting New Technologies for Competitive Advantage

REX BROWN
Rational Choice and Judgment: Decision Analysis for the Decider

WILLIAM B. ROUSE AND KENNETH R. BOFF (editors)
Organizational Simulation

HOWARD EISNER
Managing Complex Systems: Thinking Outside the Box

STEVE BELL
Lean Enterprise Systems: Using IT for Continuous Improvement

J. JERRY KAUFMAN AND ROY WOODHEAD
**Stimulating Innovation in Products and Services: With Function Analysis and
Mapping**

WILLIAM B. ROUSE
Enterprise Tranformation: Understanding and Enabling Fundamental Change

JOHN E. GIBSON, WILLIAM T. SCHERER, AND WILLAM F. GIBSON
How to Do Systems Analysis

SYSTEM OF SYSTEMS
ENGINEERING

SYSTEM OF SYSTEMS ENGINEERING

INNOVATIONS FOR THE 21st CENTURY

Edited by
MO JAMSHIDI

WILEY

A John Wiley & Sons, Inc., Publication

Published by John Wiley & Sons, Inc., Hoboken, New Jersey
Published simultaneously in Canada

For general information on our other products and services or for technical support, please contact our Customer Care Department within the United States at (800) 762-2974, outside the United States at (317) 572-3993 or fax (317) 572-4002.

Wiley also publishes its books in a variety of electronic formats. Some content that appears in print may not be available in electronic formats. For more information about Wiley products, visit our web site at www.wiley.com.

Library of Congress Cataloging-in-Publication Data:

Systems of systems engineering : innovations for the 21st century / edited by Mo Jamshidi.
     p. cm. – (Wiley series in systems engineering and management)
  Includes bibliographical references and index.
  ISBN 978-0-470-19590-1 (cloth : alk. paper)
1.   Systems engineering–Technological innovations. 2.   Large scale systems.   I. Jamshidi, Mohammad.
  TA168.S8885 2009
  620.001'171–dc22

                         2008018996

Printed in the United States of America

10 9 8 7 6 5 4 3

Contents

Preface

In the twenty-first century, information science and technology continues to be critical benefactors of systems engineering that continue to redefine the design problem in industry, energy, defense, security, environment, and so on. Systems engineering is currently undergoing a major change to extend itself beyond a single system framework. Recently, there has been a growing interest in a class of complex systems whose constituents are themselves complex. These systems are sometimes called system of systems (SoS) or federation of systems (FoS). Performance optimization, robustness, and reliability among an emerging group of heterogeneous systems in order to realize a common goal have become the focus of various applications including military, security, aerospace, space, manufacturing, service industry, environmental systems, and disaster management, to name a few. There is an increasing interest in achieving synergy between these independent systems to achieve the desired overall system performance. Critical issues that deserve attention are coordination and interoperability in an SoS. SoS technology is believed to more effectively implement and analyze large, complex, independent, and *heterogeneous* systems working (or made to work) cooperatively. The main thrust behind the desire to view the systems as an SoS is to obtain higher capabilities and performance than would be possible with a traditional system view. The SoS concept presents a high-level viewpoint and explains the interactions between each of the independent systems. However, when it comes to engineering and engineering tools of SoS, we have a long way to go. This is the main goal of this volume. Here, we have put together 22 chapters, 8 on such fundamental issues as openness, engineering, architecture, modeling, simulation, net centricity (integration), emergence, technical evaluation, and management of SoS. In addition, a set of chapters indicative of the state of the art in current or potential applications of the technology of SoS such as defense, services, commercial airlines, transportation systems, health care, space exploration, space communication, global earth oberservation, robotics, infrastructures, electric power systems, microgrid systems, and environmental impacts are all included. Experts from all over the globe have been recruited to contribute to it. The structure of the book is as follows: Chapter 1 is a brief introduction, and Chapters 2–8 examine the fundamental issue of systems engineering as outlined from SoS point of view. Application areas are covered in Chapters 8–22.

This volume in the Wiley Series on System Engineering and Management would not have been possible without the diligent work and support of the contributing authors from industry, academia, The United States, Japan, the Netherlands, Canada, and so on. The editor thanks all of them for their contributions to SoS technology and to this volume. I wish to express my sincere appreciation and thanks to Professor Andrew P. Sage, Series Editor and an author of Chapter 3, for his encouragement to make this volume a reality. I wish to thank him, among 10 other mentors, to whom I have dedicated this volume—from my days at Oregon State University (1963–1967) to University of Illinois at Urbana-Champaign (1967–1971) to the formation of my professional career after I finished my systems and control education. Last, but by no means least, I wish to thank my dear wife, Jila Salari Jamshid, for her continuous love and support in all that I have undertaken in 34 years of companionship.

Mo Jamshidi

San Antonio, Texas, USA
May 10, 2008

About the Editor

Mo M. Jamshidi (Fellow IEEE, Fellow ASME, Associate Fellow AIAA, Fellow AAAS, Fellow TWAS, Fellow NYAS) received a B.S. in Electrical Engineering from Oregon State University in June 1967 and the M.S and the Ph.D. degree in Electrical Engineering from the University of Illinois at Urbana-Champaign in February 1971. He holds three honorary doctorate degrees from Azerbaijan National University, Baku, Azerbaijan, 1999, University of Waterloo, Canada and Technical University of Crete, Greece, both in 2004. Currently, he is Lutcher Brown Endowed Chaired Professor at the University of Texas, San Antonio, TX, USA. He is also the Regents Professor Emeritus of Electrical and Computer Engineering, the AT&T Professor of Manufacturing Engineering, and founding Director of the Center for Autonomous Control Engineering (ACE) at the University of New Mexico, Albuquerque, NM, USA. He has been a consultant and special government employee with the U.S. Department of Energy, NASA Headquarters and Jet Propulsion Laboratory, and the U.S. Air Force Research Laboratory for a combined 25-year period. He has worked in various academic and industrial positions at various national and international locations including with IBM and GM Corporations. In 1999, he was a NATO Distinguished Professor in Portugal conducting lectures on intelligent systems and control. He has over 600 technical publications including 63 books (12 textbooks) and edited volumes. Six of his books have been translated into at least one foreign language. He is the founding editor or cofounding editor or editor-in-chief of many journals (including Elsevier's *International Journal of Computers and Electrical Engineering Elsevier, UK, Intelligent Automation and Soft Computing, TSI Press, USA*) and one magazine (*IEEE Control Systems Magazine*). He is editor-in-chief of the new *IEEE Systems Journal* (inaugurated in 2007) and coeditor-in-chief of the *International Journal on Control and Automation*. He has been the General Chairman of the World Automation Congress (WAC, wacong.org) from its inception. He has been active within the IEEE for 42 years. Dr. Jamshidi is a Fellow of the IEEE for contributions to "large-scale systems theory and applications and engineering education," a Fellow of the ASME for contributions to "control of robotic and manufacturing systems," Fellow of the AAAS—the American Association for the Advancement of Science for contributions to "complex large-scale systems and their applications to controls and optimization," a Fellow of Academy of Developing Nations (Trieste, Italy), Member of the Russian Academy of Nonlinear Sciences, Associate Fellow,

Hungarian Academy of Engineering, a Fellow of the New York Academy of Sciences, and recipient of the IEEE Centennial Medal and IEEE Control Systems Society Distinguished Member Award and the IEEE CSS Millennium Award. In October 2005, he was awarded the IEEE SMC Society's Norbert Weiner Research Achievement Award and in October 2006, he received the IEEE SMC Society Outstanding Contribution Award. As an OSU Alumni, he was inducted into Oregon State University's Academy of Distinguished Engineers in February 2007. He is the founding Chair and Chair of the IEEE International Conference on System of Systems Engineering since 2006–2009.

Contributors

Cyrus Azani is a Senior Systems Engineer at Northrop Grumman Corporation and an Adjunct Professor at University of Maryland, MD, USA. His areas of research are system of systems engineering, architecture, and assessment; open architecture strategy, implementation and assessment; and multicriteria decision-making models and approaches.

Kul B. Bhasin leads the architecture development team for NASA's SCaN-Constellation Integration Project at NASA Glenn Research Center in Cleveland, OH, USA. He develops communication network architectures within a system of systems environment for the upcoming exploration missions of NASA.

John Boardman graduated with 1st Class Honors in Electrical Engineering from the University of Liverpool, from where he also obtained his PhD. He is currently a Distinguished Service Professor in the School of Systems and Enterprises at Stevens Institute of Technology. Before coming to Stevens Institute of Technology he held positions at the University of Portsmouth as the GEC Marconi Professor of Systems Engineering and Director of the School of Systems Engineering and later Dean of the College of Technology. He is a Chartered Engineer and Fellow of the Institute of Engineering and Technology and the International Council on Systems Engineering (INCOSE).

Rajendra V. Boppana is a Professor of Computer Science at the University of Texas at San Antonio, TX, USA. His research interests include wireless and sensor networks, secure routing and intrusion detection techniques, and autonomic computing and communications.

Suresh Chalasani is an Associate Professor and the Chair of the Business Department at the University of Wisconsin-Parkside, WI, USA. His research interests include supply chain management, health care management, and emerging technologies.

Robert J. Cloutier is a Research Associate Professor in the School of Systems and Enterprises at Stevens Institute of Technology Hoboken, NJ, USA. His research interests include model-based systems engineering and systems architecting, reference architectures, systems engineering patterns, and model-driven architecture. Rob has over 20 year's experience in systems engineering and architecting software engineering, and project management in both commercial and defense industries.

Cihan H. Dagli, Cihan Dagli is Professor of Systems Engineering and Engineering Management and also a Professor Computer and Electrical Engineering, Missouri University of Science & Technology, USA. Dr. Dagli is also the Intelligent Systems Design Area Editor for the *International Journal of General Systems* and the director of the Smart Engineering Systems Lab (SESL) at the Missouri S&T. He received B.S. and M.S. degrees in Industrial Engineering from the Middle East Technical University and a Ph.D. Applied Operations Research in Large Scale Systems Design and Operation from the University of Birmingham, UK.

Judith S. Dahmann, Ph.D., is a principal Senior Scientist in the MITRE Corporation Center for Acquisition and Systems. Prior to this, Dr. Dahmann was Chief Scientist for the Defense Modeling and Simulation Office for the U.S. DoD, where she led the development of the High-Level Architecture for simulations, now IEEE 1516. Dr. Dahmann holds a B.A. from Chatham with a year as a special student at Dartmouth College, an M.A. from The University of Chicago, and a Ph.D. from Johns Hopkins University.

Daniel A. DeLaurentis is an Assistant Professor of Aeronautics and Astronautics at Purdue University, joining the University in 2004 under the System of Systems Signature Area. His areas of interests are system of systems modeling and analysis methodologies and advanced design techniques applied to air and space transportation systems.

Michael J. DiMario, as a Senior Program Manager at Lockheed Martin, manages System of Systems Command and Control programs and is a Ph.D. candidate in the School of Systems and Enterprises at Stevens Institute of Technology. His research interests include system of systems and interoperability of complex systems. Michael has over 25 years of experience in managing and engineering systems and software programs.

Michael Duffy, Ph.D., is the Lead Systems Engineer for the U.S. Department of Energy Hydrogen Program at the National Renewable Energy Laboratory. He has over 35 years of systems engineering experience in energy, safeguards and security, nuclear waste management, national defense, transportation, and space programs.

Liping Fang is a Professor and Chair of Mechanical and Industrial Engineering at Ryerson University, Toronto, Canada. His research interests are systems engineering, industrial engineering, multiple participant-multiple objective decision making, and decision support systems.

Bobi Garrett is the Associate Director for Strategic Development and Analysis at the National Renewable Energy Laboratory. She has 29 years of technical leadership experience, focused on advancing new technologies in the energy, environmental, defense, and health care sectors.

Alex Gorod received a B.S. in Information Systems and a M.S. in Telecommunications from Pace University. Prior to his graduate studies he held a Research Analyst position at Salomon Smith Barney. He is currently a Robert Crooks Stanley Doctoral Fellow in Engineering Management at Stevens Institute of Technology, with research interests in the area of management of complex systems. He is also the Vice President of the Stevens Student Chapter of the International Council on Systems Engineering (INCOSE).

Jeffery L. Hayden is a Space Systems Engineer and Communication Architecture Consultant for NASA and the DoD. His areas of interest include space communication system of systems and network of networks architecture development, space communication network design tools and databases, spacecraft design, exploration mission concepts of operation, and scientific instrument design.

Paulien M. Herder is an Associate Professor and holds an M.Sc. degree in chemical engineering (1994) and a Ph.D. degree in systems engineering (1999), both from Delft University of Technology. She works at the faculty of Technology, Policy, and Management and is coleader of the "Flexible Infrastructures" subprogramme within the Next Generation Infrastructures (NGInfra) programme. Her research focuses on design of large-scale networked systems.

Keith W. Hipel is a University Professor of Systems Design Engineering at the University of Waterloo in Canada. His research interests are the development of conflict resolution, multiple objective decision making, and time series analysis techniques, with applications in water resources management, hydrology, environmental engineering, and sustainable development.

Ian A. Hiskens is a Professor of Electrical and Computer Engineering at the University of Wisconsin-Madison WI, USA. His major research interests lie in the area of power system analysis, in particular system dynamics, security, and numerical techniques. Other research interests include nonlinear and hybrid dynamical systems, and control.

Mo Jamshidi is the Lutcher Brown Endowed Chaired Professor of Electrical and Computer Engineering, University of Texas, San Antonio, TX, USA. His areas of interests are system of systems simulation, architecture, and control with application to land, sea, and air rovers.

Steve D. Jolly is a Senior Systems Engineer with Lockheed Martin Space Systems and has worked many deep space missions including Mars 98, Mars Odyssey, and Mars Reconnaissance Orbiter. More recently, he is supporting the Orion Program (Crew Exploration Vehicle), the Geostationary Environmental Operational Satellite Program (GOES-R), and the Mars Science Laboratory (MSL) Program. He is also instrumental in the development of a three-course graduate series in systems engineering at the University of Colorado, Denver, CO, USA.

Charles B. Keating is a Professor of Engineering Management and Systems Engineering and Director, National Centers for System of Systems Engineering at Old Dominion University in Norfolk, VA, USA. His research interests include system of systems engineering, complex systems exploration methodologies, and R&D systems management.

D. Marc Kilgour is a Professor of Mathematics at Wilfrid Laurier University, Research Director: Conflict Analysis for the Laurier Centre for Military Strategic and Disarmament Studies, and Adjunct Professor of Systems Engineering at the University of Waterloo. His main research interest is optimal decision making in multi-decision maker and multicriteria contexts, including deterrence and counterterrorism, power sharing, fair division, voting, negotiation, and infrastructure management.

Nil Kilicay-Ergin is a Postdoctoral Research Fellow in Systems Engineering at the Missouri University of Science & Technology (Missouri S&T). She received her Ph.D. degree in Systems Engineering from the University of Missouri-Rolla. Her research interests are analysis of system of systems, complex adaptive systems, artificial life, and financial markets.

Petr Korba, Ph.D., is a Principal Scientist in the field of power and control systems at ABB Corporate Research Ltd., Baden, Switzerland. His areas of interest include robust and adaptive control, model identification and parameter estimation techniques, and their applications to power systems.

Asad M. Madni is Retired President and Chief Operating Officer of BEI Technologies Inc., and is currently the Executive Managing Director and Chief Technology Officer of Crocker Capital, San Francisco, CA, USA. His areas of interest are wireless sensor networks, miniaturized "intelligent" sensors and systems, and signal processing for aerospace and defense, automotive and transportation, and industrial and commercial applications.

José Luis Risco Martín is an Assistant Professor in Complutense University of Madrid, Spain. He received his Ph.D. from Complutense University of Madrid in 2004. His research interests are computational theory of modeling and simulation, with emphasis on DEVS, dynamic memory management of embedded systems, and net-centric computing.

Saurabh Mittal is the founder and CEO of Dunip Technologies, New Delhi, India. Previously he worked as Research Assistant Professor at the Department of Electrical and Computer Engineering at the University of Arizona, USA where he received his Ph.D in 2007. His areas of interest include Web-based M&S using SOA, executable architectures, Distributed Simulation, and System of Systems engineering using DoDAF. He can be reached at saurabh.mittal@duniptechnologies.com

Brian K. Muirhead is the Program Systems Engineer for NASA's Constellation Program, Johnson Space Center, Houston, TX, USA. He is responsible for the program architecture for the United States' human exploration of the Moon and beyond.

Amer Obeidi is a Lecturer at the Department of Management Sciences, University of Waterloo, Canada. His research interest is in the development of integrated systems of decision and conflict models that incorporate emotions with complex levels of perception and awareness, with applications in military and national security strategic and tactical planning, as well as environmental and societal concerns.

Jay S. Pearlman, Ph.D. is a Chief Engineer of NCOC&EM at Boeing and is Cochair of the GEO Architecture and Data Committee. His areas of interest are system of systems architecture, ocean studies, and information systems. He is also active in remote sensing sensors and applications and aerial observation.

Hans W. Polzer is a Lockheed Martin Fellow working for the Network Centric Integration Department within the Advanced Concepts Division of the Lockheed Martin Corporate Engineering and Technology organization. Hans conducts network-centric assessments of Lockheed Martin programs and pursuits and is the Technical Lead for the corporation's participation in the Network-Centric Operations Industry Consortium (NCOIC). He is interested in measures of diversity in perspectives, context, and scope across systems and the institutions that sponsor them.

Cynthia Riley is the Lead Systems Integrator for the U.S. Department of Energy Biomass Program at the National Renewable Energy Laboratory. She has over 30 years of engineering experience in the energy and environmental industries, focused on analysis and evaluation of emerging alternative energy technologies.

Andrew P. Sage received the BSEE degree from the Citadel, the SMEE degree from MIT, and the Ph.D. from Purdue, the latter in 1960. He received honorary Doctor of Engineering degrees from the University of Waterloo in 1987 and from Dalhousie University in 1997. He has been a faculty member at the University of Arizona, the University of Florida, and the Southern Methodist University. Following 10 years of service at the University of Virginia, where he held a named professorship and was the first chair of their systems engineering department, he became first American Bank Professor of Information Technology and Engineering in 1984 at George Mason University and the first Dean of the School of Information Technology and Engineering. In May 1996, he was elected Founding Dean Emeritus of the School and also was appointed a University Professor. He is an elected Fellow of the Institute of Electrical and Electronics Engineers, the American Association for the Advancement of Science, and the International Council on Systems Engineering. He is editor of the John Wiley textbook series on Systems Engineering and Management, the INCOSE Wiley journal Systems Engineering, and coeditor of Information, Knowledge, and Systems Management. He was elected to membership in the National Academy of Engineering in 2004. His interests include systems engineering and management efforts in a variety of application areas including systems integration and architecting, reengineering, and industrial ecology and sustainable development.

Ferat Sahin is an Associate Professor of Electrical Engineering, Rochester Institute of Technology, Rochester, NY, USA. His areas of interests are swarm robotics, multiagent systems, system of systems simulation for autonomous rovers, and MEMS-based microrobots.

Debra Sandor is the Lead Systems Engineer for the U.S. Department of Energy Biomass Program at the National Renewable Energy Laboratory. She has 18 years of experience in engineering and energy R&D focused on evaluating and reporting advances in alternative transportation fuels and renewable energy technologies.

Brian Sauser holds a B.S. from Texas A&M University, a M.S. from Rutgers, The State University of New Jersey, and a Ph.D. from Stevens Institute of Technology. He is currently an Assistant Professor in the School of Systems and Enterprises at Stevens Institute of Technology. His research interests are in theories, tools, and methods for bridging the gap between systems engineering and project management for managing complex systems. This includes the advancement of systems theory in the pursuit of a

Biology of Systems, system and enterprise maturity assessment for system and enterprise management, and systems engineering capability assessment.

Ryosuke Shibasaki is a professor and director of Center for Spatial Information Science, University of Tokyo. His research interests cover mapping/tracking technologies for mobile and immobile objects in urban environment, context-aware services based on human behavior sensing, planning and design of spatial data infrastructure (SDI) and its application to the integration of heterogeneous systems. He graduated department of civil engineering, University of Tokyo in 1980. After working for Public Works Research Institute, Minisitry of Construction for six years, he returned to Univ. of Tokyo as an associate professor. In 1998, he became a professor of Center for Spatial Information Science, University of Tokyo and since 2006 he serves as a director. In 2006, he became one of the co-chairs of ADC (Architecture and Data Committee) of GEO (Group of Earth Observation).

Prasanna Sridhar received the Bachelor of Engineering degree in Computer Science and Engineering from Bangalore University, India, in 2000, Master of Science degree in Computer Science in 2003, and Ph.D. degree in Computer Engineering in 2007, the last two from the University of New Mexico. In 2006, he joined the University of Texas at San Antonio as a Research Scientist Assistant. His current research interests are embedded sensor networks, mobile robotics, modeling and simulation, and computational intelligence. Currently, he is with Microsoft Corporation.

Wil A.H. Thissen, M.Sc. in Physics and Ph.D. in Systems and Control Engineering, is a Professor and Head of the Policy Analysis Department, Faculty of Technology, Policy and Management, Delft University of Technology. His research interests are in the development of concepts, methods, and tools to deal with the complexity of large-scale, multiactor systems, in particular infrastructure systems.

James M. Tien received the B.E.E. from Rensselaer Polytechnic Institute (RPI) and the SM, E.E. and Ph.D. from the Massachusetts Institute of Technology. He has held leadership positions at Bell Telephone Laboratories, at the Rand Corporation, and at Structured Decisions Corporation (which he cofounded in 1974). He joined the Department of Electrical, Computer, and Systems Engineering at RPI in 1977, became the Acting Chair of the department, joined a unique interdisciplinary Department of Decision Sciences and Engineering Systems as its founding Chair, and twice served as the Acting Dean of Engineering. In 2007, he joined the University of Miami as a Distinguished Professor and Dean of its College of Engineering. His areas of research interest include the development and application of computer and systems analysis techniques to information and decision systems. He has published extensively, been invited to present dozens of plenary lectures, and been honored with both teaching and research awards, including being elected a Fellow in IEEE, INFORMS, and AAAS and being a recipient of the IEEE Joseph G. Wohl Outstanding Career Award, the IEEE Major Educational Innovation Award, the IEEE Norbert Wiener Award, and the IBM Faculty Award. He is an Honorary Professor at a number of non-U.S. Universities. Dr. Tien is also an elected member of the U. S. National Academy of Engineering.

Gary D. Wells works as a Senior Systems Engineer within the federal government with over 15 year's experience in supporting the acquisition and systems engineering of national space systems. Gary is a Ph.D. candidate at George Mason University. His research interests involve management and systems engineering of systems of systems.

Nilmini Wickramasinghe is an Associate Professor and Associate Director of the Center for the Management of Medical Technologies at Stuart Graduate School of Business, Illinois Institute of Technology. Her research interests include management aspects of medical technology, e-health, and knowledge management in health care.

George F. Wilber is a Technical Fellow in the Phantom Works Research and Development Group within the Boeing Company. His areas of expertise are complex software computing algorithms and systems architecture and design for airborne computing and networking systems.

Bernard P. Ziegler is a Professor of Electrical and Computer Engineering at the University of Arizona, Tucson, and Director of the Arizona Center for Integrative Modeling and Simulation. He is developing DEVS-methodology approaches for testing mission thread end-to-end interoperability and combat effectiveness of Defense Department acquisitions and transitions to the Global Information Grid with its Service-Oriented Architecture (GIG/SOA).

Chapter **1**

Introduction to System of Systems

MO JAMSHIDI

The University of Texas, San Antonio, TX, USA

1.1 INTRODUCTION

Recently, there has been a growing interest in a class of complex systems whose constituents are themselves complex. Performance optimization, robustness, and reliability among an emerging group of heterogeneous systems in order to realize a common goal have become the focus of various applications including military, security, aerospace, space, manufacturing, service industry, environmental systems, and disaster management, to name a few (Crossley, 2004; Lopez, 2006; Wojcik and Hoffman, 2006). There is an increasing interest in achieving synergy between these independent systems to achieve the desired overall system performance (Azarnoosh et al., 2006). In the literature, researchers have addressed the issue of coordination and interoperability in a system of systems (SoS) (Abel and Sukkarieh, 2006; DiMario, 2006). SoS technology is believed to more effectively implement and analyze large, complex, independent, and *heterogeneous* systems working (or made to work) cooperatively (Abel and Sukkarieh, 2006). The main thrust behind the desire to view the systems as an SoS is to obtain higher capabilities and performance than would be possible with a traditional system view. The SoS concept presents a high-level viewpoint and explains the interactions between each of the independent systems. However, the SoS concept is still at its developing stages (Abbott, 2006; Meilich, 2006).

The next section will present some definitions out of many possible definitions of SoS. However, a practical definition may be that a system of systems is a "supersystem" comprised of other elements that themselves are independent complex

System of Systems Engineering: Innovations for the 21st Century, Edited by Mo Jamshidi
Copyright © 2009 John Wiley & Sons, Inc., Publication

operational systems and interact among themselves to achieve a common goal. Each element of an SoS achieves well-substantiated goals even if they are detached from the rest of the SoS. For example, a Boeing 747 airplane, as an element of an SoS, is not SoS, but an airport is an SoS, or a rover on Mars is not an SoS, but a robotic colony (or a robotic swarm) exploring the red planet, or any other place, is an SoS. As will be illustrated shortly, associated with SoS, there are numerous problems and open-ended issues that need a great deal of fundamental advances in theory and verifications. It is hoped that this volume will be a first effort toward bridging the gaps between an *idea* and a *practice*.

1.2 DEFINITIONS OF SYSTEM OF SYSTEMS

Based on the literature survey on system of systems, there are numerous definitions whose detailed discussion is beyond the space allotted to this chapter (Kotov, 1997; Luskasik, 1998; Pei, 2000; Carlock and Fenton, 2001; Sage and Cuppan, 2001; Jamshidi, 2005). Here we enumerate only six of many potential definitions:

> *Definition 1:* Systems of systems exist when there is a presence of a majority of the following five characteristics: operational and managerial independence, geographic distribution, emergent behavior, and evolutionary development (Jamshidi, 2005).
>
> *Definition 2:* Systems of systems are large-scale concurrent and distributed systems that are comprised of complex systems (Carlock and Fenton, 2001; Jamshidi, 2005).
>
> *Definition 3:* Enterprise system of systems engineering is focused on coupling traditional systems engineering activities with enterprise activities of strategic planning and investment analysis (Carlock and Fenton, 2001).
>
> *Definition 4:* System of systems integration is a method to pursue development, integration, interoperability, and optimization of systems to enhance performance in future battlefield scenarios (Pei, 2000).
>
> *Definition 5:* SoSE involves the integration of systems into systems of systems that ultimately contribute to evolution of the social infrastructure (Luskasik, 1998).
>
> *Definition 6:* In relation to joint warfighting, system of systems is concerned with interoperability and synergism of command, control, computers, communications, and information (C4I) and intelligence, surveillance, and reconnaissance (ISR) systems (Manthorpe, 1996).

Detailed literature survey and discussions on these definitions are given in Jamshidi (2005, 2008). Various definitions of SoS have their own merits, depending on their application. Favorite definition of this author and the volume's editor is *systems of systems are large-scale integrated systems that are heterogeneous and independently operable on their own, but are networked together for a common goal.* The goal, as mentioned before, may be cost, performance, robustness, and so on.

1.3   CHALLENGING PROBLEMS IN SYSTEM OF SYSTEMS

In the realm of open problems in SoS, just about anywhere one touches, there is an unsolved problem and immense attention is needed by many engineers and scientists. No engineering field is more urgently needed in tackling SoS problems than system engineering (SE). On top of the list of engineering issues in SoS is the "engineering of SoS," leading to a new field of SoSE (see Chapter 3). How does one extend SE concepts such as analysis, control, estimation, design, modeling, controllability, observability, stability, filtering, simulation, and so on that can be applied to SoS? Among numerous open questions are how can one model and simulate such systems (see Chapter 5 by Mittal et al.). In almost all cases, a chapter in this volume will accommodate the topic raised.

1.3.1   Theoretical Problems

In this section, a number of urgent problems facing SoS and SoSE are discussed. The major issue here is that a merger between SoS and engineering needs to be made. In other words, SE needs to undergo a number of innovative changes to accommodate and encompass SoS.

1.3.1.1   *Open Systems Approach to System of Systems Engineering*   Azani, in Chapter 2, discusses an open systems approach to SoSE. The author notes that SoS exists within a continuum that contains ad-hoc, short-lived, and relatively speaking simple SoS on one end, and long-lasting, continually evolving, and complex SoS on the other end of the continuum. Military operations and less sophisticated biotic systems (e.g., bacteria and ant colonies) are examples of ad-hoc, simple, and short-lived SoS, while galactic and more sophisticated biotic systems (e.g., ecosystem, human colonies) are examples of SoS at the opposite end of the SoS continuum. The engineering approaches utilized by galactic SoS are at best unknown and perhaps forever inconceivable. However, biotic SoS seem to follow, relatively speaking, less complicated engineering and development strategies allowing them to continually learn and adapt, grow and evolve, resolve emerging conflicts, and have more predictable behavior. Based on what the author already knows about biotic SoS, it is apparent that these systems employ robust reconfigurable architectures enabling them to effectively capitalize on open systems development principles and strategies such as modular design, standardized interfaces, emergence, natural selection, conservation, synergism, symbiosis, homeostasis, and self-organization. Chapter 2 provides further elaboration on open systems development strategies and principles utilized by biotic SoS, discusses their implications for engineering of man-made SoS, and introduces an integrated SoS development methodology for engineering and development of adaptable, sustainable, and interoperable SoS based on open systems principles and strategies.

1.3.1.2   *Engineering of SoS*   Emerging needs for a comprehensive look at the applications of classical systems engineering issue in SoSE will be discussed in this

volume. The thrust of the discussion will concern the reality that the technological, human, and organizational issues are each far different when considering a system of systems or federation of systems and that these needs are very significant when considering system of systems engineering and management.

As we have noted, today there is much interest in the engineering of systems that are comprised of other component systems, and where each of the component systems serves organizational and human purposes. These systems have several principal characteristics that make the system family designation appropriate: operational independence of the individual systems; managerial independence of the systems; often large geographic and temporal distribution of the individual systems; emergent behavior, in which the system family performs functions and carries out purposes that do not reside uniquely in any of the constituent systems but which evolve over time in an adaptive manner and where these behaviors arise as a consequence of the formation of the entire system family and are not the behavior of any constituent system. The principal purposes supporting engineering of these individual systems and the composite system family are fulfilled by these emergent behaviors. Thus, a system of systems is never fully formed or complete. Development of these systems is evolutionary and adaptive over time, and structures, functions, and purposes are added, removed, and modified as experience of the community with the individual systems and the composite system grows and evolves. The systems engineering and management of these systems families pose special challenges. This is especially the case with respect to the federated systems management principles that must be utilized to deal successfully with the multiple contractors and interests involved in these efforts. Please refer to the paper by Sage and Biemer (2007) and DeLaurentis et al. (2007) for the creation of a SoS Consortium (i.e., International Consortium on System of Systems (ICSoS)) of concerned individuals and organizations by the author of this chapter. Chapter 3 by Wells and Sage discusses the challenges of engineering of SoS.

1.3.1.3 Standards of SoS System of systems literature, definitions, and perspectives are marked with great variability in the engineering community. Viewed as an extension of systems engineering to a means of describing and managing social networks and organizations, the variations of perspectives lead to difficulty in advancing and understanding the discipline. Standards have been used to facilitate a common understanding and approach to align disparities of perspectives to drive a uniform agreement to definitions and approaches. By having the ICSoS (DeLaurentis et al., 2007) represent to the IEEE and INCoSE for support of technical committees to derive standards for system of systems will help unify and advance the discipline for engineering, healthcare, banking, space exploration, and all other disciplines that require interoperability among disparate systems.

1.3.1.4 System of Systems Architecting Dagli and Kilicay-Ergin in Chapter 4 provide a framework for SoS architectures. As the world is moving toward a networked society, the authors assert the business and government applications require integrated systems that exhibit intelligent behavior. The dynamically changing environmental and operational conditions necessitate a need for system architectures that will be effective for the duration of the mission but evolve to new system architectures as the mission changes. This new challenging demand has led to a new operational style: instead of designing or subcontracting systems from scratch, business or government gets the best systems the industry develops and focuses on becoming the lead system integrator to provide SoS. SoS is a set of interdependent systems that are related or connected to provide a common mission. In the SoS environment, architectural constraints imposed by existing systems have a major effect on the system capabilities, requirements, and behavior. This fact is important, as it complicates the systems architecting activities. Hence, architecture becomes a dominating but confusing concept in capability development. There is a need to push system architecting research to meet the challenges imposed by new demands of the SoS environment. This chapter focuses on system of systems architecting in terms of creating meta-architectures from collections of different systems. Several examples are provided to clarify system of systems architecting concept. Since the technology base, organizational needs, and human needs are changing, the system of systems architecting becomes an evolutionary process. Components and functions are added, removed, and modified as owners of the SoS experience and use the system. Therefore, in Chapter 4 evolutionary system architecting is described and the challenges are identified for this process. Finally, the authors discuss the possible use of artificial life tools for the design and architecting of SoS. Artificial life tools such as swarm intelligence, evolutionary computation, and multiagent systems have been successfully used for the analysis of complex adaptive systems. The potential use of these tools for SoS analysis and architecting is discussed, by the authors, using several domain application specific examples.

1.3.1.5 SoS Simulation

Sahin et al. (2007) have presented an SoS architecture based on Extensible Markup Language (XML) in order to wrap data coming from different systems in a common way. The XML can be used to describe each component of the SoS and their data in a unifying way. If XML-based data architecture is used in an SoS, the only requirement for the SoS components is to understand/parse XML file received from the components of the SoS. In XML, data can be represented in addition to the properties of the data such as source name, data type, importance of the data, and so on. Thus, it does not only represent data but also gives useful information that can be used in the SoS to take better actions and to understand the situation better. The XML language has a hierarchical structure where an environment can be described with a standard and without a huge overhead. Each entity can be defined by the user in the XML in terms of its visualization and functionality. As a case study in this effort (see Chapter 5 by Mittal et al.), a master-scout rover combination represents an SoS where for the first time a sensor detects a fire in a field. The fire is detected by the master rover and commands the scout rover to verify the existence of the fire. It is important to note that such an architecture and simulation do not need any mathematical model for members of the systems.

1.3.1.6 SoS Integration

Integration is probably the key viability of any SoS. Integration of SoS implies that each system can communicate and interact (control) with the SoS regardless of their hardware, software characteristics, or nature. This means that they need to have the ability to communicate with the SoS or a part of the SoS without compatibility issues such as operating systems, communication hardware, and so on. For this purpose, an SoS needs a common language the SoS's systems can speak. Without having a common language, the systems of any SoS cannot be fully functional and the SoS cannot be adaptive in the sense that new components cannot be integrated to it without major effort. Integration also implies the control aspects of the SoS because systems need to understand each other in order to take commands or signals from other SoS systems. See Chapter 6 by Cloutier et al. on network centric architecture of SoS.

1.3.1.7 Emergence in SoS Emergent behavior of an SoS resembles the slow-down of the traffic going through a tunnel, even in the absence of any lights, obstacles, or accident. A tunnel, automobiles, and the highway, as systems of an SoS, have an emergent behavior or property in slowing down (Morley, 2006). Fisher (2006) has noted that an SoS cannot achieve its goals depends on its emergent behaviors. The author *explores* "interdependencies among systems, emergence, and interoperation" and develops maxim-like findings such as these: (1) Because they cannot control one another, autonomous entities can achieve goals that are not local to themselves only by increasing their influence through cooperative interactions with others. (2) Emergent composition is often poorly understood and sometimes misunderstood because it has few analogies in traditional systems engineering. (3) Even in the absence of accidents, tight coupling can ensure that a system of systems is unable to satisfy its objectives. (4) If it is to remain scalable and affordable no matter how large it may become, a system's cost per constituent must grow less linearly with its size. (5) Delay is a critical aspect of systems of systems. Chapter 7 by Keating will provide a detailed perspective into emergence property of SoS.

1.3.1.8 SoS Management: The Governance of Paradox Sauser and Boardman, in Chapter 8, present an SoS approach to the management problem. They note that the study of SoS has moved many to support their understanding of these systems through the groundbreaking science of networks. The understanding of networks and how to manage them may give one the fingerprint that is independent of the specific systems that exemplify this complexity. The authors point out that it does not matter whether they are studying the synchronized flashing of fireflies, space stations, structure of the human brain, the internet, the flocking of birds, a future combat system, or the behavior of red harvester ants. The same emergent principles apply: large is really small, weak is really strong, significance is really obscure, little means a lot, simple is really complex, and complexity hides simplicity. The conceptual foundation of complexity is paradox, which leads us to a paradigm shift in the SE body of knowledge.

Paradox exists for a reason and there are reasons for systems engineers to appreciate paradox even though they may be unable to resolve them as they would a problem specification into a system solution. Hitherto paradoxes have confronted current logic only to yield at a later date to more refined thinking. The existence of paradox is always the inspirational source for seeking new wisdom, attempting new thought patterns, and ultimately building systems for the "flat world." It is our ability to govern, not control, these paradoxes that will bring new knowledge to our understanding on how to manage the emerging complex systems called system of systems.

Chapter 8 establishes a foundation in what has been learnt about how one practices project management, establishes some key concepts and challenges that make the management of SoS different from our fundamental practices, presents an intellectual model for how they classify and manage an SoS, appraises this model with recognized SoS, and concludes with grand challenges for how they may move their understanding of SoS management beyond the foundation.

In the previous section, a brief introduction was presented for six theoretical issues of SoS, that is, integration, engineering, standards, open and other architectures, modeling, infrastructure, and simulation. These topics are discussed in great detail by a number of experts in the field in chapters in the book.

1.3.2 Implementation Problems

Besides from many theoretical and essential difficulties with SoS, there are many implementation challenges facing SoS. Here, some of these implementation problems are briefly discussed and references are made to some with their full coverage.

1.3.2.1 Systems Engineering for the Department of Defense System of Systems
Dahmann and Baldwin, in Chapter 9, have addressed the national defense aspects of SoS. Military operations are the synchronized efforts of people and systems toward a common objective. In this way from an operational perspective, defense is essentially a "system of systems" enterprise. However, despite the fact that today almost every military system is operated as part of a system of systems, most of these systems were designed and developed without the benefit of systems engineering at the SoS level factoring the role the system will play in the broader system of systems context. With changes in operations and technology, the need for systems that work effectively together is increasingly visible. Chapter 9 outlines the changing situation in the defense department and the challenges it poses for systems engineering.

1.3.2.2 e-Enabling and SoS Aircraft Design Via SoSE   A case of aeronautical application of SoS worth noting is that of e-enabling in aircraft design as a system of an SoS at Boeing Commercial Aircraft Division (Wilber, 2007). The project focused on developing a strategy and technical architecture to facilitate making the airplane (Boeing 787, see Fig. 1.1) network-aware and capable of leveraging computing and network advances in industry. The project grew to include many ground-based architectural components at the airlines and at the Boeing factory, as well as other key locations such as the airports, suppliers, and terrestrial Internet Service Suppliers (ISPs).

Wilber (2007) points out that the e-enabled project took on the task of defining a system of systems engineering solution to problem of interoperation and communication with the existing, numerous, and diverse elements that make up the airlines'

FIGURE 1.1 A photo of the new SoS e-enabled Boeing 787 (courtesy of Boeing Company, see also Chapter 10 by G.R. Wilber)

operational systems (flight operations and maintenance operations). The objective has been to find ways of leveraging network-centric operations, to reduce production, operations and maintenance costs for both Boeing and the airline customers.

> One of the key products of this effort is the "e-enabled architecture." The e-enabling architecture is defined at multiple levels of abstraction. There is a single top-level or "reference architecture" that is necessarily abstract and multiple "implementation architectures." The implementation architectures map directly to airplane and airline implementations and provide a family of physical solutions that all exhibit common attributes and are designed to work together and allow re-use of systems components. The implementation architectures allow for effective forward and retrofit installations addressing a wide range of market needs for narrow and wide-body aircraft.
>
> The 787 "Open Data Network" is a key element of one implementation of this architecture. It enabled on-board and off-board elements to be networked in a fashion that is efficient, flexible, and secure. The fullest implementations are best depicted in Boeing's GoldCare Architecture and design.

Wilber, in Chapter 10, presents an architecture at the reference level and how it has been mapped into the 787 airplane implementation. *GoldCare* environment is described and is used as an example of the full potential of the current e-enabling.

1.3.2.3 A System of Systems Perspective on Infrastructures Thissen and Herder, in Chapter 11, touch upon a very important application in the service industry (see also Chapter 13 by Tien). Infrastructure systems (or infrasystems) providing services such as energy, transport, communications, and clean and safe water are vital to the functioning of modern society. Key societal challenges with respect to our present and future infrastructure systems relate to, among other things, safety and reliability, affordability, and transitions to sustainability. Infrasystem complexity precludes simple answers to these challenges. While each of the infrasystems can be seen as a complex system of systems in itself, increasing interdependency among these systems (both technologically and institutionally) adds a layer of complexity.

One approach to increased understanding of complex infrasystems that has received little attention in the engineering community thus far is to focus on the commonalities of the different sectors and to develop generic theories and approaches such that lessons from one sector could easily be applied to other sectors. The system of systems paradigm offers interesting perspectives in this respect. The authors present, as an initial step in this direction, a fairly simple three-level model distinguishing the physical/technological systems, the organization and management systems, and the systems and organizations providing infrastructure-related products and services. The authors use the model as a conceptual structure to identify a number of key commonalities and differences between the transport, energy, drinking water, and ICT sectors. Using two energy-related examples, the authors further illustrate some of the system of systems related complexities of analysis and design at a more operational level. The authors finally discuss a number of key research and engineering challenges related to infrastructure systems, with a focus on the potential contributions of systems of systems perspectives.

1.3.2.4 Sensor Networks
The main purpose of sensor networks is to utilize the distributed sensing capability provided by tiny, low-powered, and low-cost devices. Multiple sensing devices can be used cooperatively and collaboratively to capture events or monitor space more effectively than a single sensing device (Sridhar et al., 2007). The realm of applications for sensor networks is quite diverse, which include military, aerospace, industrial, commercial, environmental, and health monitoring, to name a few. Applications include traffic monitoring of vehicles, cross-border infiltration detection and assessment, military reconnaissance and surveillance, target tracking, habitat monitoring and structure monitoring, and so on.

Communication capability of these small devices and often with heterogeneous attributes makes them good candidates for system of systems. Numerous issues with sensor networks such as data integrity, data fusion and compression, power consumption, multidecision making, and fault tolerance all make these SoS very challenging just like other SoS. It is thus necessary to devise a fault-tolerant mechanism with a low computation overhead to validate the integrity of the data obtained from the sensors (systems). Moreover, a robust diagnostics and decision-making process should aid in monitoring and control of critical parameters to efficiently manage the operational behavior of a deployed sensor network. Specifically, Chapter 12 by Sridhar et al. will focus on innovative approaches to deal with multivariable multispace problem domain as well as other issues, in wireless sensor networks within the framework of an SoS.

1.3.2.5 A System of Systems View of Services
Tien, in Chapter 13, covers a very important applications of SoS in our today's global village — *service industry*. The services sector employs a large and growing proportion of workers in the industrialized nations, and it is increasingly dependent on information technology. While the interdependences, similarities, and complementarities of manufacturing

and services are significant, there are considerable differences between goods and services, including the shift in focus from mass production to mass customization (whereby a service is produced and delivered in response to a customer's stated or imputed needs). In general, a service system can be considered to be a combination or recombination of three essential components — people (characterized by behaviors, attitudes, values, etc.), processes (characterized by collaboration, customization, etc.), and products (characterized by software, hardware, infrastructures, etc.). Furthermore, inasmuch as a service system is an integrated system, it is, in essence, a system of systems whose objectives are to enhance its efficiency (leading to greater interdependency), effectiveness (leading to greater usefulness), and adaptiveness (leading to greater responsiveness). The integrative methods include a component's design, interface, and interdependency; a decision's strategic, tactical, and operational orientation; and an organization's data, modelling, and cybernetic consideration. A number of insights are also provided, including an alternative system of systems view of services; the increasing complexity of systems (especially service systems), with all the attendant life cycle design, human interface, and system integration issues; the increasing need for real-time, adaptive decision making within such systems of systems; and the fact that modern systems are also becoming increasingly more human centered, if not human focused — thus, products and services are becoming more complex and more personalized or customized.

1.3.2.6 *System of Systems Engineering in Space Exploration*

Jolly and Muirhead, in Chapter 14, cover SoSE topics that are largely unique for space exploration with the intent to provide the reader a discussion of the key issues, the major challenges of the twenty-first century in moving from systems engineering to SoSE, potential applications in the future, and the current state of the art. Specific emphasis is placed on how software and electronics are revolutionizing the way space missions are being designed, including both the capabilities and vulnerabilities introduced. The role of margins, risk management, and interface control is all critically important in current space mission design and execution, but in SoSE applications they become paramount. Similarly, SoSE space missions will have extremely large, complex, and intertwined command and control and data distribution ground networks, most of which will involve extensive parallel processing to produce tera-to-petabytes of products per day and distribute them worldwide.

1.3.2.7 *Communication and Navigation in Space SoS*

Bhasin and Hayden, in Chapter 15, have taken upon the challenges in communication and navigation for space SoS. They indicate that communication and navigation networks provide critical services in the operation, system management, information transfer, and situation awareness to the space system of systems. In addition, space systems of systems are requiring system interoperability, enhanced reliability, common interfaces, dynamic operations, and autonomy in system management. New approaches to communications and navigation networks are required to enable the interoperability needed to satisfy the complex goals and dynamic operations and activities of the space system of systems. Historically, space systems had direct links to Earth ground communication systems, or they required a space communication satellite infrastructure to achieve higher coverage around the Earth. It is becoming increasingly apparent that many systems of systems may include communication networks that are also systems of systems. These communication and navigation networks must be as nearly ubiquitous as possible and accessible on the demand of the user, much like the cell phone link is available at any time to an Earth user in range of a cell tower. The new demands on communication and navigation networks will be met by space Internet technologies. It is important to bring Internet technologies, Internet Protocols (IP), routers, servers, software, and interfaces to space networks to enable as much autonomous operation of those networks as possible. These technologies provide extensive savings in reduced cost of operations. The more these networks can be made to run themselves, the less humans will have to schedule and control them. The Internet technologies also bring with them a very large repertoire of hardware and software solutions to communication and networking problems that would be very expensive to replicate under a different paradigm. Higher bandwidths are needed to support the expected voice, video, and data transfer traffic for the coordination of activities at each stage of an exploration mission.

Existing communications, navigation, and networking have grown in an independent fashion with experts in each field solving the problem just for that field. Radio engineers designed the payloads for today's "bent pipe" communication satellites. The Global Positioning Satellite (GPS) system design for providing precise Earth location determination is an extrapolation of the Long Range Navigation (LORAN) technique of the 1950s where precise time is correlated to precise position on the Earth. Other space navigation techniques use artifacts in the RF communication path (Doppler shift of the RF and transponder-reflected ranging signals in the RF) and time transfer techniques to determine the location and velocity of a spacecraft within the solar system. Networking in space today is point-to-point among ground terminals and spacecraft, requiring most communication paths to/from space to be scheduled such that communications is available only on an operational plan and is not easily adapted to handle multidirectional communications under dynamic conditions.

Chapter 15 begins with a brief history of the communications, navigation, and networks of the 1960s and 1970s in use by the first system of systems, the NASA Apollo missions; it is followed by short discussions of the communication and navigation networks and architectures that the DoD and NASA employed from the 1980s onward. Next is a synopsis of the emerging space system of systems that will require complex communication and navigation networks to meet their needs. Architecture approaches and processes being developed for communication and navigation networks in emerging space system and systems are also described. Several examples are given of the products generated in using the architecture development process for space exploration systems. The architecture addresses the capabilities to enable voice, video, and data interoperability needed among the explorers during exploration, while in habitat, and with Earth operations. Advanced technologies are then described that will allow space system of systems to operate autonomously or semiautonomously. Chapter 15 ends with a summary of the challenges and issues raised in implementing these new concepts.

1.3.2.8   *Electric Power Systems Grids as SoS*   Hiskens and Korba, in Chapter 16, provide an overview of the systems of systems that are fundamental to the operation and control of electrical power systems. Perspectives are drawn from industry and academia, and reflect theoretical and practical challenges that are facing power systems in an era of energy markets and increasing utilization of renewable energy resources (see also Chapter 17 by Duffy et al.). Power systems cover extensive geographical regions and are composed of many diverse components. Accordingly, power systems are large-scale, complex, dynamical systems that must operate reliably to supply electrical energy to customers. Stable operation is achieved through extensive monitoring systems and a hierarchy of controls that together seek to ensure total generation matches consumption and voltages remain at acceptable levels. Safety margins play an important role in ensuring reliability, but tend to incur economic penalties. Significant effort is therefore being devoted to the development of demanding control and supervision strategies that enable reduction of these safety margins, with consequent improvements in transfer limits and profitability. Recent academic and industrial research in this field will also be addressed in Chapter 16.

1.3.2.9   *SoS Approach for Renewable Energy*   Duffy et al., in Chapter 17, have provided the SoS approach to sustainable supply of energy. They note that over one half of the petroleum consumed in the United States is imported, and that percentage is expected to rise to 60% by 2025. America's transportation system of systems relies almost exclusively on refined petroleum products, accounting for over two thirds of the oil used. Each day, over 8 million barrels of oil are required to fuel over 225 million vehicles that constitute the United States light-duty transportation fleet. The gap between the United States oil production and transportation oil needs is projected to grow, and the increase in the number of light-duty vehicles will account for most of that growth. On a global scale, petroleum supplies will be in increasingly higher demand as highly populated developing countries expand their economies and become more energy intensive. Clean forms of energy are needed to support sustainable global economic growth while mitigating impacts on air quality and the potential effects of greenhouse gas emissions. Growing dependence of the united states on foreign sources of energy threatens her national security. As a nation, the authors assert that we must work to reduce our dependence on foreign sources of energy in a manner that is affordable and preserves environmental quality.

1.3.2.10   *Sustainable Environmental Management from a System of Systems Engineering Perspective*   Hipel et al., in Chapter 18, provide a rich range of decision tools from the field of SE that are described for addressing complex environmental SoS problems in order to obtain sustainable, fair, and responsible solutions to satisfy as much as possible the value systems of stakeholders, including the natural environment and future generations who are not even present at the bargaining table. To better understand the environmental problem being investigated and thereby eventually reach more informed decisions, the insightful paradigm of a system of systems can be readily utilized. For example, when developing solutions to global warming problems, one can envision how societal systems, such as agricultural and industrial systems, interact with the atmospheric system of systems, especially at the tropospheric level. The great import of developing a comprehensive toolbox of decision methodologies and techniques is emphasized by pointing out many current pressing environmental issues, such as global warming and its potential adverse affects, and the widespread pollution of our land, water, and air systems of systems. To tackle these large-scale complex systems of systems problems, systems engineering decision techniques that can take into account multiple stakeholders having multiple objectives are explained according to their design and capabilities. To illustrate how systems decision tools can be employed in practice to assist in reaching better decisions for benefiting society, different decision tools are applied to three real-world systems of systems environmental problems. Specifically, the Graph Model for Conflict Resolution is applied to the international dispute over the utilization of water in the Aral Sea Basin; a large-scale optimization model founded upon concepts from cooperative game theory, economics, and hydrology is utilized for systematically investigating the fair allocation of scarce water resources among multiple users in the South Saskatchewan River Basin in Western Canada; and multiple criteria decision analysis methods are used to evaluate and compare solutions to handling fluctuating water levels in the five Great Lakes located along the border of Canada and the United States (Wang et al., 2007).

1.3.2.11 *Robotic Swarms as an SoS*

As another application of SoS, a robotic swarm is considered by Sahin in Chapter 19. Here a robotic swarm based on ant colony optimization and artificial immune systems is considered. In the ant colony optimization, the author has developed a multiagent system model based on the food gathering behaviors of the ants. Similarly, a multiagent system model is developed based on the human immune system. These multiagent system models, are then tested on the mine detection problem. A modular microrobot is designed to perform to emulate the mine detection problem in a basketball court. The software and hardware components of the modular robot are designed to be modular so that robots can be assembled using hot swappable components. An adaptive TDMA communication protocol is developed in order to control connectivity among the swarm robots without the user intervention. Details are given in Chapter 19.

1.3.2.12 *Transportation Systems*

The National Transportation System (NTS) can be viewed as a collection of layered networks composed by heterogeneous systems for which the Air Transportation System (ATS) and its National Airspace System (NAS) is one part. At present, research on each sector of the NTS is generally conducted independently, with infrequent and/or incomplete consideration of scope dimensions (e.g., multimodal impacts and policy, societal, and business enterprise influences) and network interactions (e.g., layered dynamics within a scope category). This isolated treatment does not capture the higher level interactions seen at the NTS or ATS architecture level; thus, modifying the transportation system based on limited observations and analyses may not necessarily have the intended effect or impact. A systematic method for modeling these interactions with a system of systems (SoS) approach is essential to the formation of a more complete model and understanding of the ATS, which would ultimately lead to better outcomes from high-consequence decisions in technological, socioeconomic, operational, and political policy-making context (DeLaurentis, 2005). This is especially vital as decision makers in both the public and the private sector, for example, at the interagency Joint Planning and Development Office (JPDO), which is charged with transformation of air transportation, are facing problems of increasing complexity and uncertainty in attempting to encourage the evolution of superior transportation architectures (DeLaurentis and Callaway, 2006). Chapter 20 by DeLaurentis will be addressing this application.

1.3.2.13 Healthcare Systems Under a 2004 Presidential Order, the U.S. Secretary of Health has initiated the development of a National Healthcare Information Network (NHIN), with the goal of creating a nationwide information system that can build and maintain Electronic Health Records (EHRs) for all citizens by 2014. The NHIN system architecture currently under development will provide a near-real-time heterogeneous integration of disaggregated hospital, departmental, and physician patient care data and will assemble and present a complete current EHR to any physician or hospital a patient consults (Sloane, 2006). The NHIN will rely on a network of independent Regional Healthcare Information Organizations (RHIOs) that are being developed and deployed to transform and communicate data from the hundreds of thousands of legacy medical information systems presently used in hospital departments, physician offices, and telemedicine sites into NHIN-specified metaformats that can be securely relayed and reliably interpreted anywhere in the country. The NHIN "network of networks" will clearly be a very complex SoS, and the performance of the NHIN and RHIOs will directly affect the safety, efficacy, and efficiency of healthcare in the United States. Simulation, modeling, and other appropriate SoSE tools are under development to help ensure reliable, cost-effective planning, configuration, deployment, and management of the heterogeneous, life-critical NHIN and RHIO systems and subsystems (Sloane et al., 2007). ICSoS represents an invaluable opportunity to access and leverage SoSE expertise already under development in other industry and academic sectors. ICSoS also represents an opportunity to discuss the positive and negative emergent behaviors that can significantly affect personal and public health status and the costs of healthcare in the United States (DeLaurentis et al., 2007). See Chapter 21 by Chalasani et al.

1.3.2.14 Global Earth Observation System of Systems GEOSS is a global project consisting of over 60 nations whose purpose is to address the need for timely, quality, long-term, global information as a basis for sound decision making (Butterfield et al., 2006). Its objectives are: (i) improved coordination of strategies and systems for Earth observations to achieve a comprehensive, coordinated, and sustained Earth observation system or systems; (ii) a coordinated effort to involve and assist developing countries in improving and sustaining their contributions to observing systems, their effective utilization of observations, and the related technologies; and (iii) the exchange of observations recorded from *in situ*, air full and open manner with minimum time delay and cost. In GEOSS, the "SoSE process provides a complete, detailed, and systematic development approach for engineering systems of systems

Boeing's new architecture-centric, model-based systems engineering process emphasizes concurrent development of the system architecture model and system specifications. The process is applicable to all phases of a system's life cycle. The SoSE process is a unified approach for system architecture development that integrates the views of each of a program's participating engineering disciplines into a single system architecture model supporting civil and military domain applications" (Pearlman, 2006). ICSoS will be another platform for all concerned around the globe to bring the progress and principles of GEOSS to formal discussions and examination on an annual basis. Chapter 22 by Shibasaki and Pearlman will be addressing GEOSS application. Figure 1.2 shows a number of systems in GEOSS .

1.3.2.15 Deepwater Coastguard Program   One of the earliest realization of an SoS in the United States is the so-called Deepwater Coastguard Program shown in Fig. 1.3. As seen here, the program takes advantage of all the necessary assets at their disposal, for example, helicopters, aircrafts, cutters, satellite (GPS), ground station, human, computers, and so on — all systems of the SoS integrated together to react to unforeseen circumstances to secure the coastal borders of the southeastern United States, for example, Florida Coast. The Deepwater program is making progress in the development and delivery of mission effective command, control, communications, computers, intelligence, surveillance, and reconnaissance (C4ISR) equipment (Keeter, 2007). The SoS approach, the report goes on, has "improved the operational capabilities of legacy cutters and aircraft, and will provide even more functionality when the next generation of surface and air platforms arrives in service." The key

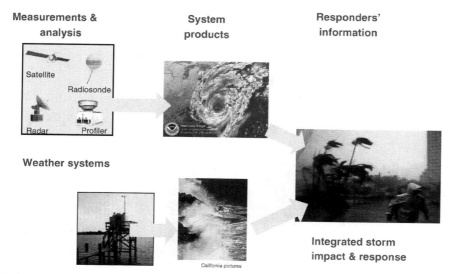

FIGURE 1.2   SoS of the GEOSS project (courtesy, Jay Pearlman, Boeing Company, see also Chapter 22 by Shibasaki and Pearlman)

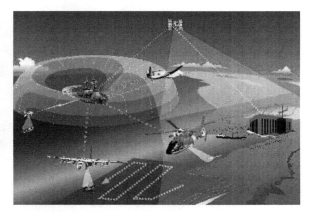

FIGURE 1.3 A security example of an SoS — deepwater coastguard configuration in United States

feature of the system is its ability to interoperate among all Coast Guard mission assets and capabilities with those of appropriate authorities at both local and federal levels.

1.3.2.16 Future Combat Missions Another national security or defense application of SoS is the future combat mission (FCM). Figure 1.4 shows one of the numerous possible configurations of an FCM. The FCM system is "envisioned to be an ensemble of manned and potentially unmanned combat systems, designed to ensure that the future force is strategically responsive and dominant at every point on the spectrum of operations from nonlethal to full-scale conflict. FCM will provide a rapidly deployable capability for mounted tactical operations by conducting direct combat, delivering both line-of-sight and beyond-line-of-sight precision munitions, providing variable lethal effect (nonlethal to lethal), performing reconnaissance, and

FIGURE 1.4 A defense example of a SoS (courtesy, Don Walker, Aerospace Corporation)

transporting troops. Significant capability enhancements will be achieved by developing multifunctional, multimission, and modular features for system and component commonality that will allow for multiple state-of-the-art technology options for mission tailoring and performance enhancements. The FCM force will incorporate and exploit information dominance to develop a common, relevant operating picture and achieve battle space situational understanding" (Global Security Organization, 2007). See also Chapter 9 by Dahmann and Baldwin for insights in this and other defense applications.

1.3.2.17 National Security   Perhaps one of the most talked-about application areas of SoSE is national security. After many years of discussion of the goals, merits, and attributes of SoS, very few tangible results or solutions have appeared in this or other areas of this technology. It is commonly believed that "systems engineering tools, methods, and processes are becoming inadequate to perform the tasks needed to realize the systems of systems envisioned for future human endeavors. This is especially becoming evident in evolving national security capabilities realizations for large-scale, complex space, and terrestrial military endeavors. Therefore, the development of systems of systems engineering tools, methods, and processes is imperative to enable the realization of future national security capabilities" (Walker, 2007). In most SoSE applications, heterogeneous systems (or communities) are brought together to cooperate for a common good and enhanced robustness and performance. "These communities range in focus from architectures, to lasers, to complex systems, and will eventually cover each area involved in aerospace-related national security endeavors. These communities are not developed in isolation in that cross-community interactions on terminology, methods, and processes are done" (Walker, 2007). The key is to have these communities work together to guarantee the common goal of making our world a safer place for all. See Chapter 9 by Dahmann and Baldwin for insights in this and other security applications.

1.3.2.18 Critical Infrastructure and Air Transportation Security   Air transportation networks consist of concourses, runways, parking, airlines, cargo terminal operators, fuel depots, retail, cleaning, catering, and many interacting people including travelers, service providers, and visitors. The facilities are distributed and fall under multiple legal jurisdictions in regard to occupational health and safety, customs, quarantine, and security.

Currently decision making in this domain space is focused on individual systems. The challenge of delivering improved nationwide air transportation security, while maintaining performance and continuing growth, demands a new approach. In addition, information flow and data management are a critical issue, where trust plays a key role in defining interactions of organizations.

SoS methodologies are required to rapidly model, analyze, and optimize air transportation systems (Nahavandi, 2007). In any critical real-world system there is and must be a compromise between increased risk and increased flexibility and productivity. By approaching such problem spaces from an SoS perspective the authors are in the best position to find the right balance (DeLaurentis et al., 2007).

1.4 CONCLUSIONS

This chapter is written to serve as an introduction to the book. The subject matter of this book is an unsettled topic in engineering in general and in systems engineering in particular. Attempt has been made to cover as many open questions in both theory and applications of SoS and SoSE. It is our intention that this book would be the beginning of much debate and challenges among and by the readers of this book. The book is equally intended to benefit industry, academia, or government. A sister volume, by the author, on the subject is under press at the present time and can give readers further insight into SoS (Jamshidi, 2008).

REFERENCES

Abbott, R., 2006, Open at the top; open at the bottom; and continually (but slowly) evolving, *Proceedings of IEEE International Conference on System of Systems Engineering,* April 2006, Los Angeles.

Abel, A., Sukkarieh, S., 2006, The coordination of multiple autonomous systems using information theoretic political science voting models, *Proceedings of IEEE International Conference on System of Systems Engineering,* April 2006, Los Angeles.

Azarnoosh, H., Horan, B., Sridhar, P., Madni, A.M., Jamshidi, M., 2006, Towards optimization of a real-world robotic-sensor system of systems, *Proceedings of World Automation Congress (WAC) 2006,* July 24–26, Budapest, Hungary.

Butterfield, M.L., Pearlman, J., Vickroy, S.C., 2006, System-of-systems engineering in a global environment, *Proceedings of International Conference on Trends in Product Life Cycle, Modeling, Simulation and Synthesis PLMSS.*

Carlock, P.G., Fenton, R.E., 2001, System of systems (SoS) enterprise systems for information-intensive organizations, *Systems Engineering,* 4(4): 242–261.

Crossley, W.A., 2004, System of systems: an introduction of Purdue University Schools of Engineering's Signature Area, *Engineering Systems Symposium,* March 29–31, 2004, Tang Center, Wong Auditorium, MIT.

DeLaurentis, D.A., 2005, Understanding transportation as a system-of-systems design, problem, *AIAA Aerospace Sciences Meeting and Exhibit,* January 10–13, 2005. AIAA-2005-123.

DeLaurentis, D.A., Callaway, R.K., 2006, A system-of-systems perspective for future public policy, *Review of Policy Research,* 21(6): 2006.

DeLaurentis, D., Dickerson, C., Di Mario, M., Gartz, P., Jamshidi, M., Nahavandi, S., Sage, A., Sloane, E., Walker, D., 2007, A case for an international consortium on system of systems engineering, *IEEE Systems Journal,* 1(1): 68–73.

DiMario, M.J., 2006, System of systems interoperability types and characteristics in joint command and control, *Proceedings of IEEE International Conference on System of Systems Engineering,* April 2006, Los Angeles.

Fisher, D., 2006, *An Emergent Perspective on Interoperation in Systems of Systems,* (CMU/SEI-2006-TR- 003). Software Engineering Institute, Carnegie Mellon University, Pittsburgh, PA.

Global Security Organization, 2007, http://www.globalsecurity.org/military/systems/ground/fcs-back.htm.

Jamshidi, M. 2005, *Theme of the IEEE SMC 2005*, Waikoloa, Hawaii, USA, http://ieeesmc2005.unm.edu/, October 2005.

Jamshidi, M., 2008, *System of Systems Engineering – Principles and Applications*, Taylor Francis CRC Publishers, Boca Raton, FL, USA.

Keeter, H.C., 2007, Deepwater command, communication, sensor electronics build enhanced operational capabilities, *US Coastguard Deepwater Prorgam site*, http://www.uscg.mil/deepwater/media/feature/july07/c4isr072007.htm.

Kotov, V., 1997, Systems of systems as communicating structures, *Hewlett Packard Computer Systems Laboratory Paper* HPL-97-124, pp. 1–15.

Lopez, D., 2006, Lessons learned from the front lines of the aerospace, *Proceedings of IEEE International Conference on System of Systems Engineering,* April 2006, Los Angeles.

Luskasik, S.J., 1998, Systems, systems of systems, and the education of engineers, *Artificial Intelligence for Engineering Design, Analysis, and Manufacturing*, 12(1): 11–60.

Meilich, A., 2006, System of systems (SoS) engineering & architecture challenges in a net centric environment, *Proceedings of IEEE International Conference on System of Systems Engineering*, April 2006, Los Angeles.

Manthorpe, W.H., 1996, The emerging joint system of systems: a systems engineering challenge and opportunity for APL, *John Hopkins APL Technical Digest*, 17(3): 305–310.

Morley, J., 2006, Five Maxims about Emergent Behavior in Systems of Systems, http://www.sei.cmu.edu/news-at-sei/features/2006/06/feature-2-2006-06.htm.

Nahavandi, S., 2007, Modeling of large complex system from system of systems perspective, Keynote presentation, *2007 IEEE SoSE Conference,* April 18, 2007, San Antonio, USA.

Pearlman, J., 2006, GEOSS- global Earth observation system of systems, Keynote presentation, *2006 IEEE SoSE Conference,* April 24, 2006, Los Angeles, CA, USA.

Pei, R.S., 2000, Systems of systems integration (SoSI): a smart way of acquiring army C4I2WS systems, *Proceedings of the Summer Computer Simulation Conference*, pp. 134–139.

Sage, A.P., Biemer, S.M., 2007, Processes for system family architecting, design, and integration, *IEEE Systems Journal*, 1(1): 5–16.

Sage, A.P., Cuppan, C.D., 2001, On the systems engineering and management of systems of systems and federations of systems, *Information, Knowledge, Systems Management*, 2(4): 325–334.

Sahin, F., Jamshidi, M., Sridhar, P., 2007, A discrete event XML based simulation framework for system of systems architectures, *Proceedings of the IEEE International Conference on System of Systems,* April 2007.

Sloane, E., 2006, Understanding the emerging national healthcare IT infrastructure, *24 × 7 Magazine*, December 2006.

Sloane, E., Way, T., Gehlot, V., Beck, R., 2007, Conceptual SoS model and simulation systems for a next generation national healthcare information network (NHIN-2), *Proceedings of the 1st Annual IEEE Systems Conference,* April 9–12, 2007, Honolulu, HI.

Sridhar, P., Madni, A.M., Jamshidi, M., 2007, Hierarchical aggregation and intelligent monitoring and control in fault-tolerant wireless sensor networks, *IEEE Systems Journal*, 1(1): 38–54.

Walker, D., 2007, Realizing a Corporate SOSE Environment, Keynote presentation, *2007 IEEE SoSE Conference,* April 18, 2007, San Antonio, TX, USA.

Wang, L., Fang, L., Hipel, K.W., 2007, On achieving fairness in the allocation of scarce resources: measurable principles and multiple objective optimization approaches, *IEEE Systems Journal,* 1(1): 17–28.

Wilber, F.R., 2007, A system of systems approach to e-enabling the commercial airline applications from an airframer's perspective, Keynote presentation, *2007 IEEE SoSE Conference,* April 18, 2007, San Antonio, TX, USA.

Wojcik, L.A., Hoffman, K.C., 2006, Systems of systems engineering in the enterprise context: a unifying framework for dynamics, *Proceedings of IEEE International Conference on System of Systems Engineering,* April 2006, Los Angeles.

Chapter **2**

An Open Systems Approach to System of Systems Engineering

CYRUS AZANI[1,2]

[1]Northrop Grumman Technical Services, Herndon, VA, USA
[2]University of Maryland, College Park, MD, USA

2.1 INTRODUCTION

The twenty-first century engineering is witnessing an unprecedented change in the way we conceive, develop, field, and sustain systems. Many of the premises underlying the traditional systems engineering (SE) strategies are no longer valid. Traditional SE has been focusing on developing stand-alone systems with stable architecture and static technology base in which improvements were slow and very costly (Azani and Khorramshahgol, 2005). These strategies incorrectly assume that all the system of systems (SoS) requirements are known in the beginning of the development process and can be frozen in time or assumed to be stable. The traditional SE strategies also wrongly assume that the concepts of operation and various technologies used for constructing today's SoS are static and are subject to minor future changes.

Organizations can no longer afford to develop SoS based on traditional systems engineering approaches characterized by frozen requirements, reactive design, poor architecture archeology, and lack of understanding and consideration of elaborative interactions and interdependencies that exist among systems, processes, practices, and stakeholder organizations. It is apparent that traditional system engineering strategies are less effective in developing complex, dynamic, and evolving SoS. Traditional system engineering strategies are slow and result in excessive total system life cycle cost and unacceptable schedule delays. Such strategies are also less responsive to evolving needs of multiple stakeholders, rapidly changing technologies, and other

System of Systems Engineering: Innovations for the 21st Century, Edited by Mo Jamshidi

challenged faced the twenty-first century systems (Azani and Khorramshahgol, 2006). In the twenty-first century, systems engineers must deliver systems with the capability to be self-organized and self-regulated, and be reconfigured affordably and very quickly in response to evolving needs and rapidly changing technologies.

Although SoSE is a new concept, the SoS development is not new. Natural systems of systems have been in existence and are subject to continuous development and evolution for millions of years. Most of the challenges faced by an engineered or synthetic SoS are indeed the same challenges encountered for millions of years by many natural SoS. To survive and sustain, natural systems of systems had no choice but to adopt and follow certain fundamental open systems characteristics and principles. For example, they had to have open interfaces (i.e., permeable boundaries) in order to interact with their environment and had to adopt modular design principles (e.g., cohesiveness, encapsulation, and self-containment) in order to multiply themselves quickly and join forces to protect themselves and survive. The natural systems of systems have also created elaborative self-government systems through homeostasis, synergism, cybernetic control systems, and reconfigurable architectures in order to adapt and evolve in their harsh environment.

This chapter contends that the SoSE can consciously mimic the natural development processes and capitalize on natural development approaches such as the open systems strategy to build evolvable, affordable, and sustainable systems. Similar to a living organism, an engineered SoS is a whole greater than the sum of its parts; is capability rather than system based; has requirements and behavior that are both emergent and prespecified; has an encapsulated, self-contained, and cohesive structure; has different motivation, priorities, values, and practices compared to other systems; is network centric; is subject to predetermined blueprints and interface standards; and depends on symbiotic relationships and open interfaces to function and evolve within an uncertain environment.

2.2 THE OPEN SYSTEM CONCEPT

Systems that can exchange energy, material, and information with the outside world are called open systems. The open system concept is not new. It was first originated in biological and physical sciences and then migrated into physical and social sciences in the early parts of the twentieth century (Azani, 2001). In the late 1960s and early 1970s, the concept began to be applied in commercial information technology. The largest natural SoS on our planet (i.e., the ecosystem) and all the organisms living in it are considered open systems (Kay, 2000). These systems have been evolved and sustained for millions of years by being open. Commercial and social entities are also considered open systems because they are constantly exchanging energy (e.g., power that illuminates offices and factories and runs machineries and computers), material (e.g., raw material, labor, tools, hardware and software products), and information (e.g., knowledge, technical and nontechnical data) among each other and with their surrounding environment. Such entities cease to exist when they become closed systems.

Besides the ability to exchange energy, material, and information, open systems also possess capability to form complex hierarchical structures enabling them to cooperate and compete at the same time. As Ben-Jacob (2003) contends, open systems behavior and structures are governed by both competitive tendencies and collaborative exchange of information. In addition to exchange of information, competition for survival also plays an important role in enhancing cooperation among open systems. In fact, complex structures emerge in open systems only when there is competition between two or more tendencies (Turing, 1952). For example, although the notions of freedom and cooperation are usually perceived as contradictory (Poundstone, 1992), bacteria solved this apparent paradox by forming complex hierarchical colonial modular structures. The bacteria have also realized, over their course of evolution, that increasing informative communication among individual bacterium will result in increased freedom and cooperation among them (Shapiro 1992; Ben-Jacob 1997). Therefore, diverse open systems, living and nonliving alike, respond to externally imposed conditions by forming complex hierarchical spatiotemporal patterns (Thompson, 1944; Stevens, 1974; Ben-Jacob and Levine, 1998, 2001; Ball, 1999). Such structural patterns simultaneously elevate the degree of freedom of the individual cells and the level of cooperation among them.

Through openness, the biotic systems are enabled to capitalize on complexity to become more flexible and adaptable. In open systems, complexity does not have a negative connotation and plays an important role in enabling systems to adapt and evolve. According to Ben-Jacob (1997, 2003), structural complexity might be a more appropriate concept instead of entropy production to describe the response of open systems to external imposed conditions especially when these conditions vary in time and/or space. This argument is centered on the notion that stability, as used for closed systems, or open systems in steady states, is not valid for the hierarchical or scale-free spatiotemporal complex patterns formed during abiotic self-organization. In such cases, higher complexity elevates the flexibility of the system, thus imparting it higher tolerance and robustness so that it can better adapt to the external variable stimuli (Ben-Jacob, 2003).

Self-organization and complexity are closely related concepts in open systems. For example, a typical bacterial colony that consists of billions of bacteria is not created by predesign, but through the process of biotic self-organization. The bacteria store the information for creating the needed "tools" and the guiding principles needed for the colonial self-organization. Additional information is cooperatively generated, in response to environmental conditions, to create an adaptable complex system that can perform many tasks and is capable to learn and change itself accordingly (Ben-Jacob and Levine, 2006). In a way, the colony can self-organize itself to cope with adverse external conditions, because each bacterium is, by itself, a biotic autonomous system with its own internal cellular informatics capabilities such as storage, processing, and evaluation of information (Ben-Jacob and Levine, 2001). New features can collectively emerge during self-organization from the intracellular level to the whole colony. The cells thus assume newly cogenerated traits and abilities that are not explicitly stored in the genetic information of the individuals (Ben-Jacob and Levine, 2006).

Open systems are also characterized by emerging behavior and evolving structure. The emerging behavior and evolving structure of an open system is a function of its permeability to outside influence, inherited guidelines, ability to self-govern, and the degree of synergistic effects created as the system interacts with other systems and with its environment. The synergistic effects produced by wholes are considered the very cause of the evolution of complexity in natural systems (Koestler, 1969). Therefore, to understand an SoS emerging behavior and evolving structure, one should study the extent of SoS openness and multiplicity of interactions among constituent parts and dynamic processes that take place as a result of such interactions. The knowledge of the individual properties of the parts and the way they are connected to and act on one another allows us to compute the behavior of the whole (Atlan, 2003).

Timely reaction to evolving needs and technology requires agile open SoS that could quickly and cost effectively be integrated and reconfigured within an SoS environment. Affordable agility and adaptable sustainment of an SoS will demand engineering of openness attributes into an SoS from early phases of development life cycle. When applied effectively, a truly open SoS can enhance the adaptability of the system to changes in requirements and technology, will reduce the total ownership costs of the system, and will improve its overall life cycle supportability. Moreover, by following an open system approach in acquisition of an SoS, the program and project managers will be in a better position to leverage investments made in commercial products, practices, and technologies by other organizations and industries to field superior capability more quickly and affordably (Azani, 2005). Furthermore, an open system strategy considers life cycle support requirements up front, facilitates

FIGURE 2.1 Open systems benefits

integration of component systems into the SoS and the SoS integration with its environment, and permits system evolution with technology development and continuing technology insertion throughout the SoS life cycle. Figure 2.1 shows the benefits of an open SoS.

2.3 OPEN SYSTEM PRINCIPLES

The open systems development approaches are among the most adaptive system development efforts known to humankind. Although the engineering approaches utilized by biotic natural system of systems and living organisms are complex and very sophisticated, they are governed by a number of fundamental open systems principles or attributes. These principles depicted in Fig. 2.2 are synergism (e.g., holistic relationships), open interfaces (e.g., permeable boundaries), conservation (e.g., least amount of waste), modularity (e.g., self-containment and encapsulation), self-governance (e.g., self-organization, homeostasis, and self-regulation), emergence (e.g., emerging structure and behavior), symbiosis (e.g., mutual self-rewarding relationships), and reconfigurability (e.g., ability to change and adapt). Through adherence to these principles, open systems maintain system efficiency and flexibility, collaborate and compete with each other, protect themselves and become cohesive, reach a steady-state equilibrium, effectively handle complexity, determine and manage their emerging behavior and evolving structure, conserve resources, and above all adapt and sustain. Let us briefly examine the concepts underlying each of these strategies or attributes and their implications for SoS engineering.

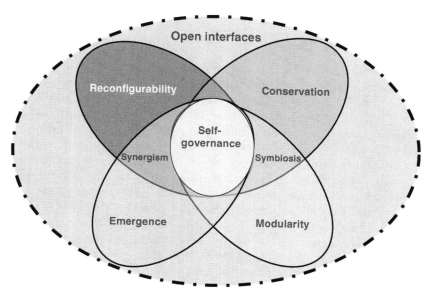

FIGURE 2.2 Fundamental open system principles

2.3.1   Open Interface Principle

In contrast to closed systems, natural open systems have permeable boundaries enabling them to exchange energy, matter, and information with other systems and with the environment surrounding them. Open systems, through permeable interfaces, exchange resources and information among themselves enabling them to learn, adapt, and evolve. Therefore, through open interfaces, for a limited time, which for most natural open systems have taken millions of years, natural open systems maintain themselves away from thermodynamic equilibrium (i.e., entropy) and live in a locally produced quasi-steady state (Kay, 2000). Such condition does not exist in closed systems, because in systems with closed interfaces, entropy is continually being maximized and disorder will ultimately rule. However, in open systems order is continually evolving from both order and disorder (Schrodinger, 1994), which postpones the ultimate state of entropy for the system.

Genetic blueprints and specifications govern the openness of interfaces within biotic systems. Such standards will enable the fittest of such systems to survive and sustain. In an engineered SoS, evolving technical baselines (functional, allocated, and product baselines), widely used standards and protocols and configuration control documents constitute the genetic makeup of the engineered system. In engineered SoS, interfaces are governed through adherence to widely supported and consensus-based standards (i.e., open standards) that are publicly available at nominal cost to all potential users (Azani, 2001a). Because these standards are reasonably stable, as well as widely distributed, competent vendors can create compliant products that work together with related products of other vendors to form a viable and evolvable SoS. The role of the builder of an SoS then changes from that of designer to that of architect and integrator (Azani, 2001b).

2.3.2   Synergism Principle

Synergism is the cooperative interaction among constituent parts of a system so that their combined effect is greater than the sum of their individual effects. Such cooperative interaction is enabled through open interfaces. In nature, system development is not performed in isolation from the system surrounding. Through open interfaces, natural systems are continually being influenced by their environment and, in turn, influence their surrounding. Based on the synergism principle, the knowledge gathered from the study of autonomous constituent systems is necessary, but not sufficient for understanding their cooperative behavior in an SoS. Therefore, one also needs to study the cooperative behavior of SoS constituent systems for full understanding of the SoS emerging behavior and the constituent systems response to SoS dynamic behavior.

The open systems approach to systems development and sustainment teaches us to always focus on systems development from a broader SoS perspective and take into consideration the interrelationships and interdependencies within a system, among systems, and between the system and its surrounding suprasystem. Such synergism implies an all-inclusive SoS design perspective. This holistic design perspective is a sustained development with minimum resource consumption and maximum return proven by millions of years of survival and sustainment. When engineered SoS are developed as closed systems (in a stove-piped manner and as end into themselves), they disturb the delicate synergistic equilibrium that exists among various types of systems within their environment. Consequently, they start to degrade their own short- and long-term functioning and sustainment through industrial waste, deteriorated ecosystem, increased pollution, obsolescence, and other intended or unintended consequences associated with closed systems.

2.3.3 Self-Government Principle

True open systems are characterized by self-governance tenets such as cybernetic control (self-control), homeostasis (self-regulation), and self-organization. Cybernetic control is the ability to predict and control a system's behavior through proactive (feedforward) as well as reactive (feedback) mechanisms to process information, react to information, and change to adapt. Homeostasis concept was first coined by Cannon (1932) to explain the mechanisms (e.g., cooperating methods or means acting simultaneously or successively) used by living organisms to maintain their constancy or steady-state operating conditions. It is the ability of an open system to self-regulate itself and maintain the system on an evolving steady state. Homeostasis does not occur by chance, but is the result of organized self-government (Cannon, 1932). In natural systems, homeostasis is the maintenance of temperature, fluid levels, blood sugar, acid–base balance, and other conditions. Self-organization is a concept with roots in physics and biology. It is a process by which the inherent order of a system increases and its internal organization becomes more complex without intervention by an outside source. In an engineered SoS, the self-governance should be built into the system through sensors, expert systems, and proactive system design that can continually monitor the environment and flow of resources into the system and adjust the emerging SoS behavior and structure accordingly and automatically.

By its very nature, self-organization determines the SoS freedom of response, which should not be completely predetermined but also regulated. Regulated freedom means that the communication biases the constituent systems toward a preferred range of responses not strictly defined by the SoS. It is possible that a component system misbehaves and produces a response that is not acceptable or even destructive to the SoS. However, the destructive ability of such individual system should be regulated by the SoS governance. It seems, on the surface, that regulation and freedom are competitive. However, as seen in the bacteria, the appropriate colonial organization can lead to an increase in both freedom and cooperation (Ben-Jacob, 2003). Similar to a biotic system, a message, for example, a chemical agent, may trigger a specific, predetermined pathway within a constituent receiving system, which in turn induces an automatic, predetermined response of the SoS. With informative communication, the message initiates in the receiving constituent system an individual interpretation process, involving internal restructuring (self-organization) based on its current internal state and previously accumulated information. The new structure enables the system the freedom to select its response to the message. Such freedom implies that the internal self-organization is connected with generation of new information.

In engineered systems, self-governance is represented by routine administrative and managerial activities contemplated to create and maintain SoS equilibrium (homeostasis) within tolerable limits. Such equilibrium will ensure SoS hygiene and will enable the SoS to create balances between the inflow and outflow of energy, material, and information into and from the system on one hand and between conflicting demands/interests of various SoS stakeholders on the other hand. Similar to living organisms, an engineered SoS can be built to have conformer, regulator, and negative as well as positive feedback control mechanisms. The regulator mechanisms (e.g., sensors) built into an SoS should maintain the SoS at a constant level over possibly wide ambient environmental variations and diverse stakeholders demands. The conforming control mechanisms built into an SoS, can enable the SoS environment to shape and sustain the SoS at a steady-state equilibrium through constant flow of information and insertion of new technology. The negative and positive feedback control loops (i.e., cybernetic control systems) should be built into an SoS in such a way as to keep conditions from exceeding tolerable limits and to push levels out of normal ranges to ensure SoS sustainment under predefined critical conditions.

2.3.4 Emergence Principle

Emergence is the occurrence of novel and coherent structures, patterns, and properties during the process of self-organization in complex systems (Goldstein, 1999). When a large number of systems or elements are assembled and linked to form a complex new system, new qualities and properties of the system as a whole, but not of its parts, emerge (Langton, 1989; Waldorp, 1993; Gell-Mann, 1994; Horgan, 1995; Badii and Politi, 1997; Goldenfeld and Kadanof, 1999; Vicsek, 2001). Knowledge of the individual properties of the modules and the way they are connected to and act on one another allows us to compute the behavior of the whole (Atlan, 2003).

The emerging behavior of an SoS is the direct result of synergy among its constituent systems, interaction of each constituent system with its immediate surroundings, and the interaction between the SoS and its environment. Based on this principle, the behavior of a system cannot be predicted by the behavior of its constituent parts taken separately. For example, no physical property of an individual molecule of air would lead one to conclude that a large collection of them will transmit sound. In addition, hurricane is the result of synergy of water, wind, and high and low pressures. Individually, these elements are generally stable, but as they interact, they can yield a tremendous force.

2.3.5 Conservation Principle

In natural open systems, materials are cycled back through the systems and everything that is produced is used in some fashion by other systems for their own benefit. Consequently, the boundary between resources and wastes does not exist in natural open systems, since output from systems is input for other systems. Therefore, the efficiency of an open systems development approach can be observed in how little natural systems waste. In building an SoS, nature utilizes multipurpose and complementary constituent systems that conserve energy and material to sustain themselves and their encompassing SoS. In open systems, there is no distinction between resources and wasteful products. For example, plants as systems absorb their necessary inputs (e.g., light, nutrients, and liquids) from the air and soil effortlessly with least amount of energy and waste. They then convert these inputs into finished rich products by delicate chemical and biological processes, and with a very high degree of efficiency, precision, and quality. Moreover, in their process of development and daily production, plants also do their huge share of contribution to the sustainment of environment and ecosystem by absorbing the harmful CO_2 and releasing the much-needed oxygen into the environment. Additionally, plants' waste is food for other plants or organisms. Animals and microorganisms for the most part (except for humans) develop their systems in a similar fashion—very effortlessly with minimum expenditure of resources and maximum possible return to themselves and to the ecosystem.

In engineering systems, man has not yet been able to remove the boundaries between resources and waste. Systems engineering and development has still a long way to go in minimizing waste and conserving energy and material. To conserve resources, SoS engineers should consider all the interrelationships and interdependencies that exist within the SoS environment and among its constituent systems to identify the sources of waste in an SoS and formulate strategies to conserve material and energy through an open systems design. An open systems approach to SoS development focuses on system life cycle supportability by considering life cycle support requirements up front, permitting system evolution with technology development, anticipating technology obsolescence in system design, and by supporting technology insertion.

2.3.6 Reconfiguration Principle

Natural open systems continually reconfigure themselves to become more adaptable to the changes in their environment. Rather than a reactive and rigid design strategy, they use proactive and flexible approaches to development and sustainment. As the environment changes, such systems reconfigure themselves to adapt to the changes without external intervention. Adaptation is a slow process caused by natural selection. Natural selection is the process by which specific traits of an open system that helps it to survive become more common in successive generations of that system. A system's genetic makeup and the environment in which the system lives determine the traits. Natural selection often results in the maintenance of the status quo by survival of the fittest or eliminating less fit variants.

Open systems adaptive reconfiguration can be structural, behavioral, or physiological. Structural reconfigurations are enabled by flexible open architecture used by the system to help it survive in its natural habitat. For example, coniferous trees have cone shape and needlelike leaves to better adapt to severe cold conditions. Other examples of structural adaptation are skin color of snakes, shape of a tree, and body covering of animals. Behavioral adaptations are unique ways a particular organism behaves to survive in its natural habitat. For example, adaptability of a bacterial colony depends on communication among bacteria (Joset and Guespin-Michel, 1993; Rosenberg, 1999; Nester et al., 2001). The communication induces changes in cell behavior by direct biochemical pathways or by triggering internal self-organization. Bacterial colonies have developed these intricate communication capabilities to evolve and sustain based on sophisticated modes of cooperative behavior and colonial self-organization (Huberman and Hugg, 1986). Physiological adaptations are innate characteristics that allow an organism to perform certain biochemical reactions (e.g., making venom, secreting slime, being able to keep a constant body temperature) to survive in its environment.

An engineered SoS adaptation to a large extent is determined by its proactive and flexible executable architecture, open standards used in defining its interfaces, and adequacy of its technical baseline and configuration control documents. One of the main differences between engineering and nature is the use of some degree of randomness—classically considered as a source of disorganization—to increase the efficiency of a system (Atlan, 2003). SoS engineers should integrate this concept into their own engineering methods to improve SoS efficiency and design an adaptive SoS. Moreover, they should strive to tailor SoSE processes and models for each SoS. For example, simulation methods used should enable SoS engineers to model SoS configuration and behavior under different scenarios to identify constituent systems that cannot adapt to the SoS requirements before they are integrated into the SoS architecture.

2.3.7 Symbiosis Principle

Symbiosis is a mutually beneficial relationship between open systems. Every organism has many internal parasites that help it to digest food and survive. For example, there is a symbiotic relationship between small fishes and sharks. Small fishes ride along the side of the shark, rub up against the shark's sandpaper-like skin to rid themselves of parasites, and not only clean the shark of parasites but also get a free meal from bits of the shark. Cross-pollination is another example of symbiotic relationship, which improves the genetic variations and allows systems to better adapt to changes in environmental conditions. Evolution is strongly based on symbiotic relationships rather than competition among species. Consequently, the organisms that cooperate with each other or members of different species often outcompete those that do not (Margulis et al., 1986).

Successful development and sustainment of an SoS is to a large extent dependent on symbiotic collaboration among stakeholders of systems comprising it. The most beneficial type of symbiosis for an engineered SoS is mutualism. With this type of relationship, all the constituent systems comprising an SoS benefit from the mutually beneficial network of relationship and neither suffers. Symbiosis is needed among diverse SoS stakeholders to create a win–win situation for all. This can be accomplished by creating overarching integrated product and process teams comprising of representatives from principal SoS stakeholders. The team needs to understand interrelationships operational within the SoS in order to prescribe changes necessary to affect the desired outcome and planned improvements that benefit all the constituent systems.

2.3.8 Modularity Principle

In the hierarchical organization of nature, each level and each system is independent, to a certain degree, of the others (Anderson, 1972). In nature, modular design is very prevalent. In fact, complex organisms have evolved and thrived because of modularity. Although galactic systems, planets, and organisms are interdependent and intertwined, they also retain modular design characteristics. They are self-contained, decoupled, scalable, and highly cohesive by design. Modularity is an important enabler to standardization and customization of natural entities as well as engineered systems. It provides some degree of autonomy to each modular element enabling modules to more easily evolve and be more affordably sustained and substituted.

Since open systems maintain their autonomy through modularity, modular design is an enabler for effective management of complexity. For example, each bacterium is, by itself, a biotic autonomous system with its own internal cellular gel that posses information processing capabilities (storage, processing, and interpretation of information). These capabilities allow the cell a certain freedom to select its response to biochemical messages it receives, including self-alteration and broadcasting messages to initiate alteration in other bacteria. Such self-plasticity and decision-making capabilities elevate the level of bacterial cooperation during colonial self-organization. Since each bacterium is an autonomous self-contained living system, colonial developments provide for even more sophisticated means of engineered self-organization (Ben-Jacob and Levine, 2006).

Modularity should play a more important role in design and development of an engineered SoS. By partitioning the SoS into constituent autonomous modules and using open standards to define the interfaces among these modules, access to multiple sources of supply becomes a reality and changes become less difficult and more affordable to make (Azani, 2001b). The constituent systems comprising an SoS must follow modular design principles so that the design decisions in each module (i.e., system) do not affect decisions in other modules. Modularity should also be used in design of engineered SoS to facilitate division of design tasks among group of designers responsible for designing each system and enable them to work independently for different firms operating at different geographical locations. In applying modular design tenets, SoS designers should watch for hidden and unexpected dependencies (e.g., unwarranted extensions to open standards) among constituent systems and analyze design options to accommodate future changes. Modular SoS architectures allow plug-and-play, mix-and-match, ability to evolve and be compatible, and capability to change as technology and needs change.

2.4 THE OPEN SYSTEMS APPROACH TO SoSE

As shown in the preceding sections, open system principles play a very important role in development and sustainment of complex natural systems and are equally important considerations for design and development of engineered SoS. Similar to a natural SoS, an engineered SoS should be designed and developed as an open system in order to continually exchange material (e.g., constituent systems, hardware and software modules, resources), energy (e.g., power and signals), and information (e.g., meaningful data) with the environment encompassing it. In addition, similar to natural systems, engineered SoS must use open architectures to ensure modularity, reconfigurability, interoperability, evolvability, and effective management of change and emerging behaviors. Open architectures should depict the decomposition of the overall functionality of an engineered SoS into a set of well-defined and synergistic functions. Such architectures should also identify the modular components that provide such functionality and define the specifications of the interfaces among components by widely supported and consensus-based standards to create an overall SoS functionality greater than the sum of the functionalities of its constituent parts.

The open systems approach to SoS engineering is an integrative methodology that will apply the fundamental open systems principles discussed earlier to build affordable, adaptable, sustainable, and interoperable SoS. It capitalizes on the synergism principle of open systems to integrate the capabilities of a mix of existing and new systems into a system of systems capability greater than the sum of the capabilities of the constituent parts. It applies the open systems principle of modularity to build expandable, adaptable, self-contained, and loosely coupled systems that comprise the portfolio of an SoS. The approach also utilizes the open interface and self-government principles to effectively manage the SoS processes and behavior, build interoperability into the SoS, enable the SoS to more easily integrate or remove constituent systems from its capability portfolio, insert new technology, and improve performance. Additionally, the approach capitalizes on the symbiosis and homeostasis principles of open systems to build a collaborative environment for engineering of an evolving SoS and regulate its internal environment to maintain a stable balance (dynamic equilibrium) of governance, resources, and operation for the system.

The open system approach to SoSE (Fig. 2.3) consists of six integrated phases. Each phase will capitalize on one or more of the open system principles identified earlier to effectively engineer, develop, and integrate affordable, sustainable, interoperable, and reconfigurable systems of systems needed for twenty-first century. The phases are as follows:

Phase 1: Establish an open system enabling environment.

Phase 2: Define portfolio of SoS capabilities and requirements.

Phase 3: Allocate capabilities to candidate constituent systems.

Phase 4: Develop an open executable SoS architecture.

Phase 5: Simulate evolving SoS structures and emerging behaviors.

Phase 6: Assemble and reconfigure the SoS as needed.

The remainder of this chapter will concentrate on specific tasks and important open system considerations at each of the above-mentioned phases.

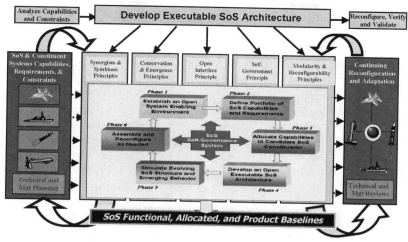

FIGURE 2.3   The open system approach to SoSE

Phase 1: Establish an Open System Enabling Environment

The first phase of the proposed methodology lays the foundation for establishing supportive business and technical practices for development of an open SoS (Azani and Flowers, 2005). Following is a list of supportive technical and business practices needed for effective conduct of this phase:

(a) Formation of a governance structure comprising an SoS Joint Engineering Team (JET) to plan, organize, and oversea all the phases and associated tasks required for SoSE. The SoS JET should foster information exchange and integrated technical action. The team members comprise technical leads from candidate constituent systems as well as representatives from other SoS principal stake-holders. The SoS JET must have sufficient authority, clearly defined areas of responsibility, and adequate resources at its disposal to successfully initiate and complete the required SoS development and sustainment activities. The team should develop and use integrated approaches to effectively share data, execute joint assessment, and make group decisions. The team members should allocate time and resources needed to conduct joint trade studies and perform reasonable analyses of the impacts at the SoS level. The team should also vigorously enforce conformance to SoS plans, policies, and interface standards. The SoS ET should also ensure that the SoS contractors are aware of those plans and interface standards as well as the responsibilities for implementing them.

(b) Presence of effective configuration management processes to effectively manage changes to SoS interfaces and their corresponding standards over the SoS life cycle.

(c) Existence of program staff with training or relevant experience in open system concept and implementation.

(d) Program management and acquisition planning efforts conducive to open systems implementation. The SoS systems engineering plan (SEP) and strategies for technology development, acquisition, test and evaluations, and product support should all be supportive of open systems implementation. The SoS SEP should encourage adherence to open systems principles identified earlier.

(e) Effective identification and mitigation of barriers or obstacles that can potentially slow down or even, in some cases, undermine compliance with open systems principles. The tendency of human participants or stakeholders (SoS owners, users, developers, suppliers, financiers, engineers, etc.) to pursue self-interest and continue to suboptimize the principal requirements and goals of an SoS create numerous formidable obstacles/barriers that must effectively be dealt with by the JET. For example, the engineering team needs to remove or at least neutralize the predicted resistance to changes brought about by a new SoS and control the tendency of stakeholders to politicize the SoS development and sustainment processes for their personal or group gain. None of such tendencies seems to exist or manifest in natural biotic systems of systems. However, in engineered SoS, they are perhaps the main contributing factors causing unpredicted emergent SoS behavior and are the principal reasons why most of manmade systems development projects fail. The JET should identify such obstacles/barriers in advance and establish plans to effectively mitigate their risks.

(f) Continuing market research and analysis to analyze commercial market capabilities and trends, collect and evaluate data on available and emerging interface standards to determine whether or not they are applicable to the particular SoS, and assess the breadth of open and de facto standard compliant products to determine if suppliers will continue to produce or support the standards selected.

(g) A list of current and future SoS stakeholders, their roles, needs, capabilities, level of expertise, and disposable resources.

(h) A comprehensive SoS database of information that among other things includes timely and sufficient information about existing and planned systems that will comprise the SoS.

Phase 2: Define Portfolio of SoS Capabilities and Requirements

During this phase, the SOS JET should identify changing needs and technologies, key performance and operational parameters, the required level of technical maturity, expected integration complexity, and imposing constraints (e.g., required interfaces, operating environment, and resources) on the SoS. The SoS capabilities and requirements are owned by SoS constituent systems and should be prioritized in collaboration with the community of users and sustainers. The identification and management of new requirements over time and the correlation and traceability between the desired capabilities and the configuration of the deployed SoS are paramount in an SoS environment characterized by evolving threats, technologies, and concept of operations. The communications and connectivity requirements such as consistency in ontology and semantic and syntactic interoperability must be managed with extreme care and fidelity. It should be noted that the traditional requirements prioritization processes may "engineer out" unnecessary redundant requirements. Within the context of an SoS, these requirements may still be valid for individual constituent systems, and removing requirements to avoid unnecessary redundancy may limit the capability of the constituent systems.

The team should use collaborative engineering tools and dynamic cost models to establish an integrated roadmap for funding, engineering, development, and evolution of the SoS. An integrated development environment (IDE) may provide bridging and integration among the individual constituent requirements systems to support collaboration and create the needed synergism. Within this phase, the SoS JET should also determine the scope of SoS, set SoS boundary, and identify operational scenarios. The team should also compile and correlate multidimensional data sets that besides supporting group decision making may also depict and explain synergistic effects. Such effects may be resulting from capability relationships (e.g., interoperability requirements), architectural relationships, and functional and programmatic interdependencies (e.g., schedule, cost, and functionality relationships) among programs/systems, and their encompassing SoS.

Phase 3: Allocate Capabilities to Candidate Constituent Systems

This phase begins with a notional or conceptual SoS framework, reference model, or architecture that depicts candidate constituent systems functionality, their relationships to each other, and candidate interface standards that might be used to govern the constituent systems behavior and evolution. In allocation of SoS capabilities to constituent systems, the SoS JET should ensure constituent systems' autonomy (e.g., independent management and operations) using modularity, symbiosis, and self-governance principles. While it is true that minor constituent program risks could be major risks to the SoS, it is also true that significant system risks may have little or no impact on the SoS functionality. Major risks associated with capability allocation may include risks associated with unwanted emergent behavior, inconsistent infrastructure, complex and costly integration, and technical risks associated with the constituent systems not meeting the SoS mission and objectives. An integrated risk management board should be established with members from constituent systems to analyze and formulate mitigation strategies for SoS that may include reallocation of SoS capabilities to substitute constituents, especially if some of the constituents are reaching their service life.

The SoS configuration will need to reflect all the available options in the available and relevant assets of constituent systems, along with their integration needs and evolution. An SoS configuration control board with representation from each of the constituent programs is essential to provide SoS configuration management and to ensure that key elements and interfaces are implemented as planned. Configuration identification will be accomplished at the SoS level through the allocation process; each SoS constituent system will likely become a configuration item.

Centralized management for key interfaces, data, and system characteristics that ensure interoperability and integration is essential. Moreover, all stakeholder organizations should participate in baseline management activities of constituent systems and vice versa. An engineering review board at the SoS level can be established to examine the impact of constituent program changes on the SoS and vice versa.

Allocation of capabilities should be done in a manner to enhance synergism and improve symbiosis. A fundamental requisite to the development of a symbiotic SoS is the self-governance process and the manner in which constituent systems' SEPs are included and integrated in the SoS SEP. The SoS governance and management processes utilized by the SoS JET should be robust enough to meet the diverse needs of large groups of stakeholders and address their multiple perspectives. The technical management processes established by the SoS JET must ensure that the views, expertise, and interests of constituent systems are factored into the processes. Technical planning for an SoS will need to augment as well as take into account the plans of those constituent systems. It must also be initiated top-down but iterated with the constituent systems until a consensus (not necessarily optimum) solution is agreed upon and resourced. Open and active exchange of information, plans, and experimentations among members of the SoS JET is important for supporting the SoS technical planning process. A key tool for accomplishing this is the development of an integrated master schedule (IMS) for the SoS. The SoS JET should consider the constituent systems analysis plans and integrate their relevant aspects (e.g., house-keeping, requirements, engineering, sustainment) into the SoS systems engineering plan.

Phase 4: Develop an Open Executable SoS Architecture

In this phase, the notional or conceptual SoS framework, reference model, or architecture, referenced in the previous phase, will evolve into an open executable SoS architecture. Such SoS architecture will capitalize on open interface and other principles of the open systems approach, will reside in executable code, and can be represented or visualized by dynamic executable and time-dependent behavior models. The SoS architecture should describe the rationale for selection of constituent systems, define synergistic relationships and symbiotic interactions among them, and delineate the overall behavior of the SoS as a whole. The SoS architecture is a living and evolving document that should represent an enterprise view rather than the structured hierarchy of constituent systems. It should set the context for design and implementation of an SoS and establish and document significant decisions about the organization of the SoS and structural elements and interfaces comprising it. The SoS architectural analysis and development is an iterative process of elaborating, partitioning, decomposing, and allocating the SoS functional and performance requirements across the constituent systems. The focal point of an SoS architecting is significant design decisions that are both structurally and behaviorally important and have a lasting impact on the SoS as well as its constituent systems' performance, evolution, and resilience. Constituent systems will likely be treated as configuration items or modules from a traditional SE perspective or as network nodes or services from an information perspective. The SoS architectural analysis involves a bottom-up assessment of available (current and legacy) systems, systems under development, and systems under planning. These assessments are merged with a top-down allocation of desired capabilities discussed earlier.

The output of the architectural analysis process is an executable SoS architecture baseline that elaborates and decomposes the SoS operational requirements, defines functional flows and open interfaces, and identifies data and control flows across the SoS. Most SoS are systems within which software is the principal, necessary, and indispensable element within the SoS itself and most of its constituent systems. Consequently, the SoS architecture should reside in executable code and be represented or visualized by dynamic executable and time-dependent behavior models of SoS processes, organization, and resources. The SoS JET should ensure that the SoS architecting will develop executable architectures with dynamic representation and reconfiguration capability enabling continuing SoS performance and effectiveness analysis under various conditions and procedures. Static architecture representations and products are not effective tools for analysis of SoS evolving structure and emerging behavior.

The SoS architecture should also enable continuous and effective exchange of resources, products, and information among the independent systems comprising the SoS and between the SoS and its environment. The SoS JET should capitalize on evolutionary development and acquisition strategies and service-oriented architectures to synchronize architecture configuration and reconfiguration. The JET may want to establish a set of descriptors to facilitate binding architecture data from different sources. One approach would be to capitalize on integrated architecture behavior models such as the one being developed by DoD Joint Systems Engineering Task Force to provide dynamic specification of component behavior as well as the emerging behavior of the SoS.

The SoS JET should ensure that open standards are used for data exchange and representations as well as for other SoS interfaces to develop a standard-based SoS architecture. Such architecture enables other systems to participate in the SoS in the future and use their existing interfaces, developed for other purposes, to support new SoS requirements. Having an open systems architecture is especially important to ensure sustainability given the number and the complexity of the interfaces. A modular and open SoS architecture creates a flexible and agile SoS by facilitating technology refreshment and insertion opportunities, while also providing for a more sustainable SoS. An open systems approach must account for varying levels of constituent system openness, as many legacy systems may have proprietary issues that can influence any middleware development.

Phase 5: Simulate Evolving SoS Structures and Emerging Behaviors

Open systems principles such as self-governance, open interfaces, reconfigurability, and modularity should provide the framework for defining the SoS architecture and federations of simulations needed to represent mission threads, communications networks, and the operational environment needed for examining costs and effectiveness associated with execution of SoS business and technical processes. The SoS openness will also facilitate the connectivity and integration of multiple models and multiple modeling tools needed for simulating emerging SoS structure and behavior. The SoS evolving structure and emerging behavior are a function of synergistic effects created as a result of dynamic interactions among diverse autonomous systems comprising the SoS and the interaction between the SoS and its evolving environment.

To facilitate simulation, an SoS should employ executable architecture models, prototyping, and advanced tools to conduct experimentation in the SoS operational context. The experimentation will assist in establishing the SoS boundary, range of feasible capabilities, and dynamic interface requirements. The boundary for an SoS should be examined periodically and modified as needed in response to evolving mission and architectures rather than taken as constraints at face value. The SoS simulation will also assist in better understanding of the constituent system characteristics, functionality, architecture, and interfaces. Testing along with simulations in an integrated simulation environment will also be needed to evaluate the SoS architecture viability and assess the technical maturity and feasibility of integrating constituent systems. The SoS JET should also establish mechanisms such as "community of stakeholders" as early as possible to provide visibility into SoS dynamic interactions and conflicting requirements. Requirement conflicts with constituent systems must be identified and adjudicated via the prototype model as early as possible.

Timely, accessible, and accurate information with regard to obsolescence, technology refreshment, compliance of current and likely constituent systems with the SoS communications and connectivity protocols and standards, and the general evolution of each of the constituent systems, particularly where such changes are likely to have an impact on the SoS capabilities, is very critical for modeling the SoS behavior. In addition, different types of data will be required for decision making at different levels (i.e., constituent system level, integration level, and SoS level). For example, there may be proprietary or restricted data tied to the constituent systems that must be managed within the SoS to protect intellectual and proprietary data. To affordably simulate the SoS emerging behavior and structure and to share and exchange program data, an integrated SoS database will be required with links to constituent systems databases. It should have syntactic correctness and retain semantic context of data across all the systems.

Phase 6: Assemble and Reconfigure the SoS as Needed

The SoS integration is not a onetime assembly of constituent components at the conclusion of design development. It is typically incremental, using an iterative process of assembling/integrating components, simulating their combined interaction, evaluating interrelationships, then reassembling, simulating, and integrating more components. SoS engineering and development is an ongoing process and will be subject to continuing refinement and reconfiguration. Through continuing assembly (integration) and reconfiguration, the SoS JET will ensure that the SoS, as developed and integrated, functions properly.

Modeling and simulation plays an important role in test and evaluation of SoS solutions. Through modeling and simulation, the SoS JET needs to ensure the readiness of integration environment and confirm that the SoS components are compliant with the SoS interface requirements. This phase also encompasses formal product verification and validation to verify component requirements, interfaces, and their integrated behavior. Assembly and reconfiguration is undertaken under the premise that the SoS constituent systems and their components (SW and/or HW) are tested and ready for integration and higher level testing. As part of this phase, the team should also establish strategies for conduct of integration and identify integration plans, procedures, and criteria. The integration strategy and reconfiguration should focus on the following factors:

- Determining whether the SoS integration begins with simulators and prototypes and steadily progresses to the "real" system or starts with building the real system.

- Identifying number and complexity of component systems: their stability, readiness, symbiosis, true openness and modularity, operational requirements, required interfaces, and management schemes.

- Identifying potential suppliers and their record of accomplishment and the extent of experience of the integration team.

- Assessing the processes, steps, complexity of the integration, and the types of risks to be mitigated before integration begins.

- The extent of COTS solutions in component systems and their future implications.

- Understanding what components must be integrated and in what order. For example, should SoS integration order be based on need, complexity, critical dependencies, or high-risk areas.

- Recognizing the requirements, test plans, and procedures approved and under configuration control. Also, understanding what simulators, tools, and COTS are under configuration control.

- Evaluating whether or not all the key players are clear on their roles and responsibilities, and what resources are required (e.g., people, facilities, test stubs/drivers, simulators, etc.).

- Establishing an integration schedule and integration entry and exit criteria.

- Establishing a list of integration activities and their sequence as well as the type of tools needed (e.g., automated test tools, debuggers, script generators, simulators, test injectors, etc.).

As the SoS integration proceeds within each iteration, there is need for continuing interface management. Interface management is a critical SoSE technical management function that provides a process for managing interface baselines and the means to identify and resolve interface issues among constituent systems and between the SoS and its environment. It includes ongoing assessment of the adequacy of the SoS interface descriptions, documentation of interface designs (e.g., interface control documents, interface design documents), management of SoS interfaces via integrated interface control working groups or other control boards, and establishing traceability matrices to ensure that SoS interface requirements and designs stay in sync.

As part of final SoS readiness review, the SoS governance should conduct assessments aimed at evaluating the composite readiness/maturity of the mix of constituent systems and their interaction, as well as the configuration of the SoS in response to the set of defined capabilities and constraints. It should also evaluate the external interfaces of the constituent systems, their compliance with the SoS connectivity and communication protocols and open standards, and their compatibility with the other constituent systems. Experimentation with the user becomes extremely important. In this context, the SoS becomes the "system of interest," and the necessary acceptance criteria and relevant technical performance measures must be developed and tracked. Technical assessments must look beyond the basics of the maturity of specific technologies and systems to their maturity with regard to integration within the context of an SoS. A multidimensional review correlation matrix may be necessary to support a final review for the total SoS. In addition, specific entry and exit criteria for technical reviews should be developed based on the required linkages or dependencies between constituent system reviews to assess the total SoS composite maturity and readiness level.

2.5   CONCLUSION

This chapter discussed an integrated business and technical methodology for development of an SoS based on open systems development principles. The chapter elaborated on a number of fundamental open system development principles; integrated these principles into an integrative and systematic methodology for engineering and development of affordable, interoperable, adaptable, and evolvable system of systems; and presented a number of guidelines for effective utilization of the proposed methodology. It should be emphasized that compliance with open system principles imposes constraints and discipline (as does any approach). For example, if the integrity of the interfaces is not maintained, or if SoS developers attach unwarranted extensions to interface standards, then system design may cease to be open, the benefits of openness are lost, and the design becomes more like a traditional point solution rather than an open system solution.

An open systems approach to SoSE considers life cycle support requirements up front, permitting system evolution with capability and technology development, and anticipates malfunctioning and obsolescence in system design as the SoS changes and evolves. Furthermore, by utilizing this approach, organizations are enabled to maintain the superiority of a modern complex SoS within constantly changing requirements, growing resource constraints, and unprecedented rate of technological change. Finally, through openness of an SoS and its constituent systems, an organization will be in a better position to continually leverage the investments made in commercial products, practices, and technologies of the global market place and adapt its behavior without external intervention.

2.6 DISCUSSION QUESTIONS

1. What are the main similarities and distinctions between natural and engineering development approaches? Why an open systems approach is needed for engineering and development of an SoS?

2. How should SoS engineers design and develop an SoS based on system development and sustainment approaches used by Nature?

3. Which open system principles should be emphasized or deemphasized in engineering and development of a typical SoS? Are these principles equally important in all circumstances?

4. What characteristics should an SoS architecture possess to make it reconfigurable, adaptable, and evolvable?

5. What causes an emerging behavior in an SoS? How could one predict the direction of such behavior and what factors should an SoS engineering team consider to proactively prepare for and manage an SoS emerging behavior?

REFERENCES

Anderson, P., 1972, More is different, *Science*, 177: 393–396.

Atlan, H., 2003, The living cell as a paradigm for complex natural systems, *Complexus*, 1: 1–3.

Azani, C.H., 2001a, The test and evaluation challenges of following an open system strategy. *The ITEA Journal of Test and Evaluation*, 22(3).

Azani, C.H., 2001b, Joint space operations via secured integrated network of modular open architectures, *Proceedings of the Joint Aerospace Weapons Systems Support, Sensors, and Simulation Symposium and Exhibition*, July 23–27, 2001, San Diego, CA.

Azani, C.H., 2005, Enabling net centric capability through secured integrated networks of modular and open architectures, *Proceedings of 8th Annual Systems Engineering Conference Sponsored by National Defense Industrial Association and International Council on Systems Engineering*, October 24–27, 2005, San Diego, CA.

Azani, C.H., Flowers, K., 2005, Integrating business and engineering strategy through modular open systems approach, *Defense AT&L Magazine*, January–February 2005.

Azani, C.H., Khorramshahgol, R., 2005, The open system strategy: an integrative business and engineering approach for building advanced complex systems, *Proceedings of the 9th World Multiconference on Systemics, Cybernetics and Informatics*, July 10–13, 2005, Orlando, FL.

Azani, C.H., Khorramshahgol, R., 2006, The open software intensive system strategy: an integrative approach for building software intensive systems, *Journal of Computer Information Systems*, 47(1).

Badii, R., Politi, A., 1997, *Complexity Hierarchical Structures and Scaling in Physics*, Cambridge University Press.

Ball, P., 1999, *The Self-made Tapestry. Pattern Formation in Nature*, Oxford University Press.

Ben-Jacob, E., 1997, From snowflake to growth of bacterial colonies: cooperative formation of complex colonial patterns, *Contemporary Physics*, 38: 205–241.

Ben-Jacob, E., 2003, Bacterial self-organization: coenhancement of complexification and adaptability in a dynamic environment, *Philosophical Transactions Royal Society of London A*, 361: 1283–1312.

Ben-Jacob, E., Levine, H., 1998, The artistry of microorganisms, *Scientific American*, 279: 82–87.

Ben-Jacob, E., Levine, H., 2001, The artistry of nature, *Nature*, 409: 985–986.

Ben-Jacob, E., Levine, H., 2006, Self-engineering capabilities of bacteria, *Journal of the Royal Society Interface*, 3(6): 197–214.

Cannon, W.B., 1932, *The Wisdom of the Body*, W. W. Norton Publication.

Gell-Mann, M., 1994, *The Quark and the Jaguar*, Freeman Publication, New York.

Goldstein, J., 1999, Emergence as a construct: history and issues, *Emergence: Complexity and Organization*, 1: 49–72.

Goldenfeld, N., Kadanof, L.P., 1999, Simple lessons from complexity, *Science*, 284: 87–89.

Horgan, J., 1995, From complexity to perplexity, *Scientific American*, 272: 104–109.

Huberman, B.A., Hugg, T., 1986, Complexity and adaptation, *Physica D*, 22: 376–384.

Joset, F., Guespin-Michel, J., 1993, *Prokaryotic Genetics: Genome Organization Transfer and Plasticity*, Blackwell Scientific, Oxford.

Kay, J., 2000, Ecosystems as self-organizing holarchic open systems, in: Jorgensen, S.E., Muller, F. (Eds.), *Handbook of Ecosystems Theories and Management*, CRS Press, Lewis Publishers.

Koestler, A., 1969, *Beyond Reductionism: New Perspectives in the Life Sciences*, Beacon Press.

Langton, C., 1989, *Artificial Life*, Addison-Wesley, New York.

Margulis, L., Chase, D., Guerrero, R., 1986, Microbial communities, *Bioscience*, 36(3): 160–170.

Nester, E.W., Anderson, D.G., Roberts, C.E., Pearsall, N.N., Nester, M.T., 2001, *Microbiology: A Human Perspective*, 3rd Edition, McGraw Hill.

Poundstone, W., 1992, *Prisoner's Dilemma*, Oxford University Press.

Rosenberg, E., 1999, *Microbial Ecology and Infectious Disease*, ASM Press, Washington, DC.

Schrodinger, E., 1994, *What is Life?* Cambridge University Press.

Shapiro, J.A., 1992, Natural genetic engineering in evolution, *Genetica*, 86: 99–111.

Stevens, F.S., 1974, *Patterns in Nature*, Little Brown, Boston.

Thompson, D.W., 1944, *On Growth and Form*, Cambridge University Press.

Turing, A.M., 1952, The chemical basis of morphogenesis, *Philosophical Transactions Royal Society of London B*, 237: 37–72.

Vicsek, T., 2001, The bigger picture, *Nature*, 409: 985–986.

Waldorp, M., 1993, *Complexity*, Simon and Schuster, New York.

Chapter **3**

Engineering of a System of Systems

GARY D. WELLS and ANDREW P. SAGE

George Mason University, Fairfax, VA, USA

3.1 INTRODUCTION

The key challenges associated with engineering a system of systems (SoS) concern needs to manage interfaces among component systems that are generally individually acquired and integrated, the distributed systems management environment (systems engineering and program management), and the challenge associated with adaptive and emergent behavior of the composite systems. Systems of systems occur on a continuum in terms of complexity and there are many variations, such as the federation of systems. The individual systems comprising a system of systems may be relatively low on a complexity continuum, or they may be large, complex systems (each one a system of systems). One trait associated with a system of systems is managerial independence of the individual systems that constitute the complete system. Managerial independence can drive the need for appropriate systems engineering groups to manage the cost, schedule, and technical aspects of the individual systems and for a separate systems engineering functional group to manage the cross-system cost, schedule, risk, and technical facets of the system of systems. In other words, there is a great need for vertical and horizontal integration (Maier, 1998). An early work addressing the characteristics of systems of systems and federations of systems is provided by Sage and Cuppan (2001). We discuss and extend some of this work here.

The engineering environment must be defined in terms of management principles and structural aspects, including definition of roles and responsibilities. In physical terms, the engineering of a system of systems often requires an organizational structure where separate systems engineering groups are responsible for

System of Systems Engineering: Innovations for the 21ˢᵗ Century, Edited by Mo Jamshidi
Copyright © 2009 John Wiley & Sons, Inc., Publication

managing/optimizing the individual systems, while other systems engineering groups are responsible for the enterprise and cross-program aspects of these systems, including the enterprise (or corporate) systems engineering function.

Guidance for establishing the structural aspect of engineering a system of systems is provided by the various systems engineering standards, as defined by the various international standards bodies (International Organization for Standardization (ISO), International Electrotechnical Commission (IEC), and Institute of Electrical and Electronics Engineers (IEEE)). Integration of effort is potentially achieved through appropriate standards (Sage and Biemer, 2007) and clear definition of roles and responsibilities, including the designation of system of interest systems engineer (SoISE), lead system integrator (LSI), enterprise systems engineer (ESE), and/or system of systems engineer (SoSE) roles and responsibilities. According to Grady (1995), an enterprise that develops and manufactures complex products should be guided by a standard, documented process for performing work. Definition of these roles and responsibilities via a standardized process allows more work to be transferred from the pool requiring new and original thinking into the pool of standard responses, and it allows more mental capacity to be applied to the really difficult problems that cannot be solved by a cookie cutter approach (Grady, 1999).

The authors of ISO/IEC/IEEE 15288, systems engineering—system life cycle process, have attempted to provide an unambiguous definition of systems engineering roles and responsibilities for each organizational level for systems of interest (Arnold and Lawson, 2004). It suggests the recursive use of processes, that is, the repeated application of the same process or set of processes applied to successive levels of detail in a system's hierarchical structure, is a key aspect of the application of this international standard (IEEE 15288, 2005). While it may very well include an all-encompassing set of processes and very useful expansion of outcomes, its treatment of roles and responsibilities above the system of interest level is incomplete for addressing a complex system. For instance, the assumption of recursive use of processes implies the system of interest can be represented as the sum of its parts (an assumption of linearity), and this is generally not a valid assumption for a complex system.

On the philosophical as well as pragmatic side, Charles Handy's (1992) new "federalist" principles provide insight into many of the considerations necessary to engineer the more complex forms of a system of systems. Handy (1992, 1995) speaks to the necessary considerations for the structuring of loosely coupled organizations. This federalist approach is necessary to obtain a systems engineering ecology—in other words, a sustainable systems engineering approach, in systems of systems and federation of systems management. The concept of federalism is particularly appropriate since it offers a well-recognized way to deal with the systems engineering management paradoxes of power and control such that the desired systems ecological balance is obtained within an enterprise.

Needless to say, an effective systems management structure must be in place to address these challenges. To be effective, the management environment and systems engineering roles and responsibilities of these systems engineering professionals must be appropriately defined and understood. Effectiveness is achieved through properly adapting and integrating the physical, functional, and philosophical domains.

3.2   BACKGROUND

There are a number of key challenges in engineering a system of systems. Here, we outline many of these.

3.2.1   Roles and Responsibilities

Engineering a system of systems is replete with challenges. One important challenge involves the increase in requirements for integrating systems with other systems while delivering them in a trustworthy manner within risk, cost, and schedule management constraints. As systems become more and better integrated, there is an accompanying need to manage more interfaces. Additionally, engineering a system of systems depends on the effective integration of technical processes with the human and organizational considerations (Brooks and Sage, 2006). Thus, the complexity of the systems engineering and enterprise management efforts is increased accordingly. As the number of systems engineering groups grows, the need to integrate their efforts increases. At best, only general systems engineering guidance presently exists to help define the roles and responsibilities for each of the higher level systems engineering groups.

3.2.2   Traditional Versus System of System Engineering

Keating et al. (2003) make several observations relevant to engineering a system of systems and why traditional systems engineering practices are often not sufficient for engineering a system of systems. Many authors have also addressed these needs, as discussed in Sage and Cuppan (2001) and Sage and Biemer (2007). Sage and Biemer (2007) define a suggested process for systems of systems engineering based on the "current" systems engineering standards. Their process flow complements the discussion on how to use ISO/IEC/IEEE 15288 to define roles and responsibilities that will be addressed later in this chapter. Keating et al. (2003) note several important realities concerning an SoS.

- Traditional systems engineering focuses on a single (often complex) system, whereas engineering a system of systems focuses on integrating multiple, asynchronous (but interdependent) complex systems.
- Traditional systems engineering focuses on optimisation, whereas engineering a system of systems focuses on satisficing to achieve realistic cost and schedule goals.
- Traditional systems engineering focuses on a final end product or enabling system; engineering a system of systems focuses on delivering an initial deployment. Systems of systems evolve over time, and the concept of a final end product may be impossible to achieve.
- Traditional systems engineering deals with requirements that remain fixed throughout development; engineering a system of systems focuses on developments where the requirements evolve over time.
- Traditional systems have well-defined boundaries; systems of systems do not.

3.2.3 ISO/IEC/IEEE 15288 Systems Engineering—System Life Cycle Processes

The authors of ISO/IEC/IEEE 15288 intend it to be broadly applicable. It applies to

- the full life cycle and to the acquisition and supply of systems, whether performed internally or externally to an organization;
- a wide variety of systems in terms of their purpose, domain of application, complexity, size, novelty, adaptability, quantities, locations, and evolutions;
- organizations in their role as acquirers and suppliers—parties can be from the same organization or different organizations; and
- creation of business environments, for examples: methods, tools, techniques, and trained personnel.

ISO/IEC/IEEE 15288 provides a high-level approach to defining organizational roles and responsibilities. The standard acknowledges that organizations typically distinguish different areas of managerial responsibility and that together these areas contribute to the organization's overall capability to trade (or effectiveness). ISO/IEC/IEEE 15288 employs a process model based on three primary organizational areas (or levels) of responsibility: enterprise, project, and technical. Within each organization, a coordinated set of enterprise, project, and technical processes contribute to the effective creation and use of systems and to achieving the organization's goals. The processes contained in ISO/IEC/IEEE 12588 are shown in Fig. 3.1 and discussed below.

At the enterprise level, different organizations and different areas of responsibility with an organization mutually establish their working relationships and acknowledge

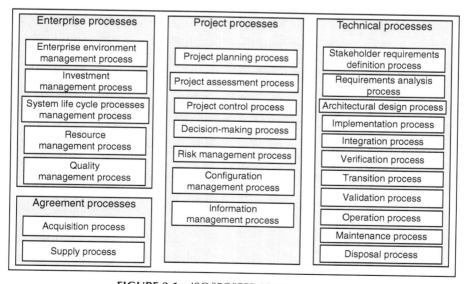

FIGURE 3.1 ISO/IEC/IEEE 15288 processes

their respective areas of responsibilities by making agreements. These agreements unify and coordinate the contributions made by different areas of responsibility in order that they can meet a common business purpose.

The enterprise processes are concerned with ensuring that the needs and expectations of the organization's interested parties are met. The enterprise processes are typically concerned at a strategic level with the management and improvement of the organization's business or undertaking, with the provision and deployment of resources and assets and its management of risks in competitive or uncertain situations. Responsibility for these processes is typically at the highest level in the organization.

The project processes are concerned with managing resources and assets allocated by enterprise management and applying them to fulfill the agreements into which that organization enters. They relate to the management of projects, in particular to the planning in terms of cost, timescales, and achievements, to the checking of actions to ensure that they comply with plans and performance criteria, and to the identification and selection of corrective actions that recover shortfalls in progress and achievement. The project processes can be employed to provide, at the corporate level, an organization's infrastructure, for example, facilities, enabling services, and technology base.

The technical processes are concerned with technical actions throughout the life cycle. They transform the needs of stakeholders first into a product and then, by applying that product, provide a sustainable service when and where needed to achieve customer satisfaction. The technical processes are applied to create and use a system, and they apply at any level in a hierarchy of system structure.

Arnold and Lawson (2004) provide an excellent characterization of this standard, and Sage and Biemer (2007) provide a comparison of it and other recent standards for use in engineering a system of systems. Two criticisms of ISO/IEC/IEEE 15288 are that the standard does not have requirements to "continually analyze and assess system of systems capabilities," and it does not "allow for incremental deployment of systems." Additionally, Sage and Biemer (2007) point out that this standard does not differentiate between the program manager's and the system engineer's responsibilities for project processes. In addition, there is some uncertainty with respect to the system engineer's responsibilities regarding enterprise processes. While it is indeed vague, a reasonable interpretation of this standard is that it includes processes for which the system engineer is directly responsible (technical processes) and those where the systems engineer acts in a supporting role (agreement, enterprise, and project processes).

3.3 DEFINITIONS

There are no commonly accepted definitions for many of the terms used in the context of a system of systems. However, establishing a set of appropriate and accepted definitions is critical to progress in this area as these establish boundaries and provide a common basis for discussion and analysis. So, it will be generally incumbent on system of systems engineering groups to establish common definitions for their specific application until truly and fully appropriate standards in this area have evolved.

3.3.1   System (System of Interest)

A system is a collection of things or elements that, working together, produce a result not achievable by the things alone (Maier and Rechtin, 2000).

ISO/IEC/IEEE 15288 defines a system of interest as the system whose life cycle is under consideration in the context of ISO/IEC/IEEE 15288. Systems of interest considered are man-made, created and utilized to provide services in defined environments for the benefit of users and other stakeholders. These systems may consist of one or more of the following: hardware, software, humans, processes, procedures, facilities, and naturally occurring entities. They can be products or services. The definition of a particular system of interest is contextual—one person's system of interest could be perceived to be a system element to another person. The hierarchy within ISO/IEC/IEEE 15288 is system of interest (uppermost), system, and system element (lowest).

A system of interest has the following attributes:

- Defined boundaries that encompass meaningful needs and practical solutions;
- A hierarchical depiction of the physical structure of the system;
- Entities at any level in a hierarchical structure can be viewed as a system;
- A fully integrated, defined set of subordinate systems;
- A set of characteristic properties at a system's boundary arising from the interactions between system elements;
- Humans, who can be users external to a system and as system elements within a system; and
- An entity capable of interacting with its surrounding environment.

3.3.2   System of Systems

Very early references to "systems within systems" or "system of systems" can be found in Berry (1964) and Ackoff (1971). Some of the early definitions compiled by Jamshidi (2005) and reiterated by Lane and Valerdi (2007) are as follows:

Early System of Systems Definitions

Source	Definition
Kotov (1997)	Systems of systems are large-scale concurrent and distributed systems that are comprised of complex systems.
Manthorpe (1996)	In relation to joint warfighting, a system of systems is concerned with interoperability and synergism of command, control, computers, communications, and information and intelligence, surveillance, and reconnaissance systems. Primary focus: information superiority.
Eisner (1993)	Systems of systems are large geographically distributed assemblages developed using centrally directed development efforts in which the component systems and their integration are deliberately, and centrally, planned for a particular purpose.

Early System of Systems Definitions *(Continued)*

Source	Definition
Shenhar (1994)	An array system (system of systems) is a large widespread collection or network of systems functioning together to achieve a common purpose.
Maier (1998)	A system of systems is a set of collaboratively integrated systems that possess two additional properties: operational independence of the components and managerial independence of the components.
Krygiel (1999)	A system of systems is a set of different systems so connected or related as to produce results unachievable by the individual systems alone.

Many characterizations of a system of systems suggest that such systems have the following properties:

1. *Operational Independence of the Individual Systems.* This suggests that a system of systems is composed of systems that are independent and useful in their own right, and if a system of systems is disassembled into the constituent systems, these constituent systems are capable of independently performing useful operations by themselves and independently of one another.

2. *Managerial Independence of the Systems.* This suggests that the component systems generally operate independently to achieve the technological, human, and organizational purposes of the individual organizational unit that operates the system. These component systems are generally individually acquired, serve an independently useful purpose, and often maintain a continuing operational existence that is independent of the larger system of systems.

3. *Geographic Distribution.* Geographic dispersion of the constituent systems in a system of systems is often very large. Often, the individual constituent systems can readily exchange only information and knowledge with one another, and not any substantial quantity of physical mass or energy.

4. *Emergent Behavior.* The system of systems performs functions and carries out purposes that may not reside uniquely in any of the individual constituent systems. The principal purposes supporting engineering of these individual systems and the composite system of systems are fulfilled by these emergent behaviors.

5. *Evolutionary and Adaptive Development.* A system of systems is never fully formed or complete. Development is evolutionary and adaptive over time, and where structures, functions, and purposes are added, removed, and modified as experience of the community with the individual systems and the composite system grows and evolves.

These were initially described in this format by Maier (1998), based on his study of works at the time, and they form the basis of his study and are the ones used by Sage and

Cuppan (2001) in their efforts. The Department of Defense (DoD) has developed an excellent and recent System of Systems Engineering Guide, which may be found at http://www.acq.osd.mil/se/to%20be%20posted/SOSE%20Guide%20Dec%2022% 20PDF.pdf.

3.3.3 Federation of Systems

Quite often, there are appropriate missions for relatively large systems of systems in situations where there is a very limited amount of centralized command-and-control authority that can be used to exercise systems engineering development options. Instead, there is a coalition of partners having decentralized power and authority and potentially differing perspectives of situations. It is useful to term such a system a "federation of systems" and sometimes a "coalition of systems." The participation of the federation partners is based upon collaboration and coordination to meet the needs of the federation.

Here, innovation includes both technological innovation and organizational and human conceptual innovation. The systems fielded to obtain desired capabilities will not be monolithic structures in terms of either operations or acquisition. Rather, they will be systems of systems, coalitions of systems, or federations of systems that are integrated in accordance with appropriate architectural constructs to achieve the evolutionary, adaptive, and emergent cooperative effects that will be required to achieve human and organizational purposes and to take advantage of rapid changes in technologies. They can potentially accommodate system life cycle change, in which the life cycle associated with use of a system family evolves over time; system purpose change, in which the focus on use of the system emerges and evolves over time; and environment change, in terms of alterations in the external context supporting differing organizational and human information and knowledge needs, as well as in the technological products that constitute constituent systems.

We can contrast system families and identify relationships between conventional systems, systems of systems, and federations or coalitions of systems with regard to three characteristics as initially noted by Krygiel (1999): autonomy, heterogeneity, and dispersion. A federation of systems, federated system of systems, or coalition of systems will generally have greater values of these characteristics than a (nonfederated) system of systems.

A system of systems generally has achieved integration of the constituent systems across communities of contractors, and sometimes across multiple customer bases, and is generally managed by more horizontally organized program management structures, such as integrated product and process development (IPPD) teams. When the IPPD team effort is well coordinated, the team is generally better able to deal with conflict issues that arise due to business, political, and other potentially competing interests.

Federations of systems face the same dilemmas identified for systems of systems but are generally much more heterogeneous along transcultural and transnational sociopolitical dimensions, are often managed in an autonomous manner without great central authority and direction such that they satisfy the objectives and purpose of an individual unit in the federation, and often accommodate a much greater geographic dispersion of organizational units and systems. Thus, the delimiters between systems of systems and federations of systems or coalitions of systems, while generally subjective, are nonetheless very principled.

The concept of federalism is particularly appropriate since it offers a well-recognized way to deal with the systems engineering and systems management paradoxes of power and control such that the desired systems ecological balance is obtained. Generally, this is accomplished by making things big by keeping them small, a goal which, in turn, is accomplished by expanding the domain of an enterprise by instantiating multiple quasi-autonomous units as opposed to acquiring mass by aggregation around a centralized command-and-control authority base, encouraging autonomy but within appropriate bounds set by process and architecture standards, and combining variety with shared purpose and individuality with partnerships at national and global levels. As we have noted, federalism is based on five principles of Handy: subsidiarity, interdependence, a uniform and standardized way of doing business, separation of powers, and dual citizenship. These systems of systems and federations of systems concepts have numerous implications for systems engineering and management.

3.3.4 System of Systems Engineering and System of Interest Engineering

System of systems (interest) engineering is the process of planning, analyzing, organizing, and integrating the capabilities of a mix of existing and new systems into a system of systems capability that is greater than the sum of the capabilities of the constituent parts. This process emphasizes the process of discovering, developing, and implementing standards that promote interoperability among systems developed under different sponsorship, management, and primary acquisition processes as described in http://stinet.dtic.mil/cgi-bin/GetTRDoc?AD=ADA429180&Location= U2&doc=GetTRDoc.pdf. (USD/AT&L, 2003).

3.3.5 Lead System Integrator

The role of the LSI can vary widely from acquisition to acquisition, especially in the role the LSI plays in the selection of the development contractor(s). In many cases, the LSI is responsible for integrating systems into a system of systems, but the LSI is generally not directly involved in source selection. Three definitions for the LSI are as follows:

- The LSIs are visionaries and leaders who can coordinate, motivate, and work closely with a set of contractors to achieve the ultimate objective in an optimal manner. The LSI seeks to perform its mission by leveraging the work that is being done by other contractors (Gupta, 2003).
- A LSI has the authority to acquire and integrate assets from a variety of potential system suppliers on behalf of an organization that is acquiring a complex software-intensive system. The LSI has the authority to contract with and

manage other suppliers on behalf of the acquirer. A primary task of the LSI is to determine early in the integration cycle whether required software assets can be mined from existing assets, can be purchased as (commercial-off-the-shelf systems) COTS components, or need to be developed from scratch (Bergy et al., 2003).

- The LSI can be government, a contractor, or a contractor team (Flood and Richard, 2005; Lane and Boehm, 2008). The LSI defines the scope of the system of systems, plans the activities to be performed, analyzes the requirements, and starts developing the system of systems architecture. As the scope, requirements, and architecture start to firm up, the LSI begins source selection activities to identify the desired system component suppliers. As the suppliers start coming on board, the LSI organization must focus on team building, rearchitecting, and feasibility assurance with selected suppliers. The LSI must periodically rearchitect to make adjustments for the selected system components that may not be compatible with the initial system of systems architecture or other selected components. Feasibility studies are conducted to better evaluate technical options and their associated risks. Many of the technical risks in a system of systems are due to incompatibilities between different system components or limitations of integrating older system components with today's technology.

A discussion of some of the issues associated with the lead systems integrator concept is presented in Brooks and Sage (2006).

3.3.6 Enterprise System Engineering

Enterprise systems engineering is focused on coupling traditional systems engineering activities with enterprise activities of strategic planning and investment analysis to enable appropriate transformation (Carlock and Fenton, 2001; Rouse, 2006). From the perspective of ISO/IEC/IEEE 15288, it is the coupling of the enterprise and, potentially, the agreement, the project, and the technical processes.

3.3.7 Complex Systems

Sage and Rouse (2009) and Rouse (2007) provide an overview of complex systems based on their study of the literature—"a system is complex when we cannot understand it through simple cause-and-effect or other standard methods of systems analysis. A system is complex when we cannot reduce the interplay of individual elements to the study of the individual elements considered in isolation. Often, several different models of the complete system, each at a different level of abstraction, are needed." One of the characteristics of large (and complex) systems is that they are often formed from a variety of component systems: newly engineered from the ground up, custom systems, potentially tailored existing commercial-off-the-shelf systems, and existing systems.

3.3.8 Complex Adaptive Systems

An understanding of complex adaptive systems provides a framework for understanding the behavior of enterprises. Complex adaptive system theory is potentially applicable to resolving a number of frustrations associated with engineering a system of systems: lack of control, where we are often unable to do what is best even though there is a clear idea of what that is; lack of predictability; a lack of willingness to cooperate; and a lack of system interoperability. Among the many contributors to complex adaptive systems, the works of Rouse (2007), Sage (2002), and Sheard (2009a, 2008b) have strong relevance to systems engineering. There is much work to be done to fully interrelate complex adaptive system concepts with the engineering of a system of systems.

3.4 MOTIVATION FOR THE ENGINEERING OF A SYSTEM OF SYSTEMS

There is much motivation for the study of the engineering of systems of systems or system families. This section will describe examples relating to space systems and summarize some of the unique challenges associated with engineering these systems of systems.

3.4.1 Government Accounting Office Studies

On April 6, 2006, the United States Government Accounting Office (http://www.gao.gov/new.items/d06776r.pdf) (GAO, 2006) provided testimony before a subcommittee on the Department of Defense's space acquisition programs. This testimony stated that "the DoD intends to spend nearly $7 billion in fiscal year 2007 to acquire space-based capabilities to support current military and other government operations as well as to enable DoD to transform the way it collects and disseminates information, gathers data on its adversaries, and attacks targets. Despite its growing investment in space, however, DoD's space system acquisitions have experienced problems over the past several decades that have driven up costs by hundreds of millions, even billions, of dollars; stretched schedules by years; and increased performance risks. In some cases, capabilities have not been delivered to the warfighter after decades of development" (GAO, 2006).

In their *Framework for Assessing the Acquisition Function at Federal Agencies*, the GAO (http://www.gao.gov/new.items/d05218g.pdf) (GAO, 2005) set forth four cornerstones—organizational alignment and leadership, policies and processes, human capital, and knowledge and information management. Clearly defining and integrating roles and responsibilities, effective communication and continuous improvement, empowering cross-functional teams, and monitoring and providing oversight to achieve desired outcomes were identified among the critical success factors.

3.4.2 Defense Science Board/Air Force Scientific Advisory Board Study on Acquisition of National Security Space Programs

In May 2003, the Defense Science Board (DSB)/Air Force Scientific Advisory Board (AFSAB) Task Force on Acquisition of National Security Space Programs issued their final report (USD/AT&L, 2003) on systemic issues affecting the acquisition of space systems, including all aspects from requirements definition and budgetary planning through staffing and execution. This task force was chartered by the Under Secretary of Defense for Acquisition, Technology, and Logistics; the Secretary of the Air Force; and the Under Secretary of the Air Force/Director of the National Reconnaissance Office. The task force provided recommendations for improvements to the acquisition process applicable to the initiation through deployment life cycle phases.

Key findings of the task force included the following:

1. U.S. national security is highly reliant on space capabilities, and this reliance will continue to grow. Requirements exist to monitor events on a global basis, to transfer enormous quantities of data, and to project force worldwide. There is no viable alternative to the unique capabilities provided by these space systems.

2. Cost has replaced mission success as the primary driver in managing acquisitions, resulting in excessive technical and schedule risk. "Space is unforgiving; thousands of good decisions can be undone by a single engineering flaw or workmanship error, and these flaws and errors can result in a catastrophe."

3. The space acquisition system is strongly biased to produce unrealistically low-cost estimates; these estimates lead to unrealistic budgets and unexecutable programs.

4. Government capabilities to lead and manage the acquisition process have seriously eroded. In many cases, the proliferation of requirements involving space systems "involve multiple systems and require a system of systems approach to properly resolve and allocate the user needs. The space acquisition system lacks a disciplined management process able to approve and control requirements in the face of these (growing user base) trends. Clear trade-offs among cost, schedule, and requirements are not well supported by rigorous system engineering, budget, and management processes."

5. While the space industrial base is adequate to support current programs, long-term concerns exist. This erosion can be traced back, in part, to the acquisition reform initiatives of the 1990s. This policy marginalized the government program management role and replaced traditional government "oversight" with "insight." This authority of program managers and other working-level (e.g., system engineers) officials subsequently eroded to the point where it reduced their ability to succeed on development programs. Policies and practices inherent in acquisition reform inordinately devalued the systems acquisition engineering workforce. As a result, today's government systems engineering capabilities are not adequate to support the assessment of requirements, conduct trade studies, develop architectures, define programs, oversee contractor engineering, and assess risk. With the growing emphasis on effect-based capabilities and cross-system integration, systems engineering becomes even more important, and interim corrective action must be considered.

3.4.3 Space Systems

The acquisition of space systems, an example of one system of systems domain, presents unique challenges. First, the system of systems concept of evolutionary development does not apply to the individual spacecraft since few spacecraft were ever designed for on-orbit maintenance or upgrade, and most orbital domains are not reachable by the space shuttle. Incremental or evolutionary development occurs at the ensemble of spacecraft level or the architecture level. It is also possible within the ground segment. Second, space architectures tend to be quite complex. Space systems require years to develop and because of their expense are often designed to be operated over a 5–10-year period. This may require that the system be validated for a period of up to two decades, for example, 10 years development and 10 years operation. Finally, space systems must be highly reliable. The quality and test programs are challenged to deliver these systems in the face of cost and schedule pressures, the lack of a highly reliable industrial base supporting space acquisitions, and the emergent behavior (the so-called unknown unknowns) associated with any complex system.

3.5 FEDERALIST MANAGEMENT PRINCIPLES

Establishing management principles is a top-down process; that is, setting organizational standards, policies, and vision are responsibilities of senior management, possibly with the help of the enterprise systems engineering group or the system of systems engineering group. Charles Handy (1992) has provided an excellent examination of the principles of federalism that are often associated with a system of systems.

3.5.1 Handy's Five Principles

Federalism is based on five principles and five axioms. These must be adopted "inside out" (Handy, 1992). A summary of Handy's five principles tailored to the domain of systems engineering and management is as follows:

1. *Subsidiarity is the Most Important of Federalism's Principles.* It means that power belongs to the lowest point within the system of systems engineering team. Subsidiarity requires that the manager enables engineering professionals by training, advice, and support to execute their responsibilities better. Subsidiarity is the reverse of empowerment in that it is not the system of systems manager who is delegating power. Instead, power is assumed to lie at the lowest point in the organization and should be taken away only by agreement between the engineering professional and the manager. The result is multiple quasi-autonomous units. Subsidiarity is an awesome responsibility because it imposes on the engineering professional or group "type-II" accountability, or accountability for solving the correct problem. Traditional engineering management structures generally operate on the basis of "type-I" accountability,

that is, solving the problem correctly by making sure that no mistakes are made. To be effective, subsidiarity has to be formalized by management through "contracts" or agreements.

2. *Interdependence is a Reality in Federalism.* The quasi-autonomous development units or teams of a system of systems stick together because they need one another. This concept is known as pluralism. Federalism encourages combination when and where appropriate but not the centralization of services. Achieving this interdependence, or pluralism, is necessary in federalism because it distributes power. While this avoids the risks of overcontrol of the typical centralized program management bureaucracy, in a system of systems program management is still the focal point for action.

3. *There is a Uniform and Standardized Way of Doing Business.* Interdependence within federated system of systems engineering organizations requires agreement on a basic set of standardized processes, common agreements for communicating, and common units for measurement of progress and quality.

4. *Separation of Powers is also Required.* Federalism requires that management, monitoring, and governing aspects of system of systems engineering projects be viewed as separate functions to be accomplished by separate bodies even if the membership of these bodies may in fact overlap. Management is the executive function responsible for delivering the discrete "engineering goods." Monitoring is the quality/judicial function responsible for seeing that the "engineering goods" are delivered in accordance with the "laws of the federation" (e.g., architectural standards, build-to requirements, and quality objectives). Governance (e.g., senior program manager or program executive officer) is responsible for overseeing the management and the monitoring functions. The governance function is also responsible for establishing the a priori strategy, policy, and direction to be followed by the system of systems engineering organization. When these three functions are combined into one body, short-term needs generally drive out long-term considerations.

5. *Dual Citizenship Exists in a Federated Engineering Development Project.* Every individual is a "citizen" in two communities—the local development group/professional group/union and the overall system of systems program. Local citizenship seldom needs much support, and the system of systems program draws its strength from the strong leadership of the local groups. It is the federated "citizenship" that requires emphasis if the benefits of subsidiarity and interdependence are to be realized by sponsors and customers of a system of systems engineering program. In a federated engineering development project, strong leadership of the local groups results in the dual citizenship needed. This is another of federalism's paradoxes and the one that ensures a strong local identity, similar to the tagging mechanism of Holland (1995,1998). This also results in the willingness to reach out and to avoid committing the "type-II" (wrong problem) errors previously discussed relative to subsidiarity. This directly leads to the psychosocial notion of shared values as initially demonstrated by successful Pacific-Rim corporations (i.e., what the Japanese would

recognize as the "spiritual fabric" of their federations). It must be recognized that accomplishing this requires more than the group's need to improve the bottom line. According to Handy (1992), it must be some modern-day equivalent of the citizens of Elizabethan England collectively acting for "the Queens Great Matter" (i.e., world stability and strategic domination over competitors).Application of these five principles creates a federalized systems engineering ecology for the system of systems.

3.5.2 Handy's Five Axioms

Once the five key principles of federalism have been incorporated, five additional guidelines, or axioms, must be implemented by the leadership and communicated throughout the distributed organization for the benefits of federalism to be realized. A brief description of Handy's (1992) leadership axioms, within a systems engineering and management context, suggests the following potential benefits associated with federalism:

1. *Authority Must Be Earned from Those over Whom it is Exercised.* Quality engineering professionals require management by consent if they are to give their best. It must be acknowledged by system of systems or federation of systems development project leadership that the engineering professional should give or withhold the consent.

2. *Engineering Professionals Have Both the Right and the Duty/Responsibility to Sign Their Work.* Subsidiarity requires that people take responsibility for their decisions by "signing their work" (e.g., configuration items and document artifacts) both literally and metaphorically. This is a key aspect of performance in other professions (e.g., doctors, lawyers, professors). A signature on one's work may be the best single guarantee of quality.

3. *Autonomy Means Managing Empty Spaces.* In a federated system of systems or federation of systems development program, groups and individuals live within two concentric circles of responsibility. The inner circle represents their minimally acceptable responsibilities, that is, everything they have to do or risk failure. The larger circle marks the limits of their authority. The in-between area is their area of discretion; this area is the space in which they have both the freedom and the responsibility to initiate action. Engineering professionals within a system of systems or federation of systems project must fill this space — it is their "type-II" accountability. Where no mistakes are tolerated (conventional "type-I" project management ideology), no professional initiative will be risked. Forgiveness, providing an individual or group "learns", is a necessary part of federalist thinking in an engineering context.

4. *Twin Hierarchies are Both Necessary and Useful.* Twin hierarchies demonstrate federalism's principle of interdependence. There is, in every organization, a clear status hierarchy. Some people are justifiably senior to others because of their proven knowledge, experience, and ability. However, in the engineering task hierarchy of a federated system of systems program, the

role dictates who is who (i.e., the junior configuration identification engineer may direct the proven senior test engineer). Distinguishing between status and task hierarchies allows organizations to be much flatter without losing effectiveness.

5. *What is Good for Me Should be Good for the Organization.* This is the twin citizenship principle brought down to the level of the individual. Successful professionals believe the necessity of self-improvement. They know that if they do not continually invest in their own learning and development, they will become obsolete. What they ask of federated system of systems program management is that it facilitates and encourages the process of self-improvement. In turn, the professional owes a loyalty to the federated entity. Federated system of systems or federation of systems program management must never take this loyalty for granted; it must be carefully nurtured. Otherwise, the professional may feel released from any sense of obligation.

Thus, we see that federalism is not just another word for restructuring conventional systems engineering management lines of authority. The ecology-based thinking behind it is simply that "autonomy releases energy; energy fuels innovation, which is required for survival in adaptive environments; people have the right to do things their own way as long as it is in the common interest of the project; people need to be well-informed, well-intentioned, and well-educated to interpret that common interest; and individuals prefer to be led rather than centrally managed."

Summarizing Sage and Cuppan's (2001) discussions on federalism, there are at least three paradoxes of power and control associated with these observations. The first paradox is that system of systems engineering organizations need to be both big and small at the same time. Federalism responds to the pressures of this paradox by balancing power among the following:

- Those at the center of the organization;
- Those at the centers of expertise; and
- Those at the center of the action, such as the line engineers and developers.

The second system of systems paradox lies in declared preferences for the equivalent to "free and open markets," which requires some level of competition among the members of the federation, as the best guarantee of efficiency, as its managers organize their own operations for centralized control. Handy (1992) observes that open markets, on their own, do not work any better than central planning. A bit of both is needed, and this circumscribes the federalist compromise.

The third paradox is best summed up by the phrase: "What you do not own you cannot command." More to the point, the third paradox is the desire by engineering leadership to run a system of systems program as if it were yours when in fact you cannot afford, or may not want, to make it yours. This leads to the notion of alliances, which are notoriously difficult to manage or lead. Each alliance is unique, to be "lived with" rather than managed, better built on mutual respect and shared interests than on administrative or legal documents and tight controls.

3.6   STRUCTURAL APPROACH—DEFINING ROLES AND RESPONSIBILITIES FOR ENGINEERING A SYSTEM OF SYSTEMS

Systems engineers will generally benefit from guidance as to how the application of traditional systems engineering practices, as defined by the ISO/IEC/IEEE 15288, must be modified in order to help an organization cope with the challenges associated with engineering a system of systems, especially the definition of the roles and responsibilities of the SoISE, ESE, SoSE or LSI charged with overseeing and ensuring the success of systems engineering and integration efforts.

Many studies addressing complex systems, federation of systems, or system of systems issues either define complex system, federation of systems, or system of systems traits or document the challenges associated with the complex system, federation of systems, or system of systems. They do not have much in the way of practical guidance for engineering a complex system exhibiting federation of systems or system of systems traits; the insight provided is more often than not a list of heuristics.

Systems engineering standards exist that document the established processes for systems engineering and systems management throughout the product life cycle. However, the authors of these standards provide only a list of processes for engineering a system, and these processes must be adapted to the complex system, federation of systems, or system of systems domain. They contain little guidance as to how this should be accomplished.

The opportunity exists to build on the established practices for systems engineering a system to create practical guidance on how to engineer a system of systems.

The focus of this section is the creation of standardized guidance, even if it must be further adapted for unique and specific situations, to assist such systems engineering professionals as the SoISE, ESE, SoSE, or LSI with the task of managing and executing the systems engineering function applicable to a system of systems. This should lead to an enterprise systems engineering and management environment that is more effective in supporting an organization's systems engineering efforts, in evaluating and managing its resource needs, in employing Handy's (1992) federalist principles, and in evaluating the proper skill mix for the effort at hand. It should also support greater cohesion within the enterprise (organizational) systems engineering function and lead to unity of purpose. Unity of purpose is an especially valuable outcome as much time can be wasted in debating roles and responsibilities, and an important role could potentially be omitted due to a lack of common understanding of roles and responsibilities, adding unnecessary risk to an already risk-filled undertaking. The following steps summarize the proposed approach to defining roles and responsibilities:

1. Define key terms and concepts; create a common data dictionary.
2. Develop a framework for the systems engineering and management functions.
3. Summarize the challenges associated with engineering and management of the system of systems.
4. Model the information flow among systems engineering groups constituting the system of systems.

5. Evaluate the degree to which Handy's (1992) federalist principles and axioms are/should be implemented.

6. Document the roles and responsibilities by adapting the agreement processes listed in ISO/IEC/IEEE 15288.

3.6.1 Define Key Terms and Concepts

A key activity underlying the integration of a system of systems is the creation of a system of systems dictionary. This dictionary is the basis for structured analysis; there will likely be overlap in terminology at the system interfaces and potentially between systems (Levis, 2009). There is no guarantee this terminology is consistently applied across the system of systems and the use of terminology can only be understood within the context of a system of systems.

3.6.2 Develop Framework for Systems Engineering and Management Functions

As stated by Grady (1999), some problems are so complex at the enterprise level that two or more enterprises (or other entities within an enterprise) must work together cooperatively. By extension, this will likely mean that systems engineering and management groups within each enterprise must work together to achieve some common purpose. The recursive use of processes, that is, the repeated application of the same process or set of processes applied to successive levels of detail in a system's hierarchical structure, is a key aspect of the application of the ISO/IEC/IEEE 15288 international standard (IEEE15288, 2005). The concept of a hierarchically decomposed system of interest is illustrated in Fig. 3.2.

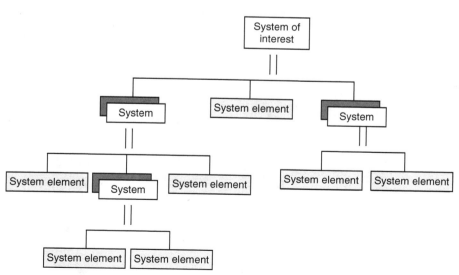

FIGURE 3.2   ISO/IEC/IEEE 15288 system of interest hierarchy

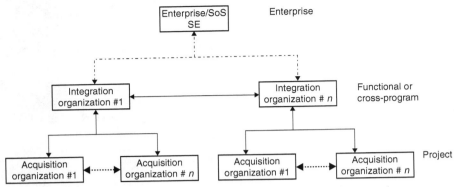

FIGURE 3.3 Systems engineering and management framework

For large monolithic systems, this is often quite adequate. For large, systems of systems, the structure no longer applies as the whole does not equal the sum of the parts. A more appropriate framework is shown in Fig. 3.3. This constitutes a suggested systems engineering and management organizational framework for engineering a system of systems. Three basic levels of the systems engineering and management function are depicted. For especially large, complex federated systems, there is the need for a systems engineering and management function at the enterprise level that supports both enterprise functions and integration at the intermediate level within the enterprise. At the intermediate level, one or more systems engineering and management groups must integrate the acquisitions within their assigned area of responsibility. At the lowest level, the systems engineering and management groups supporting the actual acquisition of products and services can be found. For smaller scale efforts, the first and second tiers may be combined to produce a two-tier framework for the systems engineering and management groups.

An exemplary, deployed view of this operational framework is depicted in Fig. 3.4. We can see that the system of interest is actually not as easily decomposed as might be implied by ISO/IEC/IEEE 15288. In Fig. 3.4, the system of interest is composed of vertically integrated systems and systems that provide a common cross-system infrastructure. Since each vertical bar represents a different acquisition organization,

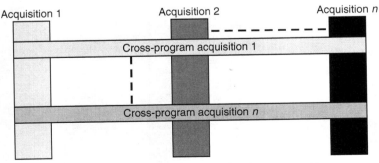

FIGURE 3.4 Operational framework for the system of interest

one can see that there are several different acquisitions and potentially several systems engineering groups (one for each system plus the integration systems engineering group and the system of systems engineering group) involved in acquisition of a complex SoS.

3.6.3 Summarize the Challenges Associated with Engineering and Management of the System of Systems

As each system of systems has its own unique terminology, it also has its own unique engineering challenges. The purpose of defining these challenges is to differentiate between traditional systems engineering and systems engineering practices that are more suitable to a system of systems. This effort will involve consideration of the trades involved in choosing the SoISE, the ESE, the LSI, or the systems of systems approach at each level within the system of systems organization. Fig. 3.4 provides an example of management and engineering complexity. This is one of the three primary drivers used in choosing and adapting the appropriate activities in ISO/IEC/IEEE 15288.

3.6.4 Modeling the Information Flow Among Systems Engineering Groups Comprising the System of Systems

Modeling the information flows is critical to establishing the relationships among the various systems engineering groups. It is a key predecessor to the task of evaluating the applicability of ISO/IEC/IEEE 15288 activities and outcomes for each systems engineering group.

3.6.5 Evaluate the Degree to Which Handy's (1992) Federalist Principles and Axioms are/should be Implemented

The principles of federalism will be used to define the nominal functional behavior of the systems engineering groups. This is the second of the three primary drivers that is used in evaluating the applicability of ISO/IEC/IEEE 15288 activities and outcomes for each systems engineering group. One important ramification of the federalist approach and placing power at the lowest point in the systems engineering framework is the vital role the integrated product team (IPT, or other unique identifier for this group) plays in the overall engineering effort.

3.6.6 Document the Roles and Responsibilities by Adapting the Agreement Processes Listed in ISO/IEC/IEEE 15288

The model for the functional behavior of the systems engineering and management professionals will be developed by adapting the processes, the associated activities, and the outcomes from ISO/IEC/IEEE 15288 based on the results of tasks 2.4.1–2.4.4. The proposed process is illustrated in Fig. 3.5. Figure 3.6, or other appropriate template, can be used to capture these roles and responsibilities. Figure 3.6 is derived from the template employed by Friedman and Sage (2004) with three exceptions. The systems engineering processes used by Friedman and Sage (2004) are replaced by

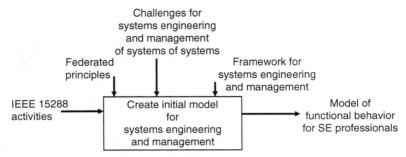

FIGURE 3.5 Development of the hypothesis

those from ISO/IEC/IEEE 15288, the roles of the systems engineering contractors are decomposed into those for the Federally Funded Research and Development Corporation (FFRDC) contractor and the Contractor Assisted Advisory Services (CAAS) contractor, and the shared roles are meant to be folded into the roles of the government, the CAAS contractor, and the FFRDC contractor, that is, the systems engineering team lead/support roles. At the conclusion of this step, the roles and responsibilities for each systems engineering group (at each level) should be defined. Inherent in the definition of roles and responsibilities at each level is the explicit recognition that the execution of the engineering function lies as much, or more, in the IPT as the more formal, centrally executed engineering functions. While the hierarchy inherent to the framework for organizing engineering groups provides the critical management and oversight of the overall engineering effort, it is the "professional structure" of the

			ESE			SoS			LSI		
			Government	CAAS	FFRDC	Government	CAAS	FFRDC	Government	CAAS	FFRDC
Process 1	Subprocess 1	Activities									
	......	.....	⋮	⋮	⋮	⋮	⋮	⋮	⋮	⋮	⋮
	Subprocess n	Activities									

FIGURE 3.6 Potential template for documenting roles and responsibilities at each organizational level and for each systems engineering professional

IPT that provides the integration of effort across organizations and projects and develops the products associated with each ISO/IEC/IEEE 15288 activity—the system of systems program draws its strength from the strong leadership of the local groups. The roles of the IPT must be explicitly documented to gain the full understanding of the engineering function.

3.6.7 Vignette

The following vignette is based loosely on a real problem encountered during the early stages (precontract award) of a government acquisition group. It is intended to provide a conceptual example and amplification of the general approach outlined in this section. Additionally, it demonstrates the amorphous boundaries that can exist within a system of systems.

Problem: The program manager approached the chief systems engineer and asked him to update the program management plan to reflect the roles and responsibilities of the entire systems engineering function. He provided an organization chart that showed a chief systems engineer at the program manger level. The chief systems engineer (synonymous with the system of systems engineer/system of interest engineer defined in Section 3.3.4) would be third in importance behind the program manager and the deputy program manager and would be responsible for technical integration of the system (actually a system of systems) and ensuring a sound technical baseline. Additionally, there was to be a systems engineering and test (SET) and two functional acquisition groups that reported to the program leadership. This effort does not include a LSI. It does include an FFRDC and a CAAS systems engineering support to the government. Higher on the organization chart, there exists another system of interest that also has a chief systems engineer (synonymous with the enterprise systems engineer definition in Section 3.3.6).

3.6.7.1 *Framework for Systems Engineering*   The framework for this vignette is shown in Fig. 3.7. One can readily see the hierarchical nature of the systems

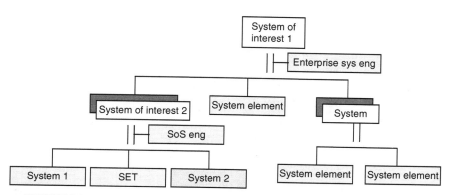

FIGURE 3.7   Framework for systems engineering—organizational perspective

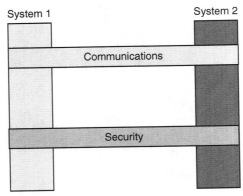

FIGURE 3.8   Framework for systems engineering—development perspective

engineering function within this system of interest 1 and system of interest 2. Not readily apparent are the systems engineering activities that are inherent in the systems comprising system of interest 2. As with the system of interest as a whole, each system must apply sound systems engineering practices in acquiring the end products and the enabling products. These systems engineering groups must be integrated, often across organizational boundaries. Figure 3.8 below shows system of interest 2 in terms of development responsibility.

As can be seen in the above simplified example, there are four different organizations involved in developing the end products and the enabling products associated with system of interest 2. In this example, there are two end products—system 1 and system 2. There are two enabling products that are common to systems 1 and 2—communications infrastructure and a security infrastructure.

3.6.7.2 Challenges   From the perspective of the definition of the characteristics of a system of systems, the key challenges are as follows:

1. Geographic distribution between systems comprising SoI 2 exists.
2. Emergent behavior manifests itself over time.
3. SoS are never fully formed; hence, they are evolutionary and adaptive.
4. There are numerous stakeholders—sponsoring and funding organizations as well as the various user communities.
5. There exist a number of parallel, independent (or not so independent) developments; the overall SoS architecture must be defined well enough up front to allow for many concurrent development efforts in parallel.
6. Multiple decision *approvers*—as the number of people involved in the decision-making process increases, the probability of getting a timely (or even any decision) often decreases.
7. There exists a number of cross-cutting risks—risks that cut across organizational boundaries and/or system components.

		TO			
		ESE	SoSE	System 1 SE	System 2 SE
FROM	ESE	N/A	• Policies/stds • Architecture • Enterprise processes	N/A	N/A
	SoSE	Inputs to • Architecture • Tech baseline • Schedule • CM/risk mgmt	N/A	• Policies/stds • Architecture • SoS tech Processes, e.g., rqmts and IV&V, •Proj processes, e.g., CM/risk mgmt, schedule	• Policies/stds • Architecture • Tech processes, e.g., rqmts and IV&V, •Proj processes, e.g., CM/risk mgmt, schedule
	System 1 SE	N/A	Inputs to • Architecture • Tech baseline • Schedule • CM/risk mgmt	N/A	Inputs to • ICDs • Analyses • Schedule
	System 2 SE	N/A	Inputs to • Architecture • Tech baseline • Schedule • CM/risk mgmt	Inputs to • ICDs • Analyses • Schedule	N/A

FIGURE 3.9   Information flows among systems engineering functions

3.6.7.3   Information Flow Among the Systems Engineering Groups   Figure 3.9 provides a high-level synopsis of the information flows between the systems engineering functional groups shown in Fig. 3.7. In accordance with ISO/IEC/IEEE 15288, the enterprise systems engineer is primarily involved in the execution of the enterprise processes shown in Fig. 3.1 across SoI 1. Although the SoS engineer is primarily involved in executing the enterprise, project, and technical processes across SoI 2, he or she is also responsible for supporting the agreement processes between the acquirers and the suppliers. The systems engineers within systems 1 and 2 are responsible for executing the agreement processes and the technical processes for their respective systems. This will be expanded upon via tailored roles and responsibilities in Section 3.6.7.5.

3.6.7.4   Applicability of Handy's Principles

1. Subsidiarity—power belongs to the lowest point within the system of systems engineering team. Training and support are provided by the enterprise and the respective SoI; IPTs are the dominant means for accomplishing the core systems engineering functions.

2. Interdependence—quasi-autonomous development units or teams of a system of systems stick together because they need one another. Each systems engineering group has a unique role and depends on the other systems engineering groups to realize the end product and enabling products.

3. It is the responsibility of a higher level (in the organizational structure) to establish a uniform and standardized way of doing business for themselves and the next lower level systems engineering group(s).

4. Management, monitoring, and governing aspects of system of systems engineering projects must be separate functions to be accomplished by separate bodies even if the membership of these bodies may in fact overlap.

5. Each system engineer is a member of their "local" systems engineering team and/or the overall SoI or system to which they belong.

3.6.7.5 Defined Roles and Responsibilities Figures 3.10, 3.11, and 3.12 provide a high-level summary of each systems engineering function's responsibilities. For illustrative purposes, only one process for each applicable process group (enterprise, agreement, project, and technical) and one to three activities per process will be shown. With the exception of Fig. 3.12, which may be performed at the SoI/SoS levels, the remaining two, Figs. 3.10 and 3.11, are meant to reflect responsibilities unique to that position given the context of the vignette.

3.7 CASE STUDIES INVOLVING THE ENGINEERING OF A SYSTEM OF SYSTEMS

Friedman and Sage (2004) investigated the role that case studies can play in systems engineering and systems engineering management, especially case studies that involve the acquisition of systems. Their efforts were driven by two main considerations: (1) a belief that case studies potentially expose realities of the world of professional practice and (2) to show that case studies may be employed to help make the distinction between the systems engineering duties and the responsibilities of the public sector and the government.

Friedman and Sage developed a two-dimensional framework for systems engineering case study research. This framework is a 9×3 matrix, where the three columns represent the responsibility domain—the systems engineering contractor responsibilities, shared responsibilities, and government system engineering responsibilities—and the nine rows represent systems engineering activities. Six of the systems engineering activities represent the systems engineering life cycle phases, and three systems engineering activities represent necessary systems engineering process and systems management support.

3.7.1 Validation

This case study effort does not conclude with the definition of these roles and responsibilities. Like the systems, they are responsible for, systems engineering roles and responsibilities are likely to evolve over time. Case studies provide an important analysis methodology for periodically assessing the continued relevance of defined roles and responsibilities. An ongoing task involves the validation of the utility of systems engineering standards for defining and predicting the functional behavior of

Enterprise processes

Enterprise environment management process

		ESE	
	Government	CAAS	FFRDC
Establish plans for each business area	Identify the short-term objectives that contribute to achieving strategic Sol 1's objectives and the development systems that will be undertaken to accomplish the strategic objectives.	Augment the government staff by providing expertise in organizational strategic planning and with specific knowledge of the organizations strategic planning processes; prepare a proposed set of objectives and projects to achieve these objectives, for government approval	Augment the government staff by providing independent analyses and advice in organizational strategic planning and with specific knowledge of the organizations strategic planning processes; provide access to and surge capability to bring world-class experts to assist, when required
Prepare system life cycle policies and procedures that implement the requirements of ISO/IEC/IEEE 15288, and are consistent with enterprise strategic plans	Prepare system life cycle policies consistent with the complexity of the work, the methods used, and the skills/training of personnel involved-- relevant policies and procedures include risk management, quality management and resource management.	Augment the government staff by providing expertise in life cycle policies; prepare policies and procedure documents for government approval	Augment the government staff by providing independent analyses and advice in organizational life cycle policies and procedures; provide access to and surge capability to bring world-class experts to assist, when required
Define criteria that control progression through the system life cycle	Establish the decision-making criteria regarding entering and exiting each life cycle stage and for other key milestones	Augment the government staff by preparing, for government approval entry and exit criteria for each life cycle stage and for other key milestones	Augment the government staff by providing independent analyses and advice; provide access to and surge capability to bring world-class experts to assist, when required

FIGURE 3.10 Enterprise systems engineer responsibilities

69

Technical processes

Stakeholder requirements definition process

	Government	SoS engineer — CAAS	SoS engineer — FFRDC
Identify the individual stakeholders or stakeholder classes	Identify the individual stakeholders or stakeholder classes who have a legitimate interest in the system of systems throughout its life cycle, including users, supporters, developers, maintainers, disposers, and regulatory bodies	Augment the government staff by supporting the identification of individual stakeholders or stakeholder classes	Augment the government staff by providing independent analyses and advice supporting the identification of individual stakeholders or stakeholder classes; provide access to and surge capability to bring world-class experts to assist, when required
Elicit stakeholder requirements	Elicit stakeholder requirements in terms of the needs, wants, desires, expectations and perceived constraints. The requirements should focus on system purpose and behavior in the context of the operational environment and conditions	Augment the government staff by providing expertise in eliciting and documenting stakeholder requirements; provide domain-specific expertise	Augment the government staff by providing independent analyses and advice by providing domain-specific expertise; provide access to and surge capability to bring world-class experts to assist, when required
Define the constraints on a system solution	Define the constraints on a system solution that are unavoidable consequences of existing agreements, management decisions and technical decisions	Augment the government staff by supporting the identification and capture of system constraints	Augment the government staff by providing independent analyses and advice by providing domain-specific expertise; provide access to and surge capability to bring world-class experts to assist, when required

Note: This role is unique to the SoS engineer

FIGURE 3.11 System of systems engineer responsibilities

		Systems engineer (system level)		
Technical processes	Verification process	Government	CAAS	FFRDC
	Define a verification plan based on system requirements	Define a verification plan for the system that accounts for the sequence of configurations defined in the integration strategy and, where appropriate, takes account of disassembly strategies for fault diagnosis. The schedule typically defines risk-managed verification steps that progressively build confidence in compliance of the fully configured product	Augment the government staff by providing expertise in verifying system requirements; provide domain-specific expertise.	Augment the government staff by providing independent analyses and advice by providing domain-specific expertise; provide access to and surge capability to bring world-class experts to assist, when required
	Identify potential constraints on design decisions	Identify verification constraints, including practical limitations of accuracy, uncertainty, repeatability imposed by the verification enabling systems, the associated measurement methods, and the availability of enabling systems	Augment the government staff by providing expertise in verifying system requirements; provide domain-specific expertise.	Augment the government staff by providing independent analyses and advice by providing domain-specific expertise; provide access to and surge capability to bring world-class experts to assist, when required
	Conduct verification to demonstrate compliance to the specified design requirements	Conduct verification in a manner, consistent with organizational constraints, such that uncertainty in the replication of verification actions, conditions and outcomes is minimized. Approved records of verification actions and outcomes are archived.	Augment the government staff by providing expertise in verifying system requirements; provide domain-specific expertise.	Augment the government staff by providing independent analyses and advice by providing domain-specific expertise; provide access to and surge capability to bring world-class experts to assist, when required

FIGURE 3.12 Systems engineer responsibilities (system level)

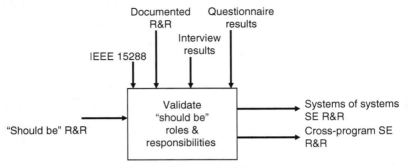

FIGURE 3.13   Proposed validation process

the systems engineering professionals and how this behavior evolves over time. These validation activities will focus on documenting "as built" practices and comparing them to the proposed "should be" practices derived from ISO/IEC/IEEE 15288. This process is diagrammed in Figs. 3.10 and 3.11. Figure 3.10 depicts the proposed approach to validation showing the key inputs and controls and the two anticipated outcomes. The "as built" practices can be derived from existing documentation, interviews, and surveys. Figure 3.6 provides a potential template for recording the "as built" roles and responsibilities based on the case study results. Figure 3.13 must be developed for each level in the framework. Figure 3.14 amplifies the approach shown in Fig. 3.13 and shows the steps involved in validating and adapting the "should be" implementations. As before, there are two emphasis areas—the roles and responsibilities of the systems engineering professionals and how these roles and responsibilities are driven, if at all, by Handy's federalist approach.

3.7.1.1   Analyze the Contributions of such Systems Engineering Professionals as the SoISE, the ESE, the SoSE, and the LSI   Within the context of the case study, it is important to document the functional behavior of such systems engineering

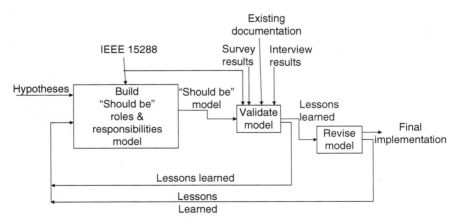

FIGURE 3.14   Validation approach

professionals as the SoISE, the ESE, the SoSE, and the LSI. This will provide invaluable insight into how these professionals are employed and the lessons learned in each case.

3.7.1.2 Evaluate the Degree to Which Handy's Federalist Principles and Axioms, as Implemented, Drive the Definition of Systems Engineering Functional Behavior In this instance, we are concerned with understanding the degree to which Handy's principles and axioms are implemented and how this influences the roles and responsibilities and functional behavior of the systems engineering professionals.

3.7.2 Benefits of Case Study Research

In addition to keeping pace with the evolving requirements for engineering the system of systems, the anticipated contributions of systems engineering case study research will be the following:

1. An extension of the applicability of exiting systems engineering standards by more explicitly treating the systems engineering functions associated with systems of systems.
2. Validation of the sufficiency of ISO/IEC/IEEE 15288 for establishing roles and responsibilities for the systems engineering of complex systems comprising a variety of enterprises and applications.
3. Documentation of the realities of the world of professional practice regarding large, complex systems.
4. Insight into the distinction between the systems engineering duties and the responsibilities of the public sector and the government.
5. Determination of the validity of the assumption of recursiveness employed by current systems engineering standards.
6. Guidance, based on established systems engineering practices, on how to organize the systems engineering functions responsible for supporting the acquisition of complex systems and system of systems, especially an expansion of our understanding of the roles and responsibilities of the SoSE or LSI charged with overseeing and ensuring the success of systems engineering and integration at the system level.
7. Recommendations for further study and activities.

3.8 SUMMARY

In this chapter, we have attempted to summarize the essential ingredients associated with the challenging and needed efforts to engineer a system of systems, especially systems operating in a federated systems management environment. The key challenges concern needs to manage interfaces among component systems that are generally individually acquired and integrated, the distributed systems engineering and program management environments, and the challenge associated with adaptive and emergent behavior of the composite systems. Additionally, these systems must be integrated with other systems while being delivered in a trustworthy manner within risk, cost, and schedule management constraints. Studies to date have clearly indicated that the lack of clearly defined, effective processes for systems engineering and management and unrealistic budgets are primary contributors to cost and schedule overruns for many systems.

The engineering environment must be defined in terms of management principles and structural aspects, including definition of roles and responsibilities. In physical terms, the engineering of a system of systems often requires an organizational structure where separate systems engineering groups are responsible for managing/optimizing the individual systems, while other systems engineering groups are responsible for the enterprise and cross-program aspects of these systems, including the enterprise (or corporate) systems engineering function. Guidance for establishing the structural aspect of engineering a system of systems is provided by the various systems engineering standards, as defined by the various international standards bodies. Additionally, engineering a system of systems depends on the effective integration of technical processes with the human and organizational considerations (Brooks and Sage, 2006).

On the philosophical as well as pragmatic side, Handy's (1992) new "federalist" principles provide insight into many of the considerations necessary to engineer the more complex forms of a system of systems. This federalist approach is necessary to obtain a systems engineering ecology–in other words, a sustainable systems engineering approach for systems of systems management.

ISO/IEC/IEEE 15288 provides a high-level approach to defining roles and responsibilities for the systems engineering professionals. ISO/IEC/IEEE 15288 employs a process model based on three primary organizational areas (or levels) of responsibility: enterprise, project, and technical. Within each organization, a coordinated set of enterprise, project, and technical processes contribute to the effective creation and use of systems and to achieve the organization's goals.

The key steps for defining the unique roles and responsibilities for each systems engineering group involve defining key terms and concepts — creating a common data dictionary, developing the framework for the systems engineering and management functions, summarizing the challenges associated with engineering and management of the system of systems being evaluated, modeling the information flow among systems engineering groups comprising the system of systems, evaluating the degree to which federalist principles and axioms are/should be implemented, and documenting the roles and responsibilities by adapting the agreement processes listed in ISO/IEC/IEEE 15288.

Periodically, the definition of roles and responsibilities will need to be revalidated. Case studies can serve this function well by extending the applicability of exiting systems engineering standards by more explicitly treating the systems engineering functions associated with engineering systems of systems or system families and identifing areas for further research.

REFERENCES

Ackoff, R., 1971, Towards a system of systems concepts, *Management Science*, Vol. 17, No. 11, Theory Series, pp. 661–671, July.

Arnold, S., Lawson, H.W., 2004, Viewing systems from a business management perspective: the ISO/IEC 15288 standard, *Systems Engineering*, 7(3): 229–242.

Bergy, J., O'Brien, L., Smith, D., 2003, Application of options analysis for reengineering in a lead system integrator environment, CMU/SEI-2003-TN-009.

Berry, B., 1964, Cities as systems with systems of cities, Papers and Proceedings of *Regional Science Association*, 13:147:164.

Brooks, R.T., Sage, A.P., 2006, System of systems integration and test, *Information, Knowledge, and Systems Management*, 5(4): 261–280.

Carlock, P.G., Fenton, R.E., 2001, Systems of systems (SoS) enterprise systems engineering for information-intensive organizations, *Systems Engineering*, 4(4): 242–261.

Checkland, P., 1999, *Systems Thinking, Systems Practice*, John Wiley & Sons, Inc., Hoboken, NJ.

Eisner, H., 1993, RCASSE: rapid computer-aided systems of systems engineering, *Proceedings of the 3rd International Symposium of the National Council of System Engineering*, NCOSE, Vol. 1, pp. 267–273.

Eisner, H., Marciniak J., McMillan R., 1991, Computer-aided system of systems engineering, *Proceedings IEEE Conference on Systems, Man, and Cybernetics*, pp. 531–537.

Flood, S., Richard P., 2005, An Assessment of the lead systems integrator concept as applied to the future combat system program, Defense Acquisition Review Journal, DAU Press, December (http://findarticles.com/p/articles/mi_m0SVI/is_4_12/ai_n16128490/pg_11).

Friedman, G., Sage, A.P., 2004, Case studies of systems engineering and management in systems acquisition, *Systems Engineering*, 7(1): 84–97.

GAO, 2005, Framework for Assessing the Acquisition Function at Federal Agencies, GAO-05-218G, Washington, DC, September (http://www.gao.gov/new.items/d05218g.pdf).

GAO, 2006, Letter to the Subcommittee on Strategic Forces, Committee on Armed Services, *Defense Acquisitions: Space System Acquisition Risks and Keys to Addressing Them*, Washington, DC, June (http://www.gao.gov/new.items/d06776r.pdf).

Grady, J.O., 1995, *System Engineering Planning and Enterprise Identity*, CRC Press, Boca Raton, FL.

Grady, J.O., 1999, *System Engineering Deployment*, CRC Press, Boca Raton, FL.

Gupta, A., 2003, Role and Importance of Lead System Integrator in Context of New Air Operations Centers, MIT Sloan School of Management White Paper.

Handy, C., 1992, Balancing corporate power: a new federalist paper, *Harvard Business Review*, 70(6): 59–72.

Handy, C., 1995, Trust and the virtual organization, *Harvard Business Review*, 73(3): 40–50.

Holland, J.H., 1995, *Hidden Order—How Adaptation Builds Complexity*, Addison Wesley, Reading, MA.

Holland, J.H., 1998, *Emergence—From Chaos to Order*, Addison Wesley, Reading, MA.

IEEE 15288, 2005, Adoption of ISO/IEC 15288:2002 systems engineering—system life cycle processes, IEEE, New York, NY.

Jamshidi, M., 2005, System of systems engineering definitions, *Proceedings IEEE Systems, Man, and Cybernetics Conference*, October, Waikoloa, HI http://ieeesmc2005. unm.edu/ SoSE_Defn.htm.

Keating, C., Rogers, R., Unal, R., Dryer, D., Sousa-Poza, A., Safford, R., Peterson, W., Rabadi, G., 2003, System of systems engineering, *Engineering Management Journal*, 15(2): 36.

Kotov, V., 1997, Systems of systems as communicating structures, Hewlett Packard Computer Systems Laboratory Paper HPL-97-124, pp. 1–15.

Krygiel, A.J., 1999, *Behind the Wizard's Curtain: An Integration Environment for a System of Systems*, CCRP Publication Series, Vienna, VA.

Lane, J., Boehm, B., 2008, System of systems lead system integrators: where do they spend their time and what makes them more/less efficient, *Systems Engineering*, 11(1): 81–91.

Lane, J., Valerdi, R., 2007, Synthesizing SOS concepts for use in cost modeling. *Systems Engineering*, 10(2): 23–32.

Levis, A.H., 2009, System Architectures, *Handbook of Systems Engineering and Management*, 2nd edition, John Wiley & Sons, Hoboken, NJ.

Maier, M.W., 1998, Architecting principles for systems-of-systems, *Systems Engineering*, 1(4): 267–284.

Maier, M.W., Rechtin, E., 2000, *The Art of Systems Architecting*, CRC Press, Boca Raton, FL.

Manthorpe, W., 1996, The emerging joint system of systems: a systems engineering challenge and opportunity for APL, *John Hopkins APL Technical Digest*, 17(3): 305–311.

Office of the Under Secretary of Defense for Acquisition, Technology, and Logistics (USD/AT&L), 2003, *Report of the Defense Science Board/Air Force Scientific Advisory Board Joint Task Force on Acquisition of National Security Space Programs,* Washington, DC, May (http://stinet.dtic.mil/cgi-bin/GetTRDoc?AD=ADA429180&Location=U2&doc=GetTRDoc.pdf).

Rouse, W.B. (Ed.), 2006, *Enterprise Transformation: Understanding and Enabling Fundamental Change*, John Wiley & Sons, Hoboken, NJ.

Rouse, W.B., 2007, Complex engineered, organizational and natural systems, *Systems Engineering*, 10(3): 260–271.

Sage, A.P., 2002, Complex adaptive systems, in *McGraw Hill Yearbook of Science and Technology*, pp. 51–53.

Sage, A.P., Cuppan, C.D., 2001, On the systems engineering and management of systems of systems and federations of systems, *Information, Knowledge, Systems Management*, 2(4): 325–345.

Sage, A.P., Biemer, S.M., 2007, Processes for system family architecting, design, and integration, *IEEE Systems Journal*, 1(1): 5–16.

Sage, A.P., Rouse, W.B., 2009, Information technology and knowledge management, chapter 35 in *Handbook of Systems Engineering and Management*, 2nd edition, John Wiley & Sons, Inc., Hoboken, NJ.

Sheard, S., 2009a, Principles of complex systems for systems engineering, *Systems Engineering*, Vol. 12, No. 2. Forthcoming.

Sheard, S., 2008b, Complex adaptive systems in systems engineering and management, in: Sage, A.P., Rouse, W.B. (Eds.) *Handbook of Systems Engineering and Management*, 2nd edition, John Wiley & Sons, Hoboken, NJ.

Shenhar, A., 1994, A New systems engineering taxonomy, *Proceedings of the 4th International Symposium of the National Council of System Engineering,* NCOSE, Vol. 2, pp. 261–276.

Chapter **4**

System of Systems Architecting

CIHAN H. DAGLI and NIL KILICAY-ERGIN

Missouri University of Science & Technology, Rolla, MO, USA

4.1 COMPLEX SYSTEM ARCHITECTING

Architecting is the process of structuring the components of a system, their inter-relationships, and their evolution over time. It is related to the structural properties of a system. Successful architecture development is important as it plays a dominating role in integration of component systems (Sage, 2005). However, classical system architecting is changing as we are increasingly becoming a networked society. This is true in industry, individuals, and all forms of government. Society is becoming increasingly dependent on these networks. It is possible to combine these systems and make them transnational; it thus provides an opportunity to respond to the dynamically changing needs imposed by global events. Consequently, this creates a need for systems architectures that will be in effect for the duration of the event, possibly necessitating the need to develop a new systems architecture for the next mission or event. This fact is important, as it complicates the systems architecting activities. Hence, architecture becomes a dominating but confusing concept in capability development. These systems are generally referred as system of systems (SoS), which is a collaborative meta-level system structure where independent complex systems are integrated to provide increased functionality and performance capabilities.

The loss of any part of the system will degrade the performance or the capabilities of the whole. They need to evolve in time to accommodate changes in requirements and technology. Hence, systems engineers need to monitor and evolve/adapt systems architectures in a timely manner. This eliminates the classical concept that was used in the past, namely, that architectures are static.

These systems evolve by adding components, and as in the case of electrical utilities, creating a potential for hidden robustness, for example, load sharing across electric utilities, and also giving rise to a potential for cascading failures, as

System of Systems Engineering: Innovations for the 21st Century, Edited by Mo Jamshidi
Copyright © 2009 John Wiley & Sons, Inc., Publication

characterized by the August 14, 2003 blackout in Northeast United States. Individual systems within the SoS may be developed to satisfy the peculiar needs of a given group, the information they share being so important that the loss of a single system may deprive other systems of the data needed to achieve even minimal capabilities.

Unfortunately, the current body of knowledge in systems research is not sufficient for effective design and operation of these types of systems. There is a need to push the boundaries of technology and systems engineering and systems architecting research in both industry and research universities to meet the challenges imposed by new demands. There is an increased uncertainty about system requirements coupled with continuous changes in technology and organization structures. Diverse spectrums of missions and operations require the development of system architectures that can adapt and evolve.

The following subsections identify the attributes of complex systems to define various forms of complex systems and then focus on distinguishing characteristics and types of system of systems. The core differences between SoS architecting and systems architecting along with SoS architecting challenges are discussed. The rest of the chapter describes evolutionary SoS architecting in terms of creating meta-architectures and methodologies for architecting these systems with specific focus on the potential use of artificial life techniques.

4.1.1 Attributes of Complex Systems

Different complex systems can be identified by analyzing the system attributes such as interdependent, independent, distributed, cooperative, competitive, and adaptive. Recent system definitions can be based on these attributes. For example, it is possible to define a family of systems (FoS) as a set or arrangements of *independent* systems that can be arranged or interconnected in various ways to provide capabilities. The mix of systems can be tailored to provide desired capabilities, dependent on the situation. Although these systems can be providing useful capabilities independently, in collaboration they can more fully satisfy a more complex and challenging capability. We can also define intelligent enterprise systems in terms of *cooperative*, *competitive*, and *adaptive* systems that evolve to respond to changing business conditions. System of systems can be defined in terms of *interdependence* attribute where a set or arrangements of interdependent systems are connected to provide a given capability. While individual systems within the SoS may be developed to satisfy the peculiar needs of a given user group, the information they share is so important that the loss of a single system may deprive other systems of the data needed to achieve even minimal capabilities.

Complexity theory is a beneficial approach to define and understand the identity of a system. It helps in understanding how complex systems are affected from their environments and how a system learns by proposing alternative ways for improvement. It also answers the question that why some good predictions and solutions can be obstructed by dynamic nature of the environment. There are some conclusions that complexity theory arrives (Levy, 2000), which are applicable to SoS:

1. *Long-term planning is impossible:* There are nonlinear relationships among components of complex systems. Therefore, long-term planning is impossible.

Systems of systems are composed of complex systems, and a metasystem behavior cannot be derived by analyzing the behavior of the component subsystems.

2. *Dramatic change can occur unexpectedly:* Complexity theory claims that small perturbations can also cause huge changes in the overall system behavior. Changes are inevitable and impact of changes is not always obvious. This property is the reason for cascading failures in system of systems. Since there is strong interdependency among systems, a small change can cause a chain reaction and result in cascading failures.

3. *Complex systems exhibit patterns and short-term predictability:* Long-term forecasting is impossible, but short-term forecasting and describing the behavioral model of systems is possible. Therefore, next time period behavior of systems can be predicted when reasonable specifications of conditions at one time period are given. System of system testing and validation is based on this characteristic. Architecture performance evaluations focus on short-term forecasting of system architecture behavior.

4. *Organizations can be turned to be more innovative and adaptive:* Complexity theory suggests that emergent order and self-organization provide a robust solution for organic networks to be successful in competitive and rapidly changing environmental conditions. The evolutionary characteristic of the SoS architecting results in emergent capabilities that individual systems are not capable of achieving. System architects can benefit from this property by designing SoS components that can self-adapt and self-organize to changing environmental conditions.

4.1.2 System of Systems

There are many definitions of SoS depending on the application area and focus (Carlock and Fenton, 2001; Sage and Cuppan, 2001; Maier, 2005.). Future Combat Systems (FCS), NATO, transnational virtual enterprises, and intelligent transportation systems are some of the networked systems that we observe in governments and commercial enterprises.

Maier (1998) highlights two principal characteristics for distinguishing SoS from other complex systems: operational independence of the components and managerial independence of the components. System of system components must operate independently when separated from the main system. The components must also maintain their existence independent of the SoS. As a result of these characteristics, SoS fulfills a common purpose as well as additional purposes of the individual subsystems. This causes inherent redundancy and increases the development cost of SoS. Maier (1998) classifies SoS into different types based on the SoS development process. If the system is developed through formal organizations to fulfill a common purpose, it is a directed SoS. If the system is developed through the collaboration of its participants, it is a collaborative SoS. Chen and Clothier (2003) classify SoS into two types based on architecture structure. If the component systems are architected so that they can be integrated to work together to fulfill a goal, it is a dedicated SoS. If subsystems are previously existing architectures that are integrated to meet an immediate mission requirement, it is a virtual SoS.

4.1.3   System of Systems Architecting Versus Systems Architecting: Architecting Challenges

SoS conceptual framework identifies three components: physical networks such as roads and power grids, information networks such as Intranets and databases, and social networks such as people, organizations, and processes. The ultimate goal of SoS is to create robust physical networks, information networks, and social networks and integrate these three main components seamlessly. This can be achieved through better networking and information sharing leading to improved situation awareness/ understanding, which enhances collaboration and interactions in social networks leading toward more agile SoS elements and effective SoS.

Even though the SoS conceptual framework outlines the steps to successful SoS formation, there are many architecting challenges when compared to classical systems architecting. System of systems architecting process is similar to systems architecting processes in terms of scoping, aggregation, partitioning, and certification, but it is at a meta-level. The meta-level architecting will be elaborated in the following sections. While the process shares common properties, there are various deviations from classical systems architecting. Classical systems architecting concentrates on optimizing individual stand-alone systems, while SoS architecting concentrates on selecting the right collection of systems to satisfy the customer requirements. Therefore, the SoS architecting process spans through multiple abstraction layers and domains to foster collaborative functions among independent systems. Collaboration requires more emphasis on interface architecting, which brings along many challenges such as interoperability, scalability, and security issues.

In the SoS environment, architecture has more influence on requirements than it does in an environment dominated by a stand-alone complex system. In a stand-alone system, architecture is the implementation solution for the client requirements. However, in SoS environment architectural constraints imposed by existing systems can have a major influence on overall capabilities and system behavior. Specifically, the amount and variety of the legacy systems impose interoperability constraints in SoS architecting processes. This is a major challenge since integration of these legacy systems to other systems depends on the abstraction level, which is not as clear as in classical systems architecting.

Different architecting tools are required at high levels of complexity. In SoS architecting, the use of many architecture and interoperability-related enablers, such as levels of information systems interoperability, architecture frameworks, technical architectures, and other elements, have missing linkages to system engineering processes (Chen and Clothier, 2003). A good balance of heuristics, analytical techniques, and integrated modeling is necessary as architecting tools for SoS. Specifically, model-centric frameworks and executable models become important tools for SoS analysis and architecting as they provide insights into SoS architecture behavior. Figure 4.1 highlights the deviations from classical system architecting for SoS.

	System of Systems Architecting	Systems Architecting
Architecting properties	• Abstract, meta-level • Fuzzy uncertain requirements • Network-centric • Softwareintensive • People-intensive • Intensive communication infrastructure • Network of various stakeholders • Collaborative emergent development • Dynamic architecture	• Domain specific systems level • Several stakeholders • Controlled development • Static architecture
Architecting constraints	• The same classical systems architecting processes, but at the meta-level • Emphasis is on interface architecting to foster collaborative functions among independent systems • Concentration is on choosing the right collection of systems to satisfy the requirements • Scalability • Interoperability • Trustworthiness • Hidden cascading failures • Confusing life cycle context	• Architecting processes at component and systems level • Monolithic system architecting (optimize individual systems) • Concentration is on the building the right physical technical architecture • Clear life cycle context
Legacy systems	• Abstraction level determines the integration of legacy systems to other systems • Large amount and variety of legacy systems	• Integration of legacy systems to system components are more clear compared to SoS
Architecting tools	• Model-centric and executable models • Balance of heuristics, analytical techniques and integrated modeling	• Document-centric frameworks • Model-centric frameworks • Pure analytical techniques • Heuristics

FIGURE 4.1 SoS architecting versus systems architecting

System of system architecting and evolution presents challenges due to changes in the system engineering context (Chen and Clothier, 2003). As different systems are integrated, the system life cycle context becomes a confusing issue for SoS. The existing component systems' life cycle and the recently added systems' life cycle complicate the SoS life cycle definition. Additionally, the diversity of the engineering focus and the engineering environment bring up additional challenges. For example, dynamically changing requirements increase uncertainty, and systems need to be designed for fuzzy attributes. Diverse spectrum of missions and operations increase complexity of architecting SoSs. Besides, the need to develop dynamic and evolving communication architecture dominates the SoS architecting processes. All these challenges open various research needs for SoS. There is a need for research to find answers to the following questions:

- How can we assure trustworthiness?
- How can we assure interoperability?
- How can we assure large-scale design along with distributed testing?
- How can we assure evolutionary growth?
- How can we deal with hidden interdependencies?
- How can we guard against hidden cascading failures?
- How can we deal with complexity?

Maier (2005) identifies several research areas specific to SoS. One area is to balance the sociotechnical equilibrium of SoS. This becomes important in social SoS such as intelligent transportation systems. Designers are challenged with explicitly incorporating interactive social and technical effects into system design. This enhances the need for incorporation of human systems into models of SoS. Another challenging research area is the adjustment of optimization techniques for identifying invariant architectures that will be useful for many design solutions rather than an optimal solution to a specific problem.

Having identified the core differences between classical systems architecting and SoS architecting and the associated challenges, the following sections elaborate on the meta-level evolutionary SoS architecting.

4.2   EVOLUTIONARY SYSTEM ARCHITECTING

Evolutionary system architecting is a distinguishing characteristic of system of systems. Complex stand-alone system architectures are static, whereas an SoS architecture is dynamic because it may not be fully formed initially. As the client of the SoS experiences with the SoS, functions and subsystems are added, removed, or modified. Therefore, while the classical system architecting is a controlled development, the architecting of SoS is a collaborative evolutionary development. As a consequence of this evolutionary architecting, SoS exhibits emergent behavior such that it achieves purposes that are not possible by its component subsystems (Sage, 2005).

SoS evolution can occur in various forms. When a subsystem is modified or improved without changing the interfaces to other subsystems, self-evolution occurs. When two or more subsystems are integrated, SoS undergoes joint evolution. Emergent evolution is observed when a new system is architected from subsystems with new capabilities (Chen and Clothier, 2003). In traditional system engineering practice, these evolutionary phases are not addressed or at most considered at the end of the life cycle of the architecture. This fact along with increased technical and managerial complexity causes problems in application of traditional system engineering practices to SoS context.

Evolutionary SoS architecting starts with addressing the business and operational requirements, the technology options, and the existing architecture interfaces. Chen and Han (2001) introduce architecture evolution environment, where they define the architecture interfaces in layered structures such as operational interfaces, application interfaces, data interfaces, software platform interfaces, and hardware platform interfaces. This layered structure is necessary due to the interoperability challenges stemming from the complexity of the subsystem architectures. Evolution requirements are identified from business requirements and technological options. Architecture gaps are identified by checking the current architecture against the new requirements. A new architecture solution can be generated by considering these gaps and the new technological options. Then the solution can be implemented (Chen and Han, 2001; Chen and Clothier, 2003) and the implemented architecture becomes the input to the evolutionary architecting process.

Another distinctive characteristic of SoS architecting is its emphasis on creating an evolvable communication architecture, which is the backbone for collaborative capabilities of the SoS. The following subsections focus on the communications architecture of SoS, the Global Information Grid (GIG), and the network-centric capabilities the communication architecture provides. Then, SoS architecting is described from another perspective as a meta-level architecture generation process.

4.2.1 Global Information Grid

Complex systems architecting is an attempt to integrate several complex systems into meta-architectures. From many potential component systems, a set must be selected to construct the meta-architecture for SoS. The selection of the set depends on the requirements, functionalities, and capabilities desired from the SoS to achieve the common mission. Since the meta-architecture operates in continuously changing environments, multiple system states and actions must be explored during the complex system architecting processes. Also, spiral development of SoS necessitates dynamically changing evolving architectures (Kilicay and Dagli, 2007). This requires the creation of a meta-architecture consisting of core components that remain unchanged for a given period as other components are evolved in time.

To achieve architecting such a metasystem, all component systems need a physical global interface to function. Initially, the Department of Defense defined the GIG as the seamless communications architecture for information superiority and the basic interface for creating meta-architectures for United States Military (Buda et al., 2001). Now, the Global Information Grid represents the system formed by the distributed collections of electronic capabilities that are managed and coordinated to support some sort of enterprise (virtual organization). It is designed to improve interoperability and reduce extra layers of redundancy and to increase the speed of exchanging data. It is like the Internet but less dependent on ground-based systems to transmit data and more dependent on mobile systems. The goal is to increase collaboration by linking users with common interests. Ultimately, sensors, business units, and military units would be tied to the GIG network serving as both users and providers of data.

Different independent systems are connected to the GIG to create a network-centric architecture such as missile systems, military forces, business systems, analysts, policy makers, and so on. The users of the GIG are at the same time data providers for the SoS. Figure 4.2 illustrates the evolving global information architecture. Mobile and ground communication networks provide data transfer capability. The servers and databases store these data to be converted to information through net-centric services and applications.

4.2.2 Network-Centric Operations

The networked systems comprise people, organizations, cultures, activities, and interrelationships. The semiautonomous systems (people, organizations) are integrated through cooperative arrangements. The network-centric operations leverage communication technologies to create synergies and emergent capabilities in a synchronized way. Net-centric operations mean rapid change, synchronization of actions, interdependent operations, and emergent capabilities. The operational environment gain information superiority, global reach, shared situational awareness, faster decision cycle, and joint operations at lower tactical levels through net-centric operations. In such environments networking, sensing, information sharing, collaboration, innovation, and agility are valued properties.

The evolving net-centric architecture is created through architecting the GIG, which has distinctive properties. The GIG is modular where portion of the ground and mobile communication networks can be expanded or removed so that the communication architecture is adaptable to changing and evolving requirements. It is robust against the cascading failures due to interdependencies. It is interoperable so that different legacy systems can be integrated into the architecture at various abstraction levels. Finally, it allows network-centric operability for sensing, networking, and collaboration. Figure 4.3 illustrates this architecture. Even though the GIG or the communication architecture is the heart of the SoS architecture, the SoS architecting is not complete as it requires different systems to operate on this communication architecture. Thus, the evolving physical SoS architecture is created by integrating different systems to the GIG. The following subsection explains the SoS architecting.

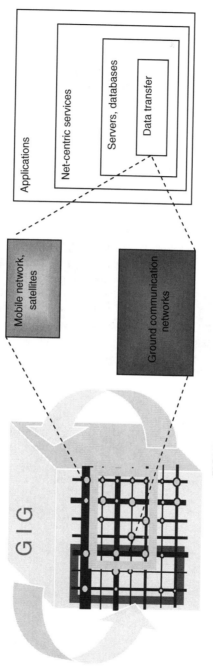

FIGURE 4.2   An evolving global physical architecture

85

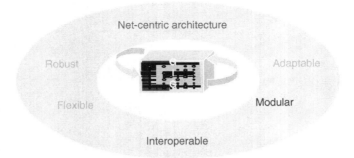

FIGURE 4.3 The evolving net-centric architecture

4.2.3 Dynamically Changing Meta-Architectures

The evolving physical architecture is created by connecting operationally and managerially independent systems to the GIG. This net-centric architecture is also evolving to meet the changes in system requirements and objectives. It is the dynamically changing architecture that creates the best net-centric systems, although data and the communication architecture are a necessity for the system to function. A dynamically changing meta-architecture for system of systems can be defined as a collection of different complex adaptive systems that are readily available to be plugged into the evolvable net-centric communications architecture. The challenge is to identify the right collection of systems that will collaborate to satisfy the client requirements. This shifts the focus from component and individual system level to meta-level architecting. Figure 4.4 illustrates this meta-architecture (Dagli and Kilicay, 2007).

This "plug and play" concept of assembling and organizing coalitions from different systems provides flexibility to respond to changing operational and environmental situations, but requires high level of interoperability in the information architecture that supports the coalition units.

The idea of creating meta-architectures can be clarified by looking at the net-centric manufacturing SoS architecture created for the production of Boeing 787. Figure 4.5

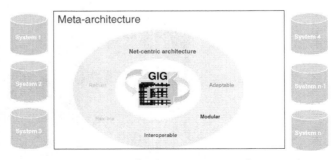

FIGURE 4.4 Dynamically changing meta-architecture for SoS

Company/Business Unit	Main Location	787 Work Statement
Boeing Commercial Airplanes (announced November and December 2003)	Washington	Airplane development, integration, final assembly, program leadership
Alenia/Vought Aircraft Industries (announced November 2003)	Italy, Texas	Horizontal stabilizer, center fuselage, aft fuselage
Boeing Fabrication (announced November 2003)	Washington, Canada, Australia	Vertical tail assembly, movable trailing edges, wing-to-body fairing, interiors
Boeing Wichita (announced November 2003; April 2004)	Kansas, Oklahoma	Fixed and movable leading edges, flight deck, part of forward fuselage, engine pylons
Fuji Heavy Industries (announced November 2003)	Japan	Center wing box, integration of the center wing box with the main landing gear wheel well
Kawasaki Heavy Industries (announced November 2003)	Japan	Main landing gear wheel well, main wing fixed trailing edge, part of forward fuselage
Mitsubishi Heavy Industries (announced November 2003)	Japan	Wing box
Hamilton Sundstrand (announced February 2004, March 2004, July 2004, September 2004)	Connecticut	Auxiliary power unit, environmental control system, remote power distribution units, electrical power generating and start system, primary power distribution, nitrogen generation, ram air turbine emergency power system, electric motor hydraulic pump subsystem.
Rockwell Collins (announced February 2004)	Iowa	Displays, communications/ surveillance systems
Smiths (announced February 2004, June 2004)	United Kingdom	Common core system, landing gear actuation and control system, high lift actuation system

FIGURE 4.5   The Boeing 787 example

provides a portion of the companies involved in the production of Boeing 787. A total of 42 national and international companies are connected to the global company information grid. Each company is operationally and managerially independent and has different organizational architectures and legacy systems. The focus of the SoS architecting for this domain is to choose the right collection of component manufacturing companies to produce Boeing 787. The emphasis is on interface architecting to achieve collaborative functions among these manufacturing companies. The Boeing information grid (GIG) provides the net-centric capabilities, and the way these independent companies are connected to the Boeing information grid creates the meta-architecture for Boeing enterprise. The information grid architecture is the backbone of the SoS architecture, but the SoS virtual manufacturing meta-architecture can achieve its mission through the independent companies connected to the information grid. As the requirements or fuzzy system attributes change or evolve over time, new manufacturing companies are added to this information grid. Hence, the meta-architecture is evolutionary. Figure 4.6 illustrates the Boeing global information grid and the various companies connected to its functional and mission networks. The architecting process spans through different abstraction layers, and in this case functional and mission layers.

System of systems is not always created with a directed mission, but can also be created by the collaboration of its users (Maier, 1998). One example of such a system is the financial market organization. The market organization and the rules for trading at the market create a meta-architecture similar to the SoS meta-architecture described in this section. Traders are connected to the market trading grid through some communication architecture, and different system dynamics are observed based on trader behaviors. Since the subsystems (traders) are connecting without any direction, the SoS architecting focuses on creating the communication grid with net-centric capabilities and designing the right price formation mechanism, the rules for trading at the market. The way the traders are connected to this architecture and their behavior form the market SoS architecture. Other SoS architectures can be created by

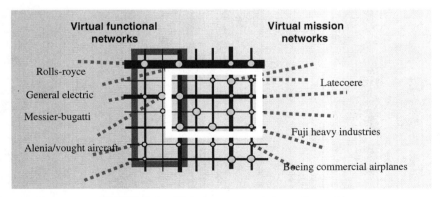

FIGURE 4.6   Global company information grid

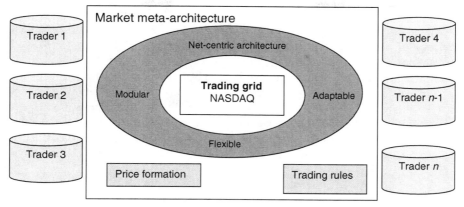

FIGURE 4.7 Meta-architecture generation for financial markets. (Adapted from Ergin, 2007)

connection of other systems such as trust funds, and so on (Ergin, 2007). Figure 4.7 illustrates the financial market meta-architecture.

4.3 SoS ARCHITECTING ENABLERS: THE ROLE OF ARTIFICIAL LIFE

There is a diversity of architecture frameworks and methodologies for enabling architecture developments. The fundamental goal of all these enablers is to capture a detailed description of the SoS architecture based on different architectural views, develop an implementation process by utilizing available technological options and knowledge, and then conduct performance evaluations. The architecture enablers can be classified into three types based on their support on the architecture design process. These are mainly enablers for static representation of the architecture, enablers for creating an executable model of the architecture, and enablers for logical, behavioral, and performance evaluation of the architecture (Wagenhals et al., 2003).

Department of Defense Architectural Framework (DoDAF) is an architecture representation standard to provide a common communication language in systems architecting for the design and operation of system of systems. It is a descriptive framework that captures operational, technical, and systems views of the architecture. The framework is important for capturing a common understanding of the architecture, but it does not provide a procedure for developing the architectures.

Zachman's Enterprise Architecture Framework (Zachman, 1987) is another enabler for static representation of the enterprise architecture. The idea of this framework is that the overall enterprise architecture is made up of other architectures that are focused on different specific areas of concern. Therefore, different architectural abstractions can be designed by different perspectives. For example, enterprise strategy architecture can be defined by the planners, an enterprise conceptual model can be created by the owner, a logical system model can be created by the designer, a physical technology model can be designed by the builder, a detailed architecture model can be created by the subcontractor, and so on. These architecture abstractions from various perspectives are integrated together taking into account the external inputs such as user requirements, concept of operations, business plans, laws, and regulations. The final architectural framework is a 6 × 6 matrix that describes overall enterprise architecture. The framework is useful in capturing the static view of the system, but does not help architects to understand how a specific change in a business process might lead to a change in business functions or software applications.

ISO Reference Model for Distributed Processing, Federal Enterprise Architecture, DoD's Technical Reference Model, Net-Centric Enterprise Services, and Net-Centric Enterprise Solutions for Interoperability are some of the other descriptive frameworks used for SoS analysis.

The object-oriented methodology is used as an alternative method for representation of architectures. Since most software systems tend to be object-oriented constructs, the transition from an object-oriented representation to system design is easier compared to structural representation frameworks (Wagenhals et al., 2003). For this purpose, UML and SysML are utilized to capture the system state models, user requirements, and other sets of complex relationships between the components of the system.

Petri nets and colored Petri nets are utilized as enablers for executable model generation. They visualize the system communications and processes through graph theoretic methods. Since these tools support simulation, architecture behavior can be analyzed to capture a dynamic view of the SoS. These tools have been successfully utilized in SoS architecture analysis. (Wagenhals et al., 2003; Madwaraj et al., 2006; Wang, 2007).

The executable models also serve as a performance evaluation enabler. The simulation models and state space analysis can identify deadlocks within the architecture, lower and upper bound system behaviors that may occur in the architecture. State models and simulation of state models have been used in performance assessments of Future Combat Systems (Campbell et al., 2005).

Heuristics techniques are utilized in system architecting to restrict the architecture search space by eliminating the past mistakes in system design. Maier (1998) highlights several well-known heuristics important for SoS architecting. *Leverage is at the interfaces* heuristic (Rechtin and Maier, 1997) becomes more important for SoS since components are highly independent. Successful SoS architecting lies in designing and leveraging the interfaces. The architecture evolution can be more rapid if *stable intermediate forms exist* (Rechtin and Maier, 1997) because there is no guarantee that all components will continuously collaborate. *The policy triage: Let the dying die; ignore the ones that will cover on their own and treat only those that would die without help* heuristic (Rechtin and Maier, 1997) puts emphasis on balancing the control and authority mechanism in SoS. Natural systems do not forecast or schedule. They respond quickly, robustly, and adaptively. SoS architecture can balance the control mechanism if the component systems are designed to have the capability of self-organization and self-adaptation.

In system of system analysis, the architecture efforts are focused on the evolution of the existing communications and processing systems, moving toward the creation of an integrated system that can provide a seamless physical, information, and social network. This brings the focus on understanding the system-level behavior emerging from these subsystems. It is feasible to understand any system of systems as an artificial complex adaptive system. The relation of SoS characteristics and complex adaptive systems (CAS) characteristics are outlined in Correa and Keating (2003). Artificial life tools have been successfully used in analysis of complex adaptive systems. Since system of systems is collection of several complex adaptive systems, we can utilize these tools for analysis of SoS behavior. The following subsections illustrate how artificial life tools can be utilized as architecting enablers.

4.3.1 Swarm Intelligence

Swarm intelligence specifically focuses on collective intelligence where many agents run concurrently performing actions that affect the behavior of other agents. There is no central communication and a well-specified task is set for the entire distributed system. The system focuses on maximizing some utility function to achieve the specified task. Bonabeau et al. 1999 provide a comprehensive survey of swarm intelligence systems that have been inspired by social insects.

Swarm intelligence is a good architecture enabler for SoS because characteristics of swarm intelligence are common with characteristics of system of systems. Figure 4.8 summarizes and compares swarm intelligence characteristics and SoS characteristics. Various applications utilized swarm intelligence for modeling and solving system of system architecture problems including collaborative robots architecture development, swarm routing in communications networks, and military swarm scenario modeling (Lambert, 2003).

Meyyappan (2006) presents a self-adaptive, swarm-based control model for real-time part routing in a flexible flow shop environment. The control model is a multiagent system that exhibits adaptive behavior, which has been inspired from the natural system of the wasp colony. The production problem consists of assigning trucks to paint booths in real time in a flexible flow shop environment with the objective of throughput maximization and minimization of number of paint flushes accrued by the production system, assuming no a priori knowledge of the color sequence or color distribution of trucks is available. The flexible flow shop environment is a dynamic environment where the product mix is not known prior to production. This is a good example of a system where changes in environmental conditions are not known and the system has to adapt quickly to changes. The self-adaptation capability is an important feature that provides evolving SoS architectures. Figure 4.9 summarizes how swarm intelligence is used for the advanced manufacturing system analysis.

For the flexible flow shop environments, three-layer software architecture is developed to conduct experiments and collect data. The software architecture includes execution layer, communications layer, and control layer. The execution layer includes the simulated flexible flow shop, developed in ARENA®. Transactions among different events, such as arrival of trucks, in the simulated model are executed by modules developed in Visual Basic (VB®) in the communications layer. An MS

	Characteristics of Swarm	Characteristics of SoS
Common goal	X	X
Local interactions	X	X
Autonomy of units	X	X
Stigmergy	X	X
Self organization	X	X
Distributed	X	X
Simple rules	X	
Flexible and robust	X	X
Large number	X	

FIGURE 4.8 Characteristics of swarm intelligence versus SoS

	Wasps	Manufacturing System
Goal	To maximize food collection, egg laying, nest building	To maximize throughput, minimize number of setups incurred, minimize average cycle time
Agents	Wasps	Machines
Work specialization	Wasps specialize to gather food or build nest or lay eggs	Machines specialize to process particular part types and avoid additional setups
Force	Force variable of wasp	Remaining processing times, setup times, wait times of machines
Threshold	Threshold of wasp	Setup requirements
Stimulus	Scent of food	Waiting times of parts

FIGURE 4.9 Manufacturing system and swarm intelligence. (Adapted from Meyyappan, 2006)

Access® database to keep track of performance measures also resides in this layer. The communications layer facilitates integration and information flow between the execution and control layers. Due to its modular structure, this software architecture easily allows replacement of one model with another and facilitates faster experimentation. This type of modular software architecture is necessary for behavior analysis of system of systems, so that various scenarios and architectures can be analyzed.

4.3.2 Multiagent Modeling

Multiagent modeling by its modular characteristic provides a valuable engineering abstraction for analysis of system of systems. There is a need for frameworks that can capture emergent behavior of system architectures. Human performance as a system parameter becomes more significant and dominant in SoS architecting. Human systems make decisions and facilitate the interactions between systems in the SoS. Therefore, there is a need to incorporate human behavior into architecture analysis. Multiagent modeling systems are used as architecting enablers for incorporating human behavior into analysis and capturing the emergent behavior of the architectures.

Multiagent studies focus on design aspects of the agents. Some of the major study areas can be described as follows (Ferber, 1999):

- Agent architectures
- Agent–system architectures
- Agent infrastructures

Agent architecture studies focus on internal architecture of agents such as perception, reasoning, and action components. Since multiagent systems are constructed without any global control, one way to prevent chaotic behavior of the system is to design perception and reasoning into agents. Therefore, cognitive architectures (Madni et al., 2005) provide the underlying foundations for a multiagent intelligent system. This component is actually a good way of incorporating humans into SoS architecture analysis.

Agent–system architecture studies analyze agent interactions and organizational architectures where agents operate and interact under specified environmental constraints. This component is analogous to the SoS meta-architecture. The way the agents are connected to the environment and their interactions create the meta-architecture for the system analysis.

Agent infrastructure studies focus on interface mechanisms of the multiagent systems, which is mainly communication aspects between agents. These studies try to achieve common agent communication language and protocols, common format for the content of communication, and shared ontology between agents. This component of the multiagent systems is analogous to the Global Information Grid of the SoS. It is the physical interface where agents interact to achieve some goals.

Multiagent systems have been successfully used in what-if scenario analysis and system structure analysis for system of systems such as Future Combat Systems.

Kewley (2004) utilizes multiagent modeling to simulate combat simulation with embedded tactical decision agents to test the effects of different command and control structures of Future Combat Systems. The model also serves as a performance evaluation enabler to estimate the effects of command and control techniques given certain information is available to the force. Dekker (2003) focuses on organizational architecture in the presence of cultural differences similar to NATO organization. The performance of a military force composed of friendly and opposing units is analyzed. Both friendly and opposing units are assumed to come from two different nations and two different services such as Air Force and Army. The performances of four different organizational architectures are analyzed in the model. Both the studies provide a good illustration of how humans can be incorporated into system architecture analysis.

4.3.3 Evolutionary Modeling

Models of elements of system of systems deploy all types of evolutionary algorithms. These algorithms provide self-adaptation capability to SoS architecture components. For example, learning classifier system is a suitable tool for SoS architecture analysis and design because it provides a good combination of exploitation and exploration. The learning classifier system has three main components: the performance component, the reinforcement component, and the discovery component (Wilson, 1995). Successful rule strategies are exploited through reinforcement component, whereas new rule strategies are discovered through the genetic algorithm discovery component. Unmanned vehicles and autonomous robots are an important component of Future Combat Systems. Learning Classifiers is utilized in Cazangi et al. (2003) to model control mechanisms of robots in unknown environments. In their study, classifier rules consist of obstacle avoidance and target capture behaviors. Their results reveal that robot controller is able to generalize properly, and thus robots are able to adapt to different environments.

Another important task in SoS environment is data analysis and prediction. There are many challenges for prediction under distributed information sharing (DIS) environment. First of all, no supervised learning technique can handle highly nonstationary evolving system's forecasting (Dagli and Kilicay, 2007). There is no good explanation why historical patterns should be expected to repeat. The enhancement for existing teacher-based learning such as Artificial Neural Network and Fuzzy Association is either too time consuming or too hard to maintain consistency. Therefore, there are a few solutions that can be applied to solve prediction problems under DIS environment. One solution is that core prediction engine can adopt supervised learning that is assisted by reinforcement learning architecture. Adaptive Critic Designs and Q-learning can be considered as potential reinforcement learning candidates. Supervised learning blocks such as ANNs and Fuzzy Association System can be used to make coarse learning, and the "normal" system patterns can be generalized from historical data. After sufficient coarse learning, fine learning is applied that employs reinforcement

DIS data environment

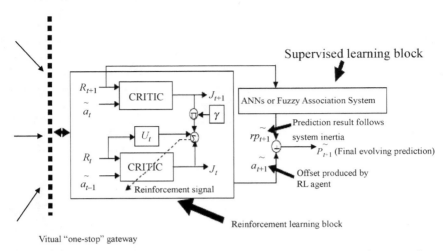

FIGURE 4.10 Supervised learning assisted reinforcement learning prediction architecture. (Adapted from Li, 2005)

learning (RL) algorithms. RL agent will mimic system evolution. The offset will be generated to show the effects of "unexpected" events (Li, 2005). Figure 4.10 outlines this architecture using Adaptive Critics method, which combines supervised learning with reinforcement learning.

4.3.4 Emergent Behavior Analysis of Architectures

The extensive complexity of the SoS architecting requires a multimethodology approach for analysis of SoS architectures. A three-step methodology can be used to capture the emergent behavior of SoS architectures (Kilicay and Dagli, 2007). The SoS architecture is first defined using a structural approach, such as a DoDAF. DoDAF defines three related views of architecture development, namely, operational view (OV), systems view (SV), and technical standards view (TV) (Umneh et al., 2007). These views are used to create a common language for stakeholders to understand the SoS. However, this framework is not sufficient to capture different state models of the SoS. Therefore, at the second step, an object-oriented approach such as UML is utilized to capture the system behavior by identifying end user's requirements, states and, sequence of events that the system can undergo (Stanilka et al., 2005). The first two steps still capture the static view of the SoS. Therefore, the third step is to convert the UML static model into an executable model so that emergent behavior of the SoS architecture can be analyzed. Finally, the architecture is modified based on the emergent behavior from the executable model. Figure 4.11 illustrates the multimethodology approach for analysis of SoS architectures.

Petri nets have been successfully used as an executable model and are easily combined with structural and object-oriented approaches (Madwaraj et al.,

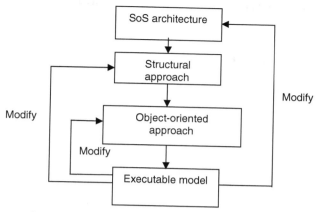

FIGURE 4.11 Multimethodology for behavior analysis of architectures. (Adapted from Ergin, 2007)

2006; Wang, 2007). The multiagent models can also be used as an executable model for emergent behavior analysis of architectures. These models provide the flexibility to incorporate evolutionary human behavior into system models. This can provide more insights during system architecting. Agent-based simulation packages such as AnyLogic™ have capabilities to convert UML constructs into executable models (Ergin, 2007).

4.4 CONCLUDING INSIGHTS

Seamless integrations and adaptive systems that can respond to changing requirements by reorganizing independent systems are the solution to today's competitive environment. This characteristic is necessary for both defense and commercial systems and can only be created with evolvable architectures. This creates a need to convert static systems architectures of the past to dynamic architectures of the future. Therefore, system of systems architecting deals with creating meta-architectures by integrating independent systems to the network-centric communication architecture. This type of architecting is different from traditional system engineering context where the architectures are static. Various forms of SoS architectures can be created to respond to the changing requirements as the users of these systems experience and change the requirements.

The SoS architecting process is challenging since the architecture evolution environment and the evolution requirements should be clarified before finding an architecture solution. Several architecture enablers such as structural architecture frameworks, object-oriented architecture representations, executable models, and simulations support this process by providing a common understanding of the SoS architecture and capturing the SoS architecture behavior.

Artificial life tools are also potential architecture enablers and aid SoS designers in critical technical analysis such as decision-aiding algorithms, development of command and control architecture, communication links, logistic infrastructure, sensor technologies, and platform capabilities. Modular architectures that utilize one or a combination of these methodologies promise more adaptability and robustness for SoS architecture design and analysis. Swarm intelligence, agent technology, and artificial intelligence can be applied wherever applicable to achieve robustness, reliability, scalability, and flexibility in SoS architecture. Therefore, new modeling and simulation algorithms based on biologically inspired approaches should be added to systems engineer's tool box to cope with modeling and analysis of emerging systems.

Evolutionary modeling and learning are essential components of system of systems. Current supervised learning techniques cannot handle rapidly evolving SoS. Reinforcement learning candidates combined with other learning mechanisms such as a supervised learning-assisted reinforcement learning architecture or classifier systems are more suitable for developing self-adaptation capabilities to the SoS architecture and its components.

Performance, risk, schedule, and cost are some of the system attributes for comparing and selecting architectures. Ability to learn and evolve new architectures from the previously generated architectures, based on system attributes' values, needs to be incorporated in modeling and simulation process.

For SoS architecting finding optimal solutions and efficiency is not as important as run-time interoperability. Flexibility and extensibility are more important compared to traditional system architecting. The dynamics of cognitive and social processes do not obey the static representations and rules of the architecture. Therefore, both structural and object-oriented analyses are required for comprehension of SoS. Simulation tools that combine various modeling paradigms (discrete, agent based, and system dynamics) should be used in analysis of SoS to capture different behavioral views.

Understanding and designing system architectures that can self-organize and adapt without any outside control is the solution to successful system of systems. How this can be achieved is the challenge that today's system engineers must face and solve.

REFERENCES

Bonabeau, E., Dorigo, M., Theraulaz, G., 1999, *Swarm Intelligence: From Natural to Artificial Systems*, Oxford University Press, New York.

Buda, G., Choi, D., Graveman, R., Kubic, C., 2001, Security standards for the global information grid, *IEEE Military Communications Conference*, 1: 617–621.

Campbell, J., Anderson, D., Lawton, C., Shirah, D., Longsine, D., 2005, System of systems modeling and simulation, *Proceedings of Conference on Systems Engineering Research 2005*, March 23–25, NJ, USA.

Carlock, P.G., Fenton, R.E., 2001, System-of-systems (SoS) enterprise systems for information-intensive organizations, *Systems Engineering*, 4(4): 242–261.

Cazangi, R., Zuben, F., Figueiredo, M., 2003, A classifier system in real applications for robot navigation, *The 2003 Congress on Evolutionary Computation*, 1: pp. 574–580.

Chen, P., Clothier, J., 2003, Advancing systems engineering for systems-of-systems challenges, *Systems Engineering*, 6(3): 170–183.

Chen, P., Han, J., 2001, Facilitating system-of-systems evolution with architecture support, *Proceedings of the International Workshop Principles Software Evolution (IWPSE)*, Vienna, Austria, pp. 130–133.

Correa, Y., Keating, C., 2003, An approach to model formulation for systems of systems, *IEEE Conference on Systems, Man and Cybernetics*, 4: 3553–3558.

Dagli, C., Kilicay, N., 2007, Understanding behavior of system of systems through computational intelligence techniques, *CD Proceedings of 1st Annual IEEE Systems Conference*, April 9–12, Honolulu, Hawaii.

Dekker, A.H., 2003, Using agent-based modeling to study organizational performance and cultural differences, *Proceedings of the MODSIM 2003 International Congress on Modeling and Simulation*, Vol. 4, Townsville, Queensland, pp. 1793–1798.

Ergin, N.H., 2007, Architecting system of systems: artificial life analysis of financial market behavior, Dissertation, University of Missouri-Rolla, July 2007.

Ferber, J., 1999, *Multi-Agent Systems: An Introduction to Distributed Artificial Intelligence* Addison-Wesley.

Kilicay, N., Dagli, C., 2007, Methodologies for understanding behavior of system of systems, *CD Proceedings of Conference on System Engineering Research*, March 14–16, Hoboken, NJ.

Kewley, R., 2004, Agent-based model of Auftragstaktik; self-organization in command and control of future combat forces, *Proceedings of the 2004 Winter Simulation Conference*, Vol. 1, No. 5–8, December 2004, pp. 930.

Lambert, J., 2003, UUV program and potential swarming applications, *Swarming Network Enabled C4ISR Conference*, January 13–14, 2003, McLean, VA, Section D, pp. 28–33.

Levy, D., 2000, Applications and limitations of complexity theory in organizational theory and strategy, *Handbook of Strategic Management*, Marcel Decker, NY, pp. 67–87.

Li, H., 2005, Financial prediction and trading via reinforcement learning and soft computing, Dissertation, University of Missouri-Rolla, December 2005.

Madni, A.M., Sage, A.P., Madni, C., 2005, Infusion of cognitive engineering into systems engineering processes and practices, *IEEE International Conference on Systems, Man and Cybernetics*, 1: 960–965.

Madwaraj, R., Ramakrishnan, S., Dagli, C., Miller, A., 2006, Modeling the global earth observation system of systems, *Industrial Engineers Research Conference*, FL, USA.

Maier, M., 1998, Architecting principles for systems-of-systems, *Systems Engineering*, 1(4): 267–284.

Maier, M.W., 2005, Research challenges of system-of-systems, *IEEE International Conference on Systems Man and Cybernetics*, 4: 3149–3154.

Meyyappan, L., 2006, Domain adaptive control architecture for advanced manufacturing systems, Dissertation, University of Missouri-Rolla, July 2006.

Rechtin, E., Maier, M., 1997, *The Art of Systems Architecting*, CRC Press, Boca Raton, Florida.

Sage, A.P., 2005, System of systems: architecture based systems design and integration, *Keynote Speech of International Conference on Systems, Man and Cybernetics*, October 10–12, Hawaaii, USA, http://ieeesmc2005. unm.edu/, accessed 2007.

Sage, A.P., Cuppan, C.D., 2001, On the systems engineering and management of systems-of-systems and federations of systems, information, knowledge, *Systems Management*, 2(4): 325–345.

Stanilka, S.P., Miller, A., Dagli, C., 2005, Object-oriented development for DoDAF system of systems, *CD Proceedings of INCOSE 2005*, July 10–14, Rochester, NY.

Umheh, N., Miller, A., Dagli, C., 2007, TOGAF vs. DoDAF: architecting frameworks for Net-centric systems, *CD Proceedings of IERC Industrial Engineering Research Conference*, May 19–23, Nashville, TN.

Wagenhals, L.W., Haider, S., Levis, A.H., 2003, Synthesizing executable models of object oriented architectures, *Systems Engineering*, 6: 266–300.

Wang, R., 2007, Executable system architecting using systems modeling language in conjunction with colored petri nets: a demonstration using the GEOSS network centric system, Master Thesis, University of Missouri-Rolla, July 2007.

Wilson, W.S., 1995, Classifier fitness based on accuracy, *Evolutionary Computation*, 3(2): 149–175.

Zachman, J., 1987, A framework for information systems architecture, *IBM Systems Journal*, 26(3): 276–292.

Chapter **5**

Modeling and Simulation for Systems of Systems Engineering

SAURABH MITTAL,[1] BERNARD P. ZEIGLER,[2] JOSÉ L. RISCO MARTÍN,[3] FERAT SAHIN,[4] and MO JAMSHIDI[5]

[1]Dunip Technologies, New Delhi, India
[2]Arizona Center for Integrative Modeling and Simulation, Electrical and Computer Engineering, University of Arizona, Tucson, AZ, USA
[3]Departamento de Arquitectura de Computadores y Automática, Facultad de Informática, Universidad Complutense de Madrid, Madrid, Spain
[4]Multi Agent Bio-Robotics Laboratory, Electrical Engineering, Rochester Institute of Technology, Rochester, NY, USA
[5]Lutcher Brown Endowed Chair, Electrical and Computer Engineering, University of Texas San Antonio, San Antonio, TX, USA

5.1 INTRODUCTION

The system of systems (SoS) concept, originally suggested as a method to describe the use of different systems interconnected to achieve a specific goal, has grown in its myriad of definitions and concepts (DiMario, 2006, 2007). Nevertheless, a common defining attribute of an SoS that critically differentiates it from a single monolithic system is interoperability, or lack thereof, among the constituent disparate systems. The plethora of perspectives on SoS problems evident in the literature (Sage and Cuppan, 2001; Morganwalp and Sage, 2004; Sage, 2007) suggests that interoperability may take the form of integration of constituent systems (e.g., element A is hierarchically superior to element B) or interoperation of constituent systems (e.g., two or more independent elements or systems with no identified hierarchy). In this chapter we focus less on SoS problems per se than on the role that modeling and simulation (M&S) can play in helping to address these problems.

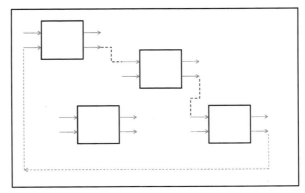

FIGURE 5.1   System composite of components systems or coupled model

Systems theory, especially as formulated by Wymore (1967, 1992), provides a conceptual basis for formulating the interoperability problem of SoS. Systems are viewed as components to be coupled together to form a higher level system, the SoS. As illustrated in Fig. 5.1, components have input and output ports (indicated with arrows) that allow couplings (indicated in dashed connectors) to be defined through which information can flow from output ports to input ports. The discrete event system specification (DEVS) formalism (Zeigler et al., 2000), based on systems theory, provides a computational framework and tool set to support systems concepts in application to SoS.

Information flow in the DEVS formalism, as implemented on an object-oriented substrate, is mediated by the concept of DEVS message, a container for port-value pairs. In a message sent from component A to component B, a port-value pair is a pair in which the port is an output port of A, and the value is an instance of the base class of a DEVS implementation, or any of its subclasses. A coupling is a quadruple of the form (*sending component A, output port of A, receiving component B, input port of B*). This sets up a path where by a value placed on an output port of A by A's output function is transmitted, in zero time, to the input port of B, to be consumed by the latter's external transition function.[1]

In systems or simulations implemented in DEVS environments the concepts of ports, messages, and coupling are explicit in the code. However, for systems/simulations that were implemented without systems theory guidance, in legacy or non-DEVS environments, these concepts are abstract and need to be identified concretely with the constructs offered by the underlying environment. For SoS engineering, where legacy components are the norm, it is worth starting with the clear concepts and methodology offered by systems theory and DEVS, getting a grip on the interoperability problems, and then translating backward to the non-DEVS concepts as necessary.

---

[1]The confluent function may also be involved; (see Zeigler et al., 2000 for details).

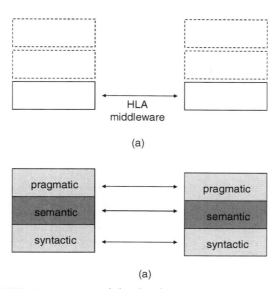

FIGURE 5.2   Interoperability levels in distributed simulation

5.1.1   Levels of Interoperability

With the conceptual basis offered by the DEVS formalism in mind, we briefly review experience with interoperability in the distributed simulation context and a linguistically based approach to the SoS interoperability problem (Carstairs, 2005; DiMario, 2006). Sage (2007) drew the parallel between viewing the construction of SoS as federation of systems and the federation that is supported by the High Level Architecture (HLA), an IEEE standard fostered by the Department of Defense (DoD) to enable composition of simulations (Dahmann et al., 1998; Sarjoughian and Zeigler, 2000). As illustrated in Fig. 5.2, HLA is a network middleware layer that supports message exchanges among simulations, called federates, in a neutral format.[2] However, experience with HLA has been disappointing and forced acknowledging the difference between enabling heterogeneous simulations to exchange data (so-called technical interoperability) and the desired outcome of exchanging meaningful data so that coherent interaction among federates takes place, the so-called substantive interoperability (Ylmaz and Oren, 2004). Tolk introduced the levels of conceptual interoperability model (LCIM) which identified seven levels of interoperability among participating systems (Tolk and Muguira, 2003). These levels can be viewed as a refinement of the *operational* interoperability type which is one of three defined by DiMario (2006). The operational type concerns linkages between systems in their interactions with one another, the environment, and with users. The other types apply to the context in which systems are constructed and acquired. They are *constructive*— relating to linkages between organizations responsible for system construction and *programmatic*—linkages between program offices to manage system acquisition.

---

[2]HLA also provides a range of services to support execution of simulations.

Linguistic level	A collaboration of systems or services interoperates at this level if:
Pragmatic — how information in messages is used	The receiver reacts to the message in a manner that the sender intends (assuming nonhostility in the collaboration).
Semantic — shared understanding of meaning of messages	The receiver assigns the same meaning as the sender did to the message.
Syntactic — common rules governing composition and transmitting of messages	The consumer is able to receive and parse the sender's message

FIGURE 5.3   Linguistic levels

Subsequently, one of the present authors coauthored a book in which the LCIM was mapped into three linguistically inspired levels: *syntactic, semantic,* and *pragmatic*. The levels are summarized in Fig. 5.3. More detail is provided in Zeigler and Hammonds (2007).

In this interoperability framework, the question to be addressed here is how M&S can help to achieve all three linguistic levels of interoperability. To discuss this question in more depth, we proceed to a brief review of the formal foundations of M&S that will allow us to frame the question and discuss the support available now and possible in the future.

5.2   REVIEW OF M&S FOUNDATIONAL FRAMEWORK

The theory of modeling and simulation presented in Zeigler et al. (2000) provides a conceptual framework and an associated computational approach to methodological problems in M&S. The framework provides a set of entities (real system, model, simulator, experimental frame (EF)) and relations among the entities (model validity, simulator correctness, among others) that, in effect, present an ontology of the M&S domain. The computational approach is based on the mathematical theory of systems and works with object orientation and other computational paradigms. It is intended to provide a sound means to manipulate the framework elements and to derive logical relationships among them that are usefully applied to real-world problems in simulation modeling. The framework entities are formulated in terms of the system specifications provided by systems theory, and the framework relations are formulated in terms of the morphisms (preservation relations) among system specifications. Conversely, the abstractions provided by mathematical systems theory require interpretation, as provided by the framework, to be applicable to real-world problems.

In its computational realization, the framework is based on the DEVS formalism and implemented in various object-oriented environments. Using Unified Modeling Language (UML) we can represent the framework as a set of classes and relations as illustrated in Figs. 5.4 and 5.5.

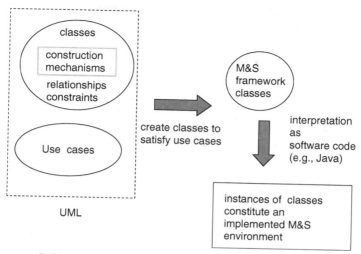

FIGURE 5.4 M&S framework formulated within UML

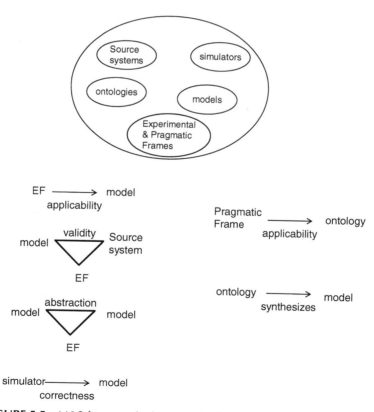

FIGURE 5.5 M&S framework classes and relations in a UML representation

Various implementations support different subsets of the classes and relations.[3] In particular, this chapter will review the implementation of DEVS within a Service-Oriented Architecture (SOA) environment called DEVS/SOA[4] (Mittal, 2007; Mittal et al., 2007b).

In the System of systems, systems and/or subsystems often interact with each other because of interoperability and over all integration of the SoS. These interactions are achieved by efficient communication among the systems using either peer-to-peer communication or through central coordinator in a given SoS. Since the systems within SoS are operationally independent, interactions among systems are generally asynchronous in nature. A simple yet robust solution to handle such asynchronous interactions (specifically, receiving messages) is to throw an event at the receiving end to capture the messages from single or multiple systems. Such system interactions can be represented effectively as discrete-event models. In discrete-event modeling, events are generated at random time intervals as opposed to some predetermined time interval seen commonly in discrete-time systems. More specifically, the state change of a discrete-event system happens only upon arrival (or generation) of an event, not necessarily at equally spaced time intervals. To this end, a discrete-event model is a feasible approach in simulating the SoS framework and its interaction. Several discrete-event simulation engines[5-8] are available that can be used in simulating interaction in a heterogeneous mixture of independent systems. The advantage of DEVS is its effective mathematical representation and its support to distributed simulation using middleware such as DoD's high level architecture (HLA).[9]

5.2.1 DEVS Modeling and Simulation

DEVS (Zeigler et al., 2000) is a formalism, which provides a means of specifying the components of a system in a discrete-event simulation. In DEVS formalism, one must specify *basic models* and how these models are connected together. These basic models are called *atomic models* and larger models that are obtained by connecting these atomic blocks in meaningful fashion are called *coupled models* (shown in Fig. 5.6). Each of these atomic models has *input ports* (to receive external events), *output ports* (to send events), set of *state variables*, *internal transition*, *external transition*, and *time advance functions*. Mathematically, it is represented as septuple system: $M = \langle X, S, Y, \delta_{\text{int}}, \delta_{\text{ext}}, \lambda, t_a \rangle$ where X is an input set, S is set of states, Y is set of outputs, δ_{int} is internal transition function, δ_{ext} is external transition function, λ is the output function, and t_a is the time advance function. The model's

[3]OMG, Object Modeling Group, www.omg.org.
[4]DUNIP: A prototype demonstration, http://www.acims.arizona.edu/dunip/dunip.avi.
[5]XDEVS webpage: http://itis.cesfelipesegundo.com/~jlrisco/xdevs.html.
[6]OMNET++, http://www.omnetpp.org/.
[7]NS-2, http://www.isi.edu/nsnam/ns/.
[8]MatLab Semulink, http://www.mathworks.com/products/simulink/.
[9]HLA, http://www.dmso.mil/public/transition/hla.

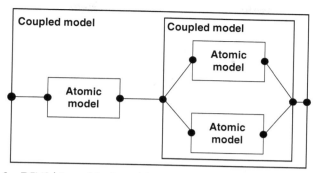

FIGURE 5.6 DEVS hierarchical model representation for systems and subsystems

description (implementation) uses (or discards) the message in the event to do the computation and delivers an output message on the outport and makes a state transition. A Java-based implementation of DEVS formalism, DEVSJAVA (Sarjoughian and Zeigler, 2000), can be used to implement these atomic or coupled models. In addition, DEVS-HLA (Sarjoughian and Zeigler, 2000) will be helpful in distributed simulation for simulating multiple heterogeneous systems in the system of systems framework.

The following section explores how Extensible Markup Language (XML) and DEVS environment can be combined in a simulation environment.

5.2.2 XML and DEVS

In DEVS, messages can be passed from one system (coupled or atomic model) to another using either predefined or user-defined message formats. Since the systems within SoS maybe different in hardware and/or software, there is a need for a unifying language for message passing. Each system need not necessarily have the knowledge (operation, implementation, timing, data issues, etc.) of another system in an SoS. Therefore, one has to work at a high level (information or data level) in order to understand the present working condition of the system. One such good fit for representing different data in a universal manner is XML. Figure 5.7 describes conceptually an SoS simulation example to demonstrate the use of XML as a message passing paradigm using DEVS formalism.

In Fig. 5.7, there are three systems in a hierarchy where systems A and B send and receive data from system C. System C sends and receives data from a higher level as described in the message of system C. The data sent by system C have data from systems A and B. In addition, it has information about system A and B being system C's subsystems.

Assuming the XML-based integration architecture and the DEVS environment exist, we can then offer solutions to robust data aggregation and fusion for system of systems that have heterogeneous complex systems such as mobile sensor platforms or microdevices.

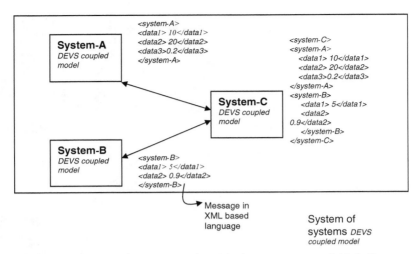

FIGURE 5.7 A SoS simulation example with three systems and XML-like message passing [SAH07a]

5.3 MODEL-BASED ENGINEERING

Model-based software engineering process is commonly referred as model-driven architecture (MDA) or model-driven engineering. The basic idea behind this approach is to develop model before the actual artifact or product is designed and then transform the model itself to the actual product. The MDA is pushed forward by Object Management Group (OMG) since 2001. The MDA approach defines system functionality using platform-independent model (PIM) using an appropriate domain-specific language. Then given a platform definition model (PDM), the PIM is translated to one or more platform-specific models (PSMs). The OMG documents the overall process in a document called MDA Guide.

MDA is a collection of various standards like the UML, the metaobject facility (MOF), the XML Metadata Interchange (XMI), common warehouse model (CWM) and a couple of others. OMG focuses MDA on forward engineering, that is, producing code from abstract, human-elaborated specifications (Wikipedia).

An MDA tool is used to develop, interpret, compare, align, and so on, models or metamodels. A "model" is interpreted as any kind of models (e.g., a UML model) or metamodel (e.g., CWM metamodel). An MDA tool may be one or more of the following types:

- **Creation Tool:** Used to elicit initial models and/or edit derived models.
- **Analysis Tool:** Used to check models for completeness, inconsistencies, or define any model metrics.
- **Transformation Tool:** Used to transform models into other models or into code and documentation.

- **Composition Tool:** Used to compose several source models, preferably conforming to the same metamodel.
- **Test Tool:** Used to "test" models. A mechanism in which test cases are derived in whole or in part from a model that describes some aspects of system under test (SUT).
- **Simulation Tool:** Used to simulate the execution of system represented by a given model. Simply speaking, is the mechanism by which model is "executed" using a programming language
- **Reverse Engineering Tool:** Intended to transform a particular legacy or information artifact into full-fledged models.

It is not required that one tool may contain all of the features needed for model-driven engineering. UML is a small subset of much broader scope of UML. Being a subset of MDA, the UML is bounded by its own UML metamodel. Progress has been made to develop executable UML models but it has not gained industry wide mainstream acceptance for the same limited scope. Potential concerns with the current MDA state of art include:

- MDA approach is underpinned by a variety of technical standards, some of which are yet to be specified (e.g., executable UML).
- Tools developed by many vendors are not interoperable.
- MDA approach is considered too idealistic lacking iterative nature of software engineering process.
- MDA practice requires skilled practitioners and design requires engineering discipline not commonly available to code developers.
- OMG sponsored CORBA project after much promises but it failed to materialize as a widely accepted standard.

Model-based testing is a variant of testing that relies on explicit behavior models that encode the intended behavior of the system and possibly the behavior of its environment (Utting et al., 2006). Pairs of input and output of the model of the implementation are interpreted as test cases for this implementation: the output of the model is the expected output of the SUT. This testing methodology must take into account the involved abstractions and the design issues that deal with lumping different aspects as these can not be tested individually using the developed model.

Following is the process for model-based testing technique (Utting et al., 2006) as shown in Fig. 5.8:

- A model of the SUT is built on existing requirements specification with desired abstraction levels.
- Test selection criteria are defined with an objective to detect severe and likely faults at an acceptable cost. These criteria informally describe the guidelines for a test suite.

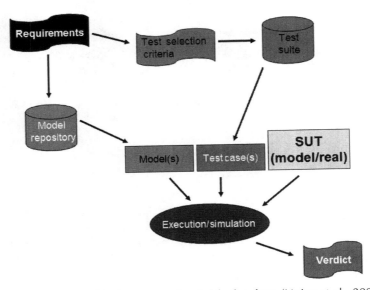

FIGURE 5.8   Graphical process extended further from (Utting et al., 2006)

- Test selection criteria are then translated into test case specifications. It is an activity where a textual document is turned "operational". Automatic test case generators fall into this step of execution.
- A test suite is "generated" that is built upon the underlying model and test case specifications.
- Test cases from the generated test suite are run on the SUT after suitable prioritization and selection mechanism. Each run results in a verdict of "passed" or "failed" or "inconclusive."

A summary of contributions to the model-based testing domain can be seen at (Utting et al., 2006).

5.4   SoS ARCHITECTURE MODELING: DoDAF, UML, AND SYSTEMS ENGINEERING PRINCIPLES

Unified Modeling Language (UML)[10]has been widely adopted by the industry as the preferred means of architecture specification due to its multimodel expressive power. However, UML constructs are not sufficient to specify the complete set of SoSE processes. A more extensive architectural framework is needed for better organization and management of SoSE artifacts. Frameworks such as Wymore's (1967, 1992), the Department of Defense Architecture Framework (DoDAF) (DoDAF Working Group, 2003, 2004), and Zachman's (O'Rourke and Fishman, 2003), are examples that may use UML as a means of presenting the concepts of SoSE. There are other representation

[10]UML, Unified Modeling Language, http://www.omg.org/technology/documents/formal/uml.htm.

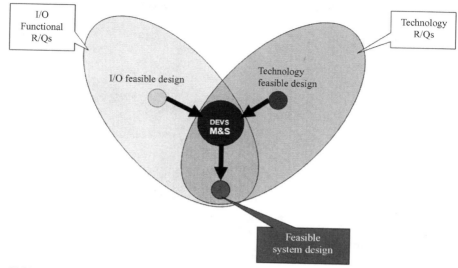

FIGURE 5.9   Role of M&S in Wymorian tricotyledon theory for systems engineering

mechanisms like IDEF[11] notation that help understand various SE perspectives. However, UML is more comprehensive in its graphical representations and these can aid SoSE frameworks in their descriptions of various portions of SE processes.

Figure 5.9 represents the integral role that M&S plays in SoSE processes with respect to Wymore's theory for systems engineering. DEVS is strategically placed between the I/O requirements and the technology requirements to provide M&S at "design" level, before a design is deemed "feasible." DEVS is based on "port" identification and is component-based modeling and simulation formalism that meets at the crossroads of the two cotyledons of Wymore's theory.

Wymore's theory of systems engineering also underlies our system theory and DEVS-based approach to SoSE, namely, integrated M&S for testing and evaluation strategy as described in sections ahead.

DoDAF is the basis for the integrated architectures mandated in DoD Instruction 5000.2 (2003) and provides broad levels of specification related to operational, system, and technical views. Integrated architectures are the foundation for interoperability in the Joint Capabilities Integration and Development System (JCIDS) prescribed in CJCSI 3170.01D and further described in CJCSI 6212.01D (Chairman, 2004, 2006). DoDAF seeks to overcome the plethora of "stove-piped" design models that have emerged. Integration of such legacy models is necessary for two reasons. One is that, as systems, families of systems, and systems of systems become more broad and heterogeneous in their capabilities, the problems of integrating design models developed in languages with different syntax and semantics have become a serious bottleneck to progress. The second is that another recent DoD mandate also intended to break down this "stove-piped" culture requires the adoption of the SOA paradigm as

[11]IDEF Functional modeling method, http://www.idef.com/.

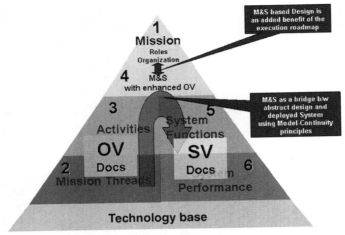

FIGURE 5.10 Role of M&S in DoDAF design process

supported in the development of Network-Centric Enterprise Services (NCES).[12] However, anecdotal evidence suggests that a major revision of the DoDAF to support net centricity is widely considered to be needed and efforts are underway toward a SOA-based DoDAF.

DoDAF consists of three views, namely, operational view (OV), systems view (SV), and technical view (TV). OV is a description of the tasks and activities, operational elements, and information exchanges required to accomplish DoD missions. DoD mission includes both the warfighting missions and business process. These are further decomposed into separate mission threads. SV is a set of graphical and textual products that describes systems and interconnections providing for, or supporting, DoD functions. SV associates system resources to the requirements of OV. TV is the minimal set of rules governing the arrangement, interaction, and interdependence of system parts or elements, whose purpose is to ensure that a conformant system satisfies a specified set of requirements.

M&S is introduced at the intersection of the modeling and technology cotyledons in Wymore's theory (recall Fig. 5.9) to help bridge the gap between the desired and the possible. Similarly, we can augment DoDAF by introducing M&S at a place where the design goes from abstraction to realism, that is, from OV specifications to SV implementations. Figure 5.10 depicts the fact that M&S can contribute to the system design process as well.

Although the current DoDAF specification provides an extensive methodology for system architectural development, it is deficient in several related dimensions— absence of integrated modeling and simulation support, especially for model continuity throughout the development process, and lack of associated testing support. To overcome these deficiencies, we described an approach to support specification of DoDAF architectures within a development environment based on DEVS-based

---

[12]DoD Metadata registry and clearinghouse, http://www.xml.gov/presentations/fgm/dodregistry.htm.

modeling and simulation. The authors (Zeigler and Mittal, 2005; Mittal, 2006) enhanced the DoDAF specification to incorporate M&S as a means to develop "executable architecture" (Atkinson, 2004) from DoDAF specifications and provided detailed DoDAF to DEVS mapping leading to simulation, and feasibility analysis. The result is an enhanced system life cycle development process that includes model-continuity-based development and testing in an integral manner.

5.5 SYSTEMS OF SYSTEMS TEST AND EVALUATION USING DEVS M&S

5.5.1 DEVS State of the Art in Test and Evaluation

Before moving onto the details of DEVS Unified Process (DUNIP), it is important to list the capabilities of DEVS technology as it stands today and how it contributes to the system of systems test and evaluation. Further, such test and evaluation (T&E) must be viewed within SOA perspective due to various DoD mandates. The following Fig. 5.11 summarizes DEVS state of the art with respect to T&E.

Desired M&S Capability for T&E	Solutions Provided by DEVS Technology
Support of DoDAF need for executable architectures using M&S such as mission based testing for GIG SOA	DEVS Unified Process (Mittal, 2007) provides methodology and SOA infrastructure for integrated development and testing, extending DoDAF views (Mittal, 2006).
Interoperability and cross-platform M&S using GIG/SOA	Simulation architecture is layered to accomplish the technology migration or run different technological scenarios (Sarjoughian et al., 2001; Mittal and Zeigler, 2003). Provide net-centric composition and integration of DEVS "validated" models using Simulation Web Services (Mittal et al., 2007a).
Automated test generation and deployment in distributed simulation	Separate a model from the act of simulation itself, which can be executed on single or multiple distributed platforms (Zeigler et al., 2001). With its bifurcated test and development process, automated test generation is integral to this methodology (Zeigler et al., 2005).
Test artifact continuity and traceability through phases of system development	Provide rapid means of deployment using model-continuity principles and concepts like "simulation becomes the reality" (Hu et al., 2003).
Real-time observation and control of test environment	Provide dynamic variable-structure component modeling to enable control and reconfiguration of simulation on the fly (Hu et al., 2003; Mittal and Zeigler, 2003, 2005; Mittal et al., 2006). Provide dynamic simulation tuning, interoperability testing and benchmarking.

FIGURE 5.11   DEVS state-of-the-art

In an editorial (Carstairs, 2005), Carstairs asserts an acute need for a new testing paradigm that could provide answers to several challenges described in a three-tier structure. The lowest level, containing the individual systems or programs, does not present a problem. The second tier, consisting of systems of systems in which interoperability is critical, has not been addressed in a systematic manner. The third tier, the enterprise level, where joint and coalition operations are conducted, is even more problematic. Although, current T&E systems are approaching adequacy for tier-two challenges, they are not sufficiently well integrated with defined architectures focusing on interoperability to meet those of tier three. To address mission thread testing at the second and third tiers, Carstairs advocates a collaborative distributed environment (CDE), which is a federation of new and existing facilities from commercial, military, and not-for-profit organizations. In such an environment, M&S technologies can be exploited to support model-continuity (Hu and Zeigler, 2005) and model-driven design (MDD) development (Wegmann, 2002), making test and evaluation an integral part of the design and operations life cycle.

The development of such a distributed testing environment would have to comply with recent DoD mandates requiring that the DoD Architectural Framework (DoDAF) be adopted to express high-level system and operational requirements and architectures (DoDAF Working Group, 2003; DoD Instruction, 5000.2, 2003; Chairman, 2004, 20062006). Unfortunately, DoDAF and DoD net-centric mandates (Atkinson, 2004) pose significant challenges to testing and evaluation since DoDAF specifications must be evaluated to see if they meet requirements and objectives, yet they are not expressed in a form that is amenable to such evaluation. DoDAF does not provide a formal algorithmically enabled process to support such integration at higher resolutions. Lacking such processes, DoDAF is inapplicable to the SOA domain and Global Information Grid (GIG) in particular. There have been efforts like in Dandashi et al. (2004) that have tried to map DoDAF products to SOA but as it stands out there is no clear-cut methodology to develop an SOA directly from DoDAF, rest aside their testing and evaluation.

5.5.2 Bifurcated Model-Continuity-Based Life Cycle Methodology

The needed solution is provided by combining the systems theory, M&S framework and model-continuity concepts that lead naturally to a formulation of a bifurcated model-continuity-based life cycle process as illustrated in Fig. 5.12. The process can be applied to development of systems using model-based design principles from scratch or as a process of reverse engineering in which requirements have already been developed in an informal manner. The depicted process is a universal process and is applicable in multiple domains. The objective of this research effort is to incorporate DEVS as the binding factor at all phases of this universal process.

The process has the following characteristics:

- **Behavior Requirements at Lower Levels of System Specification:** The hierarchy of system specification as laid out in Zeigler et al. (2000) offers well-characterized levels at which requirements for system behavior can be stated. The

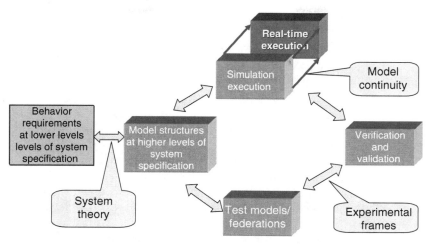

FIGURE 5.12   Bifurcated model-continuity-based system life cycle process

process is essentially iterative and leads to increasingly rigorous formulation resulting from the formalization in subsequent phases.

- **Model Structures at Higher Levels of System Specification:** The formalized behavior requirements are then transformed to the chosen model implementations, for example, DEVS-based transformation in C++, Java, C#, and others.

- **Simulation Execution:** The model base that may be stored in model repository is fed to the simulation engine. It is important to state the fact that separating the model from the underlying simulator is necessary to allow independent development of each. Many legacy systems have both the model and the simulator tightly coupled to each other that restrict their evolution. DEVS categorically separates the model from the simulator for the same simple reason.

- **Real-Time Execution:** The simulation can be made executable in real-time mode and in conjunction with model-continuity principles, the model itself becomes the deployed code.

- **Test Models/Federations:** Branching in the lower path of the bifurcated process, the formalized models give way to test models that can be developed at the atomic level or at the coupled level where they become federations. It also leads to the development of experiments and test cases required to test the system specifications. DEVS categorically aids the development of experimental frames at this step of development of test suite.

- **Verification and Validation:** The simulation provides the basis for correct implementation of the system specifications over a wide range of execution platforms and the test suite provides basis for testing such implementations in a suitable test infrastructure. Both of these phases of systems engineering come together in the verification and validation (V&V) phase.

5.6 EXPERIMENTAL FRAME CONCEPTS

An experimental frame is a specification of the conditions under which the system is observed or experimented with. As such an experimental frame is the operational formulation of the objectives that motivate a modeling and simulation project. Many experimental frames can be formulated for the same system (both source system and model) and the same experimental frame may apply to many systems. Why would we want to define many frames for the same system? Or apply the same frame to many systems? For the same reason that we might have different objectives in modeling the same system, or have the same objective in modeling different systems. There are two equally valid views of an experimental frame. One, views a frame as a definition of the type of data elements that will go into the database. The second views a frame as a system that interacts with the system of interest to obtain the data of interest under specified conditions. In this view, the frame is characterized by its implementation as a measurement system or observer. In this implementation, a frame typically has three types of components (as shown in Fig. 5.13): *generator* that generates input segments to the system, *acceptor* that monitors an experiment to see the desired experimental conditions are met, and *transducer* that observes and analyzes the system output segments.

Figure 5.13b illustrates a simple, but ubiquitous, pattern for experimental frames that measure typical job processing performance metrics, such as relate to round trip time and throughput. Illustrated in the web context, a generator produces service request messages at a given rate. The time that has elapsed between sending of a request and its return from a server is the round trip time. A transducer notes the

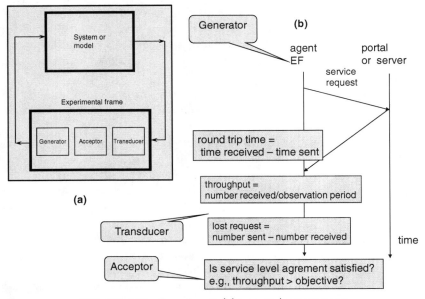

FIGURE 5.13 Experimental frame and components

departures and arrivals of requests allowing it to compute the average round trip time and other related statistics, as well as the throughput and unsatisfied (or lost) requests. An acceptor notes whether performance achieves the developer's objectives, for example, whether the throughput exceeds the desired level and/or whether say 99% of the round trip times are below a given threshold.

Objectives for modeling relate to the role of the model in systems design, management, or control. Experimental frames translate the objectives into more precise experimentation conditions for the source system or its models. A model is expected to be valid for the system in each such frame. Having stated our objectives, there is presumably a best level of resolution to answer the questions raised. It is usually the case that the more demanding the question, the greater the resolution needed to answer it. Thus, the choice of appropriate levels of abstraction also hinges on the objectives and their experimental frame counterparts.

Methods for transforming objectives into experimental frames have been discussed in the literature (Zeigler et al., 2000; Traore and Muxy, 2004). In the context of SoS engineering, modeling objectives support development and testing of SoSs. Here we need to formulate measures of the effectiveness (MOE) of an SoS in accomplishing its goal to allow us to rigorously evaluate the architectural and design alternatives. We call such measures, *outcome* measures. In order to compute such measures, the model must include certain variables; we'll call *output* variables, whose values are computed during execution runs of the model. The mapping of the output variables into outcome measures is performed by the transducer component of the experimental frame. Often there may be more than one layer of variables intervening between output variables and outcome measures. For example, in military simulations, measures of performance (MOP) are output variables that typically judge how well parts of a system are operating. Such measures enter as factors into MOEs.

5.7   EXPERIMENTAL FRAMES FOR SoS TEST AND EVALUATION

The DoD's concept of jointness, in which assets of Army, Navy, Air Force, and other military services are brought together to execute a mission, offers a progenitor for numerous SoS examples. Joint critical mission threads for a proposed SoS are intended to capture how the SoS's capabilities will be used in real-world situations. The capability to measure effectiveness of joint missions requires the ability to execute such mission threads in operational environments, both live and virtual. Experimental frame concepts offer an approach to enable simulated and real-time observation of participant system information exchanges and processing of acquired data to allow mission effectiveness to be assessed. Relevant Measures of Performance (MOPs) include quality of shared situational awareness, quality and timeliness of information, and extent and effectiveness of collaboration. Measures of Evaluation (MOEs) concern measures of the desired benefits such as increase in combat power, increase in decision-making capability, and increase in speed of command. Such metrics must be modeled with mathematical precision so as to be amenable to collection and computation with minimally intrusive test infrastructure to support

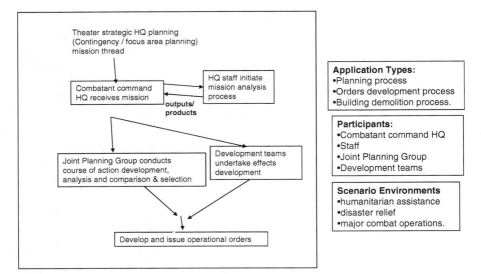

FIGURE 5.14 Illustrating a Joint Mission Thread and its dimensions for variation

rigorous, repeatable, and consistent testing and evaluation. An example of a mission thread is the Theater Strategic Head Quarter Planning at the Combatant Command Level (NSB, 2000).

As illustrated in Fig. 5.14, this thread starts when the Combatant Command HQ receives a mission that is delegated to HQ Staff. The staff initiates the mission analysis process that returns outputs and products to the commander. Then the Joint Planning Group and the development teams develop courses of action and perform effects analysis, respectively. The thread ends with the issue of operational orders. The MOP of interest might include various ways of measuring extent and effectiveness of collaboration, while the MOE might be the increase in speed of command.

These measures might be collected in assessing the value added of a collaboration support system in an operationally realistic setting. An informally presented mission thread can be regarded as a template for specifying a large family of instances. As illustrated in Fig. 5.15, such instances can vary in several dimensions, including the objectives of interest (the desired MOP and MOE), the type of application, the participants involved, and the operational environments in which testing will occur. Furthermore, mission threads can be nested, so that, for example, mission analysis is a subthread that is executed within the theater planning thread. An instance of a collaboration mission thread can be modeled as a coupled model in DEVS (recall Fig. 5.1), in which the components are participants in the collaboration and couplings represent the possible information exchanges that can occur among them.

This formulation offers an approach to capturing and implementing joint mission threads in the form of test model federations that can be deployed in a *net-centric integration infrastructure* such as the Global Information Grid/Service-Oriented Architecture (GIG/SOA) (Chaum et al., 2005). As illustrated in Fig. 5.15a, such an

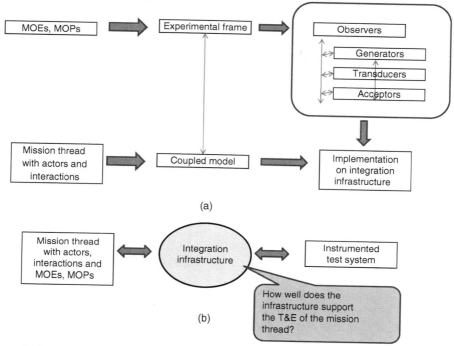

FIGURE 5.15   Mission thread implementation in an integration environment

infrastructure offers environment in which to deploy the participants in a mission thread together with network and web services that allow them to collaborate to achieve the mission objectives. As we shall see, formulation of a mission thread instance as a coupled model allows us to provide a rigorous method for realizing the thread within the integration infrastructure. At the same time, the MOEs and MOPs that have been formulated for assessing the outcome of the mission execution need to be translated into an appropriate experimental frame. In this case, the distributed nature of the execution will often require the frame to be distributed as well. The distributed frame will have such components as *observers* for the participant activities and message exchanges as well as the generators, transducers, and acceptors previously discussed. Indeed, as in Fig. 5.15b, the problem of assessing how well an integration infrastructure supports the collaboration requirements of a mission thread can be formulated as one of designing an *Instrumented Test System* where the latter is taken as an SoS and which therefore, can be addressed with M&S approach discussed here.

5.8   DEVS UNIFIED PROCESS AND ITS SERVICE-ORIENTED IMPLEMENTATION

This section describes the refined bifurcated model-continuity process and how various elements like automated DEVS model generation, automated test-model generation (and net-centric simulation over SOA are put together in the process, resulting in DUNIP (Mittal, 2007). The DUNIP is built on the bifurcated model-continuity-based life cycle methodology. The design of simulation-test framework occurs in parallel with the simulation-model of the system under design. The DUNIP process consists of the following elements:

1. Automated DEVS Model Generation from various requirement specification formats.
2. Collaborative model development using DEVS Modeling Language (DEVSML).[13]
3. Automated generation of test-suite from DEVS simulation model.
4. Net-centric execution of model as well as test-suite over SOA.

Considerable amount of effort has been spent in analyzing various forms of requirement specifications, namely, state-based, natural language-based, Rule-based, BPMN/BPEL-based, and DoDAF-based, and the automated processes, which each one should employ to deliver DEVS hierarchical models and DEVS-state machines (Mittal, 2007 and Mittal et al., 2008, Forthcoming). Simulation execution today is more than just model execution on a single machine. With Grid applications and collaborative computing the norm in industry as well as in scientific community, a net-centric platform using XML as middleware results in an infrastructure that supports distributed collaboration and model reuse. The infrastructure provides for a platform-free specification language DEVSML (Mittal et al., 2007a) and its net-centric execution using service-oriented architecture called DEVS/SOA (Mittal et al., 2007b). Both the DEVSML and DEVS/SOA provide novel approaches to integrate, collaborate, and remotely execute models on SOA. This infrastructure supports automated procedures is the area of test-case generation leading to test-models. Using XML as the system specifications in rule-based format, a tool known as Automated Test Case Generator (ATC-Gen) was developed that facilitated the automated development of test models (Zeigler et al., 2005; Mak, 2006; Mak et al., Forthcoming).

The integration of DEVSML and DEVS/SOA is performed with the layout as shown below in Fig. 5.16.

Various model specification formalisms are supported and mapped into DEVSML models including UML state charts (Martin et al., 2007), an exhibit driven state-based approach (Mittal et al., Forthcoming), Business Process Modeling Notation (BPMN)[14,15] or DoDAF-based (Mittal, 2006). A translated DEVSML model is fed to the DEVSML client that coordinates with the DEVSML server farm. Once the client has DEVSJAVA models, a DEVSML server can be used to integrate the client's model with models that are available at other sites to get an enhanced integrated DEVSML file that can produce a coupled DEVSML model. The DEVS/SOA enabled server can

---

[13]DEVSML: A Web service demonstration, http://150.135.218.205:8080/devsml/.
[14]Business Process Modeling Notation (BPMN) www.bpmn.org.
[15]Business Process Execution Language (BPEL) http://en.wikipedia.org/wiki/BPEL.

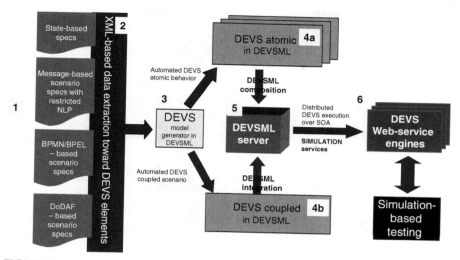

FIGURE 5.16   Net-centric collaboration and execution using DEVSML and DEVS/SOA

use this integrated DEVSML file to deploy the component models to assigned DEVS web server simulated engines. The result is a distributed simulation, or alternatively, a real-time distributed execution of the coupled model.

5.8.1   DEVSML Collaborative Development

DEVSML is a way of representing DEVS models in XML language. This DEVSML is built on JAVAML (Badros, 2000), which is XML implementation of JAVA. The current development effort of DEVSML takes its power from the underlying JAVAML that is needed to specify the "behavior" logic of atomic and coupled models. The DEVSML models are transformable back and forth to Java and to DEVSML. It is an attempt to provide interoperability between various models and create dynamic scenarios. The key concept is shown in the Fig. 5.16.

The layered architecture of the said capability is shown in Fig. 5.17. At the top is the application layer that contains model in DEVS/JAVA or DEVSML. The second layer is the DEVSML layer itself that provides seamless integration, composition, and dynamic scenario construction resulting in portable models in DEVSML that are complete in every respect. These DEVSML models can be ported to any remote location using the net-centric infrastructure and can be executed at any remote location. Another major advantage of such capability is total simulator "transparency." The simulation engine is totally transparent to model execution over the net-centric infrastructure. The DEVSML model description file in XML contains metadata information about its compliance with various simulation "builds" or versions to provide true interoperability between various simulator engine implementations. This has been achieved for at least two independent simulation engines as they have an underlying DEVS protocol to adhere to. This has been made possible with the implementation of a single atomic DTD and a single coupled DTD

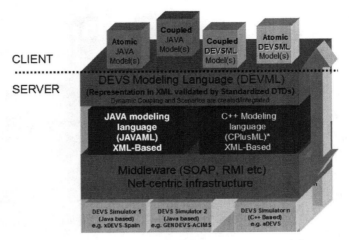

FIGURE 5.17 DEVS transparency and net-centric model interoperability using DEVSML. Client and server categorization is done for DEVS/SOA implementation

that validates the DEVSML descriptions generated from these two implementations. Such run-time interoperability provides great advantage when models from different repositories are used to compose bigger coupled models using DEVSML seamless integration capabilities. More details about the implementation can be seen at Mittal et al. (2007a).

5.8.2 DEVS/SOA: Net-Centric Execution Using Simulation Service

The fundamental concept of web services is to integrate software application as services. Web services allow the applications to communicate with other applications using open standards. We are offering DEVS-based simulators as a web service, and they must have these standard technologies: communication protocol (Simple Object Access Protocol, SOAP), service description (Web Service Description Language, WSDL), and service discovery (Universal Description Discovery and Integration, UDDI).

The complete setup requires one or more servers that are capable of running DEVS simulation service, as shown in Fig. 5.18. The capability to run the simulation service is provided by the server side design of DEVS simulation protocol supported by the latest DEVSJAVA Version 3.1 (ACIMS, 2006). The simulation service framework is two layered framework. The top-layer is the user coordination layer that oversees the lower layer. The lower layer is the true simulation service layer that executes the DEVS simulation protocol as a service.

From multifarious modes of DEVS model generation, the next step is the simulation of these models. The DEVS/SOA client takes the DEVS models package and through the dedicated servers hosting simulation services, it performs the following operations:

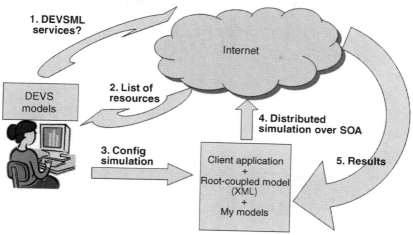

FIGURE 5.18 Execution of DEVS SOA-based M&S

1. Upload the models to specific IP locations, that is, partitioning.
2. Run-time compile at respective sites.
3. Simulate the coupled-model.
4. Receive the simulation output at client's end.

The main server selected (corresponding to the top-level- coupled model) creates a coordinator that creates simulators in the server where the coordinator resides and/or over the other remote servers selected. The DEVS/SOA web service client as shown in Fig. 5.19 below operates in the following sequential manner:

1. The user selects the DEVS package folder at his machine.
2. The top-level-coupled model is selected as shown in Fig. 5.19.
3. Various available servers are selected. Any number of available servers can be selected. Figure 5.20 shows how servers are allocated on per-model basis. The user can specifically assigned specific IP to specific models at the top-level-coupled domain. The *localhost* (Fig. 5.19) is chosen using debugging sessions.
4. The user then uploads the model by clicking the upload button. The models are partitioned in a round-robin mechanism and distributed among various chosen servers.
5. The user then compiles the models by clicking the compile button at server's end.
6. Finally, simulate button is pressed to execute the simulation using simulation service hosted by these services.
7. Once the simulation is over, the console output window displays the aggregated simulation logs from various servers at the client's end.

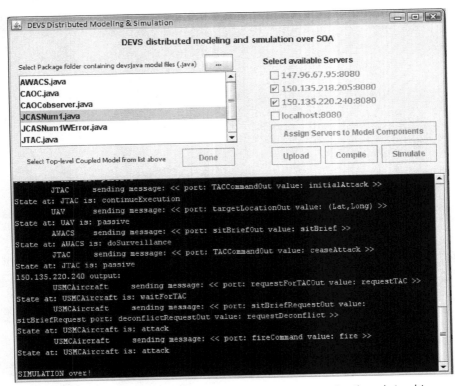

FIGURE 5.19 GUI snapshot of DEVS/SOA client hosting distributed simultion

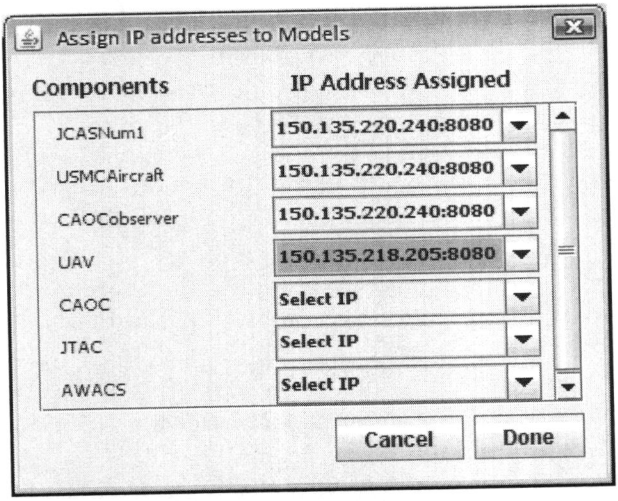

FIGURE 5.20 Server assignment to models

In terms of net-ready capability testing, what is required is the communication of live web services with those of test models designed specifically for them. The approach has the following steps

1. Specify the scenario
2. Develop the DEVS model
3. Develop the test model from DEVS models
4. Run the model and test model over SOA
5. Execute as a real-time simulation
6. Replace the model with actual web service as intended in scenario.
7. Execute the test-models with real-world web services
8. Compare the results of steps 5 and 7.

One other section that requires some description is the multiplatform simulation capability as provided by DEVS/SOA framework. It consists of realizing distributed simulation among different DEVS platforms or simulator engines such as DEVSJAVA, DEVS-C++, and so on on Windows or Linux platforms. In order to accomplish that, the simulation services will be developed that are focused on specific platforms, however, managed by a coordinator. In this manner, the whole model will be naturally partitioned according to their respective implementation platform and executing the native simulation service. This kind of interoperability where multiplatform simulations can be executed with our DEVSML integration facilities. DEVSML will be used to describe the whole hybrid model. At this level, the problem consists of message passing, which has been solved in this work by means of an adapter pattern in the design of the "message" class (Mittal et al., 2007b). Figure 5.21 shows a first approximation. The platform-specific simulator generates messages or events, but the simulation services will transform these platform-specific-messages (PSMsg) to our current platform-independent-message (PIMsg) architecture developed in DEVS/SOA.

Hence, we see that the described DEVS/SOA framework can be extended toward net ready capability testing. The DEVS/SOA framework also needs to be extended toward multiplatform simulation capabilities that allow test-models be written in any DEVS implementation (e.g., Java and C++) to interact with each other as Services.

5.8.3 The Complete DEVS Unified Process

DUNIP can be summarized as the sequence of the following steps:

1. Develop the requirement specifications in one of the chosen formats such as BPMN, DoDAF, Natural Language Processing (NLP)-based, UML-based, or simply DEVS-based for those who understand the DEVS formalism.
2. Using the DEVS-based automated model generation process, generate the DEVS atomic and coupled models from the requirement specifications using XML.

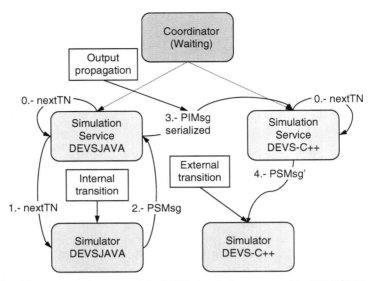

FIGURE 5.21   Interoperable DEVS simulation protocol in DEVS/SOA

3. Validate the generated models using DEVS W3C atomic and coupled schemas to make them net ready capable for collaborative development, if needed. This step is optional but must be executed if distributed model development is needed. The validated models that are which are PIMs in XML can participate in collaborative development using DEVSML.

4. From Step 2, either the coupled model can be simulated using DEVS/SOA or a test-suite can be generated based on the DEVS models.

5. The simulation can be executed on an isolated machine or in distributed manner using SOA middleware if the focus is net-centric execution. The simulation can be executed in real time as well as in logical time.

6. The test-suite generated from DEVS models can be executed in the same manner as laid out in Step 5.

7. The results from Step 5 and Step 6 can be compared for V&V process.

The basic bifurcated model continuity-based Life cycle process for systems engineering in Fig. 5.10 in light of the developments in DEVS area is summarized in Fig. 5.22 above. The gray boxes show the original process and the colored boxes show the extensions that were developed to make it a DEVS compliant process. A sample demo movie is available at http://150.135.218.205:8080/devsml.

Many case studies came about as DUNIP was defined and developed. Many of the projects are currently active at Joint Interoperability Test Command (JITC) and others are at concept validation stage toward a deliverable end. Each of the projects either uses the complete DUNIP process or a subset of it. As we shall see on a case by case basis, DEVS emerge as a powerful M&S framework contributing to the complete systems software engineering process. With the proposed DEVS-based bifurcated

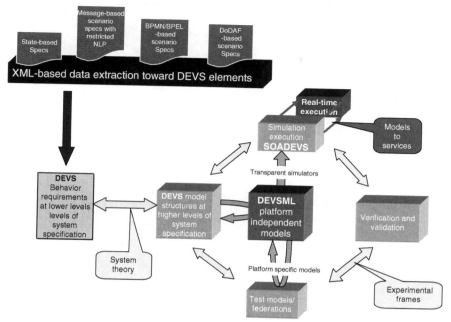

FIGURE 5.22 The complete DEVS Unified Process

model-continuity Life cycle process, systems theory with its DEVS implementation can support the next generation net-centric application development and testing.

A recent Doctoral Thesis (Mittal, 2007) provides a number of examples that illustrate how the DUNIP can be applied to import real-world simulation-based design applications. Here, we review briefly the following case studies that were developed in more detail in Mittal (2007):

- Joint Close Air Support (JCAS) model
- DoDAF-based activity scenario
- Link-16 Automated Test Case Generator (ATC-Gen project at JITC)
- Generic Network for Systems Capable of Planned Expansion (GenetScope project at JITC, 2006).

Each of the projects has been developed independently and ATC-Gen and Genet-Scope are team projects. All of the projects stand alone and each applies DUNIP (Fig. 5.22 in full or in part. Figure 5.23 below provides an overview of the DUNIP elements used in each of the projects. All of the DUNIP elements have been applied at least once in one of the projects.

The JCAS system requirements come in many formats and served as a base example to test many of the DUNIP processes for requirements-to-DEVS transformation. JCAS requirements were specified using the state-based approach, BPEL-based approach

DUNIP elements		Scenario	Project	
Requirement specification formats				X
State-based Specs	X			
Message-based Specs with restricted NLP	X			
BPMN/BPEL based Specs	X			
DoDAF-based scenario Specs		X		X
XML-based data extraction	X	X	X	
DEVS model structure at lower levels of Specification	X	X	X	
DEVS model structure at higher levels of System specification		X		X
DEVSML platform Independent models	X			
Test model development	X		X	
Verification and validation using experimental frames		X	X	X
DEVS/SOA net-centric simulation	X			

FIGURE 5.23 Overview of DUNIP application in available case studies

and restricted natural language approach. The JCAS case study describes how each of the three approaches led to executable DEVS models with identical simulation results. Finally, the simplest executable model (that specified by the state-based approach) was executed over a net-centric platform using DEVSML and DEVS/SOA architectures.

The DODAF-based activity scenario was specified using the UML-based activity diagrams. It illustrates the process needed to transform various DoDAF documents into DEVS requirement specifications. New Operational View documents OV-8 and OV-9 were proposed (Mittal, 2006) to facilitate the transformation of DoDAF requirements into a form that could be supported by DEVS-based modeling and simulation. The population of these new documents was described as well as how DEVS models could be generated from them.

The ATC-Gen project at JITC is the project dealing with automated Link-16 testing environment and the design of ATC-Gen tool. A detailed discussion and complete example are presented in Mak (2006).

The GenetScope project at JITC is another project that employs the complete DEVS software engineering process. Using automated XML data mining, a 10-year-old legacy model written in the C language was transformed to an object-oriented DEVS model with enhanced model view simulation and control paradigm (Mittal et al., 2006). The design elements of GENETSCOPE tool (GenetScope Project at JITC, 2006) were discussed and as was its relationship with the overarching DoDAF framework (Mittal et al., 2006).

Although the DUNIP was applied in part to each of the above projects, presently, there is no live case study that implements all the aspects of DUNIP elements. Recall that DUNIP was researched and developed in the context of the above active projects.

To summarize, with the DUNIP we have the capability to:

1. Transform various forms of requirement specifications to DEVS models in an automated manner.
2. Transform any DEVS model to a PIM by using DEVSML for model and library reuse and sharing, thus supporting collaborative development.
3. Simulate any valid DEVSML using the DEVS/SOA architecture, thus exploiting the abstract DEVS simulator paradigm to achieve interoperability of DEVS model execution (for models implemented in disparate languages e.g., Java and C++).
4. Transform any DEVSML model to a service component in SOA and any coupled model into a deployable collaboration of such service components.

5.9 APPLICATION: SYSTEM OF SYSTEMS SIMULATION FOR HETEROGENEOUS MOBILE SENSOR NETWORKS

In real-world systems, the system of systems concept is addressed in a higher level where the systems send and receive data from other systems in the SoS and make a decision that leads the SoS to its global goals. Let us take the military surveillance example where different units of the army collect data through their sensors trying to locate a threat or determine the identity of a target. In this type of situations, army command center receives data from these heterogeneous sensor systems such as AWACS, ground RADARS, submarines, and so on. In general, these systems are different in hardware and/or software. This will create a huge barrier in data aggregation and data fusion using the data received from these systems because they would not be able to interact successfully without hardware and/or software compatibility. In addition, the data coming from these systems are not unified, which also adds to the barrier in data aggregation.

One solution to the problem is to modify the communication medium among the SoS components. Two possible ways of accomplishing this task are (Sahin et al., 2007b):

- Create a software model of each system. In this approach, each component in the SoS talks to a software module embedded in itself. The software module collects data from the system and through the software model generates outputs and sends to the other SoS components. If these software modules are written with a common architecture and a common language, then the SoS components can communicate effectively regardless of their internal hardware and/or software architectures.
- Create a common language to describe data. In this approach, each system can express its data in this common language so that other SoS components can parse the data successfully.

The overhead that needs to be generated to have software models of each system on an SoS is enormous and must be redone for new member of the SoS. In addition, this requires the complete knowledge of the state-space model of each SoS components, which is often not possible. Thus, data-driven approach would have better success on integrating new members to the SoS and also applying the concept to other SoS application domains.

In this work, we present SoS architecture based on XML in order to wrap data coming from different sources in a common way. The XML language can be used to describe each component of the SoS and their data in a unifying way. If XML-based data representation architecture is used in an SoS, only requirement for the SoS components to understand/parse XML file received from the components of the SoS.

In XML, data can be represented in addition to the properties of the data such as source name, data type, importance of the data, and so on. Thus, it does not only represent data but also gives useful information that can be used in the SoS to take better actions and to understand the situation better. The XML language has a hierarchical structure where an environment can be described with a standard and without a huge overhead. Each entity can be defined by the user in the XML in terms of its visualization and functionality. For example, a hierarchical XML architecture like in Listing 1 can be designed for an SoS so that it can be used in the components of the SoS and also be applied to other SoS domains easily. In Listing 1, the first line defines the name of the file that describes the functionality of the user-defined keywords used to define the SoS architecture. This file is mainly used for visualization purposes so that any of the SoS components can display the current data or the current status of the SoS to a human/expert to make sure the proper decision is taken.

In Fig. 5.24, the first keyword of the XML architecture is "system-of-system" representing an SoS. Everything described after this keyword belongs to this SoS

```xml
<!--Created 11/8/2006   Author @ Ferat Sahin-->
<?xml-stylesheet type="text/css" href="genericxml.css"?>

<systemofsystem>

    <id> Id of the System of systems </id>
    <name> The name of the System of System</name>
    <system>

        <id>Id of the first system</id>
        <name> The name of the first system </name>
        <description>  The description of the first system </description>
        <dataset>

            <Output>

                <id>Id of the first output</id>
                <data>Data of the first output</data>

            </Output>
            <Output>

                <id>Id of the second output</id>
                <data>Data of the second output</data>

            </Output>

        </dataset>
        <subsystem>

            <id>Id of the subsystem of the first System</id>
            <name>The name of the subsystem</name>
            <description>This is a subsystem of the system in a SoS</description>
            <dataset>

                <Output>

                    <data> Data of the subsystem </data>

                </Output>

            </dataset>

        </subsystem>

    </system>

</systemofsystem>
```

FIGURE 5.24   An XML based system of systems architecture

131

based on the XML architecture. The following keywords, "id" and "name," are used to describe the system of systems.

Then, the keyword "system" is used to declare and describe the first system of the SoS. In addition to "id" and "name," two more keywords, "description" and "data set," are used to describe the properties of the system and to represent the data coming out of this system. Data sources are denoted by "output" keyword and data are provided with the keyword "data". After representing data from two sources, a subsystem is described by the keyword "subsystem." The subsystem and its data are presented in a similar manner. Other subsystems can be described in this subsystem or in parallel to this subsystem as well as additional systems can be described in the system of systems.

Next, we will present a simulation case study to demonstrate SoS simulation framework for threat detection in a heterogeneous mobile sensor networks. In this scenario, there is a haptic-controlled base robot (Fig. 5.25a), a swarm robot, and two sensor modules, shown in Fig. 5.25b. Before these systems are put in the field to do threat detection, the system of system concepts should be simulated using suitable simulation tools to test the performance of these SoS concepts.

In addition, a simulation environment can also assist us to evaluate our robust data aggregation, fusion, and decision-making techniques. The next section explores a possible simulation environment that can accommodate asynchronous system interactions and data exchange using XML.

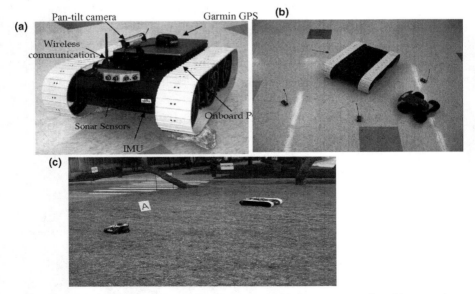

FIGURE 5.25 (a) Base robot with haptic control, (b) Components of multisensor data aggregation and fusion system of systems (Sahin et al., 2007a). (c) Base and scout robots in a field mission at University of Texas, San Antonio

5.9.1 SoS Simulation of Threat Detection (Data Aggregation)

Multiagent robotic systems can also be successfully demonstrated using the DEVS formalism and DEVSJAVA software (Sarjoughian et al., 2001). A multiagent simulation framework, Virtual Laboratory (V-Lab®), was developed for such multiagent simulation with DEVS modeling framework (Sridhar et al., 2003; Sridhar and Jamshidi, 2004).

Based on the tiered system of systems architecture (Sridhar et al., 2006, 20072006), we have simulated a data aggregation scenario where there are two robots (a base robot, a swarm robot), two sensors, and a threat (Sahin et al., 2007a). When the sensors detect the threat, they notify the base robot about the presence of a threat. Upon receiving such notification, the base robot notifies the swarm robot the location of the threat based on the information sent by the sensors.

An XML-based SoS message architecture is implemented in DEVSJAVA software (Sarjoughian and Zeigler, 2000). In this XML-based message architecture (Sahin et al., 2007a,b), each system has an XML-like message consisting of their name and a data vector. The name of each system represents an XML tag. The data vectors are used to hold the data of the systems. The length of the vectors in each system can be different based on the amount of data each system contains. For instance, the XML message of the sensors has the sensor coordinates and the threat level. The threat level is set when a threat gets in the coverage area of the sensors. However, the base robot's XML message has its coordinates, the threat level, and the coordinates of the sensor who is reporting a threat. Thus, the data vector length of base robot XML message has five elements whereas the data vector of an XML message of a sensor has three elements. Figure 5.26 presents the names and the length of the vector data of each system in the system of systems.

The data vectors are made of "double" variables in order to keep track of the positions accurately. The "threat" element in the data vector is a flag representing threat (1.0) or no threat (0.0). The elements "Xt" and "Yt" are used the destination coordinates in the XML messages of the base robot and swarm robot. These represent the coordinates of the sensor who reported a threat. The threat is named as "fire" for obvious reasons and it only has two elements in its data vector for its coordinates.

In order to generate these XML-like messages, a data structure, called "XmlEntity," is created based on the "entity" data structure in DEVSJAVA environment. This data

System	Name	Vector Data Length
Base robot	Base robot	5 (X, Y, Threat, Xt, Yt)
Swarm robot	Swarm robot	5 (X, Y, Threat, Xt, Yt)
Sensor	Sensor 1	3 (X, Y, Threat)
Sensor	Sensor 2	3 (X, Y, Threat)
Threat	Fire	2 (X, Y)

FIGURE 5.26 XML message components for the systems in the SoS

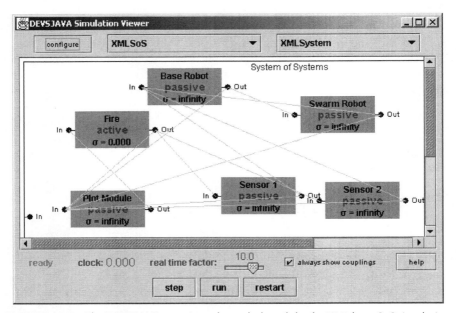

FIGURE 5.27 The DEVSJAVA atomic and coupled modules for XML base SoS simulation (Sahin et al., 2007b)

structure is used to wrap/represent the data of each system. The structures/behaviors of the systems in the SoS are created/simulated by DEVSJAVA atomic or coupled models. Figure 5.27 is a screen shot of the DEVS model of the system of systems described above.

Figure 5.28 shows a DEVS simulation step with the XML messages sent among the systems in the system of systems. As can be seen in Fig. 5.27, each system sends XML-like messages that consist of the name of the system and the data vector related to its characteristics.

Finally, Fig. 5.29 shows the simulation environment created by the "Plot Module" atomic model in DEVS. In the environment, the two sensors are located next to each other representing a border. The "threat" (Fire), red dot, is moving in the area. When the threat is in one of the sensor's coverage area, the sensor signals the base robot. Then, base robot signals the swarm robot so that it can go and check whether the threat is real or not. The behavior of the system of systems components can also be seen in Fig. 5.28 as the movements of the "threat" and the "swarm robot" are captured. The green dot represents the Swarm Robot. The base robot is not shown since it does not move in the field. When the threat enters into the coverage area of a sensor, that sensor area filled with red color to show the threat level in the coverage area. The sensor then reports the threat to the base robot. Then, the base robot issues a command to a swarm robot closest to the sensor area. Finally, the swarm robot, green dot, moves into the sensor coverage area based on the coordinates sent by the base robot to check whether the threat is real. When the swarm robot reaches the sensor coverage area, it determines the threat level with its own sensor and report it the base robot. If the

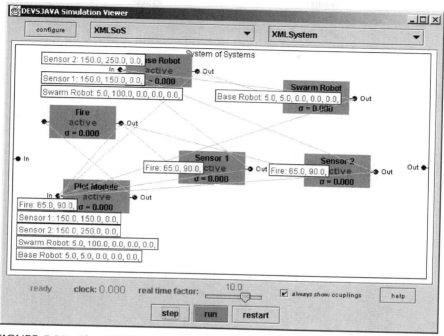

FIGURE 5.28 The DEVSJAVA simulation with XML-based messages shown at the destination (Sahin et al., 2007a)

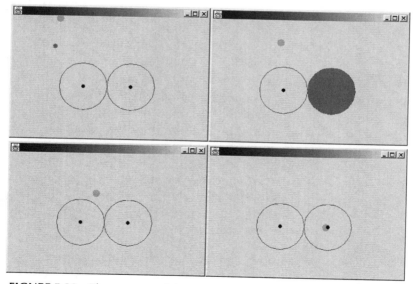

FIGURE 5.29 The progress of the DEVSJAV simulation on data aggregation

threat is not real, the swarm robot moves away from the sensor's coverage area and waits another command from the base robot .

5.9.2 Concluding Remarks

In this section, we have presented a framework to simulate a system of systems and discussed a simulation case study (threat detection) with multiple systems working collaboratively as an SoS. While DEVS formalism helps to represent the structure of an SoS, the XML provides a way to represent the data generated by each system. Together, DEVS formalism and XML form a powerful tool for simulating any given SoS architecture. To the best of our knowledge, there has been very little research directed toward the development of a *generic framework* for architectural representation and simulation of an SoS. Currently, we are working on extending the XML data representation in DEVS and making it more generic and dynamic so that when a new system is added to an SoS, it will automatically generate its XML message to send its data to other components of the SoS.

5.10 APPLICATION: AGENT-IMPLEMENTED TEST INSTRUMENTATION SYSTEM

The test instrumentation system (TIS) should provide a minimally intrusive test capability to support rigorous, ongoing, repeatable and consistent testing and evaluation. Requirements for such a test implementation system include ability to

- deploy agents to interface with SoS component systems in specified assignments;
- enable agents to exchange information and coordinate their behaviors to achieve specified experimental frame data processing;
- respond in real time to queries for test results while testing is still in progress;
- provide real-time alerts when conditions are detected that would invalidate results or otherwise indicate that intervention is required;
- centrally collect and process test results on demand, periodically, and/or at termination of testing;
- support consistent transfer and reuse of test cases/configurations from past test events to future test events, enabling life cycle tracking of SoS performance; and
- enable rapid development of new test cases and configurations to keep up with the reduced SoS development times expected to characterize the reusable web service-based development supported on the GIG/SOA.

Many of these requirements are not achievable with current manually based data collection and testing. Instrumentation and automation are needed to meet these requirements.

Before proceeding we present an example that provides an exemplar of the design approach to follow.

5.10.1 Example: Collaboration Session Timing Instrumentation

The experimental frame coupled to the activity model through observers is shown in Fig. 5.30. In the activity model, users can interact with each other through a collaboration service and also independently interact with a portal for other services, such as searching a database, checking accounting data, and so on. Information about user activity is collected by observers, shown in pink, and individually coupled with users (black arrows).

These observers are implemented as DEVS agents that are deployed in one–one fashion to user work sites. In the current example, these agents appear as applet clients with GUIs that enable users to generate status events by clicking a check box to indicate having just joined a collaboration session or having just quit one. Note that this information is obtained independently of the collaboration tool and does not rely on the latter's own status tracking. Such independent information, which might otherwise be obtained by human observers taking notes, is necessary to independently validate the collaboration tool's own logging and statistics computations. Status events generated by observers are sent to frame components, shown in yellow, for processing into performance statistics concerning the success rate in establishing collaboration sessions, the time required to establish such sessions and their duration. A somewhat different observer, that for portal services, collects user service requests sent to, and received from, the portal. This observer passes on such request messages to a frame component that computes throughput and response time statistics for portal services.

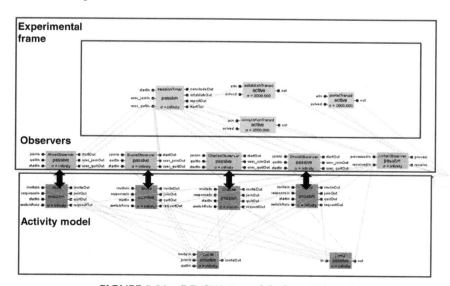

FIGURE 5.30 DEVSJAVA model of a collaboration

5.10.2 Distributed Test Federations

A DEVS distributed federation is a DEVS-coupled model whose components reside on different network nodes and whose coupling is implemented through middleware connectivity characteristic of the environment, for example, SOAP for GIG/SOA. The federation models are executed by DEVS simulator nodes that provide the time and data exchange coordination as specified in the DEVS abstract simulator protocol.

As discussed earlier, in the general concept of EF, the generator sends inputs to the SUT, the transducer collects SUT outputs and develops statistical summaries, and the acceptor monitors SUT observables making decisions about continuation or termination of the experiment (Zeigler et al., 2005). Since the SoS is composed of system components, the EF is distributed among SoS components, as illustrated in Fig. 5.31a. Each component may be coupled to an EF consisting of some subset of generator, acceptor, and transducer components. As mentioned, in addition an observer couples the EF to the component using an interface provided by the integration infrastructure. We refer to the DEVS model that consists of the observer and EF as a *test agent*.

Net-centric SOA provides a currently relevant technologically feasible realization of the concept. As discussed earlier, the DEVS/SOA infrastructure enables DEVS models, and test agents in particular, to be deployed to the network nodes of interest. As illustrated in Fig. 5.31b, in this incarnation, the network inputs sent by EF generators are SOAP messages sent to other EFs as destinations; transducers record the arrival of messages and extract the data in their fields, while acceptors decide on whether the gathered data indicates continuation or termination is in order (Mittal, 2007).

Since EFs are implemented as DEVS models, distributed EFs are implemented as DEVS models, or agents as we have called them, residing on network nodes. Such a federation, illustrated in Fig. 5.32, consists of DEVS simulators executing on web servers on the nodes exchanging messages and obeying time relationships under the rules contained within their hosted DEVS models .

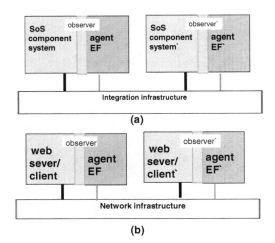

FIGURE 5.31 Deploying experimental frame agents and observers

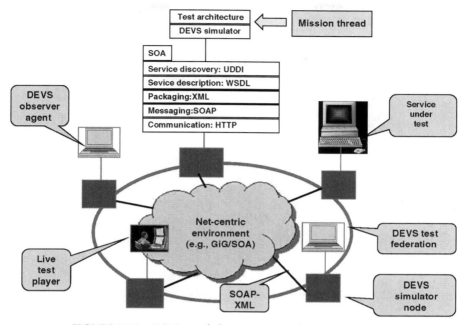

FIGURE 5.32 DEVS test federation in GIG/SOA environment

5.10.3 Distributed Multilevel Test Federations

The linguistic levels of interoperability discussed earlier provide a basis for further structuring the test instrumentation system. In the following sections, we discuss the implementation of test federations that simultaneously operate at the syntactic, semantic, and pragmatic levels (Fig. 5.33) .

5.10.3.1 Syntactic Level—Network Health Monitoring From the syntactic perspective, testing involves assessing whether the infrastructure can support the speed and accuracy needed for higher level exchange of information carried by multimedia data types, individually and in combination. We now consider this as a requirement to continually assess whether the network is sufficiently "healthy" to support the ongoing collaboration. Figure 5.34 illustrates the architecture that is implied by the use of subordinate probes. Nodal generator agents activate probes to meet the health monitoring Quality of Service (QOS) thresholds determined from information supplied by the higher layer test agents, namely, the objectives of the higher layer tests.

Probes return statistics and alarm information to the transducers/acceptors at the DEVS health layer, which in turn may recommend termination of the experiment at the test layer when QOS thresholds are violated.

In an EF for real-time evaluation of network health, the SUT is the network infrastructure (OSI layers 1–5) that supports higher session and application layers.

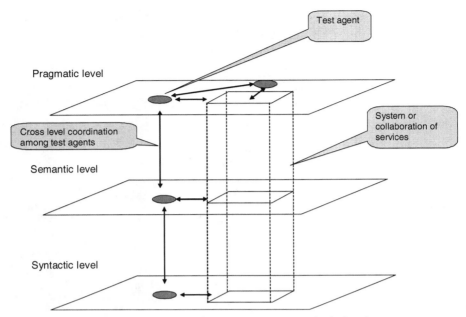

FIGURE 5.33   Simultaneous testing at multiple levels

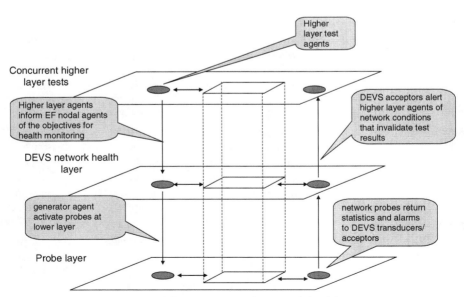

FIGURE 5.34   Multilayer testing with network health monitoring

QOS measures are at the levels required for meaningful testing at the higher layers to gather transit time and other statistics, providing quality of service measurements.

For messages expressed in XML and carried by SOAP middleware such messages are directly generated by the DEVS generators and consumed by the DEVS transducers/acceptors. Such messages experience the network latencies and congestion conditions experienced by messages exchanged by the higher level web servers/clients. Under certain QOS conditions however, video streamed and other data typed packets may experience different conditions than the SOAP-borne messages. For these we need to execute lower layer monitoring under the control of the nodal EFs.

The collection of agent EFs has the objective of assessing the health of the network relative to the QOS that it is providing for the concurrent higher level tests. Thus, such a distributed EF is informed by the nature of the concurrent test for which it monitoring network health. For example, if a higher level test involves exchanges of a limited subset of media data types (e.g., text and audio), then the lower layer distributed EF need only monitor the subset of types.

5.10.3.2 Semantic Level—Information Exchange in Collaborations   Mission
threads consist of sequences of discrete information exchanges. A collaboration service supports such exchanges by enabling collaborators to employ a variety of media, such as text, audio, and video, in various combinations. For example, a drawing accompanied by a voice explanation involves both graphical and audio media data. Further, the service supports establishing producer/consumer relationships. For example, the graphical/audio combination might be directed to one or more participants interested in that particular item. From a multilevel perspective, testing of such exchanges involves pragmatic, semantic, and syntactic aspects. From the pragmatic point-of-view, the ultimate worth of an exchange is how well it contributes to the successful and timely completion of a mission thread. From the semantic perspective, the measures of performance involve the speed and accuracy with which an information item, such as a graphical/audio combination, is sent from producer to consumer. Accuracy may be measured by comparing the received item to the sent item using appropriate metrics. For example, is the received graphic/audio combination within an acceptable "distance" from the transmitted combination, where distance might be measured by pixel matching in the case of graphics and frequency matching in the case of audio. To automate this kind of comparison, metrics must be chosen that are both discriminative and quick to compute. Further, if translation is involved, the "meaning" of the item must be preserved as discussed above. Also, the delay involved in sending an item from sender to receiver, must be within limits set by human psychology and physiology. Such limits are more stringent where exchanges are contingent on immediately prior ones as in a conversation. Instrumentation of such tests is similar to that at the syntactic level to be discussed next, with the understanding that the complexity of testing for accuracy and speed is of a higher order at the semantic level.

5.10.3.3 Pragmatic Level—Mission Thread Testing   A test federation observes
an orchestration of web services to verify the message flow among participants

adheres to information exchange requirements. A mission thread is a series of activities executed by operational nodes and employing the information processing functions of web services. Test agents watch messages sent and received by the services that host the participating operational nodes. Depending on the mode of testing, the test architecture may, or may not, have knowledge of the driving mission thread under test. If a mission thread is being executed and thread knowledge is available, testing can do a lot more than if it does not.

With knowledge of the thread being executed, DEVS test agents can be aware of the current activity of the operational nodes it is observing. This enables an agent to focus more efficiently on a smaller set of messages that are likely to provide test opportunities.

5.10.3.4 *Measuring Success in Mission Thread Executions*   The ultimate test of effectiveness of an integration infrastructure is its ability to support successful outcomes of mission thread executions. To measure such effectiveness, the test instrumentation system must be informed about the events and messages to expect during an execution, including those that provide evidence of success or failure, and must be able to detect and track these events and messages throughout the execution.

5.10.3.5 *Measuring "the Right Information at the Right Place at the Right Time"*   It is often said that the success of a mission depends on the right information arriving to the right place at the right time. Thus, an obvious measure of performance is the ability of an integration infrastructure to measure the extent to which the right information is delivered to the right place at the right time. This in turn places requirements on the test instrumentation system that it be able to gather the information needed to make such judgments. Much more than the simple ability to determine mission success for failure, such a capability would provide diagnostic capability to determine whether the final outcome was due to failure in information flow, and if so, just what information did not get received at the right time by which consumer.

Figure 5.35 graphically illustrates a formulation of the issue at hand. Given that a consumer is engaged in a particular activity during some period, there is a time window during which it expects, and can exploit in its processing, some piece of information denoted by X. As in Fig. 5.35a, if X arrives before this window opens up, it may too early and cannot be used when needed; likewise, if X arrives after the window has closed, it is too late for the information to be effectively used. Further, as in Fig. 5.35b, if the wrong information, say ~X, arrives during the window of opportunity, it may be ignored by the consumer at best, or cause a backup of messages that clog the system at worst. As in Fig. 5.35c, to make such determinations, a test agent has to know the current activity phase of the consumer, and have been informed of the expected information X and its time window of usability.

Implementing such test capabilities requires not only the right test instrumentation and protocol but also the necessary backup analysis capabilities to provide the information items needed.

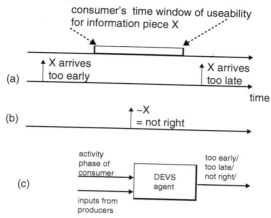

FIGURE 5.35   Instrumentation for "the right information at the right place at the right time"

5.10.4   Analysis Capabilities

A mathematical model, such as the DEVS coupled model, allows the use of analysis tools to derive useful information to support testing. As illustrated in Fig. 5.36, given a DEVS model of a mission thread, with upper and lower bounds on the times for individual activities, it is possible to derive time windows for the occurrence of events and arrival of messages. In the case of analysis, even more so than for simulation, it is important to down-select to the events and messages of particular interest, as derived from the MOPs and MOEs in the experimental frame. This is so because, unless appropriately constrained, analysis is typically intractable since it suffers from "global state explosion."

Algorithms are being developed that, under suitable restrictions of the class of models, can derive the time window specifications needed to inform the DEVS agents monitoring the related mission thread actors (Zeigler et al., 2005; Mittal, 2007).

5.10.5   Verification/Validation of the Test Instrumentation System

How do we validate/verify the TIS itself? Figure 5.37 presents the development of such a system as itself an application of the DUNIP. The overall requirements for the TIS were laid out earlier in this section and can be phrased in terms of testing for infrastructure support of a family of mission threads of interest. These requirements can be formalized in terms of DEVS-coupled models of mission threads and in terms of related measures of performance and effectiveness formulated as experimental frames. From this formal basis, the DEVS/SOA environment provides an infrastructure and tool set for implementing the TIS using DEVS test agents. An essential component of the tool set is the analytic capability to derive time windows for events and messages that determine the temporal usability of information to facilitate testing of MOPs for right place/right time arrival.

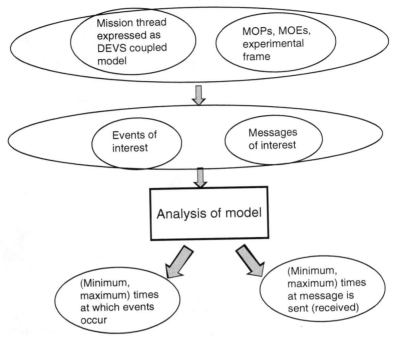

FIGURE 5.36 Analysis of mission thread models to derive test-supporting temporal information

The DUNIP process requires that in parallel, a testing process for the implementation of the TIS be developed from the same formalized basis. We could of course, apply the TIS to itself—in other words, using DEVS agents to monitor the activities of DEVS agents in their primary test federation application. However, taken literally, this could lead to an infinite regress in which another level is always needed to test the current

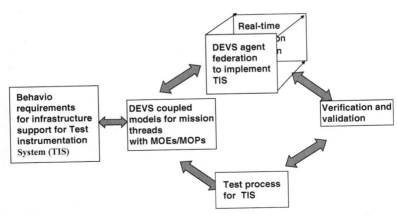

FIGURE 5.37 Bifurcated approach applied to design of test instrumentation system

level. Suitably restricted, this approach might be useful at a later stage of development after a first order level of confidence has been established. Early on, to get such confidence, the TIS should be tested on a simulated collaboration in a controlled setting where timings and behaviors of the simulated actors are controlled experimentally. For example, the collaboration model described in Fig. 5.30 could be used where faults such as communication breakdowns and unresponsive elements can be introduced to establish anomalous test cases. The extent to which the TIS responds properly to both normal and abnormal test cases is a useful measure of its performance.

5.10.6 Potential Issues and Drawbacks

Building upon and integrating earlier systems theoretic and architectural methodologies, DUNIP inherits many of the advantages that such methodologies afford and attempts to fill in the holes that they still leave. However, as with the methodologies it draws upon, and is inspired by, there are potential issues and drawbacks that may be expected to emerge in its application. Predictably, the current culture of system development still places more incentive on quickly finishing a project for an incremental development rather than on spending the extra resources needed to assure the existence of benefits such as reusability and extensibility that only provide cost-justification in the longer term. Until cultural change takes place to place greater emphasis on the fruits of well-founded methodological work, there is a risk that adopting DUNIP will create a situation in which schedules are missed and costs overrun. Particularly, personnel employed in DUNIP-based development must be adequately trained on all its supporting intellectual and technological components—system theory, DEVS, object-based software architecture, and so on. Such prior education and experience should be regarded as mandatory for DUNIP adoption and, if not present, steps must be taken to provide the necessary training, education, and experience. Further, training in one foundational element, such as software architecture, alone does not reduce the risk, and may even intensify it — in the spirit of a "little knowledge is a dangerous thing." Hopefully, books like the one in which this chapter resides and others such as those mentioned earlier (Zeigler et al., 2000; Zeigler and Hammonds, 2007) will take their place in the curricula that provide education and training the emerging SoSE field.

Further, the complexity and quality assurance issues associated with the proposed methodology need to be mitigated with the development of appropriate tools and interfaces to simplify working with the methodology. The need for such complexity-reduction tools underlies the extended discussions of tools and interfaces that have been provided here. Further quality assurance demands provision of approaches to self-checking DUNIP and its supporting infrastructures. The discussion of the test instrumentation system and its verification and validation provides an example of approaches to such assurance.

Finally, there remain many issues to resolve in the manner in which the DUNIP methodology relates to the defense-mandated DODAF. The latter is a representational mechanism, not a methodology, and does not discuss how an integrated architecture should be constructed and evolved from existing systems. DUNIP offers an approach based on systems theory and supported by DEVS-based modeling and simulation to tackle integration and interoperability issues, but the integration with DODAF remains for future consideration.

5.11 SUMMARY

In this chapter we have taken the challenge of constructing an SoS as one of designing an infrastructure to integrate existing systems as components, each with its own structure and behavior. The SoS can be specified in many available frameworks such as DoDAF, system theory, UML, or by using an integrative systems engineering-based framework such as DEVS. In this chapter, we have discussed the advantages of employing an M&S-integrated framework such as DUNIP and its supporting DEVS/SOA infrastructure. We illustrated how M&S can be used strategically to provide early feasibility studies and aid the design process. As components comprising SoS are designed and analyzed, their integration and communication is the most critical part that must be addressed by the employed SoS M&S framework. The integration infrastructure must support interoperability at syntactic, semantic, and pragmatic levels to enable such integration. We have illustrated, with an SoS consisting of heterogeneous robotic sensors and decision components, how the integration infrastructure must support interoperability at syntactic, semantic, and pragmatic levels to achieve the requisite interoperation. We discussed DoD's Global Information Grid as providing an integration infrastructure for SoS in the context of constructing collaborations of web services using the SOA. The DUNIP, in analogy to the Rational Unified Process based on UML, offers a process for integrated development and testing of systems that rests on the SOA infrastructure. The DUNIP perspective led us to formulate a methodology for testing any proposed SOA-based integration infrastructure, such as DISA's Net-Centric Enterprise Services. To support such a methodology we proposed a TIS built upon the integrated infrastructure that employs DEVS Agents to perform simultaneous testing at the syntactic, semantic, and pragmatic levels. Clearly, the theory and methodology for such SoS development and testing are at their early stages. While one book as appeared on the subject (Zeigler and Hammonds, 2007), we challenge the reader to explore the issues involved and come up with more incisive solutions that extend the very essence of systems engineering theory.

A reviewer's comments provide a well-stated summary of the chapter—"the proposed methodology can be used for integrating heterogeneous constituents of an SoS and assessing their real-time interactions and interoperability. The proposed methodology encompasses the advantages of several interrelated concepts such as the systems theory, DEVSML and DEVS/SOA, M&S framework, and the model continuity concepts. Especially, since it separates models from their underlying simulators, enables real-time execution and testing at multiple levels and over a wide rage of execution platforms, uses open standards, supports collaborative development, and has the potential to provide additional SoS architectural views."

REFERENCES

ACIMS, 2006, software site: http://www.acims.arizona.edu/SOFTWARE/software.shtml. Last accessed Nov 2006.

Atkinson, K., 2004, Modeling and simulation foundation for capabilities based planning, *Simulation Interoperability Workshop Spring.*

Badros, G., 2000, JavaML: a markup language for java source code, *Proceedings of the 9th International World Wide Web Conference on Computer Networks: The International Journal of Computer and Telecommunication networking*, pp. 159–177.

Carstairs, D.J., 2005, Wanted: a new test approach for military net-centric operations, Guest Editorial, *ITEA Journal*, 26(3).

Chaum, E., Hieb, M.R., Tolk, A., 2005, M&S and the Global Information Grid, *Proceedings Interservice/Industry Training, Simulation and Education Conference (I/ITSEC).*

Chairman, J.C.S., Instruction 3170.01D, Joint Capabilities Integration and Development System, March 12, 2004.

Chairman, J.C.S., 2006, Instruction 6212.01D, Interoperability and Supportability of Information Technology and National Security Systems, March 8, 2006, 271.

Dahmann, J.S., Kuhl, F., Weatherly, R., 1998, Standards for simulation: as simple as possible but not simpler the high Level architecture for simulation, *Simulation*, 71(6): 378.

Dandashi, F., Ang, H., Bashioum, C., 2004, Tailoring DoDAF to support a Service Oriented Architecture, White Paper, Mitre Corp.

DiMario, M.J., 2006, System of systems interoperability types and characteristics in joint command and control, *Proceedings of the 2006 IEEE/SMC International Conference on System of Systems Engineering*, April 2006, Los Angeles, CA, USA.

DiMario, M.J., 2007, SoSE discussion panel introduction, from systems engineering to system of systems engineering, *2007 IEEE International Conference on System of Systems Engineering (SoSE)*. April 16–18th, 2007, San Antonio, Texas.

DoDAF Working Group, 2003, DoD Architecture Framework Version 1.0 Vol. 3: Deskbook, DoD, August 2003.

DoD Architecture Framework Working Group, 2004, DoD Architecture Framework Version 1.0 Vol. 1, Definitions and Guidelines, February 9, 2004, Washington, DC.

DoD Instruction 5000.2, 2003, Operation of the Defense Acquisition System, May 12, 2003.

GenetScope Project at JITC 2006, (Beta Verson) *Software User's Manual, available from ACIMS* center, University of Arizona, available at:http://www.acims.arizona.edu/SOFTWARE/SCOPE/GenetScopeNetsim2_Manual.pdf.Accessed August 2008.

Hu, X., Zeigler, B.P., 2005, Model continuity in the design of dynamic distributed real-time systems, *IEEE Transactions on Systems, Man and Cybernetics—Part A: Systems And Humans*, 35(6): 867–873.

Hu, X., Zeigler, B.P., Mittal, S., 2003, Dynamic configuration in DEVS component-based modeling and simulation, Simulation: *Transactions of the Society of Modeling and Simulation International*, November 2003.

JITS, 2005, JITC reports for SCOPE command for year 2005, latest data as of October 2005.

Mak, E., 2006, Automated Testing using XML and DEVS, Thesis, University of Arizona, http://www.acims.arizona.edu/PUBLICATIONS/PDF/Thesis_EMak.pdf.

Mak, E., Mittal, S., Hwang, M.H. Automating Link-16 Testing using DEVS and XML, Journal of Defense Modeling and Simulation, Forthcoming.

Martin, J.L.R, Mittal, S., Zeigler, B.P., Manuel, J., 2007, From UML statecharts to DEVS state machines using XML, *IEEE/ACM Conference on Multiparadigm Modeling and Simulation*, September 2007, Nashville.

Mittal, S., 2006, Extending DoDAF to allow DEVS-based modeling and simulation, Special issue on DoDAF, *Journal of Defense Modeling and Simulation JDMS*, 3(2): 95–123

Mittal, S., 2007, DEVS unified process for integrated development and testing of service oriented architectures, Ph. D. Dissertation, University of Arizona.

Mittal, S., Zeigler, B.P., 2003, Modeling/simulation architecture for autonomous computing, *Autonomic Computing Workshop: The Next Era of Computing*, January 2003, Tucson.

Mittal, S., Zeigler, B.P., 2005, Dynamic simulation control with queue visualization, Summer Computer Simulation Conference, SCSC'05, July 2005, Philadelphia.

Mittal, S., Mak, E., Nutaro, J.J., 2006, DEVS-based dynamic modeling & simulation reconfiguration using enhanced DoDAF design process, special issue on DoDAF, *Journal of Defense Modeling and Simulation*, 3(4): 239–267.

Mittal, S., Risco-Martin, J.L., Zeigler, B.P., 2007a, DEVSML: automating DEVS Simulation over SOA using Transparent simulators, *DEVS Symposium*.

Mittal, S., Risco-Martin, J.L., Zeigler, B.P., 2007b, DEVS-Based Web Services for Net-centric T&E, Summer Computer Simulation Conference.

Mittal, S., Hwang, M.H., Zeigler, B.P., 2008, "XFD-DEVS: An Implementation of W3C Schema for Finite Deterministic DEVS", Demo available at: http://www.u.arizona.edu/~saurabh/fddevs/FD-DEVS.html. Forthcoming.

Morganwalp, J., Sage, A.P., 2004, Enterprise architecture measures of effectiveness, *International Journal of Technology, Policy and Management*, 4: 81–94.

Naval Studies Board (NSB), 2000, Network-Centric Naval Forces — Overview: a Transition Strategy for Enhancing Operational Capabilities.

O'Rourke, C., Fishman, N., Selkow, W., 2003, *Enterprise Architecture Using the Zachman Framework*, ISBN 0-619-06446-3, Course Technology. Available at www.eabook.info.

Sage A.P., 2007, From engineering a system to engineering an integrated system family, from systems engineering to system of systems engineering, *2007 IEEE International Conference on System of Systems Engineering (SoSE)*. April 16–18th, 2007, San Antonio, Texas.

Sage, A.P., Cuppan, C.D., 2001, On the systems engineering and management of systems of systems and federation of systems, *Information Knowledge Systems Management*, 2: 325–345.

Sahin, F., Jamshidi M., Sridhar, P., 2007a, A discrete event XML based simulation framework for system of systems architecture, *Proceedings of IEEE Internationla Conference on Systems of Systems Engineering*.

Sahin, F., Sridhar, P., Horan, B., Raghavan, V., Jamshidi, M., 2007b, System of systems approach to threat detection and integration of heterogeneous independently operable systems, *Proceedings of IEEE International Conference Systems, Man and Cybernetics*.

Sarjoughian, H.S., Zeigler, B.P., 2000, DEVS and HLA: complimentary paradigms for M&S? *Transactions of the SCS*, 17(4): 187–197.

Sarjoughian, H.S., Zeigler, B.P., Hall, S., 2001, A layered modeling and simulation architecture for agent-based system development, *Proceedings of the IEEE*, 89(2): 201–213.

Sridhar, P., Jamshidi, M., 2004, Discrete event modeling and simulation: V-Lab, *IEEE International Conference on Systems, Man and Cybernetics.*

Sridhar, P., Sheikh-Bahaei, S., Xia, S., Jamshidi, M., 2003, Multi-agent simulation using discrete event and soft-computing Methodologies, *IEEE International Conference on Systems, Man and Cybernetics.*

Sridhar, P., Asad, A.M., Jamshidi, M., 2006, Hierarchical data aggregation in spatially correlated distributed sensor networks, *IEEE Co-Sponsored World Automation Congress - International Symposium on Robotics and Applications*, Budapest.

Sridhar, P., Asad, A.M., Jamshidi, M., 2007, Hierarchical aggregation and intelligent monitoring and control in fault-tolerant wireless sensor networks, *IEEE Systems Journal*, 1: 38–54.

Tolk, A., Muguira, J.A., 2003, The levels of conceptual interoperability model (LCIM). *Proceedings Fall Simulation Interoperability Workshop.*

Tolk, A., Solick, S., 2003, Using the C4ISR architecture framework as a tool to facilitate V&V for simulation systems within the military application domain, *Simulation Interoperability Workshop*, Spring 2003.

Traore, M., Muxy, A., 2004, Capturing the dual relationship between simulation models and their context, *SIMPRA (Simulation Practice and Theory)*, Elsevier.

Utting, M., Pretshner, A., Legeard, B., 2006, A taxonomy of model-based testing, *working paper, University of Waikato , Hamilton, New Zealand,* April 2006.

Wagenhals, L.W., Haider, S., Levis, A.H., 2002, Synthesizing executable models of object oriented architectures, *Workshop on Formal Methods Applied to Defense Systems*, June 2002, Adelaide, Australia.

Wegmann, A., 2002, Strengthening MDA by drawing from the living systems theory, *Workshop in Software Model Engineering.*

Wymore, W.A., 1967, *A Mathematical Theory of Systems Engineering: the Elements*, Krieger, Huntington, NY.

Wymore,W.A.,Chapman,B.,Bahill,A.T.,1992,*EngineeringModelingandDesign*,CRCPressInc.

Ylmaz, L., Oren, T.I., 2004, A conceptual model for reusable simulations within a model-simulator-context framework conference on conceptual modeling and simulation, *Conceptual Models Conference*, October 28–31, 2004, Genoa, Italy.

Zeigler, B.P., Hammonds, P., 2007, *Modeling & Simulation-Based Data Engineering: Introducing Pragmatics into Ontologies for Net-Centric Information Exchange*, Academic Press, New York, NY.

Zeigler, B.P., Mittal, S., 2005, Enhancing DoDAF with DEVS-based system life-cycle process, *IEEE International Conference on Systems, Man and Cybernetics*, October 2005, Hawaii.

Zeigler, B.P., Kim, T.G., Praehofer. H., 2000, *Theory of Modeling and Simulation*, Academic Press, New York, NY.

Zeigler, B.P., Fulton, D., Hammonds P., Nutaro, J., 2005, Framework for M&S—Based System Development and Testing In a Net-Centric Environment, *ITEA Journal of Test and Evaluation*, 26(3): 21–34.

Chapter 6

Net Centricity and System of Systems*

ROBERT J. CLOUTIER,[1] MICHAEL J. DIMARIO,[1,2] and HANS W. POLZER[2]

[1]Stevens Institute of Technology, Hokoben, NJ, USA
[2]Lockheed Martin Company, Fairfax, VA, USA

6.1 NET CENTRICITY OVERVIEW

Although systems have connected electronically with each other for more than a century, the widespread adoption of local area networks in the 1980s and the Internet explosion of the 1990s have dramatically increased the range of possible interactions among systems and the number of systems that can interact with each other. More significantly, systems can now routinely connect with systems that belong to different enterprises, jurisdictions, and nationalities. This ubiquitous and dynamic connectivity among systems has spawned a new class of systems of systems (SoS) in which the primary interaction among the systems is via information exchanges over a data communications network.

While this property might be viewed as sufficient to label such systems of systems as being "net-centric," an important additional aspect of net centricity is that of more rapid formation of systems of systems than traditional system engineering processes would otherwise support (Wang et al., 2006). Net-centric SoS (NCSoS) incorporate significant discovery and adaptive or dynamic binding components in the participating systems, allowing them to rapidly connect to other systems that did not exist at system design time. If SoS merely use the network as a transport or connectivity mechanism among prespecified systems, they lack an important attribute of net centricity. The network conceptually represents access to an open-ended set of other systems. If

---

*The opinions and concepts expressed by the authors reflect their personal experiences and observations, and do not necessarily reflect the views or represent any position of Lockhead Martin Corporation.

---

systems have no ability to deal with this dynamism, they cannot take advantage of a key attribute that distinguishes a network from a collection of point-to-point data communications links.

To see why this is so, we need to explore the origin of the term "net-centric." The term first surfaced as "network-centric warfare" in Alberts et al. (1999) under sponsorship by the U.S. Department of Defense (DoD) Office of Force Transformation. While the reference focused on the operational warfighting impact of connecting systems and people via a network, many people objected to the term "network-centric" because it conceptually made the specific technology of networking the center of attention. This caused some to adopt the term "net-centric" as a substitute because it was perceived that the term "net" evoked the conceptual connection among participants more than the specific technology that might be used to implement that connection. It was also a useful abstraction if the participants in a net-centric interaction were not all connected to the same specific network or networking technology (e.g., an IP network and a Link-16 network). This abstraction emphasized that it was the notion of users and systems interacting and collaborating with each other over the network in response to operational needs and in operational time frames that was key to net centricity as well as the open-ended nature of possible operational needs put a premium on being able to interact with whatever people, institutions, and systems might be available to address those needs. Nondefense organizations and commercial industry are also beginning to use the term net-centric for similar reasons.

While there has been much debate about the different meanings and nuances of the terms "network-centric" and "net-centric," we think both terms can be used to denote essentially the same conceptual model of dynamic systems of systems enabled by network connectivity and some level of environmental discovery and adaptability on the part of the participating systems and users. This system-level discovery and adaptability may be augmented by network-level enabling services, such as service directories, provided by some enterprise or transenterprise entity. Other nations have used the term "network-enabled" capability to convey the same general model, using "enabled" to avoid making the network the focal point of the intended operational capabilities. While these terminology nuances reflect legitimate concerns, we will use the term "net-centric" to denote the overall conceptual model for systems of systems that form and interact with each other primarily over a network. Further discussion regarding net centricity can be found in (Polzer 2006a,b).

Additional implications of the top-level model for an NCSoS and Communities of Interest (COI), which are composed of all of the actors that can influence decisions and their respective services (Alberts, 2006), are shown in Fig. 6.1 and summarized as follows:

- Physical interaction among the systems may occur but is not required to form an NCSoS.
- Geospatial proximity of the participating systems is not required.
- The number of participating systems in an NCSoS is constrained only by network scope and capacity.

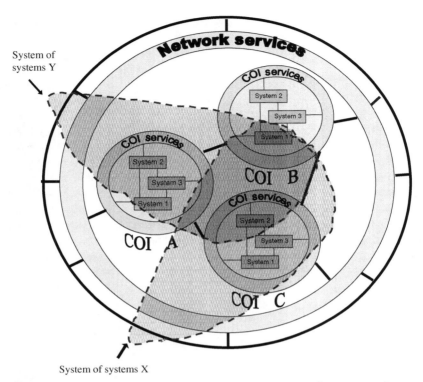

FIGURE 6.1   Net-centric community of interest system of systems ecology

- Organizational affiliation, functional domains, and operational domains of the systems in an NCSoS are constrained only by the supporting network scope.
- Systems on the network may be, and probably are, members of multiple NCSoS simultaneously.
- The network transcends any individual SoS and is a SoS in its own right. It provides services to all the participating systems such as directory and discovery services, security services, and presence awareness services.
- Systems may encounter each other on the network and begin interacting without any explicit sponsoring entity interaction or venue, legal or contractual. By contrast, physical systems interact with each other in a particular jurisdiction or institutional frame of reference, such as the law of the sea or a company's policies. For example, as yet there is no universal institutional frame of reference at the "network" level, aside from the Internet Corporation for Assigned Names and Numbers (ICANN) for the public Internet, which manages allocation of Internet domain names.
- Although systems can potentially connect with any other system on the network, in practice this is constrained by the complexity and cost associated with broad-based semantic interoperability and the law of diminishing returns regarding the

potential operational benefit from doing so. This has famously been summarized in the quip that "just because your dishwasher can talk to your refrigerator, doesn't mean that it should."

- Semantic interoperability challenges will tend to limit system interaction over the network to systems belonging to specific communities of mutual interest, or with systems in "adjacent" communities. The exception will be interaction domains that have broad applicability and widespread adoption of specific standards such as in the geospatial information domain and the financial domain.

- The network, its enabling services, and the systems connected to it create a network ecology or business/operational environment that enables new types of systems to become viable, a form of "emergent properties" thought to characterize an SoS.

- General semantic interoperability is a difficult problem and expensive for individual systems on the network to implement. Community of Interest level services devoted to brokering interactions among systems in diverse communities of interest and semantic domains will become operational enablers of increasingly net-centric systems of systems by off-loading some of this complexity from individual systems.

These characteristics have important consequences in architecting and engineering NCSoS and present a number of challenges that we will explore further in this chapter. We will start with the nature of the interactions between systems enabled by the typical network, their possible benefits, as well as their challenges and risks. We will then examine the nature of net-centric system of system architectures, with a focus on service orientation. Finally, we will examine the architecting and engineering process implications of net-centric systems of systems, as well as some of the business, governance, and policy implications of net-centric systems of systems.

6.2 NETWORK-ENABLED SYSTEMS INTERACTION

Interactions among systems over a network can range from simple ad hoc and atomic information request and responses to more complex and critical operational and business interdependencies. They can also range from specific interactions among systems defined at system requirements or design time to dynamic discovery of systems and services and dynamic binding to specific services at run time. Interactions can be unconstrained by the participating systems or constrained by an explicit business model or trust relationship with other systems or with the network provider(s). The interactions can be implicitly bounded such as Google results, which are based on publicly visible Web sites, or explicitly bounded whereby a system only accepts certain request parameter types or only certain parameter value ranges such as only locations in the United States or only certain product types. We will explore each of these interaction dimensions in turn.

6.2.1   Information on Demand

This is probably the simplest form of network-centric interaction among systems and typified by the kind of interaction provided by many organizational Web sites. It is usually constrained by the choice of interface request parameters and presentation and packaging options, as well as implicitly by the institutional scope of the site or organization offering the information. In some cases, demand is constrained by cost, with certain types or quantities of information being provided only if the demanding system is willing to pay some fee structure. The key challenge for information on demand is that there is no universal information representation standard, and the representation of most information captured by systems naturally reflects the frames of reference and perspectives of the system sponsors. The network exacerbates this challenge because it inherently exposes these many different frames of reference, scope assumptions, and representation choices to each other, usually with no explicit context representation or rationale.

Current search engine Web sites on the Internet rely on users to apply their world knowledge and context awareness to filter out extraneous responses. They also typically do not rely on such search engines to provide any notion of completeness or aggregation of data across information providers. Key potential enablers here are information broker services or aggregators. These services support mapping one frame of reference and associated identifiers to corresponding identifiers in other frames of reference and contexts, along with reasoning about scope relationships among information providers. This is essentially the tactic employed by value-added Internet service sites such as Travelocity and Priceline. Users can go to individual hotel and airline sites, for example, but exploring all possible options for a given destination at each site can be exhausting. Additionally, each site has its own way of presenting information to the user. Information broker services hide this complexity and provide "one-stop shopping" through data aggregation that general-purpose search engines and individual service provider sites cannot provide.

Another challenge in the simple information request and response interaction is that of dealing with service requests that consume a great deal of system resources to answer and with controlling or metering high volumes of service requests coming from other systems on the network. The Internet model typically deals with this problem by funding service providers through indirect revenue models not associated with information request and response by "overprovisioning" service response capacity and by simply taking longer to respond to a request. For example, the Google advertising model gives the searcher free search services by asking advertisers to pay for favorable placement of their site or product in any Google search result.

6.2.2   Services on Demand

Services may be performed by one system at the request of another system submitted via the network. The service may be a physical service, such as delivering groceries or reserving a seat in a theater, or it may be an information service, either providing information on demand or generating new information or transforming and

Service attributes	Business model issues
Generality of the service and frequency of use	Affects the size of the market for the service – the more use, the more likely a fee for service model will motivate industry
Criticality to mission success and importance of mission/capability	Decreases coupling of service cost to service delivery – fee for service models become less appropriate
Service statelessness or fungibility	Determines whether multiple service providers are feasible or practical and competition among providers is possible
Cost of providing and using the service	Fee for service incentives do not work well when a service is critical but expensive to build, deliver, and use
Service enterprise scope	Related to generality and statelessness; this can limit the market size and increase cost for delivering the service. It also places the burden on the service consumer to determine how a given service instance might need to be combined with other service requests to meet operational needs.

FIGURE 6.2   Services on demand characteristics

aggregating existing information or information representations such as providing correlations among sensor data or indexing text messages. Because the interaction for such service requests typically entails considerable resource expenditure or future commitment of assets, services on demand usually require a more explicitly defined business model or organizational relationship among the sponsors of the systems providing and consuming services. Conversely, the service requestor's business or mission operations may depend critically on the service performing as expected and within a specified time frame. This requires some means of monitoring and enforcing service quality measures, accessible to systems participating in the NCSoS, but supported by a network or COI-level service provider.

Both financial relationships and incentive models as well as authoritarian models for managing services on demand are feasible and appropriate, but under different institutional service contexts and service attributes. The service/context characteristics and their impact on possible business models for service providers and service consumers are provided in Fig. 6.2.

6.2.3   Ubiquity and Degrees of Connectivity

Depending on the physical network infrastructure deployed, an NCSoS can form between any given set of physical systems, regardless of location. However, the network connection may not be permanent or continuous, and it may be subject to physical latency due to speed of light, congestion limitations, or specific network infrastructure geospatial deployment decisions. Ubiquity of network access determines whether a given system can connect to an NCSoS depending on its current geospatial location or jurisdictional environment. The degree of connectivity may be likewise constrained by available bandwidth in a given location or the local jurisdictional environment, such as a coalition operations center or one system operating inside another system's enterprise network environment. The important point here is that physical connectivity to the network, such as wired or wireless, is not the only constraint on ubiquitous network access. The local policy, regulatory, or business environment may also limit connectivity among systems in a variety of ways. For example, privacy laws might limit what data can be exposed in services accessible over the network, and intellectual property rights might require that only systems whose sponsor have completed certain licensing or disclosure agreements can access certain data over the network.

6.2.4   Syntactic and Semantic Interoperability

Systems interacting with each other over the network are dependent on some agreement or understanding of the syntax used for representing and transporting information chunks between systems, including service requests and responses. In addition, systems depend on understanding the meaning of the interaction in institutional or specific mission or business process and language contexts referred to as the semantics. Web Services and Representational State Transfer (REST), which applies the principles of the Web to transaction-oriented services rather than publishing-oriented sites, provides a significant degree of syntactic interoperability and a modicum of semantic interoperability. However, XML and Web Services Description Language (WSDL) provide rather limited semantic capabilities. Ontology-based approaches to semantic interoperability show some promise, but automatic mapping between different ontologies is still a major challenge to research community.

There have been numerous efforts to develop formalism for characterizing syntactic and semantic interoperability among systems interacting via network technology. One of the more notable efforts was that of the U.S. Department of Defense Command, Control Computers, Communication, Intelligence, Surveillance, and Reconnaissance (C4ISR) Working Group, developer of the DoD Architecture Framework (DoDAF), in the late 1990s. The C4ISR Working Group developed the "Levels of Information System Interoperability (LISI)" model (C4ISR, 1998) to help guide the growing number of interfaces being implemented across a variety of networks among DoD information systems. The LISI model recognized the importance of syntax and semantic alignment among systems over and above the network protocols that the systems used to communicate with each other. It pointed out that the information models among the interacting systems had to be sufficiently congruent to support the "enterprise" purpose of the information exchange among the systems. However, it did not address what to do if the alignment among the information models was insufficient for the purpose at hand, or how to deal with dissimilar semantic models among the interacting systems. Nor did it address the institutional scope, context, and frame of reference issue discussed in the next section.

As discussed in the section on simple information request and response interactions, information brokers can be a key enabler of semantic interoperability among systems on the network. Generally, it is not possible, or even desirable, for every system in a SoS to implement the same information model and to utilize only that model in the service interactions it has with other systems on the network. As Fig. 6.1 illustrates, information brokers are a class of COI-level or network-level services that facilitate interaction between systems in a given COI. This ecosystem positioning also makes them a natural interface point between different COIs and with systems in other enterprises. Semantic mapping between dissimilar domains and institutions is a complex topic and costly to implement, especially in nonstatic environments such as net-centric systems of systems (DiMario, 2006). Off-loading much of this cost and complexity from individual systems and concentrating it at the COI level or higher in the net-centric ecosystem enables economies of scale and greater discoverability and adaptability for a given level of engineering investment (Polzer, 2005).

6.3 INSTITUTIONAL SCOPE AND CONTEXT REPRESENTATION IN SERVICE INTERACTION

Scope and context provide additional aids to understanding the semantics of any given set of systems via the network. The scope and context create the frame of reference of information and services along with their associated data representation provided by one system to another over the network. The scope and context provide in addition the purpose, the breadth, and the depth of coverage of any services or information that might be accessed by one system from another system or organization. Much of this may also apply to nonnetwork-connected SoS, but the lack of discernable physical attributes and interactions with inherent geospatial proximity make explicit representation of operational and organizational context and scope that much more important for network-centric systems of systems. While the representation of institutional scope, context, and frames of reference could be considered a specific aspect of semantic interoperability, we call it out separately here because it is an area that has not received a lot of attention from the semantic technology community.

Semantic technologies rightfully focus on the meaning of words, sentences, concepts, and world models, but usually do so outside any specific institutional frame of reference—a science, a topic, a body of knowledge, and a context that transcends individual institutions. However, most systems that connect to a network are commissioned by some institution to support specific institutional objectives and capabilities. The representation of concepts and entities in such systems invariably reflects the frame of reference of the sponsoring institution. For example, employee number, customer ID, product model number, and part number are all examples of data elements in an information model that is of very limited utility unless one adds the name of the enterprise to which these concepts apply. Likewise, institutions have context and scope. For example, Lockheed Martin, a large U.S. Government contractor, does not typically produce consumer products and is therefore unlikely to have an e-commerce Web site to support retail customers. Lockheed Martin builds only certain kinds of systems and certain specific platforms and thus any spare parts he makes available via a service offering on the network is unlikely to include those made by a competitor, even if the name of the ordering service and the service parameters are otherwise identical with those offered by one of the competitors.

Identities of physical and conceptual entities are the information elements most likely to be constrained by institutional context and scope, and valued within a specific institutional frame of reference. Such identities are critical to correct processing of information exchanges among systems via a network. Information brokers can help here as well, but it is also important to remember that institutional scope and context have additional dimensionalities that need to be captured and represented in system or service interfaces offered on the network. One of the early study teams to explore and document this context and scope dimensionality was that of the NATO Industrial Advisory Group, Study Group (SG)-76, on naval command and control interoperability. The final report (NATO, 2003) of the study group includes an annex that explores the dimensionality of NCSoS from an institutional scope and context perspective. This annex became the starting point for the Systems, Capabilities, Operations, Programs, and Enterprises (SCOPE) model for characterizing NCSoS, currently under development by the Network-Centric Operations Industry Consortium (NCOIC), with the institutional frame of reference, multiple perspectives, and context representation further explored in DeLaurentis (2007) suggesting that systems on the network are typically brought into initial contact with each other because some "context-shifting" event occurs that breaks the context assumptions made by the original system engineers.

6.4 INFORMATION ASSURANCE

For net-centric systems of systems, information assurance is the primary net-centric surrogate or analog for physical security, safety, and reliability measures in physical systems of systems. Systems need to be able to assure themselves of each other's identities and that they are requesting and receiving information and services from the system they think they are connected to via the network. Again, this is not unique to network-centric systems of systems (as many banking ATM physical device add-on scams have shown), but the lack of physical proximity does create new identity determination, information quality, and transport assurance/protection challenges. In addition, the degree of trust and reliance on particular systems might vary more dynamically than is typically the case in physical system interactions. This also goes beyond what has traditionally been addressed by single-system information assurance approaches in which the focus was on user authentication to the system and not vice versa.

Because the network is typically not considered to be part of a given NCSoS, the participating systems generally should not trust the network to protect the information en route between systems. The network may be monitored or compromised in some way, and even authorized users and systems on the network can be the source of information leaks or malicious actions. That is why one sees increasing emphasis on encryption at the source of information in net-centric systems of systems, as well as bilateral authentication for service invocations and responses. If any of the service request parameter values or service response elements have special protection requirements or security labels, these need to be explicitly labeled as such in the service descriptions and in the invocation and response data exchanges, over and above any encryption of the exchange content itself.

While the information assurance community tends to focus on security policy and enforcement/protection, it is worth noting that an NCSoS puts a premium on information sharing as well. However, the emphasis is on risk management more than risk avoidance. Anonymous users are not permitted, but information requested by a bono fide user will generally be made accessible upon presentation of appropriate credentials. The network information assurance (IA) mechanisms will generally capture this transaction for security monitoring and enforcement purposes. Users and service providers will be required to apply appropriate security labels to any information that leaves their system, as well as encrypting the data. Verification that the requestor meets the trust policy and authentication requirements for the requested service will generally be the role of an IA service at the network level.

6.5 ARCHITECTING AN NCSoS

6.5.1 Service-Oriented Architecture

While we made the statement that the network is typically not considered to be part of a given NCSoS, many have come to think of the DoD Global Information Grid (GIG) as the ultimate concept of an NCSoS. Others may still argue that the Internet is the ultimate NCSoS. While the GIG, like the Internet that it is patterned after (and which it uses), is generally considered a globally interconnected network, what distinguishes it from a more straightforward network such as the Internet is that it provides a more comprehensive and complete set of services that will connect personnel, organizations, and their systems with each other, supplying the right information at the right time. The Internet governance framework is pretty much limited to the Domain Name System (DNS) and the Internet Engineering Task Force (IETF). In contrast to the GIG, neither there is global user identity management and authentication services on the Internet, nor there is a set of core enterprise services provided by a central agency or a service governance framework. These core services and government structures are key enablers for net-centric operations via systems interacting with each other and with users over a network.

Most net-centric architectures are based on service-oriented architecture (SOA) concepts of service providers, service consumers, and a service registry. However, as those implementing SOA find, it is not just about technology. SOA encompasses the business, the people, the organization, and the information technology. It also includes

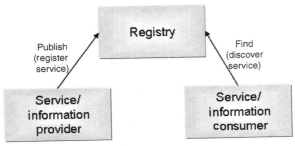

FIGURE 6.3 SOA publish and discover

business process integration, optimization and specialization, standardized business products, and application delivery across the network and around the world (Vietmeyer, 2005).

Figure 6.3 shows a simple model where there is a service or information that is published and a consumer that wants to discover whether a service or information is available. Both interact with a registry through a query. Once it is determined that the service or information is available to the consumer, the consumer will use that service using the publicly defined interface to that service or information as shown in Fig. 6.4.

To do this, the services and processes have to also provide the capability to collect, process, disseminate, and store vast amounts of information. Other services will be necessary to manage these vast stores—providing data adjudication and fusion in a near real-time environment. The right information at the right time, in the correct context, is a daunting task and will require that net-centric architecting principles be defined and that architects strictly adhere to these principles as new capability is added to the network.

The Organization for the Advancement of Structured Information Standards (OASIS) released committee specification 1 of their reference model for service-oriented architecture (OASIS, 2006). This reference model is presented as a framework for understanding the concept of service-oriented environments, based on the unifying concept of a service-oriented architecture addressing a number of critical aspects that should be considered while architecting a net-centric environment (NCE).

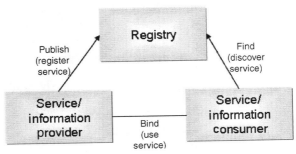

FIGURE 6.4 Using a service in an SOA

The key concepts are visibility, interaction, and effect. In the context of the previous discussion, visibility is related to the service(s) being offered by the service providers visible to the service consumers interested in using those services, interaction simply means that there is a path or vehicle for the consumer and provider to interact, and effect is the outcome of the consumer partaking of the offered service. This point is often lost in the implementation discussions of SOA, but SOA is about providing a "real-world effect" (OASIS, 2006) for the service consumer. The OASIS reference model states "SOA is a means of organizing solutions that promotes reuse, growth, and interoperability. It is not itself a solution to domain problems but rather an organizing and deliver paradigm that enables one to get more value from use both of capabilities which are locally 'owned' and those under the control of others" (OASIS, 2006, p. 9).

6.5.2   A Simple NCSoS Architecture Example

A commonly used example of an NCE is the provisioning of electricity to residential consumers. With the deregulation of electricity production and distribution around the world, the example becomes an easily understood and boundable case. In the most basic form, consider a gasoline-powered generator providing electricity to a mobile home. The generator creates electricity and offers it to a consumer who wants to turn on lights in the mobile home. The service and the consumer are connected through an extension cord. While this is a simple example, there are implied architecture decisions that would be made—would the service be a generator or a battery? What is the voltage/power level supplied by the service provider, and does it match the requirements of the service consumer? If the generator provides 220 V, and the consumer needs 110 V, then something is required to resolve the difference. Does the generator have the ability to provide enough power if the refrigerator and the water heater in the mobile home come on at the same time? Does the extension cord have the correct connectors at each end? Is it rated for the amount of power provided and to be used by the consumer? This represents what might be referred to as a simple point-to-point SOA.

While our ancestors may have lived in caves and later in grass huts, today many live in elaborate collections of single-family dwellings. Today, each of those homes requires electricity to operate our many necessities and luxuries. Early in the twentieth century, it became the goal of the U.S. Government to electrify all the homes in the United States, and power plants and transmission lines were deployed across the country. It became impossible to provide point-to-point power to each home so architects devised the concept of power transmission lines that would carry high amounts of power over large transmission lines to a collection of electric substations, which would then redistribute the electricity to the homes in a format they could use. Meters were installed at each residence so that the electric company would know how much of the power produced was consumed at each home. Some of the many architecture decisions that had to be made included power output levels, distribution means, substation location and design, and usage estimates to size the substation properly. There were also service-level agreements—implied or inferred—that the electric company would produce all the electricity a home might need, though the individual homes had no requirement to notify the electric company if they added a new appliance.

6.5.3 Birth of a Net-Centric SOA

To be able to provide power in peak load scenarios, the power system architect had to trade-off designing the power plant for some unpredictable peak load (rather than a more predictable average load), or devise a method to find more electricity in time of peak demand, without incurring the cost of additional power plants. The solution was to devise a capability to purchase more power from a neighboring power company. Thus the electric grid was born. The same concepts from the simple example are present—an electric service providing electrons across a wire to be used by a consumer to drive an electric appliance, thereby resulting in a real-world effect. Occasional power outages in extreme situations aside, many of the architecture decisions made early have made the U.S. power grid an extensible and robust system. More service consumers are added every day, and new power services are added as necessary. The ability to cross connect regions and the government deregulation have made it possible for an individual to "purchase" power from hundreds of miles away. It gives the consumer choices of service providers. Viola, we have an NCSoS—service providers are each a system—they stand alone and they satisfy their mission. Each home satisfies the IEEE and INCOSE (IEEE, 2000; INCOSE, 2006) definition of a system, and each residential house can be thought of a system—it is a collection of parts assembled to satisfy a mission, that is to provide shelter and community for a family. The electric grid enables service consumers to utilize a service offered by a service provider, without being concerned of the mechanics. The collection of electric service providers easily fits into Maier's definition of an SoS (Maier, 1996). Each plant is independent from the other, with independent missions, and each has programmatic independence and when operated as an SoS, each displays emergent behavior.

6.5.4 The Role of the Human in SOA

For the most part, without the human, the requirement for an NCE would be greatly diminished. There are some key concepts regarding human systems integration that must be taken into consideration when architecting the NCE. Some of the more important considerations include (Polzer, 2006a,b) the following:

- The NCE does not "own" the user or the complete user experience.
- Users need to access other systems/information/services on the net.
- Users have properties that transcend individual systems, for example, identity.
- Systems on the NCE may capture/convey information/actions from/to other users using other systems.
- Authorship/source/pedigree of information needs to be explicit.
- "Data" may be opinions, assessments, observations, and so on.

- Systems are authoritative only in certain contexts for certain purposes, and users need to understand those contexts and scope of applicability.
- Human resources for operating/using the system may be drawn from "foreign" organizations via the net.
- Doctrine and rules of engagement (or business processes) may be different for different users on the net.
- Business models for using systems may not be obvious to users, and resource prioritization among uses may be explicit user functionality.
- Users may have no prior training on the systems in question.
- Certain standards for user interface elements/cues to convey via the net, independent of end-user interaction means, will be needed. An example of this is that data may be coming from another user via a system.
- Systems will need to dynamically determine user context. They cannot assume user context based solely on the fact that the user is accessing the system (current default in most systems). And, users may be anywhere, using a foreign access device.
- Systems on the NCE should also consider other users as possible sources of knowledge/data or processing capability (e.g., assessment of intent).
- Discovery services need to include expertise of other users on the NCE, not just data base or document discovery.

6.5.5 Documenting Net-Centric Architectures

When architecture decisions and strategies are documented, they need to be published for all other architects to follow. This meta-architecture must support the overall organizational or business strategy in which the NCSOS is supporting. Just as if meta-data are information about the data, meta-architecture must describe information about the broad architecture—critical architectural significant strategies and architectural structures.

6.5.6 Common Parts of a Net-Centric SOA

Definitions of what constitute system services, management of these services, and accessing these services may fall into the category of meta-architecture, while not getting into the details of implementation. Other meta-architecture issues may include architecting patterns, and reference architectures for participating systems. The list below is a summary of the core services as specified by DISA[1]:

- Registry
- Discovery
- Messaging
- Mediation
- Collaboration
- Storage
- User assistant
- Metadata
- Application
- IA security
- Enterprise service management

[1]DISA, Defense Information Systems Agency, Core Enterprise Services, http://www.disa.mil/nces/enterprise_services.html.

When there are multiple systems, each based on an SOA, the question arises on how do clients on one SOA system become aware of services available on another system? A solution adopted by the DoD is to implement at federation of registry services (DoDAF[2]). That is, when a new SOA-based system joins the network, it is required to start a repository client, which can then communicate with the enterprise service registry. While the system registry contains metadata about the services offered on the system, the enterprise service registry contains meta-metadata about all of the services offered across the SoS. This allows a consumer on one SOA-based system to discover interesting available services on another available system.

The system architect must be concerned with the implementation of the definition and description of meta-metadata to ensure effective use and searching across this federation of SOA-based systems. Four key concepts that enable this federation to work are visibility, accessibility, accessibility, understandability, and trust (OASIS, 2006).

What becomes important here is that a traditional reductionism approach to systems architecting will not suffice. The traditional system architect decomposes system requirements into more understandable collections of requirements and breaks the requisite system into smaller, more understandable subsystems. The SoS architect must take a combinatorial approach of combining complete systems to construct a greater whole.

Other challenges for the SoS architect include security. Major security issues encompass multilevel security across multiple domains. This includes security level of another system attempting to connect to an NCE, as well as consumers attempting to gain access to a service for which they are not authorized. Included in this is the certification of security level of the consumer or system.

Figure 6.5 captures many of these concepts in a block definition diagram using SysML. The left side of the diagram represents the services available on the NCE. The include services such as messaging, registry, information assurance and security, and so on. There are also specialized services that we have called "COI_Services," or community of interest services. These are services that a special group may be more likely to use. An example might be a data fusion service, or a service providing access to supercomputing. Another service provided is a composeable service. Services of this type are created by aggregating multiple services into a single service to facilitate ease of access or execution. All of these services are provided to the network. Also represented are actors representing those that create and provide the services and data made available to others on the net. The right portion of Fig. 6.5 represents domains that have subscribed or attached to the NCE. They have been labeled ConsumerDomain_1 and ConsumerDomain_n. There can be 1..n domains participating in the an NCE at any given time.

6.5.7   NCE Architecting Considerations

Looking at ConsumerDomain_1, you will notice that what appear to be duplicates of some of the services we discussed are already available on the NCE. This is an

---

[2]DoDAF Ver.1.5, Vol.2, Department of Defense Architecture Framework.

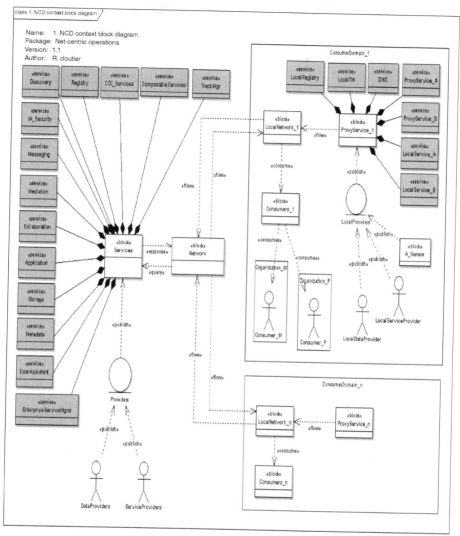

FIGURE 6.5   Net-centric environment services architecture

architecture decision that must be considered by each domain architect. If there is never going to be a condition where the domain does not have access to the broader NCE, and it is operationally acceptable to not have services available during disconnected times, then this may not be required. However, if some availability to some services is required during when the domain may be disconnected, whether it is a planned or unplanned disconnection, then the architect must consider how those services will be provided. Additionally, it is important to consider which services are more critical and whether a "lighter weight" service may be sufficient during disconnected operations. For those critical services that must be available whether connected or disconnected to the NCE, an alternative sourcing must be architected. Finally, there may be services that are required, which are not provided on the NCE. These services must be provided by local service and data providers. The architect must also consider whether to limit the access to these services to within the domain or make them available to the broader NCE.

6.5.8   The Role of the Net-Centric System Architect

Figure 6.6 represents one view of the roles of a system architect (Muller, 2004). While the original intent of Muller's work was not net-centric, nevertheless, it identifies an excellent set of activities for the net-centric architect.

- "Think, analyze" the role of each system, service, and consumer in the NCE, the operating environments, and what can go wrong.
- "Listen, talk, walk around" and understand what the architects and designers of the component pieces are doing, and how they affect the overall architecture. Do they comply with the governance model for the NCE? Do they add risk or reduce risk?
- "Design, brainstorm, explain" the architecture, what are the architecturally significant designs of the an NCE.

Some of the key goals of architecting include the following:

- Identify a collection of products and parts that work in harmony to provide a system that satisfies the customer requirements.

FIGURE 6.6   System architect role

- Identify the communication needs between those parts and ensure those needs are addressed.
- Identify the data needs of the system and ensure the right data are available at the right time.
- Analyze and synthesize architectural-level data gathered.
- Document the architecture in a manner that can be communicated to the customer and the designers.
- Document the architecture and design, including key architecturally significant decisions.

6.6 CONCLUSIONS

As has been shown, architecting an NCSoS requires many of the same skills that are necessary for architecting any complex system. However, there are differences including the recognition that the net-centric architect does not have control, or the ability to change many of the systems that are present on the NCE. This may have the result of conflicting and/or competing requirements that must be accommodated or resolved. In addition, because of the ubiquitous and dynamic connectivity opportunities, the architect must understand that the human participants may or may not be known, and may not be distinguishable from a participating system. Finally, if the NCE SoS architect is connecting a new domain to the NCE, service availability must be considered in light of connectivity and operational need. If the local domain must function during limited or no connectivity, then local services must be provided, considering the degree of functionality needed.

REFERENCES

Alberts, D.S., Garstka, J.J., Stein, F.P., 1999, Network Centric Warfare, Command and Control Research Program (CCRP).

Alberts, D.S., Hayes, R.E., 2006, *Understanding Command And Control*, CCRP, Washington DC.

C4ISR Architecture Working Group, Levels of Information System Interoperability, March 1998.

DeLaurentis, D.A., Polzer, H.W., Fry, D.N., 2007, Multiplicity of perspectives, context scope, and context shifting events, *Proceedings of the IEEE International Conference on System of Systems Engineering*, April 2007, Antanio, TX.

DiMario, M.J., 2006, System of systems interoperability types and characteristics in joint command and control, *IEEE Conference on System of Systems Engineering*, Los Angeles, CA.

INCOSE, 2006, *Systems Engineering Handbook*, Version 3.

IEEE Institute of Electrical and Electronics Engineers, 2000, *IEEE Std 1471-2000: IEEE Recommended Practice for Architectural Description of Software-Intensive Systems*, IEEE, Piscataway, NJ.

Maier, M.W., 1996, Architecting principles for systems-of-systems, *Proceeding of the 6th Annual INCOSE Symposium.*

Muller, G., 2004, System Architecting, Embedded Systems Institute, Eindhoven, The Netherlands, 2004, Version 1.3, retrieved December 20, 2004 from http://www.extra. research.philips.com/natlab/sysarch/.

NATO Industrial Advisory Group (NIAG)/SG-76, 2003, Naval C2 Interoperability Final Report, September 2003.

OASIS, 2006, Reference Model for Service Oriented Architecture, OASIS committee specification August 1, 2, 2006, http://www.oasis-open.org/committees/tc_home.php?wg_abbrev=soa-rm.

Polzer, H.W., 2005, Evolving Capabilities in a Net-Centric Eco-System, April 2005, Lockheed Martin White Paper. The Road Toward Net-centric Operations, Rob Vietmeyer, NCES Chief Engineer, DISA, March 2005, presentation to SOA Forum.

Polzer, H.W., 2006a, The Essence of Net-Centricity white paper, October 2006, Association for Enterprise Integration (AFEI), http://colab.cim3.net/file/work/SICoP/2007-08-09/HPlozer10122006.doc.

Polzer, H., 2006b, NCOIC and Industry Perspectives on Human Systems Integration in a Net Centric Environment, Net Centric Human Systems Integration Workshop, November 30, 2006.

Vietmeyer, R., 2005, The road toward net-centric operations, *Presentation to SOA Forum,* DISA, March 2005.

Wang, G., Valerdi, R., Lane, J., Boehm, B., 2006, Towards a work breakdown structure for net centric system of systems engineering and management. *INCOSE 2006—16th Annual International Symposium Proceedings, System Engineering: Shining Light on the Tough Issues*, Orlando, FL.

Chapter **7**

Emergence in System of Systems

CHARLES B. KEATING

Old Dominion University, Norfolk, VA, USA

7.1 THE SYSTEM OF SYSTEMS PROBLEM DOMAIN

The future landscape and evolution for engineering solutions to increasingly complex system of systems (SoS) requires a different level of thinking and practice for design, deployment, operation, maintenance, and transformation. There appear to be few modern age systems that practitioners would claim are not fraught with excessive complexities. Traditional systems engineering approaches (Blanchard and Fabracky, 1998; Sage, 2000; Kossiakof and Sweet, 2002; Blanchard, 2004) have proven effective in addressing complex systems problems where technical aspects dominate the solution space and boundaries are clearly discernable. However, a new class of complex systems problems has begun to emerge. This class of systems is referred to as a "system of systems" and has been receiving increased attention (Keating et al., 2003a,b). In short, these SoS represent an integration of multiple complex systems to perform a singular mission or purpose. These systems appear ephemeral, existing beyond our grasp to adequately comprehend and influence to achieve the levels of desired performance. They are problematic across the spectrum of technical, human/ social, organizational, resource, policy, political, and managerial constraints. For example, current problems in such diverse areas as integrated transportation systems, homeland security, military transformation, and critical infrastructures are all exemplars of "system of systems" problems.

The need for new approaches to deal with increasingly complex SoS is exacerbated by (1) an exponential rise in the demand, accessibility, and proliferation of information, (2) increasing requirements for interdependence between systems that have previously been conceived, developed, and deployed as independently functioning systems, (3) demands for engineering solutions willing to trade completeness for accelerated deployment, and (4) holistic solutions that exist beyond technical (hardware/software)

System of Systems Engineering: Innovations for the 21st Century, Edited by Mo Jamshidi
Copyright © 2009 John Wiley & Sons, Inc., Publication

resolution. These demands have strained the capability of traditional system engineering approaches to effectively engineer solutions to SoS problems.

A classical systems principle affecting effectiveness in SoS solutions is *emergence*. In short, emergence holds that patterns/properties in a complex system will come about (emerge) through operation of the system. These patterns/properties cannot be anticipated beforehand and are not capable of being deduced from understanding of system constituents or their individual properties. Emergence has always existed in engineering of complex system solutions, most commonly know as the "law of unintended consequences." This catchall concept somehow assures us that we are not hostage to pure mysticism in our attempts to understand apparent unexplainable behaviors as we deploy system solutions that produce unexpected consequences. However, for SoS, emergence has increased significance for two primary reasons. First, the scope of an SoS (e.g., integrated border security system) is beyond a single-system solution. This requires the integration and function of many, potentially disparate, systems to perform as an SoS. Second, the nature of an SoS solution is "holistic," with emergence occurring across the range of technical, human/social, managerial/organizational, policy, and political dimensions that characterize the SoS context.

Prior to proceeding with the discussion of emergence, the nature of the SoS problem domain bears a more critical examination. This is important since the *emergence* we seek to explore stems from the unique characteristics of the SoS problem domain.

7.1.1 Attributes of the System of Systems Problem Domain

Engineering SoS solutions is a relative new and immature field. Not every problem is an SoS problem, and therefore may not require the commitment of intellectual, fiscal, and political resources necessary to generate an SoS-based solution. However, a legitimate question is "how do we know if it is an SoS problem?" There are a set of characteristic attributes that we can use to capture the essence of the SoS problem domain and identify whether or not a particular problem fits the SoS profile. For conciseness, a set of attributes of complex system problems most appropriately addressed by SoS has been identified in Fig. 7.1. An SoS problem may have one or more of the attributes.

The attributes from Fig. 7.1 suggest that the SoS problem domain is significantly different from that of traditional systems engineering. Unfortunately, traditional systems engineering continues to be projected as a viable alternative for addressing SoS problems. The outcome is likely to result in disappointment and failure— traditional systems engineering is simply not equipped to adequately address the SoS problem domain.

Although a widely accepted approach to engineering SoS solutions does not currently exist, it is a reasonable conclusion that the significant differences in the SoS problem domain necessitate a different approach, driven from different paradigms, methodologies, and toolsets. System of systems engineering (SoSE) has been developing in response. Although it is still a relative immature field, SoSE shows promise for more effectively addressing SoS problems. There have been numerous attempts to

SoS Domain attributes	Description of attribute
Integrated primary mission/purpose	The nature of an SoS requires that existing "systems" be subordinated to the new integrated mission/purpose. This challenges existing levels of autonomy and control previously experienced by newly anointed "subsystems."
Contextual dominance	"Context," taken as the set of circumstances, factors, conditions, or patterns that influence SoS performance, is critical. Context spans technological, human/social, organizational/managerial, policy, and political dimensions of the problem domain. However, in SoS applications, arbitrary exclusion of any of these dimensions invites disappointment.
Pluralistic perspectives	The SoS environment is such that alignment of perspectives on the nature, function, and approach is subject to divergence. Pluralistic viewpoints on such issues as allocation of scarce resources, power distribution, control, performance desires, and interpersonal preferences are only exemplary of the issues that can characterize the SoS domain. These differences may be tacit or explicit and manifest in forms not easily tracked to their source.
Multiple stakeholders	SoS problem domains involve multiple stakeholders, who can have convergent or divergent views of the SoS problem domain. Each stakeholder has a "perceived" interest in the design, execution, or evolution of the SoS.
Ambiguous boundaries	Boundaries for an SoS ultimately answer the question, "What is included and excluded in the SoS, and what are the criteria for inclusion/exclusion?" Thus, boundaries are problematic in three ways. First, the inclusion/exclusion boundary criteria for an SoSE are arbitrary and necessarily qualitative in nature. Second, the nature of boundaries, and the organizing boundary paradigms, can take many forms (e.g., geography, time, conceptual, functional, physical). None of these organizing paradigms are correct or incorrect, but they are certainly problematic for the SoS, particularly if they are divergent. Third, boundaries for an SoS are not static —they may, and probably should, change over time with increased understanding of the SoS.
Uncertain resources	Resources for virtually any complex system are constrained and certainly subject to shifts and redirection. However, this is important to SoS for two primary reasons: (1) The SoS solution must be designed to be robust in response to internal or external shifts in resources, and (2) consequences of resource shifts must be understood in relationship to potential impacts.
Systemic barriers	These barriers are patterns that influence the development, deployment, maintenance, or operation of an SoS solution. Every SoS has these barriers that cross technical, managerial, contextual, and organizational boundaries. They will influence performance in the design, deployment, and evolution of SoS solutions. However, in SoS the impact of systemic barriers is more pronounced since the problem scope is much greater.
"Satisficing" solutions	More likely than not, the types of scenarios for which SoS is best suited are those that require "immediate" solutions that might not be technically complete. An immediate satisficing solution, although incomplete and potentially subject to some degree of error, may be preferable to waiting for deployment of a more complete (optimal) solution. In effect, this can invite a higher degree of emergence stemming from incomplete system solution deployment.
Ill-defined problems	SoS problems are inherently difficult to capture. Multiple (possibly divergent) perspectives, questionable ability to articulate the problem, ambiguous/shifting boundaries, and the potential for rapid change in the nature of the problem all influence the ability to develop an SoS solution.
Fluidity	SoS requirements and problem situations are subject to sudden, and potentially radical, shifts. For instance, there may be continual shifts in policy, directives, initiatives, resources, and scenarios that make addressing the situation difficult.

FIGURE 7.1 Attributes of the SoS problem domain

171

define, classify, and categorize SoSE. The following section elaborates SoSE and establishes the context for exploration of emergence as a central tenant of SoS.

7.1.2   Engineering of Systems of Systems

The difficulty in engineering solutions to complex SoS problems is well known. This is particularly true with recognition of the limitations of traditional systems engineering to address SoS problems. In fact, several early authors recognized the need to move systems engineering beyond limited perspectives to more holistic systems-based approaches (Weinberg, 1975; Beer, 1979; Checkland, 1981, 1999; Gibson, 1991; Flood and Carson, 1993). This is not meant to disparage the role of traditional systems engineering, or associated disciplines, in addressing complex problems that are well bounded, stable, and conducive to detailed analysis. However, the SoS problem domain lies beyond the grasp of traditional technical approaches (e.g., systems engineering).

The engineering of systems of systems (SoSE) is directed at a different class of problems. As we established in the preceding section, these problems are character-ized by high degrees of complexity, ambiguity, uncertainty, and potentially divergent perspectives—in other words, they are what we would classify as "messes" (Ackoff, 1999) or "wicked problems" (Rittel and Webber, 1973). These problems define the landscape of the modern technical manager, destined only to escalate in both scope and criticality in the future. For SoS problem "messes," there may be a lack of consensus concerning the nature of the problem, the strategy to proceed, or whether a problem even exists at all. Addressing these problems requires one to navigate a tangled array of interrelated issues that span technical, organizational, managerial, human, social, policy, and political dimensions—all with varying degrees of influence on the development, deployment, and effectiveness of complete SoS solution strategies. "Getting it wrong" comes at great costs, not only in wasted resources (e.g., time, manpower, money, material) but also in human costs (e.g., satisfaction, suffering, careers) and the potential catastrophic consequences stemming from the SoS vulner-abilities left from the unresolved issue(s). SoSE has been attempting to develop approaches to help guide engineering of solutions to this troublesome class of problems (Keating et al., 2003).

Multiple authors have elaborated on the nature and meaning of SoS/SoSE (Maeir, 1998; Carlock and Fenton, 2001; Sage and Cuppan, 2001; Hitchins, 2003; Keating et al., 2003a; Ring and Madni, 2005; DeLaurentis et al., 2007). Although this has led to a constructive exploration, there is currently no widely accepted perspective of SoS/SoSE. However, it appears that there are several themes of convergence in the SoS/SoSE literature (Keating, 2005). In summary, these areas include the following:

- An SoS involves the integration of multiple, potentially previously independent, systems into a higher level system (metasystem) to perform a mission/purpose for which each member plays an integral role.
- An SoS generates capabilities beyond which any of the constituent member (sub) systems is independently capable of producing.

- Integration into an SoS evokes some degree of constraint (surrender of autonomy) for previously independent systems.
- An SoS is a complex system and as such exhibits dynamic and emergent behavior, is difficult to grasp, and problematic to engineer.

Of particular interest is the fact that the primary points of convergence appear to focus on the attributes of what constitutes an SoS. This is somewhat superficial, even for a maturing field, given the range of philosophical, axiomatic, methodological, and application issues essential for a more complete field development (Keating, 2005).

A primary thrust of SoSE is establishing a "holistic" frame of reference—one that guides balanced inquiry, decision, action, and interpretation appreciative of the rich texture of the SoS problem domain. This framing is critical to bring essential structure and order to the problem domain, and above all, avoid the trap too often fallen into— *efficient development of an elegant solution to the wrong problem—destined never to be implemented, or even worse, be implemented only to cause more harm than good.* In effect, SoSE is about enabling practitioners—through development of essential methods, techniques, tools, and thinking—to be more effective in addressing complex SoS problems. SoSE has been described as: "The design, deployment, operation, and transformation of metasystems that must function as an integrated complex system to produce desirable results" (Keating et al., 2003a, p. 35). An essential aspect of accomplishing SoSE is "holistic" articulation and framing of the complex SoS problem. The correct framing of system problems is so critical that subsequent efforts are of little consequence if it is not done effectively.

Too often, it is easy for systems engineers to quickly fix attention on the technical aspects of the system being developed. This is frequently achieved through processes of reduction, often moving critical nontechnical constraints and barriers into the background or making simplifying assumptions that eliminate their consideration altogether. The assumption is that system solutions can be developed with a level of independence from their context. However, as experienced systems engineers know well, the most elegant system solution will be ineffective if it does not address the "whole" of the problem, which includes those contextual aspects that might appear to be messy. For example, it would be naive to engage in SoSE analysis of a transportation SoS without considering the prevailing political climate. However convenient it might be to place political dimensions outside the bounds of inquiry, the most novice of transportation engineers recognize the dire consequences of not considering the play of politics in the "whole" system solution development for a transportation problem.

7.2   NATURE OF EMERGENCE IN SYSTEMS OF SYSTEMS

Emergence exists in all SoS and must be dealt with to effectively address the SoS problem domain. Emergence is a principle in classical systems theory that generally suggests that system properties (patterns, capabilities, structure, behaviors) develop from interaction of system elements (Hitchins, 2003). These emergent properties are not ascribable to any of the system elements and cannot be understood/predicted from the properties of the system elements. This general concept has existed for some time, finding roots back to Aristotle who suggested that the whole is more than the sum of its parts (Checkland, 1999). However, the utility of "emergence," beyond informed thinking, continues to be questioned. Holland (1998, p. 3) stated, "Despite its ubiquity and importance, emergence is an enigmatic, recondite topic, more wondered at than analyzed. What understanding we do have is mostly through a catalog of instances, augmented in some cases by rules of thumb . . . ." Nevertheless, emergence exists in SoS and can provide insights into how we design, analyze, and transform SoS.

Emergence in complex systems generally includes the following commonly held points:

- System-level properties exist only at the system level as it functions, being different from and existing beyond the constituent element, or subsystem, properties.
- System-level properties are not held by any of the isolated elements.
- System-level properties are irreducible. They simply cannot be understood, explained, or inferred from the structure or behavior of constituent elements or their local properties.
- Understanding cause–effect relationships can only be established through retrospective interpretation. This renders traditional reduction-based analytic techniques incapable of useful predictions of emergent system-level behavior.

In addition to these characterizations of emergence, Holland (1998) makes an important point in identifying that emergent patterns are not adequately understood without the appreciation of the context within which the patterns exist. Thus, with respect to SoS, emergence has far-reaching implications for how we think, make decisions, and interpret results related to design, deployment, and transformation of SoS solutions. To effectively deal with emergence requires an appreciation of the philosophical, methodological, and axiomatic underpinnings.

A deeper examination of the nature of emergence in SoS can be centered on three central questions related to philosophical, methodological, and axiomatic predispositions. How these questions are answered determine how we think, decide, act, and interpret within the SoS problem domain. The first question focuses on how we think and drive decisions in the face of emergence—the question is, "What is our philosophical perspective regarding emergence in SoS?" Ultimately, our philosophical disposition, or worldview, will determine the range of response to SoS emergence. Therefore, it is important to take account of our philosophical disposition and understand the implications for different philosophical perspectives concerning emergence.

A second major question deals with methodological influences based on emergence in SoS. Methodology is taken, in the sense of Checkland (1999), as a broad framework that provides guidance for addressing complex systems problems. Our question becomes, "What are the methodological considerations necessary to account for emergence in SoS?" Thus methodology is general, not prescriptive, and adaptable to specific contexts and problem domains that are the subject of SoS inquiry. Methodology is critical because it provides the approach taken to address SoS problems. Thus, methodologies undertaken with respect to SoS will determine the degree to which emergence has been considered a factor in approaching the problem domain.

The third major question that must be addressed is, "What are the axiomatic principles that support a robust perspective for emergence in SoS?" Emergence is a central concept in SoS. However, emergence does not exist in isolation or mutual exclusivity from other important axiomatic principles. Instead, the perspective of emergence can be enhanced by understanding several related system principles. These principles can be instrumental in developing a more informed understanding and appreciation of the nature and implications of emergence for SoS.

In the following sections, we will deal each of these question areas in greater depth. Our objective is to provide a deeper appreciation of the nature of emergence and the implications for those tasked with operating in SoS problem domains. However, we should note that the philosophical, methodological, and axiomatic levels are (1) interrelated, (2) serve to determine the limits in the range of our decision, action, and interpretation related to emergence in SoS, and (3) exist, either explicitly or tacitly, to determine how we deal with SoS . Therefore, a more thorough accounting of the philosophical, methodological, and axiomatic implications will only strengthen our abilities to more effectively deal with emergence in the SoS problem domain.

7.2.1 Philosophical Perspective of Emergence in System of Systems

Addressing philosophical questions in SoS is at first difficult for engineers. Considering the background, education, experience, and training for most engineers, the nature and role of philosophy is frequently viewed with skepticism and considered of little relevance to the "engineering" problem at hand. However, nothing could be farther from the truth. Understanding our own philosophical leanings, as well as those of others critical to the SoS effort, is crucial. This is particularly the case when we are dealing with complex system problems that involve multiple perspectives, disciplines, and potentially conflicting values/beliefs. Ultimately, philosophical disposition will establish the acceptable range of decisions, actions, and interpretations necessary to deal with emergence inherent in SoS.

Philosophy establishes the worldview that allows us to frame a situation. Aerts et al. (1994, p. 9) present worldview as "... a worldview is a system of co-ordinates or a frame of reference in which everything presented to us by our diverse experiences can be placed." In effect, worldview permits us to give meaning to actions, decision, and events as they unfold. This worldview is what Checkland (1999) refers to as *weltanschauung*, the image or model of the world that provides meaning. A worldview is based on philosophical underpinnings (epistemological and ontological) that inform the perspective of SoS and drive purposeful decision, action, and interpretation from an internally consistent reference point. Thus, for SoS, how emergent conditions are

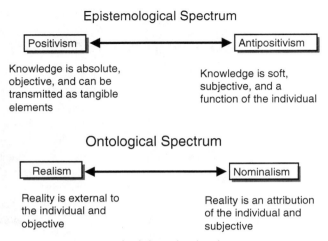

FIGURE 7.2   Philosophic-level spectrum

perceived is first based on a philosophical perspective. Following the work of Flood and Carson (1993), A brief summary of end points of the philosophical spectrum for epistemology (concerned with how knowledge of a system is gained and how that knowledge is communicated externally) and ontology (the nature of reality from which system knowledge is derived) is provided in Fig. 7.2.

Divergence at the philosophic level can result in conflict with respect to how emergence is viewed and dealt with in an SoS endeavor. For example, from an epistemological positivism worldview, emergence would quite probably be viewed as something that could/should be predicted based on absoluteness of system knowledge. There might be little tolerance of SOS approaches not based on the production of tangible products, in an objective manner, with little acknowledgement or tolerance for "emergence." On the contrary, from an antipositivism perspective, there might be more acceptance of emergent SoS conditions, since absolute knowledge of a system is not expected and multiple interpretations are anticipated. Likewise, an ontologically realist worldview would not necessarily be inclined to accept that an SoS would produce patterns not understandable through reduction-based analytical techniques. In contrast, an ontologically nominalist position might be more prone to accept that patterns, behaviors, and properties might indeed exist beyond the level of understanding capable from analysis. The point is not whether or not the ends of the spectrums are "correct" or "incorrect." Instead, it is important to understand where we place our worldview of SoS and the implications that has for how emergence is viewed. Ultimately, consideration for how we philosophically view emergence in an SoS is instrumental to the range of acceptable decisions, actions, and interpretations in addressing SoS problems. A philosophical disposition inconsistent with the ephemeral nature of emergence is likely to result in failure to meet expectations for SoS problem framing, approach, or results.

7.2.2 Methodological Perspective for Emergence in Systems of Systems

Effectiveness in dealing with SoS is largely dependent on effective systems-based approaches (methodologies). Methodology has a relationship to philosophy. Skyttner (1996) suggests that philosophy provides a broad and nonspecific guideline for action. In contrast, methodology is concerned with being a more definitive guide to action than philosophy, yet not as precise or prescriptive as a technique. In sum, a methodology is rooted in underlying philosophy and provides a general framework within which specific techniques can be integrated.

There have been a host of *systems engineering processes* developed and successfully executed (Gibson, 1991; Martin, 1996; Blanchard and Fabracky, 1998; Grady, 2000; Sage, 2000; Westerman, 2001). Although each of these processes has been successfully applied in a variety of circumstances, we suggest that they fall short of being classified as a methodology. At first glance, the distinction between "process" and "methodology" might appear trivial. However, for the practice of SoSE, it is a critical distinction with broad-ranging implications.

There are three primary distinctions to be made between a systems engineering *process* and a systems-based *methodology*. First, a systems engineering process is systematic in that it offers a sequential set of steps that must be accomplished (sometimes in strict linear fashion). Second, a process is prescriptive, providing detailed direction for application. Although many authors have not intended for their processes to take on a rigid adherence to "*n*-step" sequential application, nevertheless they are frequently deployed as such. Third, a process is specific in approach yet projected as applicable across a broad range of applications. This suggests that the process is "context" free in application. On the contrary, a systems-based methodology is not a sequential set of steps. Instead, a methodology offers a general framework with sufficient detail to guide formulation of a high-level approach. Methodology is concerned with providing guidance that is more specific than philosophy (theory) yet not as prescriptive or precise as a method/process (Checkland, 1999). Rigid process-based approaches are too restrictive to deal with the inherent emergence in SoS endeavors. Flexibility to adjust to shifting problem contexts and conditions are essential for dealing with the SoS problem domain. For complex SoS problems, it is naive to assume that a "one size fits all" *n*-step sequential process can be successfully applied to all SoS problems with equivalent results.

There are six primary conditions that suggest a methodology may be preferable to traditional systems engineering approaches (processes) to address SoS problems:

1. *Turbulent Environmental Conditions:* the environment for the effort is highly dynamic, uncertain, and rapidly changing.
2. *Ill-defined Problem Conditions:* the circumstances and conditions surrounding the problem are in dispute, not readily accessible, or lack sufficient consensus for initial problem formulation and bounding.
3. *Contextual Dominance:* the technical "hard" aspects of a problem are overshadowed by the contextual "soft" (circumstances, conditions, factors) aspects.

4. *Uncertain Approach:* the path of progression on how "best" to proceed with the effort is indeterminate. Standard processes may have failed, be failing, or likely to fail to adequately address the situation.

5. *Ambiguous Expectations and Objectives:* the ability to establish measures of success, system objectives, or requirements is questionable. This may be a result of inadequate understanding, hidden motives, or lack of technical competence to proceed with the effort.

6. *Excessive Complexity:* the bounds of the system of interest may be such that the complexities are beyond capabilities of traditional systems engineering approaches. To proceed using traditional approaches requires significant reduction, assumptions, and potential oversimplification of the problem domain.

Any or all of these conditions may support taking an SoSE perspective and approach as an useful alternative. The attributes for systems-based methodologies have been previously established (Keating et al., 2003b) and are summarized in Fig. 7.3. Although the listing is certainly not intended to be exhaustive, it does provide guidance for thinking with respect to development of methodologies appropriate to the SoSE problem domain. These attributes have been adapted for SoS and should serve as a benchmark to determine the nature of proposed approaches. Failure of any methodology to meet these attributes should call to question the legitimacy of application in the SoS problem domain and potential effectiveness in dealing with the inherent emergence for SoS.

7.2.3 Axiomatic Basis for Emergence in Systems of Systems

The axiomatic perspective involves what is accepted as source knowledge for the SoS field. For SoS, this knowledge is still in its infancy as the laws, principles, concepts, and theories are still developing. At this point in SoS, it is fair to say that there has not been a rigorous development of this base knowledge. However, there have been significant explications of systems principles, concepts, and laws that might be considered foundational to SoS (Clemson, 1994; Skyttner, 1996). Additionally, Maier (1998) has provided architecting principles for SoS. Nevertheless, the axiomatic base for SoS has not been significantly developed to provide the foundations that are necessary to provide sufficient grounding for the field. Axiomatic inquiry involves the search for a consistent set of principles, concepts, and laws that are applicable across the SoS domain. Emergence might be considered axiomatic to SoS. In addition, there are several other systems-based concepts that are supportive of the emergence perspective for SoS.

Ten systems concepts have been selected to support a developing axiomatic basis for SoS. These concepts are discussed for their implications for SoS and emergence in SoS. Consideration of these concepts provides a foundation for more effective thinking, decision, and action related to the SoS problem domain, particularly with respect to dealing more effectively with emergence.

Holism suggests that we cannot understand a complex system through reduction to the component or entity level (Skyttner, 1996). This is consistent with our prior development of emergence. The very nature of SoS suggests that traditional systems

Methodology attribute	Methodology attribute description
Transportable	Capable of application across a spectrum of complex systems engineering problems and contexts. The appropriateness (applicability) of the methodology to a range of circumstances and system problem types must be clearly established as the central characteristic of transportability.
Theoretical and philosophical grounding	Linkage of the methodology to a theoretical body of knowledge aswell as philosophical underpinnings that form the basis for the methodology and its application.
Guide to action	The methodology must provide sufficient detail to frame appropriate actions and guide direction of efforts to implement the methodology. While not prescriptively defining "how" execution must be accomplished, the methodology must establish the high-level "whats" that must be performed.
Significance	The methodology must exhibit the "holistic" capacity to address multiple problem system domains, minimally including contextual, human, organizational, managerial, policy, technical, and political aspects of a SoS problem.
Consistency	Capable of providing replicability of approach and results interpretation based on deployment of the methodology in similar contexts. The methodology is transparent, clearly delineating the details of the approach for design, analysis, and transformation of the SoS.
Adaptable	Capable of flexing and modifying the approach configuration, execution, or expectations based on changing conditions or circumstances—remaining within the framework of the guidance provided by the methodology but adapting as necessary to facilitate systemic inquiry.
Neutrality	The methodology attempts to minimize and account for external influences in application and interpretation. Provides sufficient transparency in approach, execution, and interpretation such that biases, assumptions, and limitations are capable of being made explicit and challenged within the methodology application.
Multiple utility	Supports a variety of applications with respect to complex SoS, including new system design, existing system transformation, and assessment of existing complex SoS initiatives. The methodology must provide for higher levels of inquiry and exploration of problematic situations, generating sufficient structuring and ordering necessary to move forward.
Rigorous	Capable of withstanding scrutiny with respect to (1) identified linkage/basis in a body of theory and knowledge, (2) sufficient depth to demonstrate detailed grounding in relationship to systemic underpinnings, including the systems engineering discipline, and (3) capable of providing transparent results that are replicable with respect to results achieved and accountability for explicit logic used to draw conclusions/interpretations.

FIGURE 7.3 Attributes of a systems-based methodology

engineering approaches, based on reduction to perform analysis at increasingly finer levels of detail, are ill equipped for the SoS problem domain. Senge (1990, p. 9) states, "We intuitively attempt to simplify, reduce and break problems into discrete manageable elements. While appropriate in the idealistic academic problems, this concept fails to work when applied to 'real-life'." For SoS, there is a point beyond which understanding of the whole simply cannot be accomplished by reduction. For emergence, limitations based on holism are amplified since (1) the system cannot be disassembled, analyzed, and "put back together" with any level of confidence in system-level understanding, and (2) all system-level properties cannot be known or predicted in advance of their manifestation.

The concept of *system purpose* suggests that the purpose of a system is "what it does" (Beer, 1979, 1985). Although this point seems trivial, it is not. In considering the purpose of a system, the output (tangible patterns, products, and services) and outcomes (impacts) of the system must be considered. Complex systems are designed and operated to produce outputs and achieve objectives. The systems perspective of purpose does not confuse intention with results. Regardless of the well-meaning intentions for an SoS design, purpose is based on "actual" outcomes, not those that were "intended." Therefore, it is a fundamental error to analyze an SoS based on design intentions or desires. In this sense, we conclude that every SoS is actually two systems. The first SoS is the SoS-as-designed. This system is the SoS as it was intended to operate and meet the desired performance objectives. The second SoS is the SoS-as-performed. This SoS emerges as the SoS-as-designed is deployed in an operational setting. The result is usually less than ideal as the SoS produces intended as well as unintended consequences. Thus, system purpose must be derived from the SoS-as-performed.

Although SoS purpose may be specified in advance, only after operation of the system can the achievement of the designed purpose be confirmed. Beer (1985) also adds that the system purpose is dependent on the perspective of the observer of the system. Therefore, multiple vantage points of a system may yield multiple purposes. This implies that system purpose is observer dependent and should not be considered an a priori property for an operational system. The implications of *system purpose* for emergence suggest that (1) differential between the "as-designed" and "as-performed" SoS is a function of the emergence inherent in deployment of the SoS, and (2) purpose is subject to interpretation and might "shift" over time as the SoS operates.

Complementarity suggests that any two different perspectives (or models) of a system will provide different knowledge about the system. This knowledge is neither entirely independent nor is it entirely compatible (Clemson, 1984). Each SoS perspective is correct from a particular vantage point. In addition, each SoS perspective may also be considered, to some degree, to be incorrect from an alternate system vantage point. The important argument is that there are multiple SoS vantage points, each adding to a more holistic impression of the system. Shifts in vantage points, environmental conditions, or maturing knowledge will influence perspectives of an SoS as well as its performance. It is naive to consider that there is only one SoS perspective, or representation (model) that is "correct." A representation (model) of an SoS is not the SoS but only an abstraction that represents the SoS. Therefore, it is a mistake to conduct inquiry as to which SoS perspective is "correct." Assumption of a "correct" SoS perspective encourages advocacy and competition instead of dialogue and collaboration. The conditions for complementary perspectives are amplified under high emergence in SoS.

Pluralism is a systems concept that recognizes there may be multiple purposes/ objectives in play at the individual, entity, and enterprise levels (Jackson and Keys, 1984). These differences may be tacit, undiscussable, and a source of conflict at various points in the development and execution of an SOS due to differing theories in use (Argyris and Schön, 1978, 1996) or worldviews (Aerts et al., 1994; Kovacic et al., 2007). The assumption (unitary) that we have a singular set of agreed-upon requirements and shared understandings for an SoS may be questionable. This becomes problematic for systems-based approaches that rely on rational–logical assumptions of objective/requirement alignment. In relationship to emergence, pluralism suggests that there may be differences in objectives pursued in response to patterns and properties that manifest through SoS operation.

The knowledge of a complex SoS is always *incomplete* and speculative. *System darkness* is a systems concept that recognizes there can never be complete knowledge of a system (Skyttner, 1996). Therefore, it is fallacious to think that there can ever be complete understanding of an SoS, or the context within which the SoS exists. On the contrary, as an SOS unfolds, so does our knowledge of the SoS and the emergent patterns/properties observed. In this respect, the notion that understanding will be static and capable of absolute/definitive articulation, at any point in time, is somewhat illusionary for SoS.

Boundaries in an SoS are ambiguous, fluid, and negotiable. They provide the criteria for what is included and excluded in an SoS. Therefore, they are subject to value judgments and not necessarily static. Boundaries may form around geographic, time, spatial, or conceptual delineations. SoS boundaries might very well shift radically, particularly in the initial formulation of the SoS problem domain. Ultimately, SoS boundaries—temporal, geographic, conceptual, or spatial—can and will shift throughout the performance of SoSE. This is not to suggest sloppy engineering. Instead, boundary shifts in SoS are expected and should be embraced as symbolic of our advancing understanding of the problem domain and recognized emergence in the SoS.

The *metasystem* (Beer, 1979, 1981) provides the structure of relationships that integrates the SoS. It is the metasystem that structures the appropriate balance to relieve tensions (Fig. 7.4) between (1) the autonomy of subsystems and the integration of the SoS as a whole, (2) purposeful design and self-organization, and (3) focus on maintaining stability or pursuing change . In the formation of an SoS, there are formal structural relationships that are imposed. However, through emergence, patterns, structures, and behaviors are exhibited by the SoS at the metasystem level that are not formally specified or known in advance of their development. Therefore, emergence will produce those patterns/properties that are necessary to resolve structural tensions and maintain SoS viability. At times, this may be in spite of formally specified metasystem (SoS) structures. There is not a single "correct" balance point for an SoS, and the appropriate balance in tensions might shift throughout the life of the SoS. However, a poorly performing SoS might be suspected for an inappropriate balancing in the structural tensions.

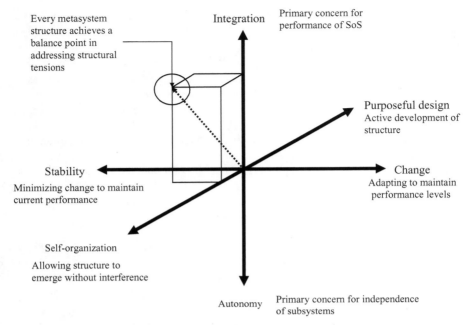

FIGURE 7.4 Balance in metasystem structural tensions

Context is the circumstances, factors, conditions, and patterns that both enable and constrain a complex system solution, deployment of that solution, and interpretation of the results of solution deployment (Keating et al., 2003). For SoS, the context can dominate the solution space and may be more important than technical aspects of a solution. As opposed to traditional systems engineering, context is critical to development of the SoS solution for SoSE. In addition to technical aspects of the SoS solution space, associated "soft dimensions" (human/social, managerial, organizational, policy, political) are also a primary consideration. Without consideration of the role of context, and its inevitable emergence, in SoS formulation, articulation, interpretation, and evolution, a significant and influential portion of the SoS problem domain is potentially ignored.

Dynamic stability holds that a system remains stable as long as it can continue to produce required performance during environmental turbulence and changing conditions. Maintenance of stability, or dynamic equilibrium (Skyttner, 1996), in complex systems is achieved through adjustments to shifts and disturbances (internally or externally generated) that impact system performance. Neither SoS nor their environments remain static and free from change. Therefore, as the SOS environment and context change, commensurate patterns/properties emerge to make the appropriate compensations necessary for maintenance of stability.

The preceding concepts for SoS are complementary to the concept of emergence. They are important in developing the system-based perspective essential to appreciate the nature of SoS. Although these concepts are drawn from systems theory, they are insightful for the SoS problem domain. However, in the future, we might expect to see SoS build new axiomatic foundations as the field matures. In the next section, we explore guidance for dealing with emergence in SoS.

7.3 DEALING WITH EMERGENCE IN SYSTEM OF SYSTEMS PROBLEM DOMAINS

Thus far we have established the nature of the SoS problem domain, explored the concept of emergence, and established related and supporting system concepts. While philosophical, methodological, and axiomatic grounding are essential for higher levels of success in SoS, they are not sufficient to guarantee success. Therefore, the need still exists to provide practitioners with some level of response to more effectively deal with emergence in SoS. In this section, direction for dealing with emergence in the SoS problem domain is offered. The material is presented as themes, organized around dealing with emergence along two facets of SoS: design of SoS and operation and maintenance of SoS.

7.3.1 Design Considerations for Emergence in System of Systems

Design plays a critical role in SoS. For SoS, it is naive to think that the effort will start from a "clean sheet," void of any predispositions. In the case of SoS, there generally already exist "legacy" systems that will be integrated to perform a new mission/ purpose. The SoS design must include the integration of the technical, organizational, managerial, human/social, policy, and political aspects of the SoS itself. In addition, it is unrealistic to think that design can be separated from deployment, particularly where high levels of emergence are characteristic of the problem domain. For many, if not most, cases SoS design and deployment are simultaneous activities. Three primary design themes are presented below for dealing with emergence in SoS.

Design Theme 1: Robustness to changes in understanding, interpretation, and context must be built into the SoS to maintain solution viability (continued existence). The very nature of the SoS problem domain hinges on the fact that perturbation is inevitable, cannot be known beforehand, and emergent patterns/properties will develop in response. Therefore, design for maintenance of stability must be built into the SoS solution design to continuously adjust and compensate for inevitable shifts. Stability of SoS must be achieved through robust designs that provide for dynamic stability over a wide range of fluctuating conditions. This occurs through system environment scanning, internal monitoring, feedforward in response to anticipated SoS deviations, and feedback to provide for continued adjustment over the mission performance of the SoS. An important distinction for emergence is that it cannot be know in advance. Unlike engaging in good risk management, where potential risk events are identified and classified, and mitigation strategies are developed, emergence is a different class, without preconception or prior probabilistic knowledge of event occurrence. Therefore, robustness in the SoS solution is essential to maintain SoS viability over a wide range of perturbations.

With respect to emergence, the actual design of an SoS can only be partially specified in advance of system operation. From the systems perspective, this explains why the most thoughtful and carefully designed SoS will have unintended consequences once the solution is deployed. In essence, SoS behavior and informal structure emerge only through system operation, regardless of the detailed design efforts conducted prior to SoS deployment. Hence, robustness in SoS solutions is essential to ensure adaptability, flexibility, and responsiveness to changing (emergent) conditions following deployment. Effective design of SoS ensures that only the minimal essential constraints, necessary to achieve the desired level of output/outcome performance, are imposed on the SoS constituent subsystems. In this sense, maximum subsystem autonomy is preserved, and therefore its maximum responsiveness at the point most capable of mounting a response to the system perturbation. In systems theory, this concept of minimal system level constraint is also known as *minimum critical specification* (Cherns, 1976, 1987). Overspecification of system-level requirements (1) is wasteful of scarce resources necessary to monitor and control system level performance, (2) reduces subsystem autonomy, which in turn restricts the agility and responsiveness of the system to compensate for environmental shifts, and (3) fails to permit subsystem elements to self-organize based on their contextual knowledge, understanding, and proximity to the operating environment . Therefore, designers of SoS should specify only the minimal SoS-level constraints necessary to achieve desired SoS-level performance.

Design Theme 2: Communications, within and external to the SoS, are essential to ensure solution viability in the face of emergence. There is generally little argument that communications are essential to performance in any complex system. However, we frequently see communications relegated to technology or high-level frameworks that provide little in the way of practical guidance. However, Beer, (1979, 1981) has provided a systems-based framework for analysis of communications in complex systems. This communication framework is based on channels. The framework is instructive for designers of SoS and can assist with higher levels of effectiveness in identifying, processing, and responding to emergence. The communication channels in an SoS are performed by "mechanisms," which are the specific vehicles to implement the channel in a complex system. The channels support design of robust SoS communications by asking two central questions: (1) are their sufficient mechanisms to fulfill the objective of each channel, based on the nature of the SoS? and (2) for existing SoS, are the individual mechanisms functioning effectively?

Drawing from Beer's (1979) communications channel description, and including extensions (Keating, 2000), the following communications are provided to guide design for effectiveness in SoS communications to deal with emergence.

- *Operations channel* provides for the direct exchange between major SoS subsystems responsible for producing the value inherent in the mission for the SoS. It is a routine transmission of information necessary to allow operations to flow smoothly. *Example*: SoS entities exchanging scheduling and resource data.

- *Coordination channel* exists to design and monitor regulatory mechanisms to ensure that sufficient SoS-level standardization (constraint) is in place to prevent unnecessary oscillation among SoS entities. *Example*: A standardized framework for verification and validation of models used within the SoS.

- *Algedonic channel* exists so that the SoS operational subsystems can have a direct link, bypassing routine communications protocols/channels, to the SoS level. The purpose of the link is to identify high-level threats that may be catastrophic to the survival of the SoS in a timely manner. In effect, other routine communications channels are bypassed when system survival is potentially in question. *Example*: Direct transmission of a breach of security at a subsystem level that could threaten the entire SoS.

- *Command channel* provides high-level direction throughout the SoS. This exists at the policy or regulatory levels, not in direction of SoS day-to-day operations, decisions, or actions. *Example*: Alert to a new external regulatory statute that must be invoked throughout the SoS.

- *Audit or operational monitoring channel* is designed to provide investigation to examine the nature of SoS disturbances or particular performance issues that impact SoS-level performance. The channel may also be used to routinely assess the health of the SoS. *Example*: An incident investigation or a routine audit of SoS financial performance.

- *Environmental scanning channel* provides for continuous monitoring of trends, patterns, and events occurring in the SoS environment that may have an immediate or future impact on the SoS. *Example*: Identification of a change in external leadership that may influence SoS funding.

- *Resource bargain – accountability channel* provides for the negotiation between the SoS and constituent subsystems concerning resource distribution. The "bargain" is developed as an explicit agreement concerning the resources provided and the expected level of performance expectation (accountability) based on those allocated resources. Thus, there is clarity in expectations for producing SoS contribution based on allocated resources. Changes in resource allocation require a renegotiation for commensurate adjustment in performance expectations. *Example*: Capital budgeting allocations for new system functionality.

- *Dialog channel* has the primary purpose of providing for examination and interpretation of SoS decisions, actions, and events. This provides for alignment of perspectives and shared understanding of SoS decisions and actions in light of the purpose and identity of the SoS. *Example*: Staff workshop to examine recent failures and SoS implications.

- *Learning channel* provides for detection and correction of SoS errors, testing of assumptions, and identification of SoS design deficiencies. This ensures that the SoS continually questions the adequacy of its design to meet mission expectations. *Example*: The examination of the continued relevance of an SoS protocol.

- *Informing channel* provides for routine transmission of information throughout the SoS. Thus, information that is not appropriate for other channels is made accessible across the entire SoS through this channel, avoiding unnecessary clutter in other channels. *Example*: Routine SoS actions that have been taken in the previous month.

- *Identity channel* provides for exploration of the essence of the SoS and the inculcation of that essence throughout the SoS. Emphasis is placed on maintaining consistency in perspective for the purpose, mission, and character that establishes the uniqueness of the SoS. Exploration of identity provides higher probability of decisions, actions, and interpretations that are consistent with the nature of the SoS. *Example*: Examination of the consistency between the SoS purpose and nature of execution.

Design Theme 3: Learning in an SoS must be designed such that higher levels of inquiry can result in continuous adjustment to emergence. Learning in SoS includes (1) identification of unexpected results, patterns, structure, or properties, (2) processing what the results mean for the SoS and constituent subsystems, and (3) mounting an appropriate response . From a systemic perspective, the work of Argyris and Schön (1978, 1996) is instructive to think about learning in SoS. They make a critical distinction in three types of learning in which systems are capable of engaging. The three basic types of learning include single-loop, double-loop, and deutero learning. These learning types are important considerations for design of SoS to effectively deal with high levels of emergence.

Single-loop learning (first-order learning) involves detecting and correcting system errors (deviations from expected results) through processes of inquiry that stay within established assumptions, constraints, boundaries, and system norms. For example, an SoS might miss a deadline for integration of a particular protocol. The response might be to simply make necessary adjustments such that the protocol will not be missed again. In contrast, *double-loop learning (second-order learning)* involves detection and correction of system error based on inquiry that calls into question operating assumptions, norms, and objectives, which may be driving the aberrant condition. For an SoS, double-loop learning challenges error in ways that call into question the ways an SoS has been designed and is operating. This is especially important where there are high levels of emergence and rapidly shifting contexts. What was once essential may no longer be appropriate for an SoS. Following the example, a double-loop learning scenario might challenge the very need for the protocol, questioning its continued legitimacy. The result of a double-loop learning episode might be abolition or redesign of the protocol requirements. Effective SoS learning must involve both types of learning at points of time and under circumstances that make them useful. A third type of learning is *deutero learning*. Deutero learning involves "learning to learn" and results in getting better at inquiry that applies the correct type of learning to the situation and finding novel responses to familiar circumstances. The important point in relationship to emergence in SoS is that each of the three levels of learning is necessary. However, designs must allow for engaging the appropriate learning level. Every occurrence of emergence in SoS represents a potential opportunity for inquiry-based learning with the potential to modify the SoS in response.

7.3.2 Operation and Maintenance Considerations for Emergence in System of Systems

Emergence is going to occur in SoS. Regardless of how much effort is placed on planning, analysis, and deployment preparation, emergence is going to occur in SoS. Emergence is not good or bad. It is simply a given for dealing with the SoS problem domain. However, there are several considerations that can ease the burden of emergence for practitioners in SoS. Three themes have been selected for operation and maintenance of SoS that are particularly relevant to more effectively dealing with emergence.

Operation and Maintenance Theme 1: SoS must continuously scan, internally and externally, to identify emergent trends, patterns, and conditions that may have an impact on performance. Scanning involves the acquisition, processing, and response to information with consequences for the SoS. In effect, scanning can provide the first indication of emergent conditions. Effectiveness in dealing with emergence will in large part be determined by how (1) quickly emergent patterns/properties are identified, (2) effective and timely analyses of the emergent conditions are conducted—at the appropriate level of depth and learning type, (3) appropriate the SoS response is, based on the results of analysis, and (4) the SoS-level follow-through is conducted . Although the time and form of emergence in SoS cannot be known in advance, the better prepared the SoS is to continuously scan, both internally and externally, to identify and deal with emergence the more effective the SoS is.

Operation and Maintenance Theme 2: Maintenance of a strong SoS identity is key to generating consistent responses to emergence. Emergence will happen at all levels of an SoS, without consideration for timing, severity, or locality. There are an incalculable number of decisions, actions, and interpretations that are necessary for an SoS to function as it maintains viability in the face of emergence. Therefore, it is necessary to establish some stabilizing force that acts as a reference point for consistency in decisions, actions, and interpretations within an SoS. Identity is the driving force that establishes the set of characteristics that is the essence of the SoS. A strong SoS identity is enduring and can provide consistency in response to emergence at all levels in an SoS. Without a strong SOS identity, the probability of consistent decisions/action on behalf of the SoS is diminished.

Operation and Maintenance Theme 3: Compatibility between SoS objectives and the technical, contextual, and conceptual/strategic support structures is essential to effectively deal with emergence. An SoS is not simply a bringing together of formerly disparate systems to function as a unity. The formation of an SoS requires that the support structures are compatible with the expectations for the SoS. Support structures include technologies consistent with expectations of the SoS, context that is not in opposition to what the SoS has been embodied to achieve, and conceptual/strategic alignment with the strategic intent and the conceptual basis for the SoS. In addition, there must be compatibility between the aims of the SoS and the resources (manpower, material, time, methods, funding) allocated to achieve those aims. Without compatible resources, it is doubtful that SoS sustainability can be achieved. However, it is important to note that without an adequate conceptual/strategic support structure, resources alone will not be sufficient to sustain the SoS through high emergence. High emergence only amplifies the need for compatibility.

7.4   CONCLUDING INSIGHTS AND CHALLENGES

Effectively dealing with emergence is critical to effectiveness in the SoS problem domain. As a concept, emergence is not difficult to grasp. In effect, we have established that emergence is simply the development of patterns, structure, or properties within a SoS that exist beyond any of the constituent subsystems, cannot be analyzed or understood as a simple manifestation from subsystem properties, and cannot be known in advance of the manifestation. However, beyond the conceptual level, dealing with emergence in SoS becomes problematic. From the preceding discussions, there are a number of insights and implications that can be taken for dealing with emergence in SoS.

First, dealing with emergence in SoS is not simply a straightforward pursuit of new techniques or methods based on traditional system development. On the contrary, the SoS problem domain was presented as a significant departure from the historically well-bounded problems that have dominated the traditional systems engineering landscape. The wider array of philosophical, methodological, and axiomatic underpinnings for emergence in SoS was presented. Understanding and effectively dealing with emergence in SoS requires appreciation of these underpinnings. They are essential to ensure development of more informed and robust responses to emergence in the SoS problem domain.

Second, design of SoS solutions requires purposeful inculcation of mechanisms to manage emergent patterns/properties as the SoS is operated. This does not assume that emergent patterns/properties are known, or can be predicted, in advance. However, the design for dealing with emergent conditions can be developed in advance. For SoS engineering, this is a necessity for generating more robust solutions. Capability to quickly identify, process, and respond to emergent conditions is to a large degree a function of effective design, which can be achieved in advance of particular emergent conditions coming to fruition. Failure to think in terms of designing for emergence in SoS is an invitation to engage solely at the reactive level. Although the specifics of the emergence in an SoS will never be routine, the robust SoS design to deal with emergence can be designed to be routine.

Third, the operation and maintenance of an SoS require that emergence be a major part of the landscape. The one consistent theme for emergence in SoS is that we must expect to be surprised. Effectiveness during SoS operation will not be so much a function of what particular emergent patterns/properties/conditions occur. On the contrary, effectiveness depends more on how the response is mounted in the light of the emergent conditions. This response must include what is necessary to address both immediate and longer term SoS needs. Long-term sustainability in the face of emergence is enhanced through the maintenance of a strong identity. This provides a reference point for action, decision, and interpretations that are consistent with the fundamental nature of the purpose of the SoS.

In closing, the following questions are provided to stimulate implications and further discussion of the material presented on emergence in SoS:

1. What is different about the nature of the SoS problem domain in contrast to more traditional problem domains?
2. How do philosophical, methodological, and axiomatic perspectives influence effectiveness in dealing with emergence in the SoS problem domain?
3. Given that emergence is going to occur in SoS, what are the implications for designing methodologies (approaches) that will enhance effectiveness in dealing with emergence?
4. What are the primary considerations that a designer of SoS should think about with respect to effectively dealing with emergence?
5. How might we assess the degree to which emergence is going to be problematic in an SoS effort?
6. What guidance for emergence might we give an engineering team getting ready to engage in an SoS problem?

REFERENCES

Ackoff, R., 1999, *Ackoff's Best: His Classic Writings on Management*, Wiley, New York, NY.

Aerts, D., Apostel, L., De Moor, B., Hellemans, S., Maex, E., Van Belle, H., Van der Veken, J., 1994, *World Views: From Fragmentation to Integration*, VUB Press, Brussels, Belgium.

Argyris, C., Schön, D., 1978, *Organizational Learning: A Theory of Action Perspective*, Addison-Wesley, New York.

Argyris, C., Schön, D., 1996, *Organizational Learning II*, Addison-Wesley, New York.

Beer, S., 1979, *The Heart of Enterprise*, Wiley, Suffolk, UK.

Beer, S., 1981, *Brain of the Firm*, Wiley, Suffolk, UK.

Beer, S., 1985, *Diagnosing the System for Organizations*, Wiley, Suffolk, UK.

Blanchard, B., Fabracky, W., 1998, *Systems Engineering and Analysis*, 3rd edition, Wiley, New York, NY.

Blanchard, B., 2004, *Systems Engineering Management*, 3rd edition, Wiley, New York, NY.

Carlock, P.G., Fenton, R.E., 2001, SoS enterprise SE for information-intensive organizations, *Systems Engineering*, 4(4): 242–261.

Checkland, P., 1981, *Systems Thinking, Systems Practice*, 1st edition, Wiley, New York, NY.

Checkland, P., 1999, *Systems Thinking, Systems Practice*, 2nd edition, Wiley, New York, NY.

Cherns, A., 1976, The principles of sociotechnical design, *Human Relations*, 29(8): 783–792.

Cherns, A., 1987, The principles of sociotechnical design revisited, *Human Relations*, 40(3): 153–161.

Clemson, B., 1984, *Cybernetics: A New Management Tool*, Abacus, Chichester, UK.

DeLaurentis, D., Dickerson, C., DiMario, M., Gartz, P., Jamshidi, M., Nahavandi, S., Sage, A., Sloane, E., Walker, D., 2007, A case for an international consotrium on system-of-systems engineering, *IEEE Systems Journal*, 1(1): 68–73.

Flood, R., Carson, E., 1993, *Dealing with Complexity*, 2nd edition, Plenum, New York, NY.

Gibson, J., 1991, *How to do Systems Analysis*, unpublished manuscript.

Grady, J., 2000, *Systems Engineering Deployment*, CRC Press, Boca Raton, FL.

Hitchins, D., 2003, *Advanced Systems Thinking, Engineering, and Management*, Artech House, Norwood, NJ.

Holland, J., 1998, *Emergence*, Plenum, New York, NY.

Jackson, M., Keys, P., 1984, Toward a system of systems methodologies, *Journal of the Operational Research Society*, 35: 473–486.

Keating, C., 2000, A systems-based methodology for structural analysis of health care operations, *Journal of Management in Medicine*, 4(3–4): 179–198.

Keating, C., 2005, Research foundations for system of systems engineering, *IEEE International Conference on Systems, Man and Cybernetics,* October 10–12, Waikoloa, Hawaii, pp. 2720–2725.

Keating, C., Rogers, R., Dryer, D., Sousa-Poza, A., Safford, R., Peterson, W., Rabadi, G., 2003a, System of systems engineering, *Engineering Management Journal*, 15(3): 36–45.

Keating, C., Sousa-Poza, A., Mun, J., 2003b, Toward a methodology for system of systems engineering, *Proceedings of the American Society for Engineering Management,* October 21–24, Norfolk, VA, pp. 1–8.

Kossiakof, A., Sweet, W., 2002, *Systems Engineering and Practice*, Wiley, New York, NY.

Kovacic, S., Sousa-Poza, A., Keating, C., 2007, Type III: the theory of the observer, *IEEE SMC International Conference on System of Systems Engineering Proceedings,* April 16–18, Los Angles, CA, pp. 107–115.

Maeir, M., 1998, Architecting principles for systems-of-systems, *Systems Engineering*, 1(4): 267–284.

Martin, J., 1996, *Systems Engineering Guidebook*, CRC Press, Boca Raton, FL.

Ring, J., Madni, A., 2005, Key challenges and opportunities in 'system of systems' engineering, *IEEE International Conference on Systems, Man and Cybernetics,* October 10–12, Waikoloa, Hawaii, pp. 973–978.

Rittel, H., Webber, M., 1973, Dilemmas in a general theory of planning, *Policy Sciences*, 4: 155–169.

Sage, A., 2000, *Systems Engineering*, Wiley, New York, NY.

Sage, A., Cuppan, C., 2001, On the systems engineering and management of systems of systems and federations of systems, *Information, Knowledge, Systems Management*, 2(4): 325–345.

Senge, P., 1990, *The Fifth Discipline*, Doubleday, New York, NY.

Skyttner, L., 1996, *Introduction to General Systems Theory*, Plenum, New York, NY.

Weinberg, G., 1975, *An Introduction to General Systems Thinking*, Dorset House, New York, NY.

Westerman, H., 2001, *Systems Engineering Principles and Practices*, Artech House, New York, NY.

Chapter **8**

System of Systems Management

BRIAN SAUSER, JOHN BOARDMAN, and ALEX GOROD

Stevens Institute of Technology, Hoboken, NJ, USA

8.1 INTRODUCTION

Currently, the attempt to understand the life cycle of modern systems has been through the engineering of systems or systems engineering (SE), and the process by which the organization, application, and delivery of systems can be managed has been called systems engineering management (Sage and Rouse, 2008; Shenhar and Sauser, 2008). A key success driver in these two fields is the process by which the integration of people, processes, problem-solving mechanisms, and information come together (Kusiak and Larson, 1999). Historically, principles of systems engineering management and systems engineering have largely been utilized and developed in the government with projects such as manned space flight, nuclear-powered submarines, communications satellites, launch vehicles, aircraft, and deep-space probes. It is the projects such as these that carry distinctions such as high complexity of the system with high technological risk, extreme design constraints, desire for complete answers, and auditability (Parth, 1998). Although the complexity of these systems, both in knowledge and connectivity, may not have been foreseeable by some of the forefathers of systems theory, like von Bertalanffy (1968) and Beer (1966), the most complex projects that have been or will be developed are of systems to solve complex engineering problems. Thus, many have argued for decades that there are fundamental differences in how we manage complex systems from simple systems (Davies and Brady, 1998; Nightingale, 1998; Shenhar, 1998; Hobday et al., 2000; Miller and Lessard, 2000; Floricel and Miller, 2001).

Opportunistically, the knowledge for the engineering of these systems is advancing rapidly but in recent years has failed to keep pace with the increasing complexity and integration of the systems themselves. While we may be able to measure and possibly predict the advancement of technology-intensive systems, we are unable to effectively measure or predict the pace at which we correlate or advance our management practices of these complex systems. We have moved to an era in which some of these complex systems are being further defined as system of system (SoS). Considering that the recognition of complex systems can be traced back to post-World War II and the complex system entitled a "system of systems" to the Strategic Defense Initiative in he early 1980s (GPO, 1989), our establishment of fundamentals is far surpassed by the realization of how to manage systems of greater complexity. Only in the last 10 years have time-honored organizations such as the Project Management Institute (PMI) (2004), the International Council on Systems Engineering (INCOSE) (2006), and the International Organization for Standardization (ISO) (ISO/IEC, 2002) begun to provide fundamentally accepted guidance on how we perform project management or systems engineering. While these fundamental bodies of knowledge have made great advancements, Shenhar and Dvir (2007) have shown through extensive empirical research that there are fundamental distinction that exist in these bodies of knowledge and how we manage systems of increasing complexity.

There is limited theory on how SoS develop and are managed. This has become even more important as there has been increasing attention in the economic activities of firms, industries, and nations to these systems. SoS differ notably from smaller, mass-produced projects and thus require different management techniques (Keating et al., 2003). Therefore, the study of SoS has moved many to support their understanding of these systems through the groundbreaking science of networks and complexity. Our understanding of networks and how to manage them may give us the fingerprint that is independent of the specific systems that exemplify this complexity. It does not matter whether we are studying the synchronized flashing of fireflies, the space stations, the structure of the human brain, the Internet, the flocking of birds, a future combat system, or the behavior of red harvester ants, the same emergent principles apply: large is really small, weak is really strong, significance is really obscure, little means a lot, simple is really complex; and complexity hides simplicity. The conceptual foundation of complexity is paradox that leads us to a paradigm shift in the SE body of knowledge.

Paradox exists for a reason, and there are reasons for systems engineers to appreciate paradox even though they may be unable to resolve them as they would a problem specification into a system solution. Till now paradoxes have confronted current logic only to yield at a later date to more refined thinking. The existence of paradox is always the inspirational source for seeking new wisdom, attempting new thought patterns, and ultimately building systems for the "flat world" (Friedman, 2006). It is our ability to govern, not control, these paradoxes that will bring new knowledge to our understanding on how to manage the emerging complex systems of SoS. For SoS, we contend that management is replaced by governance to cope with paradox.

We will present in this chapter a paradigm shift in how we think of managing systems, in this case SoS, and begin to understand the paradoxical relationship that exist in realizing them. We will establish some key concepts and challenges that make the management of SoS different from our fundamental practices, present an intellectual characterization model for how we classify and manage an SoS, appraise this model with an SoS case, and conclude with grand challenges for how we may move our understanding of SoS management beyond the foundation. First, in our pursuit to understand how we can manage this emerging era of SoS, we believe it is important to reflect back on our understanding of system theory and systems thinking to help us to distinguish what makes an SoS different and just like our understanding of systems, it must transcend domains (Boardman and Sauser, 2008).

8.2 BUILDING A FOUNDATION: AN SoS PHILOSOPHY

There is a saying: "When the going gets tough, the tough get going." For traditional systems engineering managers, this means that the classical response to a complex challenge is to knuckle down, be determined, muster your strength, and dig in until the challenge is overcome. This philosophy characterizes so much of engineering culture in the face of increasing complexity in the challenges that this community confronts. For SoS this becomes an entirely incorrect response, and more than that, it is wholly counterproductive. The tougher you become in the face of tough going, the tougher the going becomes, with only defeat and catastrophe awaiting. SoS represent a uniquely complex challenge, in terms of conception, design, development, deployment, and management. The correct response to this challenge is counterintuitive insofar as contemporary engineering culture is concerned. It is not to rise to the challenge in terms of increased complexity—of logic, strength, effort, or the usual dimensions that call for reinforcement. The correct response is essentially asymmetric: to be deliberately simple; to think counterintuitively; to behave paradoxically; to change the way we think, what people call paradigm shift; and to invert many of the traditional actions with which we have been programmed. In one vernacular, it is "to work smarter not harder." It is a matter not of reinforcement but rather reprogramming.

With increasingly inextricable links between business models and technology development, there are three forces at work: First, there is the lack of synchronicity between systems development—and their subsequent management—and knowledge development, to support that systems development and management. Second, confusion abounds as to what exactly constitutes an SoS, with intense polarization between two camps: those who attest they are "merely" systems (and so whatever is at our disposal for systems development works equally well, more or less, for these new fangled systems), and those who assert there is a fundamental difference (but who patently fail to articulate what this is, much to the amusement of the opposition and their personal chagrin). Third, there exists a huge gap between engineers who design and develop complex systems (systems engineers or complex systems engineers?) and scientists who formally study complexity and articulate complexity theory (e.g., "the groundbreaking science of networks). This gap, essentially a separation of knowledge and practice, creates a force that makes any perceptible shift in understanding of what complexity really means and how its peculiar challenge can be overcome distinctly unachievable. As we see these three forces at work, and interacting in complex ways, we are bound to produce a philosophical stance that asks and begins to explain why do we have this asynchronicity and how might it be overcome; what is the essence of a system and how can this be leveraged to explain the fundamental distinction between a system that is truly an SoS and one that is "merely" a system (of parts); and, finally, why is it imperative to shift our way of thinking and what are the pivots for accomplishing this paradigm shift? Is the answer in paradox?

8.3 PARADOX IN SoS MANAGEMENT

What is the meaning of simultaneously applicable opposites? What has such a question got to do with SoS management, especially its application to SoS and to enterprises that either execute such development programs or are themselves enabled by the SoS? We contend that this is realized through paradox.

Paradox signifies "contradictory, yet interrelated elements" (Lewis, 2000). For example, on a single sheet of paper appear two statements, one written on each side. On the first side we find, "The statement on the other side of this paper is true," and on the other side we have, "The statement on the other side of this paper is false." There is more endless circularity. Does this lead us anywhere? Cameron and Quinn (1988) proclaimed that the exploration of paradox can allow us to move outside oversimplified and polarized thinking to recognize the complexity, diversity, and ambiguity of life.

Here is what we have to say about paradox:

> A paradox is an apparent contradiction; however, things are not always as they seem. A paradox can be explained, but only by seeking wisdom from above; for the systems person, this means looking upwards and outwards, not just down and in. Paradoxical thinking is systems thinking at its best.

When we use the word apparent, we give ourselves the opportunity to introduce viewpoint or perspective that is personal or subjective to some individual (person or group), and this notion, sometimes labeled stakeholder, is germane to SoS SE. Further, we hold out the hope of resolution as opposed to the sense of despair or confusion that can befall those who get trapped in the endless circularity of evident paradoxes. Handy (1994) stated that a paradox framework shifts the model of "managing" from modern definitions based on planning and control to coping. Many apparent paradoxes that once confused the greatest minds of the day were later explained by new concepts and new ways of thinking.

The essence of paradox is tension—two statements claiming to be true and at the same time contradicting each other (Schneider, 1990). The ultimate release of that tension, not found in the resolution of the conflict within the paradox itself but rather in the recognition of the virtue of the paradox as a whole, always leads to new ways of thinking. For this reason, as much as paradox is unpalatable, especially to action-oriented people such as engineers, technologists, and business executives, it can be valued as a lever to change mindsets, to shift thinking, and a potential wellspring of new ideas leading to more effective action.

We assert that resolving this tension prematurely, for the sake of taking action, always leads to ineffective action and, more expensively, to miss an opportunity to change your way of thinking and gain breakthrough knowledge. Likewise, to ignore this tension, to pretend as it were that the paradox is unreal or irrelevant is to preserve the status quo, maintain the same old grid lines of thought and inevitably head for disaster.

Paradox is a reality of our lives. Tension is too. We have chosen to feature paradox as a significant element of SoS management because we see paradox less as a source of confusion, which at face value it certainly is, and more as a portal into new ways of thinking, new modes of working, and better ways of living. A source of tension in the development an SoS is the demand to increase efficiency, cultivate inventiveness, build individualistic teams, and think globally while acting locally—a paradox (Lewis, 2000).

We want to explore what we call "the world of both." In a paradox, there are two opposites that compete for our attention, and each demands that we make a choice. However, the very existence of the paradox itself demands that we do not choose but accept both and therefore the nature of the relationship between the parts, manifested as a conflict, contradiction, or something other that runs counter to common sense or conventional wisdom. Systemically, the parts of the paradox demand choice but the whole of the paradox requires acceptance of both, as illogical as that seems. It is as if parts and whole cannot agree, and yet parts make the whole and parts they be.

In our experience, neither engineers nor managers are comfortable with the notion of both, of holding onto a tension that must be resolved. As system engineers, we are trained to put boundaries around the system of interest. We like choice, perhaps not an overabundance of choice, and we are required to choose. It is our nature of being (Ford and Ford, 1994).

We believe that systems engineers and systems managers need to be, or learn to be, comfortable with both. Scientists have had to learn to accept subatomic matter to be both particle and wave, while technologists in pursuit of the quantum computing dream postulate the qubit, a binary digit that is both one and zero, and consequently are able to leverage computing power exponentially. Both, if not as yet "in," are on the scene, and we are offering our own thoughts and examples to make it easier for decision makers to accept simultaneous opposites (i.e., conflicting perspectives) leveraging these into richer realms of decision making.

8.3.1 The Boundary Paradox

The notion of boundary is inseparable from that of system. As problematic as it might be to locate or articulate the boundary of a system that it exists relative to the system itself is incontestable. The boundary may be defined by geography or other dimensions such as culture, organizational structure, or IT infrastructure, but howsoever it is defined it fundamentally speaks of separation, distinction, limitation, and approximation. The boundary shows who and what is in and out (Bernard, 1952).

If a system is a collection of parts and their interrelationships assembled together to form a whole for a given purpose, then the parts and their relationships are in (the system). What lies beyond the boundary is not part of the system. It may be coveted by the system and in due time be acquired and integrated into the new whole but while it lies outside the boundary, it cannot be considered part of the system and may even be considered hostile to the system.

The parts of the system can be controlled so as to serve the system, or in self-organizing style relied upon to control themselves autonomously and bring even greater well-being to the system. The externalities cannot be controlled; they may even need to be combated, if regarded as foe rather than friend. Possibly they can be influenced, thereby rendering docile an otherwise adversarial influence on the system itself.

System managers must be perpetually mindful of the boundary. The system boundary separates what belongs to the system from what does not belong. As much as the human body prefers not to be invaded by unwanted bacteria, immunization (letting a specified amount of negative bacteria in) enables the positives to get better at dealing with the negatives so when large numbers of these try to invade, the body's defenses have significantly improved. If this is a beneficial tension of a boundary, how many should belong, what boundaries do we open, how large are the openings, what does that mean to a defined system boundary?

Our boundary paradox can be stated as follows: "You have to have a boundary (in order to nurture and develop specialization in functional expertise, for example). But you must also not have a boundary (in order to allow that specialization to be rendered as a service, otherwise why has it, and to allow that expertise to be resourced via interactions with others). So the boundary must exist and must not exist—it must do *both* at the same time. The boundary must keep things out and keep things in, but it must also let things out and let things in. The boundary paradox is a realm systems engineering managers have yet to enter but to which they must journey.

8.3.2   The Control Paradox

How many people and firms do you think are involved in the end-to-end process of conceiving, making, and selling a Grand Cherokee Jeep? From initial product concept through to satisfied customers driving their new purchase of a dealer's lot. It is perhaps not a question that interests many but the answer that usually startles most.

When Thomas Stallkamp, former Vice President of Chrysler, asked this question of his line managers, it took them a little while to find the numbers. In the end, they came back with the answer: 100,000 firms and 2 million people. Stallkamp followed this question with, "Who is managing this enterprise?" The real answer is a paradox—no one is *and* lots of people are. Yet another time to make your mind up? Or a time to recognize and respect the paradox, waiting until the appropriate moment to release the tension and to achieve breakthrough thinking.

No one sits atop the Chrysler Grand Cherokee Jeep "experience." Perhaps notionally people do but in no way can they be said to be its manager. Littered throughout the management hierarchy, or network if you prefer, are hundreds of personnel, each with individual spans of care. But in what ways can this diverse collective be said to be in control of the whole experience when it is probably the case that they are largely unknown to one another? Do these managers perform like ants and somehow support excellent behavior for the Cherokee colony? And if so, understanding that the ant has no commander directing colonial affairs, are we to understand control to be just as effective, if not more so, if it is distributed rather than precisely located in a central commander? And can we really trust distributing control to a constituency that is largely unaware of the affairs and actions of its neighbors?

So when we ask "can we really trust distributing control to . . .," maybe we are asking the wrong question. Just like when Stallkamp asked who (in particular rather than plurality) is in control of this vast extended enterprise (or SoS), maybe he was asking the wrong question. We ask the wrong questions when we are in the wrong mindset. And the purpose of paradox is to confront that mindset, to force us to ask wrong questions, and to be prepared to change our mindsets, thereby releasing the tension in the paradox and moving to breakthrough thinking. So what are the right questions? Well, stepping back, What is the right mindset? Maybe, control is or at least starts with self—self-control. After all, you have to exercise self-control in response to an order, be it in the military, civil, or family domains. What is more the one issuing the order expects this, relying on this self-control and in some way is developing this in the one to whom the imperative is directed. Command assumes self-control. The question now is, "Is there an extension to command that is of a controlling nature (i.e., the communication of command carries with it or in it a controlling influence).

At this point, we are beginning to recognize the polarities. One is the command and control version by which authority located "at the top" issues directives that get resolved into executive action by a large group of people. The other is the self-organizing notion of an idea that (from the bottom) infects, propagates, and galvanizes a large group of people who then take action, as though they were a unit and had been commanded by a governing authority.

Put in other ways

- Authority must exist at the top representing order, but it must also exist at the bottom representing autonomy;
- Command must exist and orders from an external source be obeyed but so also must the power to be insubordinate operating alongside a self-will that knows its own order and orders;
- Finally, not only control must operate within a framework (a one) that grants liberty to its constituents (the many) but also control must be manifest in the self (a one) in terms of self-control and self-discipline to make a framework (for the many) work.

What are the ways in which this paradox might be resolved and its tension released? We propose that find a lesson from history in the command of Admiral Horatio Lord Nelson to be of guidance in these three concepts (Kimmel, 1998):

- *Creative disobedience* flowed naturally from his philosophy of independence of command. If his subordinates should have the freedom to deal with situations as they came up, he should be able to take the initiative as a subordinate in a battle, even if it meant ignoring orders.
- *Reciprocal loyalty* is the idea that one must give loyalty down the command hierarchy in order to gain true loyalty (as opposed to obedience through fear).
- *Servant leadership* that has been so supremely exemplified in the inspirer of the many faiths whose teachings can be incomparably paradoxical.

8.3.3 The Team Paradox

"Together each achieves more." This phrase epitomizes togetherness. It gives a sense of fulfillment that is somehow eluded by the mindset of going it alone. It seems to make selfishness redundant and self-achievement more rewarding because self is being helped by others and self is helping others simultaneously. It also conjures the notion of being coached or mentored or somehow developed as a consequence of which life is more rewarding, learning is gained, and transferable skills acquired. It is almost so engrained nowadays that you cannot be in a team without realizing that while more is the goal of each and everyone, it comes at the expense of being together.

A team simply has to be a system. It may be a poorly performing team and therefore a failing system, but a system nonetheless. It is worth our while spending some time looking at what a team is and what it means, as an example of a system, to discover yet another interesting and rather fundamental paradox that can so easily go unnoticed by system designers and operators, as a consequence of which we have more bad systems than we need.

In a team, we find both sameness and differentiation. Sameness is exhibited in uniformity. Everybody on the team and associated with the team identifies with the unifying themes and artifacts. They become recognizable and identifiable. They help to define the personality of the team distinguishing it from other teams. It is an emergent oneness that covers the many identically. Sameness is also exhibited in the common aim of success, of achieving more. A team has no room for mavericks, loners, rebels, dissidents, and the like, regardless of individual expertise, no matter how exceptional. A member not committed to togetherness cannot be on the team. Indeed, different team members have differing views on what constituted team success. Ironically, this apparent sameness can be highly distinctive, though improbably divergent. But this distinctiveness is unimportant because at a deep-rooted level the sameness, the single commitment to team success and achieving more together, is overpoweringly unifying.

From time to time, the sharp distinctions in subjective interpretations of together-ness can be a powerful disintegrating force, that is, the price of differentiation operating simultaneously with sameness. The tension between the two is, in this instance, disruptive in the most unhealthy fashion. However, patterns exist to provide early warning signs, detect the potential demise, and make timely interventions. But differentiation is essential for many reasons. Sameness is needful but not singularly so. A team is a blend of many skills. In an engineering team, you need people from different disciplines: electrical, electronic, mechanical, and software. You also need people with different project experiences—in leadership, in work package management, in test, and in manufacturing. Teams need different skills, knowledge bases, and experiences. The team becomes a pool for blending these differences together, so that each achieves more.

A team simply has to be a system because it has to have requisite variety, that is differentiation, parsimony—a meanness that culminates in the single-mindedness of each member to put the team first, and harmony—that makes togetherness feasible. What is true for a team of individuals is also true for a collection of systems gathered together to form a new system (or SoS)—there is differentiation of both functionality and sameness—performing the functions of achieving SoS purpose, of fulfilling the SoS's mission.

So what of paradox? There is an indisputable duality about the individual (or system). One aspect is that of distinctive individuality (or autonomy), and the other is that of membership, of belonging, of being a part. The thing belonged to the team, benefits from that individuality, but only when it is brought into play as part of the team via membership, so that individual has to maintain individuality and at the same time sacrifice it via membership, indicating the primacy of the team over the individual. But if that sacrifice is wasted, the team primacy is immediately suspected. Howsoever the belonging is expressed, it is very real and not without expense to that individual.

Let us also realize that there is not one single individual but many. This duality is replicated many times with these instances being highly varied. It is through this variability that the system has its being, not merely in the existence of members, nor in the distinctive roles that they play—helping the team toward dynamism—but in the diversity of expressions of this very duality of the maintenance and rendition of many distinctive individualities. Diversity is more than difference. Diversity is the measure of the paradox of the one and the many.

Here is how we express the paradox of diversity. First, we recall that a system is a collection of entities and their interrelationship assembled in such a way that the whole is greater than the sum of the parts. This notion of "greater than" has been summed up as more is different (Anderson, 1972). The parts belong so as to serve the purpose of the whole. Yet this purpose is not well served if the parts belong for that reason alone. The homogeneity of partness is good, it ensures that each and every individual is signed up as a part. But homogeneity is not something we want of the system itself. In order for it to survive and prosper, heterogeneity is required. How does the system inherit this heterogeneous quality? Our understanding is that this occurs when each and every part expresses simultaneously its individuality and its partness in manifold and diverse ways. This diversity is what gives the system its heterogeneity. Thus a tension is set up for each of the many between autonomy, maintaining individuality, and belonging, rendering this distinctiveness to serve the many. This aggregated tension creates the whole. The system is continually in tension, produced by this paradox of diversity.

8.4 BUILDING A CONTEXT: SoS CHARACTERISTICS

We are striving to understand the paradox that resonates in the realization of an SoS management. Thus, in this section, we propose a context based on a characterization of an SoS so we may identify its paradoxical challenges. In our first cognition of what an SoS was by definition, we were enlightened by others who had previously defined SoS, such as Bar-Yam (2004), Carlock and Fenton (2001), DeLaurentis (2005), Maier (1998), Sage and Cuppan (2001) to name a few of the over 40 definitions we discovered (Boardman et al., 2006). While many have pursued a definition of SoS, like Sage and Cuppan and Bar-Yam, we have pursued a characterization of SoS. In an earlier paper, we presented these characteristics that enable the differentiation between an SoS and a system and then used qualitative techniques in pattern formation of text to validate cross-references from our literature search where we believe others were articulating our chosen discriminating characteristics (Boardman and Sauser, 2006). We will describe these characteristics in the realization of managing an SoS in the following sections. Figure 8.1 is a summary of the five characteristics, their definition, their motivation, and the origin of their understanding.

Characteristic	Definition	Motivation	Provenance
Autonomy	The ability to make independent choices; the right to pursue reasons for being and fulfilling purposes through behaviors.	Legacy systems are indispensable to an SoS; the SoS has a higher purpose than any of its constituent systems, independently or additively.	Managerial and operational independence
Belonging	Happiness found in a secure relationship.	Legacy systems may need to undergo (radical) change in order to serve in an SoS.	Shared mission
Connectivity	The ability of a system to link with other systems.	Legacy systems targeted for an envisioned SoS are very likely highly heterogeneous and unlikely to conform to a priori connectivity protocols; the SoS places a huge reliance on effective connectivity in dynamic theaters of operations.	Interdependence, distributed, networked, multiple solutions, interoperability
Diversity	Noticeable heterogeneity, having distinct or unlike elements or qualities in a group; the variation of social and cultural identities among people existing together in an operational setting.	Legacy systems were most unlikely to have been purposed to work together prior to targeting the envisioning of the SoS; the SoS can only achieve its higher purpose(s) by leveraging the diversity of its constituent systems.	Independence, diversity, heterogeneous
Emergence	The appearance of new properties in the course of development or evolution.	A boundary is indispensable to a system; all systems are emergent; emergence requires a well-defined boundary; an SoS has dynamic boundaries but always clearly defined; ergo, an SoS should be capable of developing an emergence culture with enhanced agility and adaptability.	Evolving, intelligence, synergy, dynamic, adaptive

FIGURE 8.1 Foundations and descriptions of the SoS characteristics

8.4.1 Autonomy

The reality of legacy systems relative to an envisioned SoS is inescapable just as an individual freedom of choice is incontestable. For human beings, autonomy is defined as a person's ability to make independent choices. What of a system? Each legacy system that is envisaged to become a constituent system in the SoS must be accorded autonomy, the right to pursue reasons for being, and to fulfill purposes through behaviors. Respect for this autonomy is paramount, and it is a respect that the SoS itself must pay. That is, not to argue that the legacy systems cannot be migrated or morphed to more aptly serve the SoS, but it is to argue that such transformation must be out of respect for that constituent system's autonomy. To do otherwise is to imperil that constituent system's functionality and essence of being that might then be lost to the SoS, a foolish thing since it is these features that are wanted for inclusion. We argue that the capabilities of the SoS are enhanced by the exercising of constituent systems' autonomy and that the opposite is true of a system that is not an SoS, whereby its parts must cede whatever autonomy they might have had in a totally subservient act of granting autonomy to the system.

Smuts (1926) introduced the term "holon," which later was explained in more detail by Koestler (1990) as being both whole and part. This term aptly fits constituent systems relative to an SoS. However, it is proposed that an SoS cannot be so called on the basis of structure alone, including hierarchies and holarchies. It must also qualify on the basis of dynamics, for which the remaining distinguishing characteristics provide further explanation.

8.4.2 Belonging

Just as legacy systems are a reality so also is the problematique, which goes unsolved by these systems, singly and additively. By the same token, the envisioned SoS is a reality, if only in concept. Someone or some persons see that the SoS, by making use of the constituent systems via a new framework, will in a real sense deal with the problematique. So there are two new realities: the problematique and the envisioned SoS. This makes the second differentiating characteristic, belonging, a key one. The SoS cannot translate from conceptual reality into physical reality without the constituent systems belonging. But why should they? What is in it for them? Who can make them belong? What will become of them once they do belong, given that they will not lose autonomy? How will they belong? Will they continue to belong, come what may, or will their belonging be strictly conditionally? Can they exit without hurting the SoS and/or themselves?

The parts of a system (that is not an SoS) have no choice in the matter of belonging since they have no reason for existence and no dynamics to contribute without belonging. Parts in such a system are integral, and the system cannot function without them. In an SoS the parts, also wholes and therefore holons, are integrable, that is, capable of being integrated. It is proposed that for an SoS there must be negotiation between it and each constituent system about the latter's belonging and the former's acceptance. There will be manifestations of the problematique when it is better for a constituent system to unbelong or for it to be believed that they do not belong when they actually do. We must continually bear in mind that the existence of the SoS is to confront a perpetual problematique for which no single point solution, no single system, is adequate. It is not about the system as such but about the SoS capabilities for resolving or addressing the problematique. Hence "belonging" becomes a core competence or stratagem available to the SoS for dealing with the problematique.

8.4.3 Connectivity

For the U.S. military, interoperability translates into net-centricity (Stenbit et al., 2003). They want the same powers of connectivity between their warfighters, commanders, and others who "need to know" that global commerce has acquired, via the Internet and the World Wide Web, instruments that have transformed business models. No surprises there, except there is an irony considering that the Department of Defense's (DoD) chief concern is with an enemy that is organized as a network, testimony to the maxim "fight fire with fire?" Later we will get into the practical application of the "connectivity" distinguishing characteristic, but for now we want to explain its central importance.

Most designed systems require the relationships between elements to be designed simultaneously with the design of the elements themselves. Hence, connectivity between components is considered alongside the design of these components, regardless of the topology of the connections be this integrated, distributed, hub and spoke, or whatever. This design pattern normally leads to hierarchies (or holarchies) and a valued stability in development, whereby parts or subsystems are themselves stable enabling a gradual build up of the designed whole, which of course must also be stable. However, many such wholes or systems (that are not of the SoS kind) have designed connectivity to their environment, and this is fixed; it cannot emerge. The problematique that confronts an SoS will ensure that such limited, presciently designed connectivity leads to inevitable system failure.

Therefore, we argue that a distinguishing feature of an SoS is that the internal connectivity of the SoS is not presciently designed but emerges as a property of present interactions among holons. Net-centricity is a form of prescient design, enabling full connectivity by supporting interactions and connections between all the elements, according to defined protocols. Further, it supports extension as more holons are added to the SoS, provided that these holons conform to the protocols. In our scheme for an SoS, this connectivity is itself adapted as holons enter and exit the SoS. And this takes place in a way that enhances the connectivity or interactivity of the SoS with its environment, that is, dealing with the problematique. In the context of this discussion, connectivity has to do with a lot more than just topologies and protocols and interoperability standards, although it does address these practical matters and is more concerned with the agility of structures for essential connectivity in the face of a dynamic problematique that defies prescience.

8.4.4 Diversity

Imagine soldiers who are not soldiers but who wage war that is not war. Citizens who are loyal to no nation state to which they notionally belong, but who really belong to the vision of an integrated, faith-based, global-wide superpower governed by a single ruler headquartered in the Middle East. Imagine warriors who are not trained in their country of origin but in foreign lands including that of their enemy and trained by that enemy in skills needed for battle. Fighters who have neither armor nor weapons to speak of save the legacy systems of their enemy, namely, Internet, cell phone technology, Boeing aircraft, air transport infrastructure, up to a point, and box cutters. Can you imagine that? If we had, could 9/11 have been averted? Our problem in perceiving these threats to an extent lies in our inability to cope with diversity. Ashby (1956) posited a law of requisite variety asserting that for a system to be sustained, it must have at least the same number of degrees of freedom as the environment in which it operates. To paraphrase, interior diversity must match exterior diversity, or the boundary that separates them is futile. Post 9/11 efforts have largely concentrated on the boundary—understanding it, strengthening it, and in one sense extending it, for example, by military occupation of some nation states. Greater attention is now being given to increasing interior diversity and reducing exterior diversity a role, we argue, that falls to SoS thinking and acting.

Engineers have a problem with diversity, summarized in the maxim "keep it simple, stupid (KISS)." In an age when complex systems give rise to simple patterns and simple systems produce complex behavior (Waldrop, 1992), perhaps it is time for diversity to be seen less as a problem and more as an opportunity. There is still ample scope to apply KISS and this will undoubtedly continue in traditional systems engineering. Given that legacy systems ab initio present a given and possibly great diversity, what should the SoS designer do? The purpose of the interoperability framework is to get the legacy systems, holons, to work together, and to do so not additively as in the current underachieving case but synergistically. Does this mean reducing diversity, and if so, how can the SoS match the huge diversity in the problematique it faces? The opportunity for the SoS is to increase connectivity, which probably translates into standard protocols and specific architectures or topologies, an imperative for uniformity, and increase diversity. This respects the autonomy of the holons, allowing them to maximize their contributions to the SoS but within the context of the SoS.

Increasing diversity is not a license for anarchic design, but it is a spur to realizing resilient capability. Situational awareness is enhanced by multiple perspectives. But in the end a common operating picture that informs command decision is just that: a final conclusion. But no one wants to make decisions based on a conclusion that is not richly informed, that is, lacking a vital piece of data, information, knowledge, or wisdom. Diversity, through a variety of viewpoints, processes, technologies, and functionalities, ensures richness, and the SoS must be able to leverage this, in an unencumbered fashion.

8.4.5 Emergence

The terms emergent and system are inseparable. By definition, when parts and their relationships are assembled together what emerges is the system. All systems are emergent. Simon (1996), a Nobel Prize winner, said this another way when he argued that complex systems will evolve from simple systems much more rapidly if there are stable intermediate forms than if there are not; the resulting complex systems in the former case will be hierarchic.

The properties, behaviours, and purposes attributed to systems can also be said to be emergent. Some of these, for designed systems including the engineered variety, are intended. For example, it is intended that an automobile serves the purpose of transporting goods and people across reasonable distances and terrains safely, comfortably, and in timely fashion. This is an emergent or resulting property of that system. The same emergent property cannot be attributed to any of the parts therein, although every one of these will have its own emergence. Each one is engineered to a specific purpose to deliver an emergent property, for example, the power train to provide propulsion, the wheels to provide traction, and the steering to provide guidance control. With this example in mind, one can move up and down the scale of systems enumerating specific emergent properties for each part, subsystem, and system.

Some emergent properties are unintended, and of these, some are undesirable and others serendipitous. Relative to the auto, perhaps the chief undesirable and unintended behavior is atmospheric pollution most acutely experienced in city traffic. At that level, traffic jams are another example of unintended emergence: not a single vehicle is responsible for a traffic jam, and it takes a bunch of interacting autos to make one of these. Yet a desirable emergent property at that level is a personal mass transit system, highly convenient if not altogether rapid, one that obviates the need for investment in alternatives such as subways (for cities) and rail networks (for intercity travel).

The question arises, if all systems are emergent, is there anything different or special about an SoS? An SoS must match the agility of the problematique, which calls for greater emphasis on strategic capability than on rigid tactical measures. The exact nature of the SoS is often determined in real time and indeed at higher clock speed than that of the environment (or the threat within that environment). The simplest way this can be further explained is to draw a comparison between a system and an SoS.

A system provides a response to a set of predetermined "requests," that is, threats or opportunities arising from the environment in which it operates. By contrast, an SoS is an anticipatory responder having an a priori undetermined and unknowable range of responses subordinated to auxiliary mechanisms for anticipation, including disturbing the ability of the environment to pose threats or limit opportunity. In the next section, we will use a case example of a proclaimed SoS to show how these characteristics may define and realize an SoS.

For an SoS, we contend that the realization of these characteristics or the life cycle of a SoS is an evolutionary, self-synchronization by which the SoS will define itself when it has reached a state of optimum performance. This differentiates SoS from other systems because it is not the manager of the system that governs the life cycle. In an evolutionary state, the rates at which change will occur, the order of change, the legacy information upon which changes are imposed, and the environmental challenges during the periods of the change are all unknown. The error in the design of an SoS is that change may be seeking a specific goal, and basic to the assumption of evolution is that no goal exists. The lack of a goal limits the analysis and ability to calculate a likelihood or failure, and thus most of the assumptions in an SoS are not verifiable.

8.4.6   Characteristics as Paradoxes

Glass' chaos theory (1996) explains that most organizations believe that there is a state of equilibrium that management tries to balance around. Fundamental to Glass' theories and chaos theory is that the environment is not inherently stable, and there is a fine balance that produces order from disorder. This becomes a balance on the edge of chaos while letting a certain level of disorder bring order and direction. Brown and Eisenhardt (1998) refer to chaos theory in management as *structured chaos* (figuring out what to structure and not to structure, on both an organizational level and a managerial level).

Likewise, for each of the five characteristics presented in Section 8.4, there are opposing forces or paradoxes that are influenced by fluxes in realizing or recognizing a system (Sauser and Boardman, 2007a; Sauser and Boardman, 2007b). While this balance may be considered reversible, the reversible is conditions under which the forces are so nearly balanced that an infinitesimal change in one or the other would reverse the realization of the system. In any system, we seek ideal conditions that the realization of the system is carried out reversibly. Under these conditions, the realization of the system yields the maximum possible performance, although reversibility does not hold true in practice. The flow of these forces and their relationship work in distinguishing types of systems and determine the togetherness of a system that fortifies its realization.

We want to establish a fundamental construct that provides a scaffold for paradoxical reasoning that further distinguishes systems from SoS. Our reason for doing so is twofold. First, it is important to develop frameworks that provide a holistic view of the subject matter since we know that in any whole the parts are interconnected, and thus the effects produced by however small a change can be highly significant and lasting. Interconnectedness, interdependency, and for that matter interoperability demand a holistic approach. Second, the principles that have served as well before complexity theory are being assailed by uncertainty, indeterminacy, and paradox. This is not to say that they are rendered useless but that they have to be reviewed with these dynamics in mind.

By comparison, we believe the characteristics presented in Section 8.4 have a paradoxical nature that governs their expression and potential architecture. Figure 8.2 shows the paradoxical tensions that exist for each of the characteristics and what it means to describe types of systems. That is, suppressing or avoiding one side of a polarity intensifies pressure from the other (Hofstadter, 1979).

Likewise, the expression of any one of these characteristics is interrelated to the expression of another that exemplifies the tensions. For example, the level of

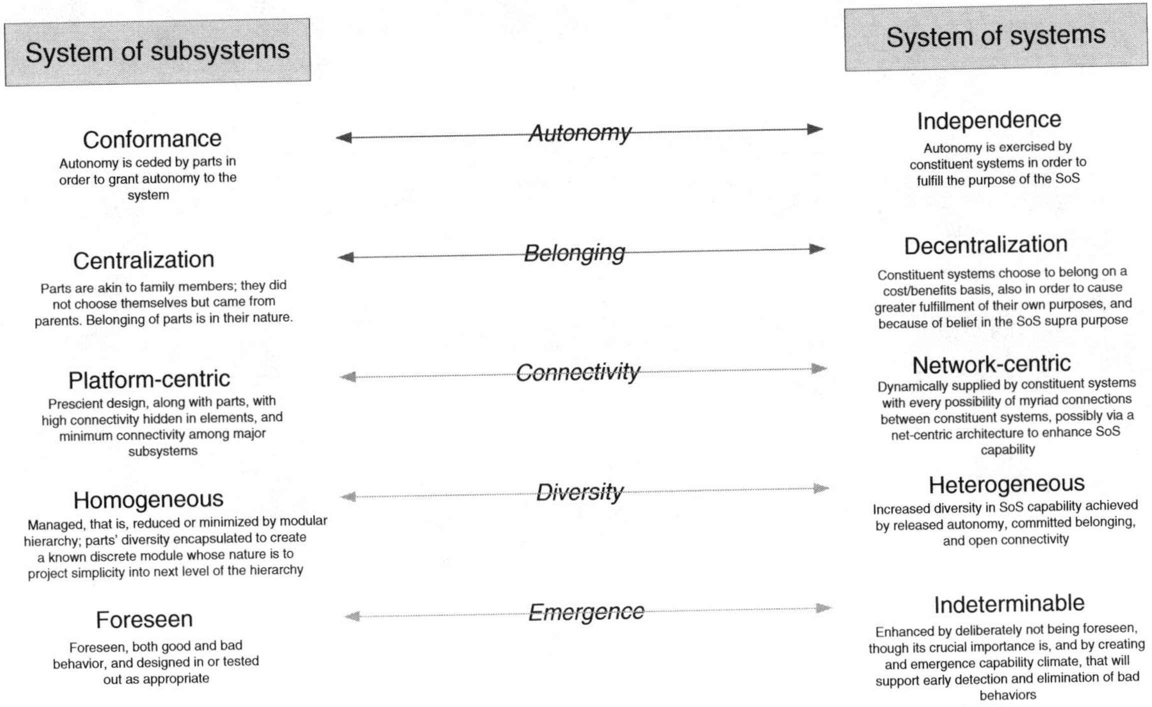

System of subsystems

System of systems

Autonomy

Conformance
Autonomy is ceded by parts in order to grant autonomy to the system

Independence
Autonomy is exercised by constituent systems in order to fulfill the purpose of the SoS

Belonging

Centralization
Parts are akin to family members; they did not choose themselves but came from parents. Belonging of parts is in their nature.

Decentralization
Constituent systems choose to belong on a cost/benefits basis, also in order to cause greater fulfillment of their own purposes, and because of belief in the SoS supra purpose

Connectivity

Platform-centric
Prescient design, along with parts, with high connectivity hidden in elements, and minimum connectivity among major subsystems

Network-centric
Dynamically supplied by constituent systems with every possibility of myriad connections between constituent systems, possibly via a net-centric architecture to enhance SoS capability

Diversity

Homogeneous
Managed, that is, reduced or minimized by modular hierarchy; parts' diversity encapsulated to create a known discrete module whose nature is to project simplicity into next level of the hierarchy

Heterogeneous
Increased diversity in SoS capability achieved by released autonomy, committed belonging, and open connectivity

Emergence

Foreseen
Foreseen, both good and bad behavior, and designed in or tested out as appropriate

Indeterminable
Enhanced by deliberately not being foreseen, though its crucial importance is, and by creating and emergence capability climate, that will support early detection and elimination of bad behaviors

FIGURE 8.2 System characteristics and their paradoxes

Characteristic	Determines the degree of...	Interdependent with...
Autonomy	Perspectives of systems are maintained and stay independent from each other	*Diversity*
	Mistakes may become correlated with each other	*Emergence*
	Portfolio of systems is balanced between unique and general	*Belonging*
Belonging	Conformity is forced upon requisite systems	*Autonomy*
	Responsibility is formalized for controlled belonging	*Diversity*
	Capability allows for distinguishing the good solutions from the bad, "Survival of the Fittest"	*Emergence*
Connectivity	Self-organization and decentralization	*Belonging*
	Self-interest and independence	*Autonomy*
	Who, what, where, when, and how—allowing the coordination of activities on one hand while solving different problems on the other	*Diversity*
Diversity	Preservation of independence is permissible	*Autonomy*
	Collectivism becomes a destructive characteristic (e.g., "group think")	*Belonging*
	Legacy of new systems is not redundant to that of the existing systems	*Connectivity*
Emergence	Knowing and knowing that you know are different skills	*Diversity*
	Trying to find the optimal systems will lead you astray (and trying to find the true systems will not)	*Connectivity*

FIGURE 8.3   The interdependence of the characteristics (Sauser and Boardman, 2007a)

autonomy may determine the degree of belonging, which will affect the extent of connectivity and possibly restrict the diversity (of elements) and maybe the emergent properties (of the system). But equally, a shift in diversity may have an effect on belonging and hence connectivity, leading to an increased autonomy level and consequential effect on emergence. Not only the coding in these pathways depends on the levels in the characteristics to which they connect, but it also has an effect on these levels being maintained or changed, up or down. Figure 8.3 shows the interdependence of these characteristics. In the following sections, we will further describe the paradoxes that exist in the realization of these characteristics.

8.5   DESCRIBING A SoS—A CASE STUDY OF THE NYC YELLOW CAB SYSTEM

The New York City Yellow Cab SoS is an essential part of the NYC transportation network. It consists of over 13,000 licensed medallion taxicabs (Commission, 2007). It serves about 240 million passengers a year, which creates a 1.82 billion dollar industry (Schaller Consulting, 2006d). NYC Yellow Cabs bring in around 30% of all fares paid by passengers for all trips in NYC and approximately 45% of fares paid for trips within Manhattan (Schaller Consulting, 2006a). On average, Manhattan residents hail a cab 100 times a year (Schaller Consulting, 2006a). The taxicabs are privately operated by independent companies or individuals. At the same time, they are closely regulated by the New York City Taxi and Limousine Commission (TLC). In contrast to

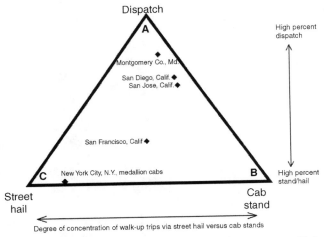

FIGURE 8.4   Schematic diagram of taxi customer market segments (Schaller, 2006)

other major cities of the United States, NYC Yellow Cabs operate predominantly on a street hail basis and do not operate on a dispatch basis as illustrated in Fig. 8.4.

Using the five characteristics and their paradoxes, we will describe the New York City Yellow Cab SoS. We will touch upon the opposing forces/paradoxes within each characteristic and then heuristically indicate a point where those forces meet.

8.5.1   Autonomy

All of the yellow cabs are operated by independent entities. Although they are strictly regulated by the TLC, they make their own decisions on how to operate. There are two opposing forces within the autonomy characteristic, which are as follows:

1. *Conformance*: The TLC is a government agency that was created to regulate and improve taxicab service within NYC. The main objectives of the TLC are to set the regulations such as the taxi fares, inspect the vehicles for safety reasons, issue licenses, and establish the local rules and regulations. The TLC enforces that each taxicab is inspected three times a year. Also, it receives and investigates passenger complaints. It has a right to impose fines, issue summons, and has the authority to suspend or revoke licenses for rule violations (Commission, 2007b).

2. *Independence*: Although the yellow cabs are closely regulated by the TLC, each medallion cab operates on an independent basis. They make independent choices in regard to schedule, routes, breaks, service, and so on. According to the survey done by Schaller Consulting (2006b), one of the main reasons licensed drivers want to drive a taxicab is because of the independence/ flexibility aspect of operating a taxicab. On the basis of the same survey, data (in percentage) shows that independence is more important then making money.

As a result, the point where conformance and independence meet leans toward the independence side within autonomy.

8.5.2 Belonging

The entire NYC medallion cab SoS shares the same mission, which is to serve NYC customers. It has to be done in a safe, efficient, and affordable manner. Licensed operators have the option, ability, and the rights to either belong to the SoS or not. For example, NYC ridership took a downfall between 1963 and 1977 as New York City went through a deep economic decline in the 1970s, and many taxicabs were sold or operated one shift instead of two shifts per day. It triggered a sharp decline in medallion prices. They went from $30,000 in 1960s to $10,000 in 1971 (Schaller Consulting, 2006c; Schaller Consulting, 2006g). Today, the medallion price can be as high as $420,000 for an individual and almost $600,000 for a corporate (fleet owned) license (TLC, 2007a). Giving individual cab owners/operators incentives based on the overall yellow cab SoS performance encourages cooperation among them toward a common goal. The opposing forces for belonging are

- *Centralization*: NYC Yellow Cabs are regulated by one single agency, which is the TLC. It acts as a centralized body that sets the overall transportation policy governing NYC Yellow Cab services.
- *Decentralization*: Taxicabs are privately operated and cannot be prearranged for pickup. They provide transportation exclusively via street hails/cab stands, because there is no central dispatcher.

Overall, there is a minimum set of constraints and maximum freedom for Yellow Cab drivers and/or operators to make decisions. As a result, the meeting point for the opposing forces is close to the decentralization side.

8.5.3 Connectivity

Yellow cabs that operate in NYC are connected through the transportation and social networks. The New York City infrastructure, which includes roads, bridges, tunnels, gas stations, parking lots, and repair shops, enables the yellow cab SoS to work in a continuous manner without any interruptions. The social networking provides access to the workers' alliances to protect drivers' rights, traffic news reports, hints on how to avoid traffic, and tips where most customers are (i.e., events, concerts, games, etc.). While there is a strong net-centricity in the structure of the yellow cab SoS, each driver pursues his/her own interests. Therefore, the opposing forces for the connectivity are

- *Platform-Centric*: Driver's income, which averaged $158 per shift in 2005 (Schaller Consulting, 2006e) depends on the cab mileage driven with passengers. It is also referred to as "paid mileage." In 2005, only 61% of the total cab mileage was "paid mileage" (Schaller Consulting, 2006l). Consequently, there is a fierce competition among the yellow cab drivers for customers.

- *Network-Centric*: On the other hand, all of the yellow cabs have the same power of connectivity. All have the same equal access to the infrastructure. Each yellow cab within the SoS is a whole on its own and also a part of the SoS network. Operating cabs form a taxi network and share the same goal, which is to serve NYC customers, while individually optimizing different success criteria.

Therefore, the opposing forces of the connectivity intersect in close proximity to the network-centric side. The heterarchy of the NYC Yellow Cab structure combines the hierarchy and the network:

8.5.4 Diversity

Although all NYC taxicabs are yellow in color, there is noticeable heterogeneity in a number of aspects such as:

- *Diversity of Cars* (i.e., hybrid, van, SUV, old, new, etc.): The majority of the taxicabs are Ford Crown Victoria. They accounted for 92% of all taxicabs in 2005 (Schaller Consulting, 2006j). The rest of the vehicles constitute Toyota, Mercury, Saturn, Lexus, Honda, and others. Some of them are hybrid, vans, and SUVs (TLC, 2007b). As of April 2005, 72% of all cabs were cars of year 2003 or newer, 19% were models produced in year 2001/2002, and 9% were of year 2000 or older (Schaller Consulting, 2006i). Of all the cabs, only 52% passed the initial inspection in 2005 (Schaller Consulting, 2006h). Recently, Mayor Bloomberg has proposed a law that states that all NYC operating taxicabs must be hybrid by 2012 (Mayor, 2007).
- *Diversity of Drivers* (i.e., background, ethnicity, age, experience, etc.): In 2005 48% of the drivers were identified as Asian, 25% as Black, 18% as White, and 7% as Hispanic (Schaller Consulting, 2006f). During the same year, the average age for licensed driver was 44 (Schaller Consulting, 2006f). According to the 1991 data, only 44% of the new applicants indicated that they had previous taxi driving experience (Schaller Consulting, 2006f). Another interesting fact is that only 9% of the drivers were born in the United States (Schaller Consulting, 2004; Schaller Consulting, 2006p).
- *Diversity in Operators*: NYC taxicab operators can be divided into three groups (Schaller Consulting, 2006o):
  1. *Owner Operators*: The owner of the medallion is also the driver.
  2. *Fleet-Type Operators*: Leased to the drivers by the shift.
  3. *Long-Term Leasing*: Leased to drivers for a period of months.

Mileage differs substantially between operators. Usually fleet-type operators utilize two shifts per day 7 days a week averaging 72,000 miles per year. Long-term leased cabs average 62,000 miles per year. Finally, owner operators use their cars only one shift a day five to six days a week averaging 42,000 miles per year (Schaller Consulting, 2006i). Thus, the tensions for the diversity are as follows:

- *Homogeneous*: Medallion cab operators are predominantly male. Only 1% of drivers were female in 2005 (Schaller Consulting, 2006f). Standards that are set by the TLC (i.e., requirement for all medallion taxicabs to be painted yellow, driver licensing requirements, safety requirements, etc.) also contribute to the level of homogeneity of the NYC Yellow Cab SoS.
- *Heterogeneous*: The diversity of cars, drivers (aside from gender), and operators makes NYC Yellow Cab SoS very heterogeneous.

As a result, the point where paradoxes within the diversity characteristic meet is very close to the heterogeneous side.

8.5.5 Emergence

As we have mentioned earlier, all systems are emergent, and this emergence is classified as intended and unintended. The intended emergent property of a NYC Yellow Cab SoS is to safely, comfortably, and efficiently transport NYC passengers. A single cab cannot achieve this result on its own. In the yellow cab SoS, the removal of one cab will not change the overall performance of the SoS. When a single cab is taken out of service, another one will emerge to take its responsibility.

In 1998, a New York Taxi Workers Alliance emerged to protect the interest of the taxicab workers. The formation of this union was the result of the April 1998 TLC's declaration that involved passing of new rules to increase the summons to the taxicab drivers for small violations. Thus, an organized strike with about 40,000 taxicab drivers was formed lasting for 24 hours (York, 2007; Press, 1998). This is an example of unintended emergent behavior.

Also, as we write this case, another unintended development emerged. This issue concerns the use of technology within the cabs. Recently, the TLC has announced plans to require the global positioning system (GPS) to be installed in every cab. The New York Taxi Workers Alliance opposed this because the taxicab drivers believe that the new device will track their whereabouts, and it will be very costly to install (Jones, 2007). This new law could potentially affect the emergence and all the other characteristics of the yellow cab SoS. Therefore, the union has threatened to strike against this law.

Finally, the tensions for the emergence are

- *Foreseen*: Taxi riders expect from NYC Yellow Cabs personal security, comfort, cleanliness inside the vehicles, fast and reliable service, and being charged the correct fee.
- *Indeterminable*: As opposed to the foreseen factor, in reality, customer satisfaction of NYC Yellow Cab service is not always met. In 2004, the New York City Transit Transportation Panel Survey revealed that in comparison to the other means of transportation, the New York City's overall taxicab customer satisfaction rating is below private cars, subway, car service, and local buses (Panel, 2004; Schaller Consulting, 2006k; Schaller Consulting,

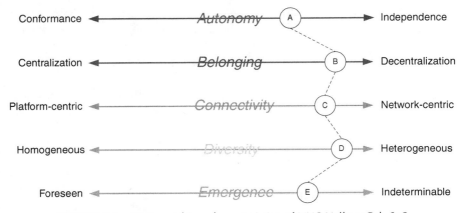

Conformance	←	Autonomy	Ⓐ	→	Independence
Centralization	←	Belonging	Ⓑ	→	Decentralization
Platform-centric	←	Connectivity	Ⓒ	→	Network-centric
Homogeneous	←	Diversity	Ⓓ	→	Heterogeneous
Foreseen	←	Emergence	Ⓔ	→	Indeterminable

FIGURE 8.5   Distinguishing characteristics of NYC Yellow Cab SoS

2006n). Also, passengers filed over 18,000 individual complaints with the TLC in 2004 (Schaller Consulting, 2006m).

As it is evident from the above analysis, the NYC Yellow Cab SoS is highly adaptable to the environment, whether it is intended or unintended. As a result, the NYC Yellow Cab SoS leans toward the indeterminable side within the emergence.

Figure 8.5 is a qualitative representation of the yellow cab SoS as we have described in the five characteristics and their paradoxes. Due to the high number of constituent systems, highly dynamic and complex environment, and the involvement of the human factor, this picture can be viewed differently in the future. The NYC Yellow Cab SoS is constantly attempting to find a point of balance where the opposing forces meet and find equilibrium. A nonprofit group called "Design Trust for Public Space," in collaboration with the TLC, is working on the taxi system report called "Taxi 07: Roads Forward" (Space, 2007). The report examines the present taxi system situation and accesses the developmental and enhancement strategies for the next decade.

8.6   CONCLUSION

There exists a lack of development as being a product of two entities: first, because knowledge developers are looking in one direction, namely reinforcement, the addition of knowledge is essentially incremental, whereas the growth in complexity of systems is very real, driven by exponentially rising customer expectations coupled with unbridled innovation in companies; second, because knowledge developers are looking in one direction and not in other directions, such as what the complexity science knowledge developers have discovered and are continuing to explore, there is inevitably a paucity of thinking and of thought processes, which could otherwise provide an immensely valuable asset for the systems engineering community.

We believe that there is a fundamental distinction to be made between a system (of parts) and a system of systems (Boardman and Sauser, 2006; Boardman and Sauser,

2008), in keeping with the spirit of complexity that significant difference is to be found in the obscure, that is, the word *of* is what matters most, rather than the systems (or parts) that make up the system (of parts or systems) itself. We have explained this significance and provided a framework for identifying how, when, and why a system takes on one manifestation or another, and therefore provides an opportunity for managing the system appropriately.

Our perception that there exists a lack of inclination to shift the line of thinking among the engineering (and business communities) and what can be done about this, that is, to accomplish a paradigm shift is grounded in the notion of complexity and how it applies to the different communities—engineering and complexity science.

The term complex derives from the Latin word *complexus* meaning to entwine. Its meaning today as an adjective is "consisting of interconnected or interwoven parts." As a noun, it means "a whole composed of interconnected or interwoven parts." A complex really is a system, and being complex means being interconnected or interwoven, as for example, a tapestry. The fact that the Bayeux Tapestry is not really a tapestry, is probably incomplete and inaccurate, in addition to being anomalous, adds spice to what we have to say about complex, because we now understand complex to be replete with paradox.

The science of complexity is relatively new. A scintillating account of its origins, historical and technical, is provided by Waldrop (1992). A comparative newcomer to the science family, it is nevertheless possible to demarcate phases in our understanding of complexity. The initial phase was occupied by understanding the forces of self-organization, not realizing that self-organization was a "system" in its own right for example Smith (1759, 17761968), Engels (1884), Darwin (1859), and Turing (1950). This was followed by seeing self-organization as a problem that transcended local disciplines and solving that problem, partially by comparing behavior in one area with that of another, for example, comparing slime molds and ant colonies. We might well consider to have now entered a new phase that concerns the creation of self-organizing systems and the invention of artificial emergence: systems built with a conscious understanding of what emergence is, and systems designed to exploit those laws, for example, software that makes book recommendations on Amazon, does voice recognition on our cell phone, and finds mates over the Internet.

The entire process of complexity understanding is driven by the paradoxical theme of unraveling that which is interwoven in order to understand the parts, their interactions, and their interweaving, while keeping it in its whole state since the unraveling will fail to produce the extraordinary emergent behaviors attributable solely to the existence of that whole and to that whole's properties meaningful only to it and remarkably different from properties attributable to the interwoven parts.

In summary, the paradoxes we have observed from our own studies and experiences of complexity include the following:

- Complexity is much simpler than it first appears;
- Simple things exhibit very complex behavior;
- Little things mean a lot;

- Myriad things are closer than we think;
- Significant things are both vital and obscure;
- Weak relationships bring strength and security;
- To those who have yet more shall be given and yet to those who have little even this will mean less; and
- perhaps the ultimate paradox, the reality that a complex is both a one and a many simultaneously.

Therefore, our philosophy can be found in our treatments of the meaning of *of* and in the reestablishment of the virtue of paradox as a pivot for shifting thinking (Boardman and Sauser, 2008). In our treatment of *of*, we try to establish the significant architectures that draw whole systems together to produce not simply emergence but an emergent culture, one that relies on emergence, the element of surprise, and the risk of chaos to produce new patterns of simplicity and elegance about which we could not know or foresee.

In our treatment of paradox, we try to reestablish the significance of wisdom and where it might be found:

- In the multiplicity and simultaneity of viewpoints;
- In respect for competing perspectives;
- In the duality of the one and the many; and
- In the conflict and the confusion that arise when we refuse to accept simple choices and with wisdom find a simplicity unforeseeable by mere resolution— something that is hidden and can only be found in the unthinkable.

So we conclude with these future challenges:

1. How can we design and manage autonomy, belonging, diversity, connectivity, and emergence?
2. How do enterprises build project capabilities needed for SoS when managing paradox?
3. How do we learn about "best practices" or attributes in the management of SoS?
4. How can we measure the breadth and depth of the capabilities of an SoS?
5. What frameworks can best explain the definition, initiation, planning, execution, and controlling of an SoS?
6. What kind of product life cycle models fit an SoS?
7. How does a system relate to the wider technological SoS in which they are embedded and vice versa?
8. How can the notion of network be applied to the levels of process and organization in an SoS?
9. Can we provide a context or system through which to interpret, define, and establish the boundaries of an SoS?

REFERENCES

Anderson, P.W., 1972, More is difficult, *Science*, 1777(4047): 393–396.

Ashby, R., 1956, *Introduction to Cybernetics*, Chapman Hall, London.

Bar-Yam, Y., 2004, The characteristics and emerging behaviors of system of systems, *NECSI: Complex Physical, Biological and Social Systems Project*, pp. 1–16.

Beer, S., 1966, *Decision and Control: The Meaning of Operational Research and Management Cybernetics*, Wiley & Sons, New York.

Bernard, C., 1952, *Introduction a la Medecine Experimentale*, Flammarion, Paris.

Bertalanffy, L.V., 1968, *General System Theory: Foundations, Development, Applications*. New York, George Braziller.

Boardman, J., Sauser, B., 2006, System of systems: the meaning of, *IEEE International System of Systems Conference*.

Boardman, J., Pallas, S., Sauser, B.J., Verma, D., 2006, Report on system of systems engineering, Final Report for the Office of Secretary of Defense, Stevens Institute of Technology, Hoboken, NJ.

Boardman, J., Sauser, B., 2008, *Systems Thinking: Coping with 21st Century Problems*, CRC Press/Taylor & Francis Group, Boca Raton, Florida.

Brown, S.L., Eisenhardt, K.M., 1998, *Competing on the Edge: Strategy as Structured Chaos*, Harvard Business School, Boston.

Cameron, K.S., Quinn, R.E., 1988, Organizational paradox and transformation, In: Quinn, R.E. and Cameron, K.S. (Eds.), *Paradox and Transformation: Toward a Theory of Change in Organization and Management*, Cambridge, MA, Ballinger, pp. 12–18.

Carlock, P.G., Fenton, R.E., 2001, System of systems (SoS) enterprise systems engineering for information-intensive organizations, *Systems Engineering*, 4(4): 242.

Commission, N.Y.T.C.T.L., 2007, Current Medallions—sorted by TLC Medallion Number, current_medallions.xls (Ed.), TLC.

Commission, T.N.Y.C.T.a. L. 2007b, About TLC, Vol. 2007.

Schaller Consulting, 2006a, *The New York City Taxicab Fact Book*, pp. 2.

Schaller Consulting, 2006b, *The New York City Taxicab Fact Book*, pp. 60.

Schaller Consulting, 2006c, *The New York City Taxicab Fact Book*, pp. 41.

Schaller Consulting, 2006d, *The New York City Taxicab Fact Book*, pp. 1.

Schaller Consulting, 2006e, *NYC Taxicab Fact Book*, pp. 37.

Schaller Consulting, 2006f, *NYC Taxicab Fact Book*, pp. 55.

Schaller Consulting, 2006g, *NYC Taxicab Fact Book*, pp. 25.

Schaller Consulting, 2006h, *NYC Taxicab Fact Book*, pp. 44.

Schaller Consulting, 2006i, *NYC Taxicab Fact Book*, pp. 42.

Schaller Consulting, 2006j, *NYC Taxicab Fact Book*, pp. 43.

Schaller Consulting, 2006k, *NYC Taxicab Fact Book*, pp. 29.

Schaller Consulting, 2006l, *NYC Taxicab Fact Book*, pp. 35.

Schaller Consulting, 2006m, *NYC Taxicab Fact Book*, pp. 13.

Schaller Consulting, 2006n, *NYC Taxicab Fact Book*, pp. 12.

Schaller Consulting, 2006o, *NYC Taxicab Fact Book*, pp. 31.

Schaller Consulting, 2006p, *NYC Taxicab Fact Book*, pp. 57.

Darwin, C., 1859, *On the Origin of Species by Means of Natural Selection, or the Preservation of Favoured Races in the Struggle for Life*, John Murray, London.

Davies, A., Brady, T., 1998, Policies for complex product system, *Futures*, 30(4): 293–304.

DeLaurentis, D., 2005, Understanding transportation as a system of systems design problem, *43rd AIAA Aerospace Sciences Meeting*.

Engels, F., 1884, The Origin of the Family Private Property, and the State.

Floricel, S., Miller, R., 2001, Strategizing for anticipating risks and turbulence in large-scale engineering projects, *International Journal of Project Management*, 19: 445–455.

Ford, J.D., Ford, L.W., 1994, Logics of identity, contradiction, and attraction in change, *Academy of Management Review*, 19: 756–795.

Friedman, T.L., 2006, *The World is Flat: A Brief History of the Twenty-First Century*, Farrar, Straus and Giroux, New York.

GPO, U.S., 1989, Restructuring of the Strategic Defense Initiative (SDI) Program, Services, U.S.C.S.C. o. A. (Ed.), United States Congress, pp. 16.

Glass, N., 1996, Chaos, non-linear systems and day-to-day management, *European Management Journal*, 14(1): 98–106.

Handy, C., 1994, *The Age of Paradox*, Harvard Business School, Cambridge, MA.

Hobday, M., Rush, H., Tidd, J., 2000, Innovation in complex products and system, *Research Policy*, 29(7–8): 793–804.

Hofstadter, D.R., 1979, *Godel, Escher, Bach: An eternal golden braid*, Norton, New York.

INCOSE 2006, INCOSE Systems Engineering Handbook, Vol. 3.

ISO/IEC 2002, Systems Engineering—Systems Life Cycle process, Vol. ISO/IEC-15288.

Jones, K.C., 2007, New York Cabbies may strike over GPS tracking, InformationWeek, July 25, 2007.

Keating, C., Rogers, R., Unal, R., Dryer, D., Sousa-Poza, A., Safford, R., Peterson, W., Rabadi, G., 2003, System of systems engineering, *Engineering Management Journal*, 15(3): 36.

Kimmel, L., 1998, Lord Nelson and Sea Power, http://www.geocities.com/Athens/3682/nelsonsea.html.

Koestler, A., 1990, *The Ghost in the Machine*, Penguin, London.

Kusiak, A., Larson, N., 1999, Concurrent engineering, in: Sage, A.P. Rouse, W.R.(Eds.), *Handbook of Systems Engineering and Management*, Wiley & Sons, New York, pp. 327–370.

Lewis, M.W., 2000, Exploring paradox: toward a more comprehensive guide, *Academy of Management Review*, 25(4): 760–776.

Maier, M., 1998, Architecting principles for system-of-systems, *Systems Engineering*, 1(4): 267–284.

Mayor, N.O.o.t., 2007, Mayor Bloomberg announces taxi fleet to be fully hybrid by 2012.

Miller, R., Lessard, D.R., 2000, *The Strategic Management of Large Engineering Projects*, MIT Press, Boston.

Nightingale, P., 1998, A cognitive model of innovation, *Research Policy*, 27: 689–709.

Panel, N.T.T., 2004, New York City Transit Transportation Panel Survey.

Parth, F.R., 1998, Systems engineering drivers in defense and in commercial practice, *Systems Engineering*, 1: 82–89.

PMI 2004, (Ed.), *Guide to the Project Management Body of Knowledge*, Project Management Institute, Newtown Square, PA.

Press, T.A., 1998, "New York cabbies stage boycott over tougher rules," *U. S. News*.

Sage, A., Cuppan, C.D., 2001, On the systems engineering and management of systems of systems and federations of systems, *Information, Knowledge, Systems Management*, 2(4): 325–345.

Sage, A., Rouse, W.B. 2008, An Introduction to systems engineering and systems management, in: Sage, A., Rouse, W.B. (Eds.), *Handbook of Systems Engineering and Management*, Wiley & Sons, Hoboken, NJ.

Sauser, B.J., Boardman, J., 2007a, Taking Hold of System of Systems Management, Engineering Management Journal, Vol. forthcoming, pp.

Sauser, B.J., Boardman, J., 2007b, Complementarity: In Search of the Biology of Systems, *IEEE International Conference on Systems of Systems Engineering, IEEE*, pp. 1–5.

Schaller Consulting, 2004, NYC Taxicab Fact Book.

Schaller, B., 2006, Entry Controls in Taxi Regulation: Regulatory Policy Implications of U.S. and Canadian Experience, pp. 1–17.

Schneider, K.J., 1990, *The paradoxical Self: Toward an Understanding of Our Contradictory Nature*, Insight Books, New York.

Shenhar, A.J., 1998, From Theory to Practice: Toward a Typology of Project Management Style, *IEEE Transactions on Engineering Management*, 45(1): 33–47.

Shenhar, A.J., Dvir, D., 2007, *Reinventing Project Management: The Diamond Approach to Successful Growth and Innovation*, Harvard Business School, Boston.

Shenhar, A.J., Sauser, B.J., 2008, Systems engineering management: the multidisciplinary discipline, In: Sage, A.P., Rouse, W.R. (Ed.), *Handbook of Systems Engineering and Management*, Wiley & Sons, Hoboken, NJ, pp. 113–136.

Simon, H.A., 1996, *The Sciences of the Artificial*, MIT Press, Cambridge.

Smith, A., 1759, The Theory of Moral Sentiments, Library of Economics and Liberty.

Smith, A., 1776, An Inquiry into the Nature and Causes of the Wealth of Nations, MetaLibri Digital Library.

Smuts, J., 1926, *Holism and Evolution*, Macmillan, London.

Space, T.D.T.f.P., 2007, The Design Trust for Public Space.

Stenbit, J.P., Wells, L., Alberts, D.S., 2003, Complexity Theory and Network Centric Warfare, DoD Command and Control Research Program.

TLC, 2007a, Monthly Medallion Sales—Average Prices & Number of Transfers.

TLC, 2007b, Saturn Vue and Toyota Camry Hybrids Join List of Approved Taxicab Models.

Turing, A.M., 1950, Computing machinery and intelligence, *Mind*, 59: 433–460.

Waldrop, M., 1992, *Complexity: The Emerging Science at the Edge of Order and Chaos*, Simon & Schuster, New York.

York, F.f.t.C.o.N., 2007, New York Taxi Workers' Alliance.

Chapter **9**

Systems Engineering for Department of Defense Systems of Systems

JUDITH S. DAHMANN

Center for Acquisition and Systems Analysis, MITRE Corporation, McLean, VA, USA

9.1 BACKGROUND

The Department of Defense (DoD) recognizes systems engineering (SE) as an important enabler to successful defense systems acquisition. In the face of rising acquisition costs and growing time to deliver, there has been a push to revitalize SE with a focus on major defense acquisition programs (which account for about 40% of DoD acquisition investment funding). The Department has enacted a set of SE policies that emphasize technical planning, technical authority, and substantive technical reviews. All of these policies are designed to give weight to the influence of systems engineering in the design and management of new defense systems development. Systems engineering processes are closely tied to the defense processes and procedures for acquisition of military systems, with systems engineering reviews playing an increasingly important role in oversight of large systems acquisition. The Defense Acquisition Guidebook (DoD, 2004a) lays out the core SE technical and management processes (Fig. 9.1) and aligns these with the phases of acquisition of a defense systems.

At the same time that the role of SE in individual acquisition programs has been strengthened, other changes have been taking place in the Department of Defense that contribute to an increased focus on the broader capabilities needed by the military as

System of Systems Engineering: Innovations for the 21ˢᵗ Century, Edited by Mo Jamshidi
Copyright © 2009 John Wiley & Sons, Inc., Publication

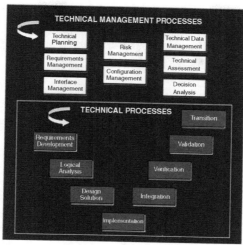

FIGURE 9.1 Systems engineering technical and technical management processes (DoD, 2004a)

basis for investment and acquisition decisions. Changes have been made in the defense requirements definition process (CJCS, 2007) to focus on desired battle space effects, capabilities needed to achieve these effects, and gaps in those capabilities that need to be addressed to meet DoD objectives. Acquisition Reform Initiatives (DoD, 2003; DoD, 2004b; Durham, 2006, 2007) call for an assessment of alternatives with broader trade space, weighing affordability, and capability needs with technical feasibility. This evaluation of options for investment is made in light of the broader portfolio of capabilities in place and considering the Department's strategic direction. Recognition of the power of information exchange in the battle space (DoD CIO, 2004; DoD CIO, 2005) is high on the list of priorities supported by policy to enable net-centric operations through battle space networks and broad-based, flexible information sharing (DoD CIO, 2003; DoD, 2004c). This net-centric strategy calls for investment in common communication, networking, and data sharing capabilities that will allow for cooperation among users and systems across the battle space. These capabilities are needed to support joint and coalition operations, which require a higher degree of coordination and synchronization of forces.

As a consequence, the Department is increasingly aware of the need for more focused attention on the design and development of systems based on the way those systems function as part of the larger environment and the joint capabilities desired. This growing awareness has led to increased attention on the management and engineering of solutions at a systems of systems (SoS) level.

In particular, in 2006 the DoD Deputy Under Secretary of Defense for Acquisition and Technology (DUSD A&T) initiated an effort to develop a guide on systems engineering for systems of systems. In the Department of Defense, SoS is defined as "a set or arrangement of systems that results when independent and useful systems are integrated into a larger system that delivers unique capabilities" (DoD, 2004a). This guidebook effort serves as a vehicle to raise awareness of the need for SE beyond individual systems and has provided an opportunity to identify and begin to address issues in this area of growing interest and importance for DoD.

9.2   DEFENSE CONSIDERATIONS IN SoS SE

With the focus on joint and coalition warfighting capabilities, many DoD capability needs will be satisfied by groupings of legacy systems, new programs, and technology insertion, in short by SoS. A number of considerations affect how DoD approaches SoS and SoS SE.

First and perhaps most important is the current governance and management structure of defense systems. Today, individual systems are owned by the military services or agencies that define their own approaches to addressing user needs and to equipping and training their forces. Although there are broad joint and DoD-wide crosscutting processes, as a practical matter, each of the services has its own processes, lines of authority, and funding mechanisms for systems. Systems of systems, when recognized by the DoD, are addressed as an overlay to these service processes, which continue to be the dominant drivers in defense acquisition management.

In addition, the DoD has a large inventory of legacy systems that will be part of the defense inventory for the long term and will need to be factored into any approach to SoS. Defense budgets are stagnant or declining. Time and costs of new developments are staggering, reinforcing the need to leverage current investments. As a result, there will be a strong push to look for ways to use the systems we have today in different combinations to meet new needs, all of this at a time when the operational environment is calling for greater agility and creativity in addressing changing threat scenarios.

DoD is responding in a number of ways. The Department is moving toward increased network dependence and recognizing the criticality of software as an enabler in constructing cooperative or distributed SoS. There is also a move to align the Department's investment into a series of capability portfolios, forcing a linkage to strategic and joint warfighting priorities. Investment decisions can be made in the context of related capabilities and systems. Systems engineering provides the technical base for selecting components of the systems needed to support portfolio objectives.

These considerations taken together contribute to a sizable management challenge to SoS systems engineers that has a direct impact in terms of technical issues for systems engineering as illustrated in Fig. 9.2. The issues of coordination and synchronization are present in all situations when you are bringing together multiple existing systems, In the DoD, when these systems are owned, managed, and funded by a single service or agency, there is a common funding authority to set priorities. In SoS incorporating systems from multiple services or agencies, the problems are compounded because of the multiple funding authorities.

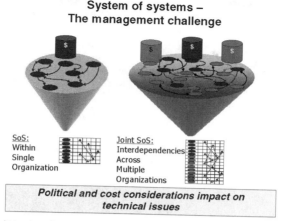

FIGURE 9.2   Political and management considerations impact on SoS SE (Dahmann and Baldwin, 2007)

9.3   SoS IN THE DoD TODAY

Most military systems today are part of an SoS whether or not explicitly recognized. Operationally the DoD acts as an SoS, but DoD development and acquisition have focused on independent systems. Most systems are created and evolve without explicit SE at the SoS level.

When we look at the SoS in the DoD today, we see that a formal SoS comes into existence only when something occurs that is important enough to trigger recognition of the SoS and bring into play management and governance processes that cut across established individual system boundaries. Reasons can vary. In some cases it is the recognition of the criticality of an SoS area, such as the Air Force Recognition that the suite of systems that work together to support the Air Operations Center (AOC) come together without benefit of coordinated preplanning and integration, and hence put at risk a critical military operational asset. An SoS may be created in response to the operational problems in which new needs are identified that cannot be supported without cooperative efforts of multiple systems (e.g., Single Integrated Air Picture (SIAP)) (OUSD AT & L, 2008).

Once recognition of the need for an SoS occurs, an organization is identified as "responsible for" the SoS "area" and the objective of the SoS is broadly defined. Typically, however, ownership of the systems in the SoS is not changed or nor are the objectives of each of the individual systems. The SoS objective is often framed in terms of improved "capabilities" and not as a well-specified technical performance objective.

The SoS is then structured. Membership is defined starting with identification of systems in the SoS. Processes and organizations are established for the SoS, including SE. In systems acquisition the processes are well established along with the SE roles and responsibilities. However, in an SoS, the organizational structures and processes that bridge the systems and stakeholders need to be created and socialized.

SoS are not typically new acquisitions. Rather they usually take the form of an overlay to an ensemble of existing systems with the objective of improving the way the systems work together to meet a new user need. This runs counter to the prevailing notion that the prime example of a DoD SoS is the Army Future Combat Systems (FCS). FCS is a mammoth effort to build 18 new systems concurrently to create a broad new Army combat capability. In fact, FCS is a unique undertaking and is not typical of the defense realization of SoS. Furthermore, as FCS matures, it has evolved to an incremental technology insertion program. FCS leverages new technologies and capabilities into the stable of legacy systems currently in the Army inventory. This process resembles the more common form of an SoS, that of an evolution or upgrade to a group of existing or planned systems.

Under these circumstances, Defense SoS managers, when designated, typically do not control the requirements or funding for all of the individual systems in the SoS and consequently find themselves in a position of influencing rather than directing as they work with systems to meet the SoS need. This affects the SE approach for the SoS, which has to accommodate the fact that the SoS needs may not be able to influence the individual systems development.

The focus of the SoS can be on the evolution of capability over time. Initial efforts may enhance the way current systems work together, anticipating change in internal or external effects on SoS. Only eventually, new functionality may be added through new systems or changes in existing systems. In some cases, the aim may be to eliminate systems or reengineer systems to provide better or more efficient capability. The latter is often problematic when the redundant systems features have been created to meet specific user needs beyond the reach of the SoS.

Because of the prevailing system-level environment, the startup of a functioning SoS can take time. Identifying the right systems and stakeholders, gaining their attention given the other demands on their time and resources and focusing them productively on their role in the SoS can be challenging. Individual systems will have their own user needs and development schedules, contractual issues can take time and synchronization can be difficult. Once formed and in a steady state, functioning SoS seem well suited to incremental development, allowing for improvements in different parts of the SoS on independent schedules with full benefit to the SoS awaiting the full mix of changes over time.

9.4 COMPARING SYSTEMS ENGINEERING AT A SYSTEM LEVEL WITH SYSTEMS OF SYSTEMS

There are some clear differences between systems and SoS in areas that influence systems engineering. (Fig. 9.3)

First, from the perspective of *community involvement*, systems differ from SoS in the nature and mix of stakeholders and in their governance. For individual systems, there is typically a single program manager with authority across the elements of the acquisition program, a funding line to support system acquisition, and a set of stakeholders committed to that system. With an SoS however, the structure is rarely if ever that clean. There is often a program manager (PM) for the

	System	System of Systems
Management /Oversight		
Stakeholder Involvement	Clear set of stakeholders	Added levels of complexity; stakeholders at both system level and levels with competing interests and priorities; in some cases the system stakeholder has no vested interest in the SoS
Governance	Single PM and funding	May have management and funding for the SoS but also have management and funding of for individual systems
Operational Environment		
Operational Focus	The systems are designed and developed to meet operational objectives	SoS is called upon to meet a set of operational objectives using systems whose objectives may or may not align with the SoS objectives
Implementation		
Acquisition/Test & Validate	Established process aligned to ACAT milestones, specified requirements, SE with a systems engineering plan (SEP)	No established process across multiple system lifecycles across acquisition programs, involving legacy systems, developmental systems, and technology insertion
	Testing or validating the system is possible	Testing is more challenging due to the difficulty of synchronizing across multiple systems life cycles; testing all permutations, given the complexity of all the moving parts, is not possible
Engineering		
Boundaries and Interfaces	Focuses on boundaries and interfaces for the single system	In SoS the focus is on identifying the systems that contribute to the SoS objectives and enabling the flow of data, control and functionality of the SoS within the constraints of the systems
Performance & Behavior	Optimize performance of the system to meet performance objectives	Provide end-to-end performance across the SoS that satisfies user capability needs within the context and constraints of the systems

FIGURE 9.3 Comparing a system with a system of systems (OUSD AT & L, 2008)

SoS who may have fun- ding for SoS development and a set of stakeholders with vested interests in the SoS.

However, in an SoS, there are also PMs for the individual systems with their own set of requirements and independent funding for system development as well as stakeholders focused on the capabilities of the individual systems who may not have any interest in the larger SoS. This mix of stakeholders at both the systems and SoS levels means that the SoS SE must forge a plan for the SoS in light of these different sets objectives and priorities as well as mandates, plans, and funding to accomplish them.

Changes to ownership of individual systems supporting a newly "formalized" SoS usually are not done because those individual systems often support other SoS. Balancing the autonomy and multiple needs of the individual systems against those of the multiple larger SoS that they support is a core issue for SoS.

Second, from the perspective of the *employment environment*, systems differ from SoS in the complexity and variability of the mission environment and operational focus of the system as compared with an SoS. While faced with the inevitable change that characterizes defense today, for systems the target mission environment is relatively stable with clear operational focus on priority capabilities. However, the SoS mission objectives tend to be stated in terms of broad capabilities. Beyond this, the SoS is charged with meeting this new set of mission objectives at the same time that the systems continue to meet their own objectives. These sets of objectives may or may not be mutually supportive or even consistent.

Finally, when considering implementation, there are significant differences between the individual systems of the DoD and the SoS. Individual systems have well-defined SE processes aligned to the DoD acquisition process. These systems have system-level requirements under the auspices of a single PM and a chief engineer. The acquisition process requires the PM to deliver a Systems Engineering Plan (SEP) and Test and Evaluation Master Plan (TEMP). However, an SoS typically involves legacy and new systems with multiple PMs and operational and support communities. SoS efforts seek to improve the ability of the systems to work together to meet a broad user capability need over time, often without any clear completion. SoS synchronization across program development paths can be hard and testing is more difficult. In fact with SoS, the line between development and maintenance tends to blur.

9.5   CORE ELEMENTS OF SoS SE

The core elements of SoS SE provide the context for the application of SE processes. By understanding the tasks facing the SoS systems engineer, we can better appreciate how basic SE processes are applied in an SoS environment and identify some emerging principles for SoS SE.

9.5.1   Translating SoS Capability Objectives into High-Level Requirements over Time

From the outset of the formation of an SoS, the SE is called upon to understand and articulate the technical level expectations for the SoS. SoS objectives are typically couched in terms of needed capabilities, and the systems engineer is responsible for translating these capabilities into high-level requirements that can provide the foundation for the technical planning to improve the capability over time. Unlike the experience of an individual systems for which the technical requirements are understood up front and the systems engineer is responsible for assessing alternative approaches to meeting these requirements, with SoS the SE has an active role in the process of translating capability needs into technical requirements. For an SoS, this process is an and needs to reflect changes in needs and options as the SoS evolves over time.

9.5.2 Understanding the Components of the SoS and Their Relationships over Time

One of the most important aspects of the SoS SE role is the development of an understanding of the systems involved in the SoS and their relationships and inter-dependencies. In an individual system acquisition, the SE is typically able to clearly establish boundaries and interfaces for the new system. In an SoS, the problem is more of understanding the ensemble of systems that affect the SoS capability and the way they interact and contribute to the capability objectives. Definition of what is "inside" the SoS is somewhat arbitrary because there are typically key systems outside of the control of the SoS management that have large impacts on the SoS objectives. What is most important here is understanding the players, their relationships, and their drivers so options for addressing SoS objectives can be identified and evaluated, and impacts of external changes can be anticipated and addressed. The SoS SE needs to identify the stakeholders, including users, of SoS and systems, including their organizational context as a foundation for their role as the SoS SE.

9.5.3 Assessing Extent to Which the SoS Meets Capability Objectives over Time

In an SoS environment, there may be a variety of ways to address objectives. This means that independent of the alternative approaches, the SoS SE needs to establish metrics and methods for assessing performance of the SoS in terms of objective capabilities. Since SoS are often fielded suites of systems, feedback on SoS perfor-mance may be based on operational experience and issues arising from operational settings. By monitoring performance in the field or in exercise settings, areas for attention can be identified and impacts of unplanned change in constituent systems can be assessed.

9.5.4 Developing, Evolving, and Maintaining a Design for the SoS

Once an SoS has clarified the high-level technical objectives of the SoS, identified the systems key to SoS objectives, and assessed the current performance of the SoS, a technical plan is developed, beginning with a design for the SoS. The SoS design addresses the concept of operations for the SoS; the systems' functions, relationships, and dependencies; and the end-to-end functionality, data flow, and communications. The SoS design (or architecture) provides the technical framework for assessing changes needed in systems or options for addressing requirements. In the case of a new system development, the systems engineer can begin with a "clean sheet" approach to design. In an SoS, on the contrary, the current state of the individual systems is an important factor in the design process.

9.5.5   Monitoring and Assessing Potential Impacts of Changes on SoS Performance

An SoS is comprised of multiple independent systems. These systems evolve independently of the SoS, possibly in ways that could affect the SoS. Consequently, a big part of SoS SE is anticipating change that will impact SoS functionality or performance. These may be changes in the systems within the SoS as well as changing external demands on SoS. By understanding impacts of proposed or potential changes, the SoS SE can either intervene to preclude problems or develop strategies to mitigate the impact on the SoS.

9.5.6   Addressing New Requirements on SoS and Solution Options

In an SoS, requirements reside both at the level of the SoS and at the level of the individual systems. Depending on the circumstances, the SoS systems engineer may have a role at one or both levels. At the SoS level, as with systems, a process is needed to collect, assess, and prioritize user needs, and then to evaluate options for addressing these needs. It is important for the SoS systems engineer to understand the individual systems and their technical and organizational context and constraints so they can anticipate the affect of SoS options on the systems. This can be difficult because of the multiple requirements and acquisition stakeholders that are engaged in an SoS. The SoS design, if done well, will provide the framework for both identifying and assessing alternatives, and will provide stability as different requirements come to fore for consideration, hopefully moderating the impact of changes in one area on other parts of the SoS.

9.5.7   Orchestrating Upgrades to SoS

Once an option for addressing a need has been selected, it is the SoS systems engineers' role to work with the SoS PM and the system PMs and systems engineers to plan, facilitate, integrate, and test upgrades to the SoS. The actual changes are made by the systems themselves and it is the role of the SoS SE to orchestrate this process, taking a lead role in the synchronization, integration, and test across the SoS.

9.6   EMERGING PRINCIPLES FOR SoS SE

Given these core elements of SoS SE, what are some approaches that seem to be well suited to SE in this environment? We conducted reviews with a set of pilot programs, programs that were nominated by the military services as examples of SoS. The systems engineers in these SoS took the time to review their current approaches to SE and their views about how the current SE processes applied to their experiences.

Based on these reviews, there were several common approaches that were generally useful to the systems engineers in executing their SE role in the SoS environment.

First, SoS SE addresses organizational as well as technical issues in making SE trades and decisions. When assessing how to support SoS functions, it is important to develop a solid technical understanding of the functionality, interrelationship, and dependencies of the constituent systems. In an SoS, however, it is equally important to understand the objectives, motivations, and plans of systems, since these factors play a large role in the SE trades in an SoS. In many cases, decisions about where to implement a needed function are based on development schedules or funding as much as on optimized technical allocations. When a needed function is aligned with the longer term goals of a systems owner, it is often logical to select that system to host the function even if there are other more technically favorable alternatives. Funding is more likely to be available for development and maintenance, and the program sponsor will be more motivated to adjust schedules and make alterations if the SoS functionality benefits the system's owning organization in the long term.

One of the big issues in an SoS is the need to acknowledge the role and relationship between the SE conducted at the systems versus the SoS level. Systems engineers of SoS find it is important for them to focus on those areas that are critical to the SoS success and leave the remainder of the SE to the systems engineers of the systems. The systems engineers at the system level have the knowledge and responsibility to address implementation details, and they are in the best position to do this. The biggest challenges are determining the areas that need to be addressed at the SoS level and focusing the limited SoS SE attention on those areas. SoS systems engineers typically focus on risk, configuration management, and data as they apply across the SoS. For SoS, a key area of concern is the synchronization across development cycles of the systems. The SoS integrated master schedule (IMS) focuses on key intersection points and dependencies across the SoS rather than focusing on individual systems schedule details. In general, the more SE the SoS can leave to the systems engineers of the individual systems the better.

Technical management of the SoS, particularly the level of participation required on the part of the systems, can be a challenge. Particularly during the early formative stage of an SoS, the tendency can be to ask the systems engineers of the systems to participate in all aspects of the SoS SE process. Given the workload of these systems engineers, this amount of support is simply not sustainable. A successful SoS technical management approach reflects the need for transparency and trust along with focused active participation by the system-level systems engineers. Once a level of understanding and trust has been developed, then a pattern of participation can be created and sustained.

Given the tension between the needs of systems themselves and the demands of the SoS, there is a real advantage to an SoS design based on open systems and loose coupling that impinges on the systems as little as possible. This type of design approach provides systems with maximum flexibility to address changing needs of original users, and permits engineers to apply technology best suited to those needs without affecting the SoS. SoS design trades hence may place a greater emphasis on approaches that are extensible, flexible, and persistent over time and that allow the addition or deletion of systems and changes in systems without affecting other systems or the SoS as a whole.

Specific attention needs to be focused on the design strategy and trades both upfront in the formation of the SoS and throughout the SoS evolution. A traditional systems acquisition program benefits by focusing analysis upfront in the design process. An SoS, however, benefits by conducting this type of analysis on an ongoing basis, since the SoS SE's success depends on a robust understanding of internal and external sources of change. Having understood the sources of change, the SE is then able to anticipate changes and their effects on the SoS.

9.7   FUTURE DIRECTION

So what does this mean for the current DoD guidance for SE for SoS?

As stated earlier, the DoD has identified 16 technical and technical management processes for DoD SE (see Fig. 9.1). These processes are drawn from international standards for SE (International Organization for Standardization, 2002). Given the state of SoS in the DoD and the elements and emerging principles for SoS SE described in the preceding sections, do these basic SE processes still apply in the DoD SoS SE environment?

The processes themselves are fundamental and at the level at which they are specified it is hard to see how these would not apply to SE for SoS. What is different for SoS is the context in which these processes are conducted (Fig. 9.4). Decision analysis, for example, is a basic process in SE. In an SoS context, the decisions are somewhat different and the SoS context means that decisions for the SoS need to be considered in light of the impact on the systems themselves. Likewise areas such as configuration management and data management may be needed at the SoS level but only to address aspects of the SoS not addressed in the SE of the individual systems.

The elements and principles of SoS discussed here are intended to augment current DoD SE practice to account for the SoS challenges.

How much these processes are affected by the SoS context and what added areas the SOS SE needs to address are open questions. It will be important to address these questions as we get more SoS experience in the SE community. This will be particularly important if, as is the case in the DoD, SoS becomes a more prevalent aspect of systems development and employment and systems engineers are called upon to provide the type disciplined technical foundation to SoS that SE has traditionally provided to systems.

In particular:

- What are the real constraints to testing in an SoS environment? How do we address the objectives of testing in ways that acknowledge the complexity and asynchronous development and deployment patterns that characterize SoS environments?
- What are the risk and cost drivers in SoS? How do we identify these, accommodate them, and where possible mitigate against their adverse effects on SoS evolution?

Technical Processes | **Technical Management Processes**

SoS Core Elements	Rqts Devl	Logical Analysis	Design Solution	Implement	Integrate	Verify	Validate	Transition	Decision Analysis	Tech Planning	Tech Assess	Rqts Mgmt	Risk Mgmt	Config Mgmt	Data Mgmt	Interface Mgmt
Translating Capability Objectives	X											X	X	X	X	
Understanding Systems and Relationships		X											X	X	X	X
Assessing Performance to Capability Objectives							X		X		X		X		X	
Developing and Evolving an SoS Architecture	X	X	X						X	X		X	X	X	X	X
Monitoring and Assessing Changes									X							X
Addressing Requirements and Solution Options	X		X						X	X		X	X	X	X	X
Orchestrating Upgrades to SoS				X	X	X	X	X	X		X	X	X	X	X	X

FIGURE 9.4 Technical and technical management processes as they apply to the core elements of SoS SE (OUSD AT & L, 2008)

- Finally, what are the enablers to allow systems engineers to better operate in SoS environments? Are there additional processes that need to be considered as part of the SoS SE toolbox when operating in an SoS environment? Are their new ways to implement current processes that are amenable to SoS characteristics? What incentives could be established for systems to participate in SoS developments? Are there management tools, new contracting methods, or new models of governance that need to be addressed to enable the effective conduct of SE in this new and changing environment?

The questions will form the basis for the investigation of future directions for SE in defense, as the Department works to better support the warfighter capability needs by addressing SoS issues in developing and engineering defense systems as well as in defining capability needs and supporting warfighting operations.

ACKNOWLEDGMENT

The authors would like to express their appreciation to Ralph Lowry, Jo Ann Lane, George Rebovitch, and Bebe Hollingshead for their work conducting the pilot SoS reviews and analysis of the pilot data, which provided the basis for this chapter. They also thank the programs and community stakeholders for sharing their time and experiences that have helped shape this research.

REFERENCES

Chairman of the Joint Chiefs of Staff (CJCS), 2007, CJCSI 3170.01F, Joint Capabilities Integration and Development System, May 1, Pentagon, Washington, DC.

Department of Defense (DoD), 2003, DoD Instruction 5000.2, Operation of the Defense Acquisition System, Ch. 3.5, Concept Refinement, May 12, Pentagon, Washington, DC.

Department of Defense (DoD), 2004a, Defense Acquisition Guidebook, Ch. 4.2.6., System of Systems Engineering, October 14, Pentagon, Washington, DC.

Department of Defense (DoD), 2004b, Defense Acquisition Guidebook, Ch. 10.1, Decisions, Assessments, and Periodic Reporting, October 14, Pentagon, Washington, DC.

Department of Defense (DoD), 2004c, DoD Directive 8320.2, Data Sharing in a Net-Centric Department of Defense, December 2, Pentagon, Washington, DC.

Department of Defense Chief Information Officer (DoD CIO)/Assistant Secretary of Defense for Networks and Information Integration, 2003, DoD Net-Centric Data Strategy, May 9, Pentagon, Washington, DC.

Department of Defense Chief Information Officer (DoD CIO)/Assistant Secretary of Defense for Networks and Information Integration, 2004, Net-Centric Checklist, V2.1.3, May 12, Pentagon, Washington, DC.

Department of Defense Chief Information Officer (DoD CIO)/Assistant Secretary of Defense for Networks and Information Integration, 2005, Net-Centric Operations and Warfare Reference Model (NCOW RM), V1.1, November 17, Pentagon, Washington, DC.

Durham, J.M., 2006, Defense Acquisition University Presentation, AT&L Hot Topics in Acquisition, August 9, https://acc.dau.mil/CommunityBrowser.aspx?id= 120571&lang=en-US.

Durham, J.M., 2007, Defense Acquisition University Presentation, Acquisition Reform Initiatives Concept Decision: a Strategic Choice, February14, http://dodcas.org/DoDCA-S2007presentations/Track2/Session1.1Durham.pdf.

International Organization for Standardization, 2002, ISO/IEC 15288, Systems Engineering – System Life Cycle Processes, http://www.iso.org/iso/iso_catalogue/catalogue_tc/catalogue_detail.htm?csnumber=27166.

Under Secretary of Defense for Acquisition, Technology and Logistics (OUSD AT&L), 2004a, Memorandum on Policy for Systems Engineering in DoD, February 20, Pentagon, Washington, DC.

Under Secretary of Defense for Acquisition, Technology and Logistics (OUSD AT&L), 2004b, Memorandum on Policy Addendum for Systems Engineering, October 22, Pentagon, Washington, DC.

Under Secretary of Defense for Acquisition, Technology and Logistics (OUSD AT&L), 2008, System of Systems System Engineering Guide: Considerations for Systems Engineering in a System of Systems Environment, V1.0, Pentagon, Washington, DC.

Chapter 10

Boeing's SoSE Approach to e-Enabling Commercial Airlines

GEORGE F. WILBER

Boeing Company, Seattle, WA, USA

10.1 BOEING e-ENABLING INTRODUCTION

Alan Mullaly, the then Boeing Commercial Airplane Company President, challenged Boeing leadership to develop a comprehensive approach to "e-enabling" the Boeing commercial airplane fleet. While concrete objectives were yet to be defined, the intent was to accelerate progress in developing a network-centric vision and implementation plan for the commercial airplane fleet and its connectivity into the vast terrestrial-based commercial aviation support infrastructure.

Boeing launched an "e-Enabled Airplane Project." The project focused on developing an enterprise strategy and a technical architecture. These documents were to serve as the framework for making the airplane more network aware and more capable of leveraging industry advances in computing and network technologies. Shortly after launching this project, it became apparent that providing the real value would require broadening the scope of the project to include the entire airline perspective. We needed to view the airplane as only one key component of a much larger system. As a result the e-Enabled Airlines Project was initiated. The project grew to include many ground-based architectural components at the airlines and at the Boeing factory, as well as other key locations such as the airports, suppliers, and terrestrial Internet Service Suppliers (ISPs).

The e-Enabled Airlines Project took on the task of defining a system of systems engineering (SoSE) solution to the problem of interoperation and communication with and between the existing numerous and diverse elements that make up the airlines'

System of Systems Engineering: Innovations for the 21st Century, Edited by Mo Jamshidi
Copyright © 2009 John Wiley & Sons, Inc., Publication

operational systems. The focus was flight operations and maintenance operations, the areas of Boeing expertise. Since that time, the goal has been to find ways of leveraging network-centric operations and systems of system engineering principles to reduce production, operations, and maintenance costs for Boeing, airlines, and key partners, and to leverage e-enabling to provide additional value for the Boeing fleet of aircraft.

One of the key products of this effort is the "Boeing e-Enabled Technical Architecture." The architecture is critical to developing the right technical solutions and unlocking the value of truly network-centric or e-Enabled airline. The architecture is defined at multiple levels of abstraction. At the top, there is a single "reference architecture" that is necessarily highly abstract. Below this there are multiple "implementation architectures" that are very precise and accurate. The reference architecture maps directly to the enterprise strategy. The implementation architectures map directly to real-world airplane and airline implementations. Together they provide a family of physical solutions that have common attributes that exhibit the behavior necessary for the systems to work together and allow reuse of systems components across dissimilar airplane platforms.

The 787 is the first commercial airplane designed from the bottom up to be e-Enabled and implemented network-centric principles from inception. The 787 open data network (ODN) provides a network-centric infrastructure based on the Rockwell-Collins core network. It is one implementation of the e-Enabling architecture. The 787 ODN enables onboard and offboard elements to be networked in a fashion that is efficient, flexible, and secure. The fullest implementations of network-centric principles are best depicted in Boeing's Gold Care Architecture and design. Boeing's "Gold Care" solution, for the 787, leverages global network-centric and SoSE principles and practices. It implements a state-of-the-art solution for airline maintenance operations and support. It is based on emerging onboard and onground computing and networking technologies, combined with real-time air-to-ground communications via satellite and VHF links.

The following lays out Boeing's e-Enabled Airlines Project and its objectives. The focus is on the technical aspects of the problem and chosen solutions. The e-Enabled Technical Architecture is presented at the reference level and is mapped to the 787 airplane implementation. The Gold Care maintenance solution and top-level design is described and is used as an example of the potential of current Boeing e-Enabling capabilities.

10.2 BOEING'S e-ENABLED AIRLINE PROGRAM

Boeing's e-Enabled Airline Program evolved from a small airplane-centric project into a very robust and well-organized program with dedicated leadership, technical staffing, and a deep support infrastructure. The program is housed in the Commercial Aviation Services (CAS) organization with the Commercial Airplane Group. Placing the program in CAS facilitates an understanding of airlines' perspectives since CAS routinely supports airline operations. This adds flight operations and maintenance operations perspectives to the already existing production and development point of view. Much of what follows has a strong operational flavor, in part as a result of the organization structure and deep domain knowledge.

A stated objective of e-Enabling is "the strategic connection and integration of business processes, people, airplanes, information, assets, and knowledge into a single focused business system," and the focus is "*centered on breaking through operational constraints.*"

Today's commercial airline environment is built on a paper-based communication infrastructure and does not rely on data communications between the airplane and the ground or between the ground elements to keep airplanes in the air. Planes routinely fly with voice only communications and carry paper documents onboard. Some planes have narrowband data communications using VHF data links over land and satellite data-links over the ocean. However, these communications links are not efficiently leveraged into the infrastructure at this time. Some airlines have begun to leverage IP-based ground communications between airline, airport, and maintenance sites. e-Enabling is about bringing these technologies together using SoSE approaches to provide "the real-time connectivity of the airplane to the ground, the delivery of information across the enterprise . . .."

Business value, cost reductions, increase in passenger comfort, and overall efficiency are gained when the entire enterprise is connected in real time. Information is created and disseminated in real time from the data that are shared between applications, services, and organizations involved in airline operations, including flight operations, maintenance operations, dispatch, and airport operations. Figure 10.1 shows the e-enabling environment.

10.2.1 The Binding Thread

Initial e-enabling efforts focused on bringing together various existing Boeing programs and projects. Each of these projects was funded and developed independently. However, it was determined that only by bringing them together could we achieve the synergistic value we desired.

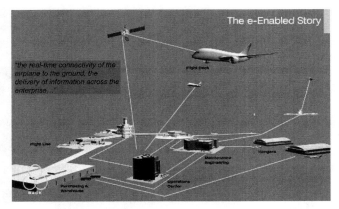

FIGURE 10.1 The e-enabled story

The first step in enabling network-centric airline operations is developing the communications infrastructure. A communication infrastructure based on Internet protocols, Internet infrastructure, LAN-based computing, wireless communications, and low-cost, pervasive computing provides a basis for sharing of data over a global, mobile, ad hoc, and dynamic network. The next step is developing airline applications and services and layering them on this emerging infrastructure. The applications convert the newly available data into useful information. They then provide the information to the appropriate decision makers. e-Enabling provides the binding thread between the application and service-oriented programs. Figure 10.2 depicts the "binding thread" of e-enabling.

One example of new emergent capability is performing real-time maintenance. Bringing together networked avionics, onboard maintenance systems, pilot communications with ground-based maintenance teams, automated sparing, and inventory management systems using air-to-ground data links and the e-enabling infrastructure has opened the door for a more efficient paperless solution to the flight line maintenance problem. We simply needed to identify the final ingredient, the creation of an electronic logbook. The electronic logbook replaces the paper for documenting, managing, and sharing the list of maintenance defects on the airplane. Making it electronic allows it to be shared quickly and easily with all of the organizations and systems involved in the maintenance support process. While the additional effort was small, the emergent value was huge.

The result is the capability to monitor airplane health, determine corrective action, and position maintenance resources in real time, while the airplane is still in flight. This emergent capability relies primarily on existing capabilities: applications, airplane functionality, maintenance, and operational support processes. We leveraged existing airplane capabilities and those of emerging maintenance support projects that were already in place with independent business cases. Each function was independently designed to provide a unique solution to a piece of the airplane maintenance problem. Linking them together resulted in additional capabilities and more efficient solutions to the maintenance support problem than originally anticipated. It created tremendous new value for the airlines. This is possible because the airplane is a very high-value resource and small increases in efficiency provide large increases in value. Figure 10.3 shows the real-time maintenance example of e-enabling emergent behavior.

10.3   BOEING e-ENABLED TECHNICAL ARCHITECTURE

One of the major products developed by the e-Enabled Airline Program is the "Boeing e-Enabled Technical Architecture." This architecture has been developed to provide a common technical approach and framework for the enterprise. It has been designed to map the enterprise strategy and vision into the technical world that builds, flies, and maintains the airplanes.

The architecture embodies the enterprise strategy and vision by ensuring all projects, features, and capabilities are included and leveraged. Further, it contains

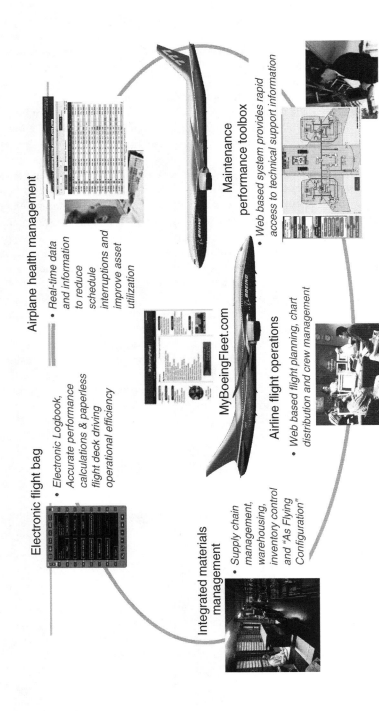

Electronic flight bag

• Electronic Logbook,
  Accurate performance
  calculations & paperless
  flight deck driving
  operational efficiency

Airplane health management

• Real-time data
  and information
  to reduce
  schedule
  interruptions and
  improve asset
  utilization

Maintenance
performance toolbox

• Web based system provides rapid
  access to technical support information

MyBoeingFleet.com

Airline flight operations

• Web based flight planning, chart
  distribution and crew management

Integrated materials
management

• Supply chain
  management,
  warehousing,
  inventory control
  and "As Flying
  Configuration"

Integration and connectivity

FIGURE 10.2   The binding thread

236

Achieve competitive advantage through integration and connectivity

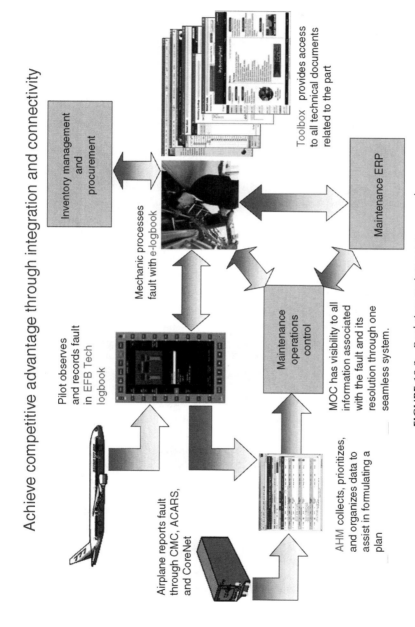

Pilot observes and records fault in EFB Tech logbook

Mechanic processes fault with e-logbook

Inventory management and procurement

Toolbox provides access to all technical documents related to the part

Airplane reports fault through CMC, ACARS, and CoreNet

AHM collects, prioritizes, and organizes data to assist in formulating a plan

Maintenance operations control

MOC has visibility to all information associated with the fault and its resolution through one seamless system.

Maintenance ERP

FIGURE 10.3 Real-time maintenance example

interconnections, data links, applications, and support services deemed essential by each individual program for that program's independent success. It functions across the enterprise, yet embodies the needs and expectations of each individual program. It provides the glue that allows these individual programs and projects to function as a whole. Each program is its own independent entity with its own system. Together they form a system of systems.

The architecture is designed to facilitate the implementation of network-centric and SoSE principles and to remove SoSE roadblocks. Roadblocks arise when individual programs are unable to achieve the greater value of the enterprise due to suboptimal pressures on the individual program. There is usually a cost associated with joining a system or project into a larger system or system of systems. The architecture is designed to show the value of the investment, reduce integration costs, maximize interoperability, and maximize reuse of the already developed and emerging components. The benefits must outweigh the cost of implementation for the architecture to be adopted. If the costs are not balanced against the overall and program-level values, then some seed money must accompany the architecture to ensure its adoption and compliance.

Projects adopting the architecture have access to new, already developed capabilities and features built-in to the architecture. The value obtained by any emergent behavior from adopting the architecture is immediately available to the participating projects. The most significant benefits of adhering to the architecture are reuse, reduced development time, reduced application complexity, and increased design stability. These benefits result from efficient use of existing capabilities, services, and resources. There are many forms of reuse, including direct reuse of a component (software or hardware), reuse of services, reuse of infrastructure, reuse across airplane types, reuse across airlines, reuse of certification basis and related paperwork, and reuse of the interfaces.

The architecture is made up of components and their relationships. The components are reusable and their relationships are both logical and physical. Components may be hardware elements or software services and applications. Physical relationships are based on off-the-shelf protocols, standards, and communication mediums.

The architecture needs to be simple to understand and convey key principles, yet it needs to be comprehensive to be readable by the detailed technical team members who will implement it. The resulting architecture has a single reference architecture component, with several different views. It also contains implementation drawings that map the simple, high-level reference architecture to real-world, complex implementations. It is viewed from the top-down by the reference drawings and from the bottom-up by the implementation drawings.

The reference drawings are developed, maintained, and controlled in a single organization for the entire enterprise. The implementation drawings are developed and maintained by coordinated architecture and project teams and are then controlled and maintained by the program or project team that is implementing the actual equipment, airplane, or system.

The Boeing e-Enabled Technical Architecture top-level reference drawings include an operational view of the key components and domains in the architecture's scope. Note that there are potentially an infinite number of domains and components that could be addressed. The current architecture focuses on Boeing production and airline flight and maintenance operations. This is necessary to manage the scope of the effort and maintain focus on the areas that provide the biggest return on investment. The domains considered essential to the architecture include Boeing, airlines, industry partners, airports, Internet, communication services, and of course the airplanes themselves. Figure 10.4 shows the top-level operational view of the e-Enabled Technical Architecture.

10.3.1 Onboard Imperative Architecture

While the overall technical architecture combined ground and airborne components, there is a natural break between the onboard and ground-based systems and systems of systems. The ground-based systems are generally developed and managed using information technology (IT) methods and standards. The onboard systems and offboard communication elements are usually managed using avionics development processes and commercial aviation standards. This section focuses on the onboard subset of the overall technical architecture.

The onboard elements of e-enabling make up a system of systems themselves. There are a variety of onboard domains, each with their own set of systems, development rules, operational constraints, and interfaces. The onboard architecture challenge is to bring these all together using network-centric principles to leverage their capabilities and to reduce equipment development and support costs, while maintaining aviation safety and security standards. Onboard considerations differ from the ground-based ones as they include power, weight, cooling, size, hazardous effects, and many more unusual aspects. Figure 10.5 depicts many of the existing onboard domains that must be connected.

The first attempt at developing an onboard architecture resulted in a formal document that was an inch thick. The second version was more than double that size. When presented to the program's chief engineers, they disparaged the size of the document and even the cost of just reading and understanding it. Boeing took the challenge of augmenting the architecture with an "imperative" overlay. This is a simplified readers' digest version that is technically accurate but only 12 pages in length. Developing a 12-page architecture that is technically accurate and complete was an exciting challenge that we accomplished in less than a year. The result is known as the "Boeing e-Enabled Onboard Imperative Architecture."

The intent was to focus on the key elements of the architecture that bring big value, drive the enterprise to more affordable solutions, and can be accomplished in the very near future. The imperative architecture would be refreshed from time to time as new business needs or technologies emerge. This approach provides for low-cost under-standing of key architecture attributes and for updating and sharing of the architecture with all of the parties involved in the various air and ground implementations. The limited page count forced a concerted effort on focusing the message. It provided for a crisp definition of components and their relationships. Finally, it provided a lexicon for communications between the individual programs and the architecture team.

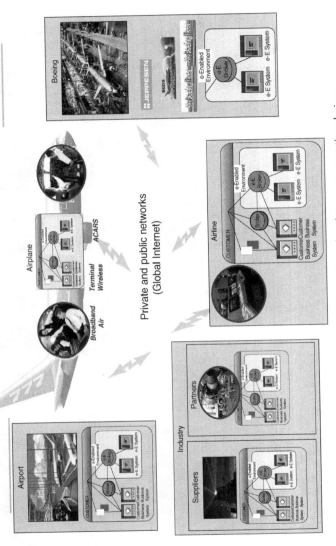

FIGURE 10.4 Boeing e-enabled technical architecture operational view

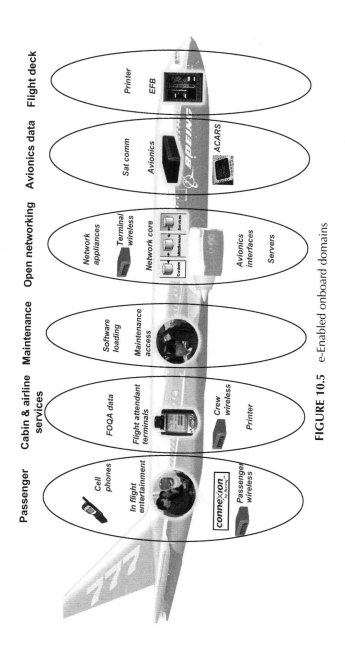

FIGURE 10.5 e-Enabled onboard domains

Passenger | Cabin & airline services | Maintenance | Open networking | Avionics data | Flight deck

Cell phones
In flight entertainment
Passenger wireless

FOQA data
Flight attendant terminals
Crew wireless
Printer

Software loading
Maintenance access

Network appliances
Terminal wireless
Network core
Avionics interfaces
Servers

Sat comm
Avionics
ACARS

Printer
EFB

The decision was made to develop 12 drawings with one view graph and one page of notes for each drawing. Three drawings would represent the reference architecture, six drawings would describe the most strategic components, and three drawings would map common implementations to the reference architecture. Figure 10.6 shows the Onboard Imperative Architecture context.

10.3.2   Onboard Imperative Architecture Elements

The imperative architecture contains only 12 drawings (with notes).

Three reference architecture drawings include (1) a connectivity drawing that defines the reference components, domains and interconnections; (2) a layers charts that shows components and hierarchical relationships with each other; and (3) a reference-to-implementation names drawing that provides the lexical conversions between naming conventions of the reference components and the real-world equipment in the implementations .

The top six strategic components are represented in drawings. These are the key components for today. The components were selected from the larger set, as those with strategic value. They are important to more than one project, need interproject management attention, and/or critical to deployment of the overall architecture. Each component drawing describes the component, includes a development roadmap for component implementations, provides links to owners and developers, and contains mapping to detailed descriptions in the larger technical architecture document(s).

Three common implementation drawings are also provided. The initial attempt at getting working level buy-in from implementation team engineers made it clear that high-level abstracted architectures do not map quickly to real-world implementations since the working level architectures are mired in real-world designs, issues, and drawings. These drawings provide a bridge between the high-level abstraction and the plethora of detailed implementations and drawings. A close look at current and near-term production technologies revealed the need for three classes of hardware implementations. So, we determined to develop one common and reusable drawing for each of these classes of hardware. These represent the current state of the art in onboard aviation design and packaging. The three drawing include (1) an integrated modular solution with a version one system, (2) the 787's second-generation integrated modular solution, and (3) a federated or individual component-based implementation suitable for very small airplane platform types and retrofit implementations. Figure 10.7 shows the 12 drawings in the Onboard Imperative Architecture.

The connectivity drawing provides a graphical representation of all of the key onboard reference components. They are organized by onboard network domains. Each domain has unique handling characteristics from other domains. Intercomponent communication is shown by connecting lines. Most of the intercomponent communication is by IP over Ethernet; however, some connectivity is accomplished using native avionics busses or wireless techniques. Figure 10.8 shows the reference architecture Connectivity Drawing.

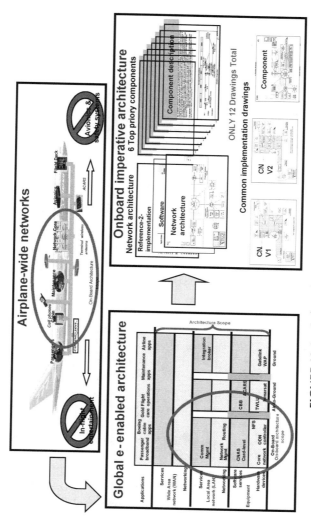

FIGURE 10.6 Onboard imperative architecture context

243

Onboard imperative architecture

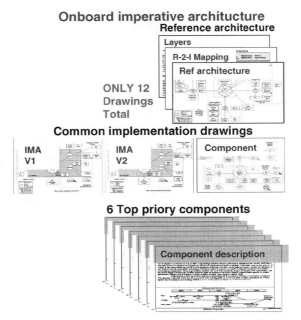

FIGURE 10.7   Onboard imperative architecture drawings

The domains provide for common security management within the domain. The private airplane domain contains applications and services that augment the avionics systems. These components are relatively stable over time. They also require some high-level security protections and even safety-related evaluations. The cabin and airline services domain contains components designed to serve the flight attendants and cabin crew. They are moderately stable applications and require moderate security to protect data but are not safety critical. The network control domain contains all of the network operations, management, and control elements and must not be tampered with by outside forces. The passenger services domain contains in-flight entertainment access and passenger business services. This domain is very fluid and includes passenger computing and telecommunications equipment, such as laptops, PDAs, and cell phones.

The layer drawing shows the reference components and their relationships based on architecture layers. This is similar to protocol layers, except that strict communication between layers is not implied. The objective is to migrate from vertical solution-based designs. Today's designs are implemented such that the applications and services are tied to the specific hardware equipment. Services are rarely exported from the specific hardware component in today's architecture. Moving to a service-oriented approach facilitates software reuse, minimizes redevelopment costs, and isolates the impacts of hardware and application changes. The transition to a service-oriented open architecture requires significant changes in our system engineering and procurement processes as it affects all aspects of avionics design, development, procurement, and supplier relations.

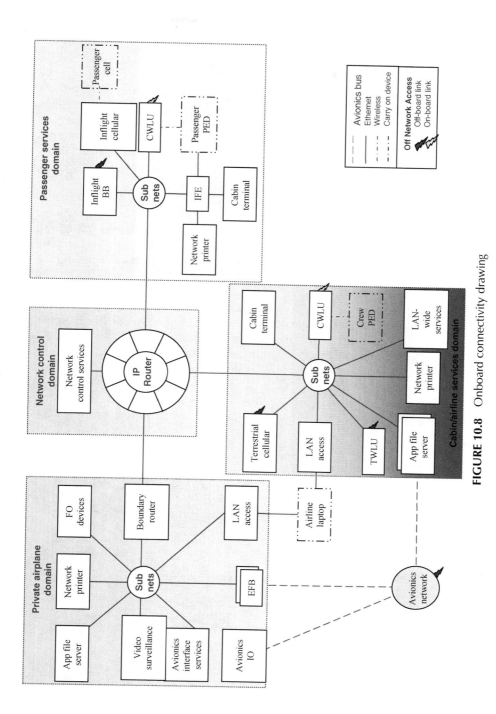

FIGURE 10.8 Onboard connectivity drawing

245

We are taking a multistep approach at implementing the layered architecture. The first step is separating the physical hardware equipment from the applications. Next we defined the reference architecture using a hardware-neutral construct. Then, we added common services in middleware layers and removed much of the duplication from the individual applications, reducing their size and complexity. We are now in the process of adding an information services layer to provide more value through even higher level services. Finally, we will modify our procurement and supplier relationships to allow development and procurement of separate hardware equipment and software services and applications. This allows reuse of common applications, data, and services across differing airplane platforms, and physical implementations. Figure 10.9 shows the reference architecture layer drawing.

Name Mapping Drawing: One problem with abstracted architectures is what to name the components. Create a new abstract name and nobody knows what you are talking about. Reuse an existing name and everyone understands, but adds their own individual meanings. So, we added a mapping from the reference names to existing implementation names as a lexicon for translation between reference discussions and real-world implementations. Figure 10.10 shows the reference architecture name mapping drawing.

Common Implementation Drawings: The reference architecture drawings are all necessarily abstract. They are logical and functional but not physical. Many project engineers cannot understand or interpret these abstract drawings. Making a cognitive link between them and a really messy wiring diagram for a specific airplane platform required the creation of a new set of drawings that show both physical and logical in one drawing.

We determined to take the reference components and overlay them on a physical background of real-world airplane hardware-based designs. We did not want to build

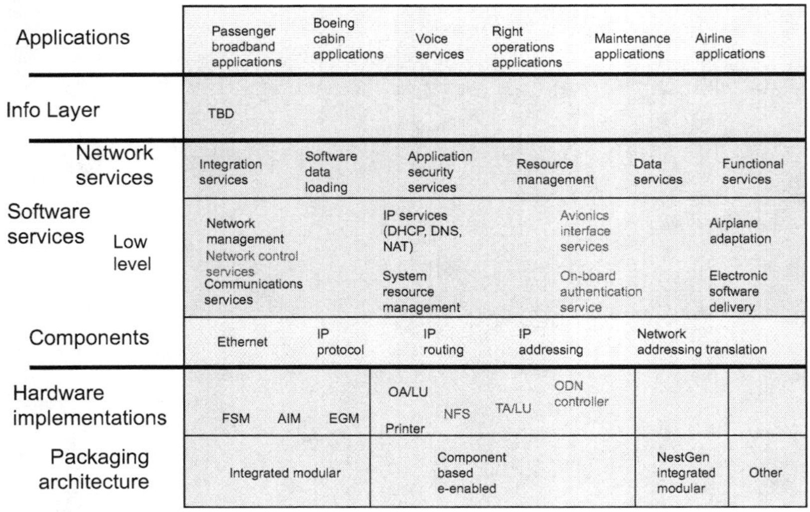

FIGURE 10.9 Onboard layer drawing

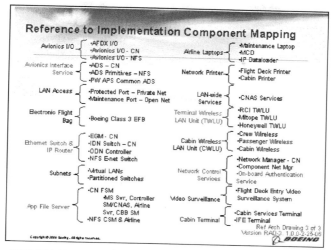

FIGURE 10.10 Reference to implementation name lexicon

one for each unique airplane model and submodel but chose to build the drawings based on common topologies and architectural solutions. We have two types of physical architectures. The first is based on highly integrated solutions, with few big boxes. The second is based on highly federated solutions with many small boxes. The first is more efficient for a new larger airplane, while the later is more efficient for smaller and older aircraft.

We created three reusable or common implementation drawings based on the three existing hardware architectures used in Boeing airplanes. The first is based on Connexion by Boeing first-generation integrated solution. The second is based on 787's second-generation hardware implementation. The third is based on a federated solution of small ARINC 600 hardware boxes like that required for the 737 NG. Figure 10.11 shows the three reference architecture common implementation drawings.

10.3.3 Mapping to Implementations

The architecture with reference and common implementation drawings are provided to all Boeing commercial airplane project teams. Project teams builds a set of implementation drawings for their particular project based on their specific needs. We, e-Enabled Airlines Project Architecture Team, work with the BCA project team to ensure the project team's implementation maps to the overall architecture. There are no formal mapping procedures at this time. The manual review process is a key part of getting agreement on the architecture being adopted by the projects and understood by the working-level engineers across the enterprise. It facilitates discussion and communication about the architecture, critical strategies, and driving principles, allows us to create a common set of terms and descriptions of baseline functionality, and allows us to level set expectations for the common reuse of applications and services.

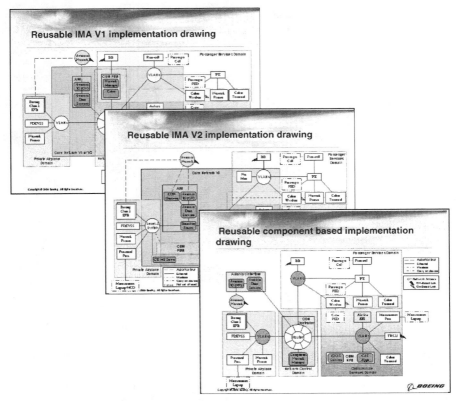

FIGURE 10.11   Common implementation drawings

Figure 10.12 shows how the On-Board Imperative Architecture maps to program implementation drawings.

10.3.4   Infrastructure Implementations

The architecture is designed in layers to allow applications and services to be abstracted from the hardware. This facilitates the continual rollout of equipment and support for the life span of the airplane, which is in excess of 30 years. Working hardware is the key to delivering the systems. There are two approaches to developing flight hardware. The first is using integrated modular techniques, putting everything in a box. The second is a federated or component-based solution that leverages many smaller and reusable boxes.

The core network component developed by Rockwell-Collins is an integrated modular solution that is used on airplanes equipped with the Connexion by Boeing and on the 787. It is a network in a box. Core network provides a single box solution to an IP network designed from the ground up to operate in the hostile and unique airplane environment with a broad set of interfaces developed specifically to link to the variety

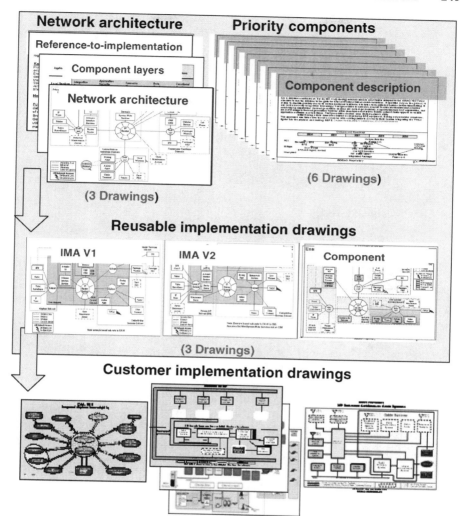

FIGURE 10.12   Projects drawing relationships to reference architecture

of existing and emerging avionics architectures in the fleet today. One box has five computers, 24-port Ethernet/IP router, firewalled avionics interfaces, internal fire-walls and border routers, and a robust power supply. Usually, it is much less expensive to buy and install one bigger box than numerous small ones. Figure 10.13 shows the Core Network version 2.

Federated solutions use many smaller components and offer a broader scale of solutions. Small-size aircraft often find it easier to fit in several smaller boxes than one large one. Federated solutions are essential to fill the needs of minimally equipped e-enabled airplanes and provide the physical attributes necessary to retrofit e-enabling into the older, existing fleet.

FIGURE 10.13   Rockwell-Collins core network equipment

10.4   e-ENABLED APPLICATIONS

While the first steps in implementing the e-enabled airline solution are developing the architecture and implementing the network, computing, and communications infrastructures, the real value exists in the applications that are built on them. The e-Enabled Airline Program is responsible for developing the infrastructure and application enablers, not the applications themselves. The e-Enabled Airlines Program does, however, provide seed funding and evangelization to the projects, program, and application teams to energize them to become part of the broader enterprise-wide e-enabling team. Each application or program is independent, each is a system that stands alone, each must support its own business case, and each must meet its internal development plans and schedules. Like any system of systems development, the end product results by bringing numerous independent, yet cooperative applications to the end users where the emergent behavior provides additional business capabilities.

10.4.1   Electronic Flight Bag

The electronic flight bag (EFB) is a critical element of the e-Enabled vision. The Boeing EFB is a class 3 device that provides the flight crew, both pilot and copilot, color graphic displays and controls. It consists of four computers and hosts numerous applications.

The EFB has many built-in applications and displays, including en route charts, terminal charts, video surveillance, and access to electronic documents and specifications. One emerging application is the electronic logbook. This is the application that enables real-time maintenance capabilities by keeping maintenance record in electronic format. Figure 10.14 shows the Boeing Class 3 EFB installed in captain's side of cockpit. Figure 10.15 shows some of the EFB applications displays for the flight crew.

FIGURE 10.14 Pilots EFB

10.4.2 Airplane Health Management

Boeing's Airplane Health Management (AHM) system provides "real-time fix or fly decision support." AHM is a hosted software application that combines data and information into an integrated decision support tool. It is designed to improve airplane availability while providing maintenance and engineering efficiencies. Its users include maintenance controllers, real-time fleet management, engineers, and mechanics. AHM provides in-depth analysis and trending of maintenance problems and solutions and advance awareness of inbound airplane health conditions. Figure 10.16 shows AHM data flow concepts.

10.4.3 Maintenance Performance Toolbox

Boeing's maintenance performance toolbox is an online, browser-delivered suite of services that combine intelligently linked documents with visual navigation methods to support efficiency and effectiveness of maintenance operations. Toolbox services are accessed through MyBoeingFleet.com to support airplane system troubleshooting and research, structural repair record management, job card management, customer content authoring, and document and hosting. Modules are combined to form solutions for aircraft maintenance operations. Solutions are designed to dramatically improve the efficiency and capabilities of the maintenance and engineering staffs; find critical information faster; collaborate internally, with partners and with Boeing; control airline content; and raise level of proficiency through on-demand training. Figure 10.17 shows two maintenance performance toolbox displays.

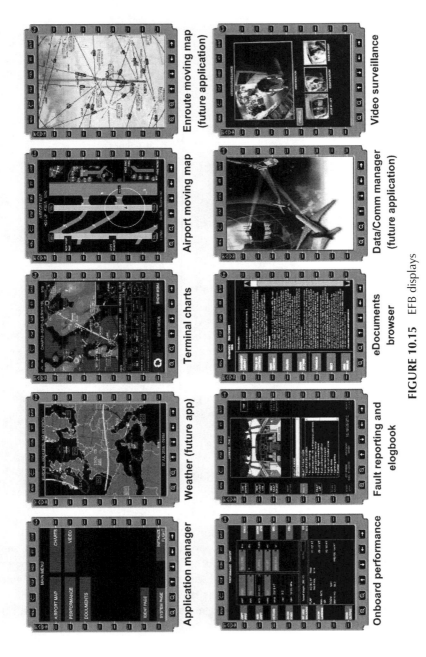

FIGURE 10.15 EFB displays

Enroute moving map (future application)

Video surveillance

Airport moving map

Data/Comm manager (future application)

Terminal charts

eDocuments browser

Weather (future app)

Fault reporting and elogbook

Application manager

Onboard performance

FIGURE 10.16 Airplane health management

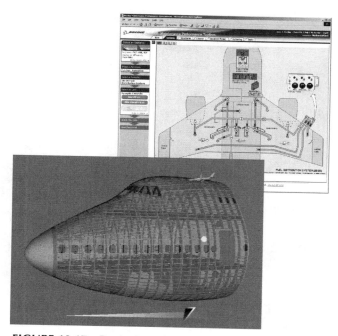

FIGURE 10.17 Boeing's maintenance performance toolbox

10.5   THE e-ENABLED 787

The Boeing 787 is the first airplane designed to be e-enabled from initial development. Many of the key e-enabled capabilities exist in the 787. The airplane has been designed to operate on the network or with intermittent network access. It has onboard e-enabled applications built-in. The capabilities give the 787 a competitive advance. They provide improved operational efficiencies, reduced maintenance costs, and dramatically improved traveling experience for passengers.

Boeing is seeking to deploy new technologies and products on the Dreamliner, including electronic documents, high-bandwidth connectivity, electronic flight bag, wireless Ramplink, Jeppesen electronic charts, Boeing Digital Technical Documents, and online support via MyBoeingFleet.com. These independent products are highly integrated with the onboard maintenance, dataload, and crew information systems. Together they set a new standard for network-centric services in a commercial airplane. Figure 10.18 shows a mock-up of the e-Enabled 787 cockpit where the displays and controls for the integrated products are brought together.

BOEING GRAPHIC

The flight deck in the 787 Dreamliner will include the Electronic Flight Bag, which features an electronic logbook and leverages the airplane's satellite connections. The logbook automatically provides reports to airline maintenance teams, allowing mechanics to better plan repairs and pre-position parts prior to airplane arrival.

FIGURE 10.18   e-Enabled 787

10.6   BOEING'S GOLD CARE MAINTENANCE FOR 787

Many of the e-enabled applications are designed to provide value to the airlines, and Boeing prides itself in bringing value to the airlines. However, the cost of development of applications and infrastructure is often high. It was through the development of the Boeing Gold Care offering that we were able to implement many of the more complex e-enabled capabilities. Gold Care is the ultimate in e-enabling today. Gold Care requires a fully e-enabled airplane that is fully connected to the Internet and a fully developed global network-centric solution for the ground side.

Boeing's Gold Care solution is about providing full maintenance for the 787. Boeing takes on many of the maintenance chores and allows the airlines to focus on operations and customer service. Basically, the airplane can be leased to the airlines by the hour, much like the airplane engines are today. Gold Care simplifies aircraft ownership by reducing cost and risk throughout the 787s life cycle. It provides benefits in acquisition, operation, and transition. Boeing takes on many maintenance responsibilities currently required for the airlines. By managing the activities from a central location, overhead costs are reduced, saving the airlines money and increasing efficiency.

Gold Care is possible because of e-enabled technologies, yet Gold Care is an excellent example of what is possible when airplanes and ground systems become network aware and leverage system of systems engineering principles. Figure 10.19 shows the reduction in organizations possible using Gold Care.

Gold Care's approach is based on a globally connected airplane and maintenance organizations. Gold Care manages the maintenance aspects of the business and leverages the reduced overhead. Figure 10.20 shows how Gold Care is based on IT infrastructure.

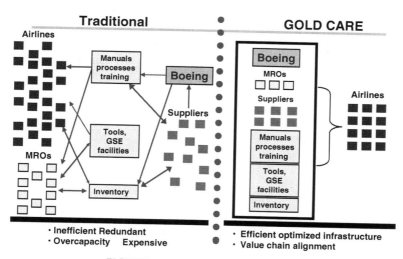

FIGURE 10.19   Gold Care approach

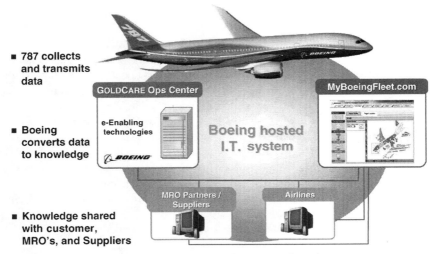

- 787 collects and transmits data

- Boeing converts data to knowledge

- Knowledge shared with customer, MRO's, and Suppliers

FIGURE 10.20   Gold Care is IT based

10.7 SUMMARY

The Boeing e-enabled journey began with a simple directive from our president. Along the way, we learned to comprehend and embrace network-centric principles and ultimately wound up describing and developing it as a system of systems engineering effort. While we did not set out to develop a system of systems solution, it soon became evident that this was the only way to unlock the hidden values and the only way we could afford to develop the new capabilities.

Boeing developed the e-Enabled Technical Architecture and its subarchitecture the Onboard Imperative Architecture to capture the salient and essential elements of the technical designs. Boeing developed several common implementation designs for current aircraft platforms and then launched the 787 based on these architectures and designs. New applications, services and, value propositions emerged along the way, and the ultimate outcome is the 787 Gold Care offering that is the state of the art in commercial airplane systems of systems engineering solutions.

Chapter 11

System of Systems Perspectives on Infrastructures

WIL A.H. THISSEN and PAULIEN M. HERDER
Delft University of Technology, The Netherlands

11.1   INTRODUCTION

The functioning of modern society to a large degree depends on the quality of the infrastructural facilities available. Physical infrastructures such as networks for transport, water and energy provision, waste removal, and telecommunications provide the basic conditions for the functioning of individuals and businesses. Technological progress and the shift to a predominantly service-oriented economy have brought the information, telecommunication, and knowledge infrastructures to the forefront.

Over time, infrastructures have become increasingly critical to the functioning of society, as economic and social processes to a large extent rely on the services provided by such systems. Failure of these infrastructures is one of the most important vulnerabilities of modern society as has been demonstrated by the detrimental effects of interrupted water provision, electricity blackouts, or failing telecommunications in the past.

Historically, infrastructures have been physically and logically separated systems. Advances in technology, and especially in information technology, however, have led to an integration and interlinking of the various infrastructures. Many of the critical infrastructures depend upon a stable and reliable electricity supply. Failure in the electricity infrastructure consequently harms many aspects of society. Similarly, information and communication technology (ICT) systems have become crucial for

System of Systems Engineering: Innovations for the 21st Century, Edited by Mo Jamshidi
Copyright © 2009 John Wiley & Sons, Inc., Publication

the functioning of most if not all other infrastructure systems (and for many other sectors as well).

In addition to the inherent vulnerability of the interconnected infrastructures, many of them are or have been in institutional change processes, such as privatization and liberalization processes. This seriously complicates the matter of reliable and safe design and operation of the infrastructures, because many more actors are involved in the system, each with their own goals and objectives. Especially the interplay between technological, behavioural, and institutional factors and sometimes the rapid expansion of the number of public and private parties involved increase the complexity of the infrasystem. Further, the increasing transboundary dependencies require analysis and design at the international level for technological and institutional innovations. Finally, the tensions between market concerns addressing (short term) economic gains and quality of service, issues of equity and fairness, and long-term sustainability requirements complicate the design and operation of infrasystems.

These changes pose formidable challenges to policy makers, business innovators, system designers, infrastructure operators, and scientists alike. The infrasystem research field therefore covers a broad spectrum of subjects, which can be viewed from different angles, exist in different contexts, give rise to different questions, and may lead to different solutions. There are, however, a number of common aspects among the infrasystems and their associated research (Herder and Thissen, 2001):

- Infrasystems display spatial, networked characteristics.
- Ongoing change in technology and institutions, such as privatization and liberalization in many sectors.
- There is a need to consider sometimes conflicting performance indicators. These include reliability of infrasystems, quality of service, costs, public access, and environmental and other impacts Van de Riet, (2003).
- The efficacy of infrasystems depends on the complex interplay of technological, institutional, and human factors.
- Infrasystems converge, that is, they become more and more intertwined and interdependent. Technologically, many if not all infrasystems depend on the energy system and, increasingly, on the ICT infrastructure. Institutionally, more and more companies through mergers or takeovers exploit different infrasystems such as energy and water provision (CMIIP, 1995 and GAO, 2003).
- Infrasystems display all characteristics of systems of systems (Sage, 2005): they consist of multiple systems that are governed independently, contain strong legacy elements, contain both technological and behavioral elements, display emergent behavior, and so on.

Because of these commonalities, lessons learned in one infrastructural sector may be transferable to other sectors and the exchange of research approaches, findings, and solutions may be rewarding. However, this is not an easy task. First, because numerous differences exist among the sectors, an adequate framework for identification of commonalities and differences is virtually missing. Second, because research has traditionally been focused on a single sector, on a part or on a single aspect only. Few, if any, attempts have been made at comparative research. While an understanding of the complex interactions of the variety of system components and aspects is needed, infrasystem complexity is so large as to make integrated system analyses including all relevant aspects virtually impossible.

In this chapter, we will explore ways in which a system of systems perspective on infrasystems can assist in capturing the intricate interplay of technology and institutions at various system levels. First, we will present a simple conceptual system of systems model that can be used to:

1. identify the key aspects and elements of different infrasystems,
2. compare commonalities and differences among the sectors.

Next, we will describe two more specific examples of energy-related analysis and design issues to illustrate, on the one hand, the potential utility of systems of systems decompositions, and, on the other hand, the variety of possible decompositions. Finally, we will discuss some determinants of infrasystem performance and show how these performance indicators tie in with the various system of systems model levels. We conclude our chapter by setting a research agenda.

11.2   GENERAL CONCEPTS AND MODEL

11.2.1   Toward a Reference Model: the Infrasystem Layer Model

Owing to the multidisciplinary character of the infrasystems research area, many different models exist. The plethora of models, each developed from a monodisciplinary perspective, and with a specific goal in mind, does not facilitate comparison of findings, let alone smooth the way for cross-sectoral learning. In this section, we will introduce a generic multilevel model and illustrate the use of the model.

We, first, note a fundamental disciplinary difference in modeling approaches. Authors rooted in the engineering disciplines generally adopt a layer modeling approach. This fits the established engineering paradigm of dividing large systems into smaller subsystems. The system of systems modeling approach (e.g., DeLaurentis and Callaway, 2004; Sage, 2005.) fits in this modeling paradigm. Authors with a nonengineering background, such as political science, economics, or public management, primarily use a model in which the different institutions, actors, or parties of relevance to a certain phenomenon are analyzed. To tackle the multidisciplinary problems that are posed by the infrasystems, however, we need to address both engineering and nonengineering aspects (De Bruijn et al., 2005; De Bruijn and Herder, 2008).

We propose a conceptual infrasystem reference model that is based upon a systemic layered view on infrasystems (see also Koolstra, 2000; Thissen and Herder, 2003). Each layer comprises physical/technical, operational, institutional, and regulatory components, actors and their interactions. We distinguish three generic layers: a

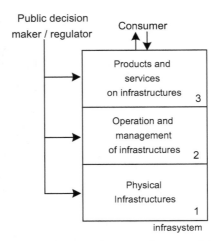

FIGURE 11.1 Infrasystem layer model

physical infrastructure layer, an operational layer, and a services layer. The distinction between the layers is based on functionality: the lower layer systems provide the conditions necessary for the existence and proper functioning of the higher layer systems. This infrasystem layer model is depicted in Fig. 11.1.

The bottom layer of the three layers covers the physical infrastructure, that is, the links (roads, pipelines, cables, etc.) and nodes (power plants, waste incinerators, water buffers, etc.). The processes concerning the design, construction, and maintenance of the infrastructure are also included in this layer, and executed by actors such as physical infrastructure owners, designers, project commissioners, contractors, and the like. This bottom layer lays the foundation for the "higher level" processes that take place in layer two and three.

The second layer of the general infrasystem model encompasses the network operation and management processes and actors. Processes in this layer deal with network control, capacity management, and routing on the network. For example, the stability of the national electricity grid, traffic control at airports and on roads, and Internet routing algorithms are processes that are executed in this layer by the coordinated actions of network operators and capacity managers. The processes in this layer interact systematically with the top layer via the capacity assigned to various service providers acting in the top layer.

The third and top layer of our layer model concerns the supply and use of infrastructure-based products and services, such as energy provision, public transportation, communication, drinking water provision, or wastewater removal. In some infrastructure sectors, such as in the electricity or water infrastructure, specific tangible products, produced by a single or a few large producers, can be identified. Other infrasystems, such as the transportation and ICT sectors, do not include producers of tangible, physical products. In these sectors, service providers supply services instead of concrete products to the consumers, for example, train companies that run transport services on railroads, or telecommunication service providers.

Processes in the model's top layer interact with other societal processes that determine service demand.

Finally, a public decision maker/regulator is introduced in the model. This actor is represented in all levels of the model. The visibility and influence of this actor on the systems and processes within the infrastructure varies among the infrastructures. Some have strong regulators that regulate the physical infrastructures as well as the operation and services levels, whereas other infrastructures have regulators that mainly operate on the services level.

11.2.2 Sector Comparison

Using the layer model introduced in the previous section, we are able to identify a number of fundamental commonalities and differences between the sectors. Figure. 11.2 (adapted and extended from Thissen and Herder, 2003) provides an indicative overview of the way key elements in different sectors fit in the three-layer model. For each layer and sector, we indicate roughly:

- the key processes and/or services,
- the key actors/roles, and
- the key components.

Concerning the bottom or physical layer, not all infrasystems appear to have a well-defined physical network structure, nor do actual physical goods always flow through the links and nodes of the infrastructure. In addition, some infrastructures comprise "active" nodes, where a physical good is processed, changed, or converted, whereas other infrastructures comprise mainly "passive" nodes, where only buffering or (re-)routing activities take place. Finally, the physical layer contains differences with respect to the existence of sources and sinks: the energy infrastructure has well-defined large-scale sources, while other infrastructures, such as the telecommunications infrastructure contain no large-scale sources or sinks, as consumers and providers play both small scale roles. Therefore, some problems associated with infrasystems, such as the vulnerability of an infrasystem to large-scale source outages are not found in every infrasystem, due to the sheer physical structure and workings of some infrasystems.

In the middle layer, fundamental differences are noted with respect to the ways the network is operated and with respect to the ways the capacity is managed on the networks. Some infrasystems have few means of network control or management, for example, road infrastructures, whereas other infrastructures are strongly controlled, such as a rail infrastructure. A general development recognized in all infrasystems is the penetration of ICT into the infrasystem's control and management layer. ICT enables more efficient data capturing and processing, enabling a more effective and efficient use of the infrasystem.

Furthermore, ICT has a strong impact on the service layer of the infrasystem. It enables more diverse services, and better service, by allowing the consumer to buy

Model layer	Sector	Transport	Energy	Drinking water	ICT
Products & services	function	delivery of freight/persons from A to B	delivery of energy, for example, gas, electricity, heat, fuel	delivery of water	delivery of communication services, provision of access to text, images, sounds, movies
	processes	driving, flying, sailing, etc and so on	gas, electricity, heat, and fuel distribution	water distribution	content production, signal distribution
	actor roles	transporters, for example, by means of bus, train, taxi, truck, airplane	energy distributor	water distributor	content provider, mobile tel. provider, fixed tel. provider, cable content provider
Operation & management	processes	traffic control	network and plant operation	network and plant operation	network operation, providing bandwidth
	actor roles	traffic control centre, road administration, rail operator	network operator, plant operator	network operator, plant operator	network operator
Physical infrastructure	nodes	airport, train station, bus station, car park, harbour	power plant, gas plant, refinery	water production plant, reservoir	satellite, antenna, telephone, television, radio
	links	rail roads, roads, water ways	power lines, gas pipes	water pipes	wires, cables, frequencies
	carriers	bus, train, car	–		–
	typical topology	roads: meshed air: star (hub-and-spoke)	main grid: ring local grid: star	tree or ring	meshed
	processes	design, construction, maintenance, repair	design, construction, maintenance, repair	design, construction, maintenance, repair	design, construction, maintenance, repair
	actor roles	(rail-)road administrator, contractor; airport authority	energy supplier, network owner, contractor	plant owner, network owner, contractor	network /cable owner, contractor

FIGURE 11.2 Structured comparison of differences and commonalities among infrastructure sectors

tailor-made solutions. The trend toward the branching out of services appears to be more pronounced in some sectors than in others. For example, tailor-made services can be bought quite easily in the transportation sector, whereas customized services are rare in the electricity sector. Reasons for these differences can be found in the nature of the service provided, and in the extent to which the sector has been liberalized. Multiple service providers are present in highly liberalized sectors, whereas only one service provider may be available to a customer in an oligopolistic or monopolistic setting.

The multilayer system of systems model can also be used to analyze and compare the various changes that are taking place. Change and innovation occur along various relevant dimensions simultaneously: technological development, institutional change, and changes and innovations in the product spectrum. Technological change may occur at all levels, but primarily at the physical and operational (control systems, ICT) levels. Technological change has been most visible in the ICT sector.

Institutional change takes place at all three levels: Privatization and liberalization have transformed a largely monopolistic situation to a situation where many actors compete. For the ICT sector, this has happened at all three levels. In the energy sector, competition has developed with respect to energy production (infrastructure layer) and delivery (products and services layer), but not with respect to the energy distribution networks and their operation. A similar situation occurs in transport, where the physical networks and nodes have largely remained under monopolistic control while, for example, in the rail sector, experiments have been introduced with competition at the service level. Relatively little institutional change has occurred in the drinking water sector, where, typically, all layers have remained in the same hands (but in some countries have been privatized).

Changes in the products delivered/markets served have also been most pronounced in the ICT sector, while being virtually absent in the drinking water sector. Figure 11.3 provides a comparative overview of the rates and types of change in the four sectors.

Two important changes related to increased interaction and interdependency between sectors are not included in this figure: the increased dependency on energy provision and ICT of all sectors, and the institutional convergence. The latter means that utility companies have started to broaden their operations to different sectors, mostly by takeovers and mergers, resulting in large corporations active in, notably, both the energy and water sectors (like, e.g., SUEZ).

The differences in the rates and types of change have implications for the choice of strategies for infrastructure designers, operators, and the like, in view of future

Speed/intensity of change	ICT	Energy	Water	Transport
Technological	+++	+	+	+
Institutional	+++	++	+	++
Product/market	+++	++	0	+

FIGURE 11.3 Rates and types of changes compared across sectors

uncertainties. For example, the stability in the market (type of product) in the drinking water sector suggests that a robustness strategy (essentially designing the system such that it will be able to adapt to changes in the amounts of demand) may be the most appropriate in designing new infrastructure elements for the long term. However, in the ICT sector, the fast changes at all levels rather suggest a flexibility strategy, where the system is designed in such a way that it can easily be adapted to changes in technology, type and size of demand, and so on.

11.3   SUSTAINABLE RESIDENTIAL ENERGY INFRASYSTEM

We now illustrate the application of systems of systems concepts with reference to a more specific case: a possible transition of the energy system toward sustainability (see also Pruyt and Thissen, 2007). For the sake of simplicity, we limit ourselves to a small but important component of the energy system, namely energy use for residential heating. In the Netherlands, this accounts for roughly 20% of total energy use and related CO_2 emissions. The question is what kind of government policies would be needed to significantly reduce CO_2 emissions related to residential heating.

Taking a model-based, systems analytic approach, one quickly faces difficult questions regarding the most appropriate model boundaries. Although energy use for residential heating is only part of the overall energy system, its evolution is codetermined by many factors and mechanisms related to the choice of energy resources and energy production technology as a whole, to investment in and exploitation of dwellings, to technology innovation, and so on. Including all these in a detailed modeling effort will quickly lead to a very complex model covering multiple levels, processes operating at very different timescales, and so on. But excluding many of these mechanisms from the model may easily lead to leaving important if not crucial processes out of the analysis. A conceptual systems of systems approach can be of help to put modeling efforts in perspective, and make deliberate choices with respect to model boundaries. Key to the approach is the identification of different systems that, in mutual interaction, determine CO_2 emissions from residential heating. Here, we limit ourselves to a simplified presentation of the approach followed in Agusdinata and Dittmar (2007), Agusdinata, (2008) and Agusdinata and DeLaurentis (2008), and distinguish the following systems that codetermine residential emissions:

- The *physical heat consumption system* is, of course, at the core of the analysis. It represents actual heat consumption in dwellings, as influenced by, among other things, tenant behavior and preferences, and physical dwelling properties such as the degree of insulation, and the type of heat generation system used.
- The *dwelling investors system* represents the decisions and actions by dwelling owners and investors related to dwelling refurbishment, investment in newer and/or more efficient heat generation technology, and the like. These decisions, of course, are determined by the business/economic interests and position of the dwelling owners, which only to a limited extent will be energy emissions related.

Costs of refurbishment and costs and availability of new technology will, of course, affect these decisions.

- The *heat technology innovation system* represents the mechanisms and actors working on R&D in heat technology, leading to the availability on the market of new products and installations for heat generation, such as Micro-CHP installations, more efficient heat pumps, solar collectors, and the like.
- The *heating infrastructure system* consists of the available hard- and software for distribution and delivery of heat or energy carriers to the individual dwellings. Key actors include local, regional, and perhaps national energy companies. Components may include local heat distribution systems, networks for gas and electricity delivery, the fuel distribution system, and so on. Important attributes from the perspective of CO_2 emissions include the mix of energy carriers, the type of primary energy source used, conversion efficiencies, and — related to the energy sources — CO_2 emission per unit of energy delivered.
- The *national energy system* consists of the total of energy sources, carriers, conversions and distribution systems, including energy provision for industry, agriculture, and transport. Major players here are the national and international energy companies, network owners and operators, and the like.

Figure 11.4 illustrates some of the key linkages between these systems. We note that each of the identified systems is managed or governed by a different set of actors, each having their own goals and objectives, many of which are not or only partially related to CO_2 emissions from residential heating. This makes governance of the system of

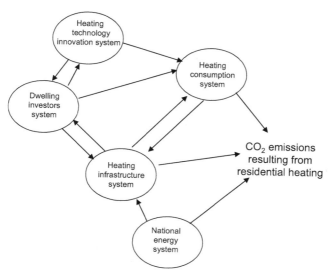

FIGURE 11.4 Simplified representation of residential housing emission system of systems

systems a nontrivial task. The conceptualization as illustrated here, however, assists in identifying the points of leverage a (national) policy maker may have. For example,

- Research budgets may be set aside, and subsidies may be given for R&D in the heating technology innovation system.
- subsidies or other mechanisms (such as the requirement of a heating efficiency report for each dwelling being sold or rented) may be used to trigger dwelling owners and investors to adopt new technology, thus creating a market for new products.
- public campaigns may be held to sensitize the public to the potential consequences of CO_2 emissions in hope of affecting consumer behavior; environment taxes (such as a tax on CO_2 emissions or a CO_2 emission permit trading system as currently under development in the EU) may be installed to create cost differences between different types of energy carrier of energy sources, and this also may affect consumer behavior as well as the market for more efficient and/or more sustainable technology.
- Incentives may be given and regulations installed at the national level to stimulate development and use of renewable energy sources (such as solar, wind, tidal).

The conceptualization described here shows the system of systems character of even limited parts of the energy system. First, illustrates how such a conceptualization can help identify the variety of levers available to policy makers. It, thereby, points to the necessity of concerted policy development, such that an appropriate mix of complementary policy measures targeted at the different component systems is chosen. This is, at present, not always the case, as recently illustrated by the Netherlands government when, for strictly budgetary reasons, subsidies for development of sustainable energy sources at national level were stopped, eliminating an important element from an otherwise reasonably coherent set of policy measures.

A second, important feature of the system of systems conceptualization is that it puts the multiactor character of the system of systems in the spotlight. Different systems are controlled by different actors being guided by different values and objectives. Such a system of systems cannot be controlled in a technocratic way. Knowledge of markets, motives for actor behavior, and so on are required, and policy makers may also benefit from entering into negotiations with different parties (such as large scale dwelling exploiters, energy companies, etc.) about their commitment to contributing to sustainability targets.

When compared to the three-layer system of systems conceptualization presented in Section 11.2, the decomposition in this example shows remarkable differences. First, the distinction between the two bottom layers in Fig. 11.1 is not made in Fig. 11.4: the heating infrastructure system contains elements of both. Second, in Fig. 11.4, a distinction is made between a specific infrasystem (heating infra) and a broader national energy infrasystem — while such a distinction has not been made in Fig. 11.1. Third, the heating technology innovation system is a very relevant system when considering transitions to more sustainable heating, while in Fig. 11.1 technological innovation systems are not included.

These differences illustrate a more general principle: it is the specific purpose of analysis that determines what scope and what decomposition and selection of systems is relevant.

11.4 FLEXIBLE SYNTHESIS GAS INFRASYSTEM

The second example to illustrate the application of different system of systems modeling techniques concerns the possible design and implementation of a flexible multifuel energy provision in The Netherlands (Herder et al., 2008a,2008b). The current energy infrastructure in the Netherlands is rigid and inflexible: capital-intensive units like power stations and oil refineries convert one specific source of energy into just a few products; most coal becomes electricity, crude oil is turned into transport fuels, and natural gas is used to produce power and heat. At none of the plants involved it is possible to change the energy source easily, if at all, in the short term. Moreover, even long-term modification opportunities are limited and expensive.

Given the uncertainty in available energy resources mix for the Netherlands in the near and far future, it is strategically desirable that the national energy provision be made more independent of specific fuels. An idea to realize this is by using synthesis gas (or syngas, a mixture of hydrogen and carbon monoxide) as the key energy carrier. We have used a systems of systems modeling approach to explore such a multifuel syngas infrastructure system. It consists of a distribution network of large but flexible multifuel gasifiers producing raw synthesis gas of various qualities, of units to refine that gas to set specifications, and of industrial processes that convert it into useable products. A multifuel syngas plant can turn solid, liquid, and gaseous raw materials into synthesis gas.

Figure 11.5 illustrates this possible future energy infrastructure. Key advantages as compared to the present situation are (1) the significantly increased flexibility of switching between crude energy sources, and (2) the increased flexibility in usage functions toward which the intermediate energy carrier (syngas) can be applied .

The syngas infrasystem comprises the following systems:

- The *physical syngas production system* is the central physical system (Level 1). It concerns syngas production in a network of gasifiers, including the selection of feedstock for the gasifiers. Other relevant variables are gasifier properties and efficiencies, with feedstock market behavior, and actual syngas consumption, as constraints that are influenced by the demand for chemical and energy products.
- The *syngas infrastructure* (Levels 1 and 2) is the physical network of compressors, pumps, and pipes that transport and distribute the syngas. In addition, it represents the control and safety mechanisms, which are operated by controllers, and rules and regulations concerning the physical connections to the main infrastructure.

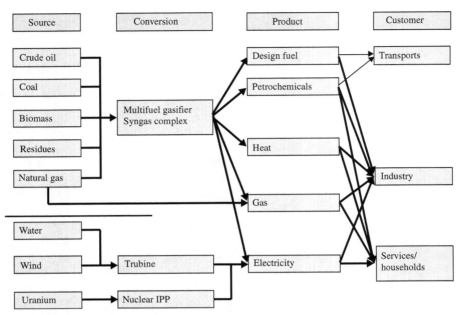

FIGURE 11.5 The system elements of the suggested energy infrastructure for the Netherlands in 2040

- The *syngas market system* (Levels 2 and 3) overlaying the physical system deals with issues such as the economics of a flexible multifuel syngas complex compared to those of specialist conversion businesses. It also comprises the internal syngas market, contractual arrangements, ownership issues, pools, and exchanges. It interacts with the conventional oil and gas markets worldwide.

- The *technology innovation system* (Level 3) represents the mechanisms and actors working on R&D in gasifier technology, leading to the availability on the market of new products and installations for syngas generation and use, and on R&D in syngas applications, leading to new technologies and products, and to more efficient existing technologies.

- The *worldwide energy and syngas products market* (Level 3) serves as the system in which the syngas infrastructure is embedded. It represents the interactions of the syngas infrastructure with its physical and economic surroundings.

For the syngas infrastructure system, a more detailed example of a system model is depicted in Fig. 11.6. It consists of a double bus network: one pipeline contains pressurized "high-quality" (HQ) syngas with a high CO/H_2 ratio, enough to satisfy the most demanding production processes. The other bus contains syngas with a lower ratio that supplies to users who do not require high-ratio syngas.

By means of illustration, Fig. 11.7 represents a more detailed system description of the syngas market system. For suppliers and users on the HQ line, two options are

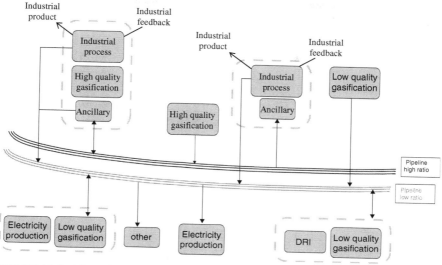

FIGURE 11.6   The infrastructure system design: a double bus network (Apotheker et al., 2006)

included: transactions through bilateral contracts or syngas trading on a syngas spot market. For actors on the low-quality (LQ) pipeline bilateral deals and a syngas pool are included, from which users can buy the quantities they need. In addition, market balancing tasks are assigned to specific actors (see Fig. 11.7).

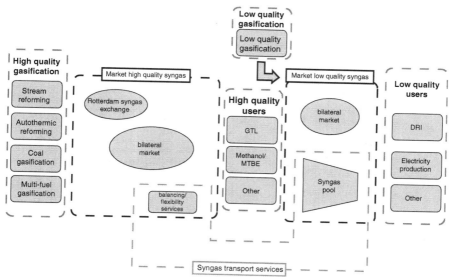

FIGURE 11.7   The market system (Apotheker et al., 2006) GTL— Gas to liquids processes; MTBE— Methyltertbutylether production process; DRI–Direct reduction of iron process

The transition over time of the current cluster into a syngas cluster like the one illustrated here is far from trivial. Creating it will cost several billion euros in all. The development and completion process will last decades, and will cover a large region. A highly developed cluster of capital-rich companies is probably needed to attract the necessary funds. Many physical as well as social and economic systems have to be designed or shaped. This design is complicated by the combination of many systems incurring emergent behavior, by deep uncertainty, and because of strong interaction between the various physical and socioeconomic systems. The establishment of a large-scale syngas infrastructure and the investment required are far beyond the means of any single company or local government.

A system of systems description of the syngas infrasystem is helpful in this respect because it shows that systems and system layers (as indicated in Fig. 11.1) are intricately interwoven, requiring an integrated modeling and design approach. By initially separating the analysis and design tasks into tasks for the different systems and for the three different layers, more adequate models can be made. The impact of other system's and other layer's variables can initially be accounted for in the form of constraints that befit the system model. For example, variables from an economic market model can be translated into relevant variables or inputs in different physical or chemical models, without losing the strengths of either modeling technique.

11.5   RESEARCH AGENDA

Without any pretension of completeness, we will conclude this chapter with a selection of subjects for a research agenda into infrasystems as systems of systems. Most research on infrasystems concentrates on one sector, one layer, and a single aspect. Many of the contributions have been restricted to one layer, focusing on layer structure, or at best to two layers, focusing on layer interaction, as the intricate inclusion of three layers into one research project is considered to be very complex. Yet, efforts to increase understanding at the overall system of systems level are much needed, in view of the fact that the key performance indicators of infrasystems are in the end determined by the interplay of most, if not all the component systems. Modeling and researching the interplay of the various component systems to fully study the determinants of overall system performance will require the integration of different disciplinary perspectives, different timeframes, and different cultures in research and practice.

There is, for example, a need to understand better the interplay of technological, institutional, and behavioral factors (Weijnen et al., 2003). This requires extensive efforts to analyze the effects of particular changes and the interplay of changes in different dimensions. Such efforts cannot succeed except in a multidisciplinary setting that combines qualitative and quantitative approaches, recognizing both social and technical complexity and combines multiactor with physical network perspectives. This will also require novel research concepts and methods that are able to include behavioral elements, for example, gaming simulations (Mayer and Veeneman, 2002) and agent-based modeling (Nikolic, 2008).

In addition to these general questions and requirements for infrasystem research, we identify five important prominent topics requiring a system of systems perspective, all related to the present changes and transitions in some of the sectors.

11.5.1   Reliability and Criticality of Infrasystems

Little insight still exists into the consequences of unbundling and other rapid changes for an infrasystem's reliability. A key unresolved question is whether and how the system's reliability can be guaranteed in a context where independent actions of interacting organizations, many with competing goals and interests, instead of single organizations, are determining infrasystem reliability. What factors are crucial? Can technological solutions compensate for the loss of central control? Can smart incentive and tariff structures guarantee sufficient attention for overall system integrity in times of rapid change?

11.5.2   Designing Infrasystems Amidst Uncertainty

Infrastructures are large-scale systems with long lifetimes. Many uncertainties therefore pertain during the system's life. New ways to deal with such inherent uncertainties both in the design stage and in the operational stages need to be developed, such as flexible designs, real options and adaptive strategies, or even designing evolutionary, self-learning systems.

11.5.3   How to Safeguard Public Objectives?

The current focus on increasing infrasystem efficiency and innovative power may not only put system reliability, but also other public objectives such as equal access, sustainability, health, and so on at risk (De Vries, 2004). Similar research questions come up as for the case of reliability: What institutional arrangements, incentive structures, regulations, and the like are required? Under what conditions and technologies will they work efficiently?

11.5.4   How to Manage Transitions?

Policy makers have introduced the market paradigm with a view to overall system efficiency. While the envisioned end situation, open competition with a sufficient number of market players, would indeed seem effective from the economic perspective, experiences show that the transition to such a situation is not self-evident. At longer term, and from a different perspective, transitions to more sustainable infrasystems (notably transport, energy) are called for. Research is needed to understand further these mechanisms and design appropriate regulatory frameworks for transition periods.

11.5.5   Cross Sectoral and Generic Questions

Last but not least, there is cross-sectoral and generic infrasystem research. Key questions include:

- What are the commonalities and differences across sectors, as illustrated in Figs. 11.2 and 11.3, perhaps extended to other sectors than the four sectors mentioned, for example, waste? How can commonalities and differences be explained?
- To what extent is it possible to transplant lessons from one sector to other sectors?
- To what extent, and for which subjects, is it possible to develop generic theories about infrasystems? Which subjects cannot be adequately dealt with at the generic level, requiring deep studies while taking the peculiarities of the individual sectors into account?

11.6   CONCLUSIONS AND DISCUSSION

Infrastructure systems are crucial to the functioning of modern society. Significant changes and developments in technology, institutional setup, and economic position are taking place in many infrastructures, yet little is known how these and other changes do and will affect key infrastructure performance indicators such as reliability, costs, quality of service, public accessibility, and environmental impacts. Understanding the mechanisms and factors affecting these performance indicators will require a true system of systems approach.

We have illustrated the application of system of systems concepts with reference to a comparison of different infrastructure sectors, introducing a three-layer decomposition distinguishing between systems at the physical infrastructure level, at the operation and management level, and at the products and services level. Next, we have illustrated a conceptual system of systems approach for analysis and design in two more specific energy-related cases.

We conclude that systems of systems approaches can significantly contribute to (see also Pruyt and Thissen, 2007):

1. describe and analyze different levels in infrasystems, and the interactions (within and between the different levels) before (re-) designing a system of systems;
2. clearly uncover the multiactor nature of infrasystems
3. reflect on positioning, omissions, boundaries, interfaces, and interactions with other component systems at the same or at different levels;
4. put models/analyses insight/results of component system analyses into perspective;
5. describe and communicate clearly the structure, dynamics, and insights developed within infrasystems and among infrasystems; and
6. reflect on the differences and commonalities between different infrastructure sectors.

There exist, however, also potential traps related to the use of system of systems concepts (and other conceptual decomposition frameworks) in infrasystem modeling. First, there are many different ways to decompose infrasystems into systems of systems. Boundaries could, for example, be drawn on the basis of the internal governance structure, the physical system boundaries, the geographic boundaries, the policy system boundaries, the technological boundaries, the market boundaries, or the outputs of interest. But the basic assumptions of the decomposition actually lead to a specific perspective on the infrasystem that influences the outcomes of the analyses.

Second, the noncritical use of such frameworks could also be dangerous because each one of these frameworks contains a set of basic assumptions that lead to a specific perspective (of many possible perspectives) on the issue. And many (pre-) analytic choices — such as the choice of the level of analysis — are in fact implicit (ethical) choices: the level chosen partly determines the perspectives and goals adopted. It is, therefore, important to critically reflect on the decomposition and -if possible- to explore several alternative decompositions to put them into perspective and to generate an understanding of their influence on the design choices and policy recommendations.

Many publications illustrate the potential of crosscutting, generic analyses in the areas of evolutionary economics, political science, law, and public management. A similar body of knowledge on infrastructures and infrastructure behavior is virtually absent in the realm of science and engineering. Present technological research and development generally focus on a single sector. Generic, formal specification methodologies, design methods, or models are lacking. We believe that a system of systems paradigm could aid in drawing lessons and sharing these lessons among different infrasystems.

REFERENCES

Agusdinata, D.B., 2008, Exploratory modeling and analysis: A promising method to deal with deep uncertainty, PhD Thesis, Delft University of Technology, The Netherlands.

Agusdinata, D.B., DeLaurentis, D., 2008, Specification of systems of systems for policy making in the energy sector, *Journal of Integrated Assessment.* Forthcoming.

Agusdinata, D.B., Dittmar, L., 2007, System-of-systems perspective and exploratory modeling to support the design of adaptive policy for reducing carbon emission, *Proceedings of System of Systems Engineering, 2007. SoSE '07. IEEE International Conference*, April 16–18, 2007.

Apotheker, D., van der Elst, D.J., Gaillard, M., van den Heuvel, M., van Lelyveld, W., 2006, Design of a Syngas Infrastructure, Report of MSc SEPAM Design Project, Delft University of Technology.

CMIIP, Committee for Measuring and Improving Infrastructure Performance, 1995, Measuring and Improving Infrastructure Performance, National Academy Press, Washington, DC.

De Bruijn, J.A., Herder, P.M., Priemus, H., 2005, Systems and actors, *Proceedings of Foundations of Engineering Systems, University Council on Engineering Systems*, December 14th, 2005, Georgia University of Technology, Atlanta, USA.

De Bruijn, J.A., Herder, P.M., 2008, Systems and actors, *IEEE Transactions on Systems Man and Cybernetics Part A-Systems and Humans*, Accepted for Publication.

De Vries, L.J. 2004, Securing the public interest in electricity generation markets, The myths of the invisible hand and the copper plate; PhD thesis, Delft University of Technology, available online via www.tbm.tudelft.nl/webstaf/laurensv.

DeLaurentis, Dan, Callaway Robert, K., 'CAB' 2004, A system of systems perspective for public policy decisions, *Review of Policy Research*, 21(6): 829–837.

General Accounting Office: *Critical Infrastructure protection. Challenges for Selected Agencies and Industry Sectors. report to the Committee on Energy and Commerce*, GAO-03-233, US General Accounting Office, Washington, DC., 2003-07-06

Herder P.M., Thissen, W.A.H., (Eds.) 2001, *Proceedings Critical Infrastructures, Delft 2001, 5th International Conference on Technology, Policy and Innovation* (Den Haag, June 26–29, 2001), Lemma Publishers, Utrecht, The Netherlands.

Herder, P.M., Stikkelman, R.M., Dijkema, G.P.J., Correlje, A.F., 2008a, Design of a syngas Infrastructure, *Proceedings of Escape-18, European Symposium on Computer-Aided Process Engineering*, Lyon, France.

Herder, P.M., Weijnen, M.P.C., Bouwmans, I., 2008b, Designing Infrastructures from a Complex Systems Perspective, *Journal of Design Research*, 7(1): 17–34.

Koolstra, K., 2000, Slot allocation in different transport sectors, in: Weijnen et al., (Ed.), Walking a Thin Line in Infrastructures, *Balancing Short Term Goals and Long Term Nature*, Delft University Press, Delft, The Netherlands.

Mayer I., Veeneman, W.W., (Eds.), 2002, *Games in a World of Infrastructures – Simulation-games for Research, Learning and intervention*, Eburon Publishers, Delft, The Netherlands.

Nikolic, I., Dijkema, G.P.J., Van Dam, K., 2008, Understanding and shaping the evolution of sustainable large-scale sociotechnical systems. towards a framework for action-oriented industrial ecology, in: Ruth, M., Davidsdottir, B., *Dynamics of Industrial Ecosystems*, Edward Elgar. Forthcoming.

Pruyt, E., Thissen, W., 2007, Transition of the european electricity system and system of systems concepts, *Proceedings of IEEE Conference on System of Systems Engineering*, April 2007, San Antonio, Texas.

Sage, A.P., 2005, Presentation at Systems of Systems Engineering IEEE SMC2005 Conference: http://ieeesmc2005.unm.edu/smc_keynote_sage.pdf.

Thissen W.A.H., Herder, P.M., (Eds.) 2003, *Critical Infrastructures - State of the Art in Research and Application*, Kluwer Academic Publishers, Boston/Dordrecht/London, USA, pp. 283–300.

Van de Riet, O.A.W.T., 2003, *Policy Analysis in Multi-Actor Policy Settings, Navigating Between Negotiated Nonsense and Superfluous Knowledge*, Eburon, Delft.

Weijnen, M.P.C., ten Heuvelhof, E.F., Herder, P.M., Kuit, M., 2003, Next Generation Infrastructures, Bsik Research Programme Proposal, Delft University of Technology, The Netherlands.

Chapter **12**

Advances in Wireless Sensor Networks: A Case Study in System of Systems Perspective

PRASANNA SRIDHAR,[1] ASAD M. MADNI,[2] and MO JAMSHIDI[1]

[1]University of Texas, San Antonio, TX, USA
[2]Crocker Capital, San Francisco, CA, USA

12.1 SYSTEM OF SYSTEMS OVERVIEW

As seen from previous chapters, there is no universally accepted definition of system of systems (SoS). Therefore, we will describe rather than define some important properties that will constitute an SoS. We will extend and refine some of the properties that are already found in literature (Maier, 1998, 2005; Keating, 2005). The word SoS and system of systems engineering (SoSE) is beginning to gain some visibility in aerospace and defense research (Defence Acquisition Guidebook.[1] DiMario, 2006; IEEE, 2006; Keating et al., 2003; Meilich, 2006; Wojcik and Hoffman, 2006). The components in the SoS, which are themselves systems, are sufficiently *complex* as shown in Fig. 12.1. Added to this complexity, the systems can be *heterogeneous* in nature. Heterogeneity of individual systems could be architectural or functional or a combination of both. Therefore, SoS can be a combination of functionally and/or architecturally different systems. Consider, for example, a car (or any land vehicle) that is equipped with different sensors that serve different purposes. The functionality of each sensor is very specific. However, when operations of all these functionally independent sensors are considered, they ensure *safety*, *efficiency*, and *situational awareness* (in this case traffic). Therefore, a car

---

[1]Defense Acquisition Guidebook, System of Systems Engineering, Chapter 4, Section 4.2.6., Electronic Book at http://akss.dau.mil/dag/Guidebook/Common_ InterimGuidebook.asp

---

System of Systems Engineering: Innovations for the 21st Century, Edited by Mo Jamshidi
Copyright © 2009 John Wiley & Sons, Inc., Publication

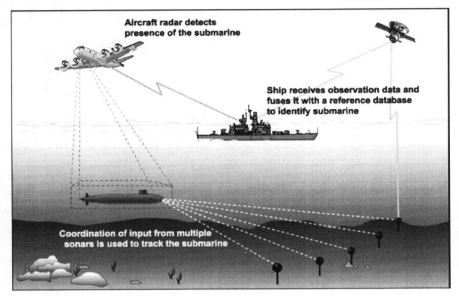

FIGURE 12.1 A classical system of systems application (Image courtesy of Halls and Llinas (2001)

can be seen as a SoS in a *functional* perspective. A car can also be seen as SoS in an *architectural* perspective. A car is generally equipped, for example, with air bag sensors, gyroscopes, accelerometers, CCD camera, IR sensors, tire pressure sensors, steering sensors, and so on. These sensors are architecturally different devices, yet they have a common talk mode as seen in controlled area network (CAN)-BUS protocol. Such protocols ensure that different (both architecturally and functionally) sensing devices work in synergy to address higher level mission objectives, that is, safety and reliability of a given car.

Combination of such complex systems (thus forming an SoS) can also deliver better *situation awareness* of a given task as opposed using individual systems. The systems within a given SoS are operationally and managerially *independent* in nature. The *degree of independence* (*DoI*) will help us to further classify SoS. The composition of systems to work as SoS should ensure that the SoS is *scalable*.

A system that scales well does not necessarily mean that it performs well. A system is said to be scalable if its *performance*, *availability*, and *fault tolerance* are in the acceptable thresholds when the load or subsystems are increased in the system. All the three parameters signify the usability of the system. Ideally, if a system is scalable, then its performance increases (or remains same), its availability increases, and its mean time to failure (MTTF) increases. Performance of a system is based on the output behavior of the given system. Availability signifies that there is no single point of failure in the system. The system shows graceful degradation with an increase in fault tolerance.

A system can be

- Geographically scalable—usability is still good no matter how geographically far the system is located from the users. However, usability also depends on the system's sensitivity, range, power, and other parameters.
- Load scalable—usability is still good even with the increase in load on the system.

SoS also exhibits *emergent* and *evolutionary* properties (DeLaurentis, 2005). Some behaviors are not seen in individual systems (components of SoS) but exhibit or emerge new behaviors when seen within an SoS framework.

An important feature that is to be considered is the application that requires several complex large-scale systems to work in synergy to accomplish a common goal. The combination of two or more systems is entirely necessitated by the application under consideration. Therefore, a single scalable system would not have accomplished the task efficiently as compared to multiple scalable systems working in tandem.

12.2 SENSOR NETWORKS AS SYSTEM OF SYSTEMS

Each of the sensor nodes has multiple sensors onboard along with limited processing and storage capabilities. Therefore, each of them is sufficiently complex and operationally independent in nature. The hardware characteristics provided by these sensor nodes make it feasible to load software (and operating system) on these nodes. These software tools along with operating system make the nodes managerially independent. The type of sensors onboard a given node determines if the node is functionally different from other nodes. Within a given sensor network, the sensor nodes can have different processors (ARM, Intel XScale, etc), different RF technologies (ZigBee, 802.11, etc), and even different software layers on top of hardware. These characteristics make them architecturally different. Having multiple sensor nodes clearly demonstrates the advantage of increased situational awareness. For example, multiple sensors deployed in a vast geographical area can provide better coverage than a single sensor.

As described earlier, the system of sensors (or simply sensor network) is said to be scalable if its performance, availability, and fault tolerance increase as the sensor nodes in the system increase. As a motivating example, we consider a system of *in-situ* sensing devices and a colony of mobile sensor nodes as two interacting systems in the system of systems platform. In sensor networks, we identify coverage criteria as an output behavior or a metric for performance. We can consider an idealistic situation where the sensor nodes are placed in such a way that sensing regions of two or more sensor nodes do not overlap (dead region). In such situations, an increase in sensor node deployment will increase the coverage of the area, thereby increasing the performance. However, this might not be a linear increase due to the failure of some of the already deployed nodes (dead nodes). The availability and failure of the entire network also increase due to increase in node density (as some nodes can act as redundant backup nodes). This method of scaling is sometimes referred to as scale-out, where more load (nodes) is added to a given system. In scale-up, one can add more resources (memory, processor power, low power usage, etc.) to a single sensor node so as to increase its performance (coverage), which in turn can affect the scalability of the entire network. In a mobile sensor network, each of the sensor nodes is capable of navigating and sensing the region from a given point A to a given point B. For these systems, we impose a hard-real time constraint of achieving the sensing task from point A to point B within a specified time T. This system can scale out by increasing the number of mobile nodes to sense a given area faster (acceptable time limit) than fewer sensor nodes.

By simply introducing another system to cooperate in achieving a common task will change overall system performance. The interaction among the systems influences each system by changing the acceptance threshold for performance, availability, and fault tolerance which the system designer would not have considered prior to the introduction of the new system. From a systems engineering (SE) perspective, the systems would have been optimally designed, developed, and deployed to meet the requirement specifications (RS) and to ensure increased performance and capacity of the system. SE is a management technology to assist and support policy making, decision making, scheduling, and resource allocation (Sage, 1992). SE consists of *formulation*, *analysis*, and *interpretation* as various ingredients in its life cycle. However, due to the introduction of new systems generally with different dynamics and different timescales for a given application, we obtain an assemblage of systems that interact with each other. This results in emergent behavior that is normally not formulated during the SE design process.

A generic solution then would be to *centralize* the complex *decentralized SoS*. To bring in the effect of interaction among systems, we centralize the control and information flow between the systems through a central command called *coordinator*. The coordinator acts as a central manager to make decisions on the acceptable threshold levels for the SoS in the given application scenario.

The role of coordinator can influence the classification of SoS as follows:

1. *Central or Closed SoS*: The coordinator has complete control of collaboration among the systems. Generally, all information and control flow among systems pass through the coordinator.

2. *Decentralized or Open SoS*: The coordinator exists, but it does not have complete control over all the components (systems). Each of the systems generally has a local controller, and the systems can directly interact with each other. Only critical and advisory information is controlled by the coordinator.

3. *Virtual SoS*: The coordinator does not exist in such SoS. The systems interact with other. The elements of SoS exhibit self-configuration and self-organization to fulfill the high-level objectives.

In this chapter, we will investigate two parameters related to *scalability*—fault tolerance and performance improvement of a sensor network within SoS context. We will consider decentralized form of SoS in order to propose our algorithms.

12.3 FAULT-TOLERANT DESIGN

The principle idea behind fault tolerance is the system's ability to perform or operate correctly even in the presence of faults. The problem of fault identification and isolation is generally a hard task in sensor networks due to the very nature of their construction and deployment. Moreover, due to the low computation and communication capabilities of the sensor nodes, the fault-tolerant mechanism should have a very low computation overhead. It is an intractable problem to actually detect whether a sensor is faulty by looking at the raw data acquired from the sensors. A simple mechanism to detect and isolate faults in sensors is to use range tests. A sensor is deemed to be faulty if its reading exceeds the threshold limits (minimum and maximum values specified). Such tests are commonly used to capture "hard faults" that occur rapidly and are termed as built-in self test (BIST) (Madni and Costlow, 2003). Although this threshold mechanism proves useful in capturing faults, they do not necessarily capture the performance degradation of a given sensor. We can refine such threshold scheme by a simple windowing scheme.

The design of threshold limits for a given sensor should be meaningful. If the limits (min and max values) are tight, we obtain high false alarms. If the limits are relaxed, then we will capture very few faults. We will assume that each sensor within a node is assumed to work within a *usable threshold window* (min, max). A built-in test is said to have passed if the sensor reading r is within this window. That is, reading r should follow the equation min $< r <$ max, where *min* and *max* are chosen appropriately for a given sensor and a given application, which constitute the operational behavior of the sensor. Every sensor is guaranteed to work "correctly" within a given operating range specified by the manufacturer. For example, a manufacturer could specify an operating range for a temperature sensor as from -25 to $125°C$. Similarly, a chemical sensor (such as carbon monoxide sensor) can have an operating range from 0 to 500 ppm (parts per million). We call this operating range as *guaranteed window*. This window is usually obtained from the sensor manufacturer. The usable threshold window (min, max) will incorporate the guaranteed window for a given sensor. That is, min of the usable threshold window will be lesser than the minimum range of operation defined by the guaranteed window and max will be greater than the maximum operating range of the sensor. For example, we can define from -40 to $+150°C$ as usable threshold window for a temperature sensor that is guaranteed to sense temperature within the range from -25 to $+125°C$ with a specified accuracy. The sensor might still work outside this guaranteed window, however, with a much lesser accuracy. We can assign a weighting factor to each of the sensors in a sensor node based on the reading r. Depending on how close the reading is to the usable window boundaries, the weight can be adaptively decreased. As described earlier, each sensor can operate within a *guaranteed window*, and the weight is set to 1 if the output of the sensor is within this window, that is, $w_{ij} = 1$, where $i = (1, \ldots, m)$ signifies the sensor node number and $j = (1, \ldots, n)$ signifies the sensors onboard the ith sensor node. If the sensor reading goes beyond the guaranteed window and approaches the usable window boundary, the weights are decreased. By using the concept of an added usable window versus a guaranteed window alone, we are trading off the performance optimization of the individual sensor versus optimizing (maximizing) the performance of the sensor network.

Threshold tests prove useful in detecting hard faults. "Soft faults" by nature occur when sensors are working within the threshold limits but might still be deemed faulty. For example, when the sensor reading is within the manufacturer's guaranteed window, it is generally difficult to classify whether the reading is an environmental stimuli captured by the sensor, if it is an intermittent transient (short-lived oscillation due to sudden voltage or current change), or if it is an intermittent fault (soft fault). To detect such faults, we will take advantage of the nature of applications in which the sensor nodes are often deployed. These applications often require dense deployment of sensor nodes, thereby making the sensor nodes *spatially correlate* with neighboring nodes within the same event region as shown in Fig. 12.2. This means that a given sensor would read the same event value (with minimal variations) as neighboring k sensors which are closely deployed. For example, consider three sensor readings a, b, and c from three redundant sensors. Two sensor readings are averaged ($(a+b)/2$, $(a+c)/2$, and $(b+c)/2$) at each predetermined time interval. The actual reading is set to the value at which the majority of the three averaged values agree upon—a *plurality voting principle*. This helps to eliminate the faulty sensor in the group of the three sensors under consideration.

A more comprehensive mechanism based on the plurality voting principle is to decrease the contribution of the faulty sensor and increase the contribution of the nonfaulty sensors. This can be achieved by simple *weighting factor*.

Each sensor node has a weighting factor at any instance of time t, given by $w_i(t)$. In the event of sensor failure, the proposed algorithm adaptively decreases the weight for sensors that have failed or demonstrate likelihood to fail. At the same time, weighting factors for the neighboring sensor nodes are increased. Hence, every reading from each sensor is weighted at each predetermined time interval t, and weight updates are computed as follows:

$$wi(t+1) = wi(t) \pm \Delta wi(t) \qquad (12.1)$$

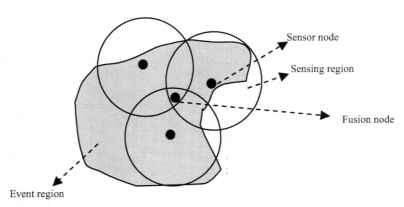

FIGURE 12.2   Spatially correlated sensor nodes

A coordinator in SoS framework (a cluster head in sensor network) can simply query the nodes for sensor reading in the event region. On the basis of a specific timeout, the fusion node performs weighted average aggregation based on the data it has currently received from the sensors.

In traditional neural networks, the change in weights $\Delta w_i(t)$ is a function of the error estimate, which is based on the difference between the expected reading and the actual reading. However, in sensor nodes, we do not know the expected or desired reading a priori. To estimate $\Delta w_i(t)$, we use the concept of spatial correlation as explained earlier. To estimate $\Delta w_i(t)$, we propose the following model:

$$\Delta w_i(t) = |\tau_i| \times \varepsilon \qquad (12.2)$$

where, τ, the *adaptation parameter* is given by

$$\tau_i = \frac{r_1 + r_2 \ldots r_{i-1} + r_{i+1} \ldots r_{k+1}}{k} - r_i \qquad (12.3)$$

r_i is the reading from the ith sensor, k is the number of neighboring sensors, and ε, the *scaling factor*, is a small value $0 < \varepsilon < 1$ chosen appropriately for a given application. The scaling factor ensures that $0 < \Delta w_i(t) < 1$. A complete implementation can be found in (Sridhar et al., 2006a,b).

To validate our algorithm, we aggregated the data based on the obtained weighting factor. We compared this weighted average data to a simple average of three sensory data without fault correction as shown in Fig. 12.3. Detailed real-world experimentation can be found in (Sridhar et al., 2007a,b).

12.4 DECISION MAKING

Fault-tolerant calibration mechanism when done at a local system level (sensor node level) can have tremendous impact on a global SoS level (sensor network level). Certain critical control, policy, and decision making need to be carried out at the coordinator level. Such control and decision making help to improve the performance of a sensor node and of the network in general. For example, a fuzzy logic-based controller at a coordinator (base station) for sensor network has been proposed in (Sridhar et al., 2006a,b). In this section, we will propose an innovative mechanism for decision making at the coordinator level based on several competing and/or contradicting criteria that exists in systems within an SoS.

Consider the application of monitoring a large structure such as a bridge. Using sensor networks would ideally require sustaining the lifetime of the deployed sensors for a long time, since the redeployment generally can be difficult, in terms of both ease and cost of deployment. In this case, network lifetime is more important than criteria such as accuracy of data, and hence we assign network lifetime a higher weighting factor. Consider another instance, an application such as habitat monitoring. High network lifetime is desired but not a required behavior. However, more importance or

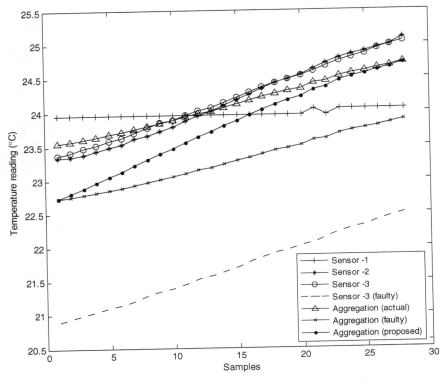

FIGURE 12.3 Comparison of aggregation under faulty conditions

priority needs to be given to efficient communication from the habitat to a command center. Consider yet another application of monitoring chemical or nuclear spill in a region. Such applications have high demands for larger node deployment in order to capture and localize all critical events in the region. Each application thus has varying demands or requirements that need to be satisfied by properly prioritizing the behavior or properties of sensor networks.

In order to prioritize the system behavior, we will need to establish criteria for prioritizing. For example, if there were three different tasks that need to be completed, a human might prioritize them based on time, cost, or importance. However, if the criteria are of equal importance with a certain kind of interaction between them, prioritizing the system behavior becomes a multicriteria decision-making problem.

The problem of decision making has been extensively studied in the field of economics. The problem of selecting an action among set of alternatives becomes harder when the decision-making process involves several *criteria* rather than a single criterion. Such problems are referred to as multicriteria decision making (MCDM) problems (Belton and Steward, 2002). MCDM is the study of discrete decision making involving two or more criteria (sometimes conflicting) or objectives. In MCDM problems, the goal is to select an alternative (choice or a system) from a set of relevant alternatives by evaluating a set of criteria. For example, consider the problem of selecting a car from a given set of three cars $S = \{A, B, C\}$. This set S represents our set of *alternatives*. Selecting a car of our choice is the *action*. The sample set of *criteria* to be evaluated can be $C^S = \{$Fuel efficiency, Luxury, Price$\}$. A conventional methodology to select a car is based on prioritizing the criteria for selection. Such priorities are generally user dependent. A simple weighting factor for each criterion can prioritize the selection process.

Let us now generalize the problem of MCDM by taking finite number of actions and criteria. Let $\Omega = \{s_1, s_2, \ldots, s_m\}$ and $X = \{x_1, x_2, \ldots, x_n\}$ be set of alternatives and set of criteria, respectively. The decision-making process proceeds by formulating a matrix $\mathbf{A}$ with set of criteria and set of alternatives given by

$$
\mathbf{A} = \begin{matrix} s_1 \\ s_2 \\ \vdots \\ s_m \end{matrix} \begin{pmatrix} a_{11} & a_{12} & \cdots & a_{1n} \\ a_{21} & a_{22} & \cdots & a_{2n} \\ \vdots & \vdots & & \vdots \\ a_{m1} & a_{m2} & \cdots & a_{mn} \\ x_1 & x_2 & \cdots & x_n \end{pmatrix} \tag{12.4}
$$

Each entry a_{ij} denotes the degree to which the criterion x_j is satisfied by the alternative s_i. The idea is to now reduce the multicriteria problem into a single global criterion problem by aggregating all the elements of matrix $\mathbf{A}$, given by $a = \mathbf{H}(a_{1j}, a_{2j}, \ldots, a_{mj})$, where $\mathbf{H}$ is the aggregation operator. Most common aggregation operator is the weighted arithmetic mean. We will now investigate the necessity of MCDM in sensor networks, the pitfalls of common aggregation operators (such as weighted mean) and provide a countermeasure for aggregating criteria without using common aggregators.

12.4.1 Interacting Criteria

A common method as discussed earlier to evaluate set of criteria is to use aggregator operator to reduce the multicriteria problem to a single global criterion problem by aggregating all the elements of matrix $\mathbf{A}$. A tradition method is to use weight sum (or weighted mean) on the row of matrix $\mathbf{A}$ given by

$$
\sum_{i=1}^{n} w_i \times a_{1i} \tag{12.5}
$$

While this is a simple approach, however, it has drawback in that it assumes that the criteria are independent. The criteria can interact with each other, which requires the replacement of weighting factor w by a more comprehensive nonadditive set function on set X (set of criteria) that considers not only weighting factor on each criterion but also weighting on each subset of criteria. Marichal (2000) gives an overview of different types of interaction among criteria that could exist in the decision-making problem. Three kinds of interaction defined and described in (Marichal, 2000) are as follows: correlation, complementary, and preferential dependency.

Correlation can be further divided into positive correlation and negative correlation. Positive correlation is existent when two or more criteria present some form of redundancy. For example, consider again the problem of evaluating a given car based on three criteria {fuel efficiency, luxury, and price}. A highly luxurious car generally comes with a higher cost. In this case, luxury and price form positive correlating criteria, and the evaluation will be an overestimate. As discussed before, this problem can be overcome by using weighting factor on subset of criteria, such that $w(ij) < w(i) + w(j)$, where i and j are two criteria, and subadditive feature overcomes the overestimate during the criteria evaluation. In the reverse case (negative correlation), weighting factor $w(ij)$ will be superadditive, given by $w(ij) > w(i) + w(j)$.

In complementary type of interaction, one criterion can replace the effect of multiple criteria. This means that importance of criteria pair (ij) is close to the importance of having single criterion i or j. Clearly, when such criteria pair exists, a weighted sum cannot be helpful during the evaluation process. A more complex weighting factor needs to be considered.

The third type of interaction is the preferential dependence. In this type of interaction, the decision maker's preference for selecting an alternative is simply given by a logical comparison; that is, if there exists a function M such that, for any two alternatives a_1 and a_2, the decision maker selects one of the alternatives (say a_1) if $M(a_1) > M(a_2)$.

Clearly, when such complex interactions exist among criteria, it is necessary to use a well-defined weighting function on subset of criteria rather than a single criterion during global evaluation. One such methodology for evaluation is the Choquet integral with the use of fuzzy measure as weighting function.

12.4.2 Choquet Integral

A fuzzy measure (Grabisch, 1995) on a set of criteria (X) is defined as a mapping function $\mu: 2^X \rightarrow [0,1]$, where 2^X is the power set of X. Additionally, μ should satisfy the following properties:

1. $\mu(\emptyset) = 0$ and $\mu(X) = 1$, where $\emptyset$ represents the null set
2. If A is a subset of B, then $\mu(A) \le \mu(B)$

For example, consider a set $X = \{x_1, x_2\}$. Power set of X is given by, $P(X) = \{\emptyset, \{x_1\}, \{x_2\}, \{x_1, x_2\}\}$. The fuzzy measure on the elements of set P, for example, can be defined as $\mu(\emptyset) = 0, \mu(\{x_1\}) = 0.4, \mu(\{x_2\}) = 0.5$, and $\mu(\{x_1, x_2\}) = 1$. If μ is the fuzzy measure on X (set of criteria), then the Choquet integral (Denguir-Rekik et al., 2006; Grabisch, 2000) of a function $f: X \rightarrow [0,1]$ with respect to μ is defined as

$$C_\mu(f(x_1) \ldots f(x_n)) = \sum_{i=1}^{n} (f(x_{(i)}) - f(x_{(i-1)})) \times \mu(Y_{(i)}) \qquad (12.6)$$

where $x_{(i)}$ indicates that the indices have been permuted such that $f(x_{(1)}) < f(x_{(2)}) < \ldots < f(x_{(n)})$ and $Y_{(i)} = \{x_{(i)}, \ldots, x_{(n)}\}$. If the fuzzy measure μ is additive (i.e., $\mu(xy) = \mu(x) + \mu(y)$), then C_μ represents discrete Lebesgue integral (Wang et al., 1995). The μ above equation for discrete Choquet integral can also be given as

$$C_\mu(f(x_1) \cdots f(x_n)) = \sum_{i=1}^{n} f(x_{(i)}) \times (\mu(Y_{(i)}) - \mu(Y_{(i+1)})) \qquad (12.7)$$

12.4.3 Motivating Example

We propose a case study for multicriteria decision making in mobile robot path planning in an environment deployed with sensor nodes. We can generalize such a decision-making process to a more complex system management. We develop an efficient data collection and sensor node replacement scheme for sensor network in a cluttered environment. The autonomous sensor nodes embedded in the environment are generally low-powered devices. High events in the environment usually require constant monitoring and dense deployment for precisely localizing the threat events. To capture all important events, we would ideally want more nodes deployed in the region of event compared to other regions in the environment. Any dying nodes would also require a replacement (redeployment) to sustain the lifetime of entire network. This is a novel methodology for a mobile robot to collect data, replace any dying node, and deploy more nodes in the region of higher events.

The decision-making problem for the robot is to efficiently navigate through the sensor field to reach all the nodes. In the event of multiple paths available to the robot, the robot path planning algorithm would intelligently decide which node to reach first. The robot is challenged with equally "important" paths to navigate in order to fulfill its goal. The goal is to collect data and/or to deploy a node. With advanced technology, robots may be able to even recharge the battery on the sensor node. However, due to low cost in sensor node construction, we assume that it is economical to redeploy a node instead of recharging the battery. The importance of a given path is based on several parameters relating to the sensor nodes in the field. Given a deployed embedded network of sensors, the task of the robot is to reach the sensor nodes based on several competing criteria. For example, a sensor could have critical data that need to be collected. At the same time, another sensor node may be dying due to low battery power, requiring immediate attention.

We formulate the above problem by defining the set of criteria, alternatives, and the goal as follows:

Criteria: $X = \{x_1, x_2, \ldots, x_n\}$ set of criteria

$X = \{$distance, battery power, scheduling, data criticality$\}$

Alternatives: $\Omega = \{s_1, s_2, \ldots, s_m\}$ set of systems on which criteria is to be evaluated

$\Omega = $ set of sensor nodes

Goal: Evaluate the set of systems/alternatives $\{s_1, s_2, \ldots, s_m\}$ based on set of criteria $\{x_1, x_2, \ldots, x_n\}$.

$G =$ Select a sensor node to be reached first.

The criteria and alternatives are organized in a tabular fashion as shown in Fig. 12.4. Distance represents how far the node is to the base-station or the robot. Battery power represents the voltage remaining in the sensor node's battery. At the sensor node level, to conserve battery power, the sensors can be *scheduled* to sense the environment at different samples or time intervals and sleep as much as possible. Although, some information might be lost, this is an effective way to optimize energy consumption. A simple way to represent criticality is to look at the threshold of the sensed value. Generally, if the sensed value is beyond the specified threshold, the criticality will be high.

C-1···C-m shown in Fig. 12.4 are evaluation results based on the current criteria and interaction among the criteria. The methodology used to obtain C-1···C-m is by using Choquet integral. A simple pairwise comparison between two evaluation items can help to determine the *preference* for selecting a particular system (sensor node). For example, if C-1 $>$ C-2, then sensor s_1 is preferred over s_2.

As discussed before, there are three types of interaction among criteria identified *correlation*, *complementary* and *preference dependency* as three different forms of interaction among criteria. In our case study on sensor network, criteria such as power level and capturing events are correlated and complementary. For example, to capture

Criteria x Sensors s	Distance	Battery power	Scheduling	Data criticality	Evaluation
sensor 1 (s_1)	d-1	b-1	e-1	cr-1	C-1
sensor 2 (s_2)	d-2	b-2	e-2	cr-2	C-2
sensor 3 (s_3)	d-3	b-3	e-3	cr-3	C-3
...	...	...	...	...	...
sensor m (s_m)	d-m	b-m	e-m	cr-m	C-m

FIGURE 12.4   Evaluation of Alternatives

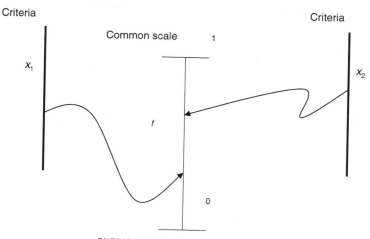

FIGURE 12.5 Mapping criteria

critical environmental events, a deployed sensor should ideally have a low sleep time and high sampling frequency. This means that power consumed by the sensor is high, suggesting that power consumption and events are correlated and complementary. Recall that Choquet integral is defined over the function $f: X \rightarrow [0,1]$. This function f is often called the *utility function* or *score* (Huédé et al., 2006). The utility function is required to make the criteria comparable, since criteria generally are not measured on a common scale. By using utility function, we map the criteria to a common scale, making them *commensurable criteria* as shown in Fig. 12.5.

Given the three criteria/attributes related to a sensor node, we can generate the utility function based on the defined goal as follows:

From Fig. 12.6, for example, a shorter distance to a given node generates a high score or utility function. For example, a distance of 1 m will generate a score $f(d) = 0.9$.

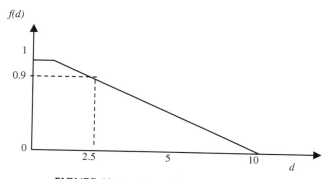

FIGURE 12.6 Generating utility function

The overall evaluation of different alternatives (sensor nodes) is obtained by aggregating the utility functions using Choquet integral with appropriate fuzzy measure (which acts as a weighting factor).

We use criticality, distance from node to base station, and sleep time (scheduling) as three inputs to the Choquet integral decision comparator. We set λ-fuzzy measure (or sometimes called Sugeno measure (Tzeng et al., 2005) to -0.9, suggesting a *positive interaction*. A *positive interaction* or *positive synergy* between two criteria i and j represents some degree of opposition between two criteria, and the fuzzy measure then becomes *subadditive*, that is, $\mu(ij) < \mu(i) + \mu(j)$. $\mu(ij)$ is calculated using the formula $\mu(ij) = \mu(i) + \mu(j) + \lambda\mu(i)\mu(j)$. Therefore, if $\lambda = 0.0$, then fuzzy measure is just additive, $\mu(ij) = \mu(i) + \mu(j)$, and the Choquet integral reduces to weighted average with fuzzy measures acting as weighting factors. Figure 12.7 gives the fuzzy measure on each criterion and fuzzy measure on subset of criteria calculated using λ-fuzzy measure.

Given input values for distance, criticality, and scheduling as 0.9, 0.6, and 0.5, respectively, we obtain Choquet integral value of 0.8096 (refer Fig. 12.7 for computation). We can then formulate a table for different input values for three criteria as shown in Fig. 12.8.

In this case, C-1 > C-2 > C-3 and sensor s-1 is selected.

Sridhar et al. (2007a,b) compare Choquet integration with other aggregation operators and discuss decision-making approaches (such as max–min) when criteria are fuzzy variables.

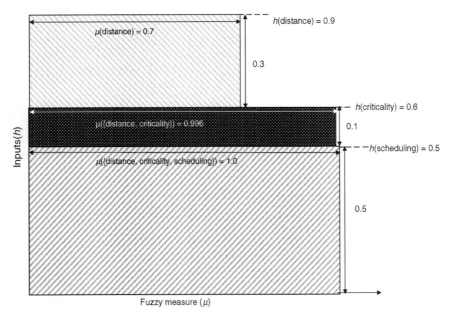

FIGURE 12.7 Fuzzy measures for criteria

No.	Distance	Criticality	Scheduling	Choquet-Integrated Values
1	0.9	0.6	0.5	C-1 = 0.8096
2	0.5	0.9	0.1	C-2 = 0.8184
3	0.1	0.1	0.9	C-3 = 0.5800

FIGURE 12.8 Inputs and computer Choquet integrated values

12.5 CONCLUDING REMARKS

The concept of SoS arises from the need to more effectively implement and analyze large, complex, independent, heterogeneous, and autonomous systems working cooperatively. The driving force behind the desire to view these systems as an SoS is to achieve higher capabilities and performance than would be possible with a traditional stand-alone system. However, when systems interact (and often cooperate and coordinate) with each to address the defined high-level objectives, the performance of the entire SoS could degrade due to scalability, optimization, autonomy, and so on.

We have addressed the scalability issue in sensor networks as case study by proposing innovative solution to fault tolerance and policy/decision making. The solution provided is generic in nature and can be extended to any system of systems applications. The coordinator takes critical managerial operations (for performance improvement), whereas each system takes local hard-real-time operations (for fault isolation in this case study). Other factors that could influence the performance of SoS, such as optimization, can also be carried out at a coordinator level. For example, Azarnoush et al. (2006) address and formulate an optimization problem when system of sensors works cooperatively with system of robots with varying degree of autonomy for a specific application. Such analysis and modeling can be generalized to any SoS application.

Although this chapter presented a case study given a specific SoS (sensor network), it is generally hard to identify whether a given application is truly an SoS application. Moreover, when issues such as scalability, optimization, and so on are considered for a given SoS and application, the ability to control information and decision becomes more complex.

REFERENCES

Azarnoush, H., Horan, B., Sridhar P., Madni, A.M., Jamshidi, M., 2006, Towards optimization of a real-world robotic-sensor system of systems, *Proceedings of the World Automation Congress*, Budapest, Hungary, 2006.

Belton, V., Steward, T.J., 2002, *Multiple criteria decision analysis: an integrated approach*, Kluwer Academic Publishers, 2002.

DeLaurentis, D., 2005, Understanding transportation as a system-of-systems design problem, *43rd AIAA Aerospace Sciences Meeting and Exhibit*, Reno, Nevada, 2005.

Denguir-Rekik, A., Mauris, G., Montmain, J., 2006, Propagation of uncertainty by the possibility theory in choquet integral-based decision making: application to an e-commerce website choice support, *IEEE Transactions on Instrumentation and Measurement*, 55(3): 721:728.

DiMario, M.J., 2006, System of systems interoperability types and characteristics in joint command and control, *Proceedings of the IEEE Conference on System of Systems Engineering,* April 2006, Los Angeles.

Grabisch, M., 1995, The application of fuzzy integral in multicriteria decision making, *European journal of operational research*, 89: 445–456.

Grabisch, M., 2000, A graphical interpretation of the Choquet integral, *IEEE Transactions on Fuzzy Systems*, 8(5): 627–631.

Hall, D., Llinas, J., 2001, *Handbook of Multisensor Data Fusion*, CRC Press, 2001.

Huédé, F., Grabisch, M., Labreuche, C., Savéant, P., 2006, Integration and propagation of a multi-criteria decision making model in constraint programming, *Journal of Heuristics*, 12(4–5): 329–346.

IEEE, 2006, Proceedings of the IEEE International Conference On System of Systems Engineering, April 2006 Los Angeles.

Keating, C., Rogers, R. Unal, R., Dryer, D., Sousa-Poza, A., Safford, R., Peterson, W., Rabadi, G., 2003, System of systems engineering, *Engineering Management Journal*, 15(3): 36–45.

Keating, C., 2005, Research foundations for system of systems engineering, *Proceeding of the IEEE System, Man and Cybernetics*, 3: 2720–2725.

Madni, A.M., Costlow, L., 2003, Common design techniques for BEI GyroChip® quartz rate sensors for both automotive and aerospace/defense markets, *IEEE Transactions on Sensors Journal*, 3(5): 569–578.

Maier, M., 1998, Architecting principles for system-of-systems, *Systems Engineering*, 1: 267–284.

Maier, M., 2005, Research challenges for system-of-systems, *Proceedings of the IEEE System, Man and Cybernetics*, 4: 3149–3154.

Marichal, J-L., 2000, An Axiomatic approach of the discrete choquet integral as a tool to aggregate interacting criteria, *IEEE Transactions on Fuzzy Systems*, 8(6): 800–807.

Meilich, A., 2006, System of systems (SoS) engineering & architecture challenges in a net centric environment, *Proceedings of the IEEE Conference on System of Systems Engineering*, April 2006, Los Angeles.

Sage, A. 1992, *Systems engineering*, Wiley IEEE, 1992.

Sridhar, P., Madni, A.M., Jamshidi, M., 2006a, Hierarchical data aggregation in spatially correlated distributed sensor network, *Proceedings of the World Automation Congress*, 2006.

Sridhar, P., Madni, A.M., Jamshidi, M., 2006b, Intelligent monitoring of sensor networks using fuzzy logic based control, *Proceeding of the IEEE International Conference on Systems, Man and Cybernetics*, 2006.

Sridhar, P., Madni, A.M., Jamshidi, M., 2007a, Hierarchical aggregation and intelligent monitoring and control in fault-tolerant wireless sensor networks, *IEEE Systems Journal (Inaugural Issue)*, 1(1): 38–54.

Sridhar, P., Madni, A.M., Jamshidi, M., 2007b, Multi-criteria decision making and behavior assignment in sensor networks, *Proceedings of the First Annual IEEE Systems Conference*, April 2007, Hawaii.

Tzeng, G-H., Yang, Y-P., Lin, C-T., Chen, C-B., 2005, Hierarchical MADM with fuzzy integral for evaluating enterprise intranet web sites, *Journal of Information Sciences—Informatics and Computer Sciences*, 169(3–4): 409–426.

Wang Z., Klir, G.J., Wang, W., 1995, Determining fuzzy measures by Choquet integral, *Proceedings of the 3rd International Symposium on Uncertainty Modeling and Analysis*, p. 724.

Wojcik, L.A., Hoffman, K.C., 2006, Systems of systems engineering in the enterprise context: a unifying framework for dynamics, *Proceedings of the IEEE Conference on System of Systems Engineering*, April 2006, Los Angeles.

Chapter **13**

A System of Systems View of Services

JAMES M. TIEN

University of Miami, Miami, FL, USA

13.1 SERVICE SYSTEM

Before viewing a service system as an integrated system of systems in Section 13.3, it is helpful to define the individual systems or system components that comprise a service system in Section 13.2. Moreover, it is critical to first define a service system in this section. Some concluding insights are included in Section 13.4.

As detailed in Tien and Berg (1995, 2003, 2006, 2007), the importance of the services sector cannot be overstated; it employs a large and growing proportion of workers in the industrialized nations. As reflected in Fig. 13.1, the services sector includes a number of large industries; indeed, services employment in the United States is at 82.1%, while the remaining four economic sectors (i.e., manufacturing, agriculture, construction, and mining), which together can be considered to be the physical "goods" sector, employ the remaining 17.9%. Alternatively, one could look at the distribution of employers for graduates from such technological universities as Rensselaer Polytechnic Institute (RPI); not surprisingly, as indicated in Fig. 13.2, there has been a complete flip of employment statistics within the past twenty years—from 71% being hired into manufacturing jobs in 1984–1985 to 69% entering the services sector in 2004–2005.

Clearly, the manufacturing sector provides critical goods or products (e.g., autos, aircrafts, satellites, computers.) that enable the delivery of effective and high-quality services; equally clear, the services sector provides critical services (e.g., financial, transportation, design, supply chain.) that enable the production, distribution and consumption of efficient, and high-quality products. Moreover, such traditional

System of Systems Engineering: Innovations for the 21st Century, Edited by Mo Jamshidi
Copyright © 2009 John Wiley & Sons, Inc., Publication

Industries	Employment (M)	Percent
Trade, Transportation & Utilities	26.1 M	19.0%
Professional & Business	17.2	12.6
Health Care	14.8	10.8
Leisure & Hospitality	13.0	9.5
Education	13.0	9.5
Government (except education)	11.7	8.5
Finance, Insurance & Real Estate	8.3	6.1
Information & Telecommunication	3.1	2.2
Other	5.4	3.9
Services Sector	112.6	82.1
Manufacturing	14.3	10.3
Construction	7.5	5.5
Agriculture	2.2	1.6
Mining	0.7	0.5
Goods Sector	24.7	17.9
Total	137.3	100.0

FIGURE 13.1 Scope and size of U.S. employment. *Source*: Bureau of Labor Statistics, April 2006

manufacturing powerhouses like GE and IBM have become more vertically integrated and are now earning an increasingly larger share of their income and profit through their services—including maintenance—operation. For example, in 2006, IBM's pretax income was $13.3 billion (based on a total revenue stream of $91.4 billion) and it was divided into three parts: 23% from systems and technology, 40% from software (which can be considered to be a service activity), and 37% from global services. Thus,

Economic sector	1984–1985 Graduates	2004–2005 Graduates
Services	29%	69%
Manufacturing	71%	29%
Agriculture	0%	0.0%
Construction	0%	2%
Mining	0%	0%
Total	100%	100%

FIGURE 13.2 Reported jobs by graduating students. *Source*: Career Development Center, Rensselaer Polytechnic Institute

IBM earned 23 and 77% of its net revenues from goods and services, respectively; as a result, IBM no longer considers itself a computer company. Instead, it offers itself as a globally integrated innovation partner, one which is able to integrate expertise across industries, business processes, and technologies.

Yet, as Tien and Berg augur (2006), university research and education have not followed suit; the majority of research is still manufacturing- or hardware-related and degree programs are still in those traditional disciplines that were established in the early 1900s. As a consequence, Hipel et al. (2007) maintain that services research and education deserve more attention and support in this twenty-first century when the computer chip, the information technology, and the Internet and the flattening of the world (Friedman, 2005) have all combined to make services—and services innovation—the new engine for global economic growth.

What constitutes the services sector? It can be considered "to include all economic activities whose output is not a physical product or construction, is generally consumed at the time it is produced and provides added value in forms (such as convenience, amusement, timeliness, comfort or health) that are essentially intangible . . ." (Quinn et al., 1987). Implicit in this definition is the recognition that services production and services delivery are so integrated that they can be considered to be a single, combined stage in the services value chain, whereas the goods sector has a value chain that includes supplier, manufacturer, assembler, retailer, and customer.

The following subsections consider, respectively, the emergence of electronic services, the relationship of services to manufacturing, and the movement toward mass customization of both goods and services.

13.1.1 Emerging Electronic Services

Prospectively, it is perhaps more appropriate to focus on emerging e(lectronic)-services. e-Services are, of course, totally dependent on information technology; they include, as examples, financial services, banking, airline reservation systems, and consumer goods marketing. As discussed by Tien and Berg (2003) and detailed in Fig. 13.3, e-service enterprises interact or "co-produce" with their customers in a digital (including e-mail and Internet) medium, as compared to the physical environment in which traditional or bricks-and-mortar service enterprises interact with their customers. Similarly, in comparison to traditional services, which include low-wage jobs, e-services typically employ high-wage earners—and such services are more demanding in their requirements for self-service, transaction speed, and computation.

In regard to data sources that could be used to help make appropriate service decisions, both sets of services rely on multiple data sources; however, traditional services typically require homogeneous (mostly quantitative) sources, while e-services increasingly require nonhomogeneous (i.e., both quantitative and qualitative) sources. Paradoxically, the traditional service enterprises have been driven by data, although data availability and accuracy have been limited (especially before the pervasive use of the Universal Product Code (UPC) and the more recent deployment of radio frequency location and identification (RFLID) tags); likewise, the emerging e-service

Issue	Service Enterprises	
	Traditional	Electronic
• Coproduction medium	Physical	Electronic
• Labor requirement	High	Low
• Wage level	Low	High
• Self-service requirement	Low	High
• Transaction speed requirement	Low	High
• Computation requirement	Medium	High
• Data sources	Multiple homogeneous	Multiple nonhomogeneous
• Driver	Data-driven	Information-driven
• Data availability/accuracy	Poor	Rich
• Information availability/accuracy	Poor	Poor
• Size	Economies of scale	Economies of expertise
• Service flexibility	Standard	Adaptive
• Focus	Mass production	Mass customization
• Decision time frame	Predetermined	Real-Time

FIGURE 13.3   Comparison of traditional and electronic services

enterprises have been driven by information (i.e., processed data), although information availability and accuracy have been limited, due to a data rich, information poor (DRIP) conundrum (Tien, 2003).

Consequently, while traditional services—like traditional manufacturing—are based on economies of scale and a standardized approach, electronic services—like electronic manufacturing—emphasize economies of expertise or knowledge and an adaptive approach. Another critical distinction between traditional and electronic services is that, although all services require decisions to be made, traditional services are typically based on predetermined decision rules, while electronic services require real-time, adaptive decision making; that is why Tien (2003) has advanced a decision informatics paradigm, one that relies on both information and decision technologies from a real-time perspective. High-speed Internet access, low-cost computing, wireless networks, electronic sensors, and ever-smarter software are the tools for building a global services economy. Thus, in e-commerce, a sophisticated and integrated services system combines product (i.e., good and/or service) selection, order taking, payment processing, order fulfilment, and delivery scheduling into a seamless system, all provided by distinct service providers; in this regard, it can be considered to be a system of systems.

13.1.2   Relationship to Manufacturing

The interdependences, similarities, and complementarities of services and manufacturing are significant. Indeed, many of the recent innovations in manufacturing are relevant to the service industries. Concepts and processes such as cycle time, total quality management, quality circles, six-sigma, design for assembly, design for manufacturability, design for recycling, small-batch production, concurrent engineering, just-in-time manufacturing, rapid prototyping, flexible manufacturing, agile manufacturing, distributed manufacturing, and environmentally sound manufacturing can, for the most part, be recast in services-related terms. Thus, many of the engineering and management concepts and processes employed in manufacturing can likewise be used to deal with problems and issues arising in the services sector.

Nevertheless, there are considerable differences between goods and services. Tien and Berg (2003) provide a comparison between the goods and services sectors. The goods sector requires material as input, is physical in nature, involves the customer at the design stage, and employs mostly quantitative measures to assess its performance. However, the services sector requires information as input, is virtual in nature, involves the customer at the production/delivery stage, and employs mostly qualitative measures to assess its performance. Of course, even when there are similarities, it is critical that the coproducing nature of services be carefully taken into consideration. For example, in manufacturing, physical parameters, statistics of production, and quality can be more precisely delineated; however, since a services operation depends on an interaction between the process of producing the service and the recipient, the characterization is necessarily more subjective and different. Consequently, since services are to a large extent subject to customer satisfaction and since, as Tien and Cahn (1981) postulated and validated, "satisfaction is a function of expectation," service performance or satisfaction can be enhanced through the effective "management" of expectation.

A more insightful approach to understanding and advancing services research is to explicitly consider the differences between services and manufactured goods. As identified in Fig. 13.4, services are, by definition, coproduced; quite variable or heterogeneous in their production and delivery; physically intangible; perishable if not consumed as it is being produced or by a certain time (e.g., before a flight's departure); focused on being personalizable; expectation related in terms of customer satisfaction; and reusable in its entirety. However, manufactured goods are pre-produced; quite identical or standardized in their production and use; physically tangible; "inventoryable" if not consumed; focused on being reliable; utility related in terms of customer satisfaction; and recyclable in regard to its parts. In mnemonic terms and referring to Fig. 13.4, services can be considered to be "chipper," while manufactured goods are a "pitirur."

Another critical difference between manufacturing and services concerns their intellectual property (Tien and Berg, 2006). More specifically and in contrast to a manufactured product, services are based on intellectual property that is rarely protected by any patents belonging to the service provider. Usually, the service provider uses physical technologies or products that belong to outside suppliers who protect their intellectual property by patents. However, the use of the intellectual property, either by-product purchase or by license, is available non-exclusively to all competing service providers. Examples abound: the airline

Focus	Services	Manufactured Goods
Production	Coproduced	Preproduced
Variability	Heterogeneous	Identical
Physicality	Intangible	Tangible
Product	Perishable	Inventoryable
Objective	Personalizable	Reliable
Satisfaction	Expectation-related	Utility-related
Life cycle	Reusable	Recyclable
Overall	Chipper	Pitirur

FIGURE 13.4   Services versus manufactured goods

industry uses jet airplanes, which technology is protected by patents owned by the aircraft manufacturers and other suppliers; Wal-Mart, as part of its vaunted supply chain leadership, relies on point-of-sales cash registers developed and sold by IBM, which holds the intellectual property for those devices; and Citibank, the leader in employing the automated teller machine (ATM) innovation, does not hold the ATM-related patents—Diebold does.

Although the comparison between services and manufacturing highlights some obvious methodological differences, it is interesting to note that the physical manufactured assets depreciate with use and time, while the virtual service assets are generally reusable, and may in fact increase in value with repeated use and over time. The latter assets are predominantly processes and associated human resources that build on the skill and knowledge base accumulated by repeated interactions with the service receiver, who is involved in the coproduction of the service. Thus, for example, a lecturer should get better over time, especially if the same lecture is repeated.

In services, automation-driven software algorithms have transformed human resource-laden, coproducing service systems to software algorithm-laden, self-producing services. Thus, extensive manpower would be required to manually coproduce the services if automation were not available. Although automation has certainly improved productivity and decreased costs in some services (e.g., telecommunications, internet commerce.), it has not yet had a similar impact on other labor-intensive services (e.g., health care, education.). However, with new multimedia and broadband technologies, some hospitals are personalizing their treatment of patients, including the sharing of electronic records with their patients, and some institutions are offering entire degree programs online with just-in-time learning capabilities (Tien, 2000).

13.1.3 Toward Mass Customization

Tien et al. (2004) provide a consistent approach to considering the customization of both goods and services—by first defining a value chain and then showing how it can be partitioned into supply and demand chains, which, in turn, can be appropriately managed. Of course, the key purpose for the management of supply and demand chains is to smooth out the peaks and valleys commonly seen in many supply and demand patterns, respectively. Although only depicting a simple two-by-two, supply versus demand, matrix, Fig. 13.5 provides an insightful understanding of supply chain management (SCM, which can occur when demand is fixed and supply is flexible and therefore manageable), demand chain management (DCM, which can occur when supply is fixed and demand is flexible and therefore manageable), and real-time customized management (RTCM, which can occur when both demand and supply are flexible and thereby allowing for real-time mass customization).

Figure 13.5 identifies several example SCM, DCM, and RTCM methods. The literature is overwhelmed with SCM findings (especially in regard to manufacturing), is only recently focusing on DCM methods (especially in regard to revenue management), and is devoid of RTCM considerations, except for a recent contribution by Yasar (2005)—he combines two SCM methods (i.e., capacity rationing and capacity extending) and two DCM methods (i.e., demand bumping and demand recapturing) to deal with the real-time customized management of, as examples, either a goods problem concerned with the rationing of equipment to produce classes of products or a services problem concerned with the rationing of consultants to coproduce classes of services.

The shift in focus from mass production to mass customization (whereby a service is produced and delivered in response to a customer's stated or imputed needs) is intended to provide superior value to customers by meeting their unique needs. It is in this area of customization—where customer involvement is not only at the goods

Supply	Demand	
	Fixed	**Flexible**
Fixed	Unable to manage Price established (at point where fixed demand matches fixed supply)	Demand chain management (DCM) Product revenue management Dynamic pricing Target marketing Expectation management Auctions
Flexible	Supply chain management (SCM) Inventory control Production scheduling Distribution planning Capacity revenue management Reverse auctions	Real-Time customized management (RTCM) Customized bundling Customized revenue management Customized pricing Customized modularization Customized coproduction systems

FIGURE 13.5 Research taxonomy for demand and supply chains

design stage but also at the manufacturing or coproduction stage—that services and manufacturing are merging in concept (Tien and Berg, 2006). It should be noted that customization is both an enabler and a driver for services innovation. After a detailed review and analysis, Tien (2006) suggests that innovation in the services area— especially in e(lectronic)-services—are facilitated by nine major innovation enablers (i.e., decision informatics, software algorithms, automation, telecommunication, collaboration, standardization, customization, organization, and globalization) and motivated by four innovation drivers (i.e., collaboration, customization, integration, and adaptation). Not surprisingly, all four drivers are directed at empowering the individual—that is, at recognizing that the individual can, respectively, contribute in a collaborative situation, receive customized or personalized attention, access an integrated system of systems, and obtain adaptive real-time or just-in-time input.

Increasingly, customers want more than just traditional or electronic services; they are seeking experiences (Pine and Gilmore, 1999) that they can customize to their liking. Customers walk around with their iPods, drink their coffee at Starbucks while listening to and downloading music, dine at such theme restaurants as the Hard Rock Cafe or Planet Hollywood, shop at such experiential destinations as Universal CityWalk in Los Angeles or Beursplien in Rotterdam, lose themselves in such virtual worlds as Second Life or World of Warcraft, and vacation at such theme parks as Disney World or the Dubai Ski Dome, all venues that stage a feast of engaging sensations that are provided by an integrated set of services and goods or products. There is, nevertheless, a distinction between services and experiences; a service includes a set of intangible activities carried out for the customer, whereas an experience engages the customer in a personal, memorable, and holistic manner, one that tries to engage all of the customer's senses. Obviously, experiences have always been at the heart of entertainment, from plays and concerts to movies and television shows; however, the number of entertainment options has exploded with digitization and the Internet. Today, there is a vast array of new experiences, including interactive games, World Wide Web Sites, motion-based simulators, 3D-movies, and virtual realities. Interestingly, one may well ask: will experiences accelerate the commoditization of services, just as services—especially electronic services—have accelerated the commoditization of goods?

13.2 SYSTEM COMPONENTS

In general, a service system can be considered to be a combination or recombination of three essential components—people (characterized by behaviors, attitudes, values, etc.), processes (characterized by collaboration, customization, etc.), and/ or products (characterized by software, hardware, infrastructures, etc.). These components and subcomponents are summarized in Fig. 13.6. In particular, people can be grouped into those demanding services (i.e., customers, users, consumers, buyers, organizations.) and/or those supplying the services (i.e., suppliers, providers, servers, sellers, organizations.); processes can be procedural (i.e., standards, evolving, decision-focused, network-oriented.) and/or algorithmic (i.e., data mining,

System Sub components	System Components (Elements)		
	People (Behaviors, attitudes, values, etc.)	**Processes** (collaboration, customization, etc.)	**Products** (software, hardware, infrastructures, etc.)
Demand	Customers, users, consumers, buyers, organizations, etc.		
Supply	Suppliers, providers, servers, sellers, organizations, etc.		
Procedural		Standards, evolving, decision-focused, network-oriented, etc.	
Algorithmic		Data mining, decision modeling, systems engineering, etc.	
Physical			Facilities, sensors, information technologies, etc.
Virtual			e-commerce, simulations, Web 2.0, e-collaboration, etc.

FIGURE 13.6 Services system: components and subcomponents

decision modeling, systems engineering.) in structure; and products can be physical (i.e., facilities, sensors, information technologies.) and/or virtual (i.e., e-commerce, simulations, e-collaboration.) in form.

13.2.1 People

Given the coproducing nature of services, it is obvious that people constitute the most critical component of a service system. In turn, because people are so unpredictable in their behaviors, attitudes, and values, they can raise the complexity of a service system. Moreover, the multistakeholder—and related multiobjective—nature of such systems serve to only intensify the complexity level and may render the system to be indefinable, if not unmanageable. Human performance, social networks, and interpersonal interactions combine to further aggravate the situation. People-oriented, decision-focused methods are discussed in Section 13.3.

The U.S. health care system is a good example of a people-intensive service system that is in disarray. It is the most expensive and, yet, among the least effective system for a developed country; a minority of the population receives excellent care, while an equal minority receives inadequate care (National Academies, 2006). This situation is not due to a lack of well-trained health professionals or to a lack of innovative technologies; it is due to the fact that it is based on a fragmented group of mostly small, independent providers driven by cost-obsessed insurance companies—clearly, it is a nonsystem. As a consequence, a coordinated and integrated health care system must evolve, one requiring the participation and support of a large number of stakeholders (i.e., consumers, doctors, hospitals, insurance companies.). For example, patients must take increased responsibility for their own health care in terms of access and use of validated information.

13.2.2   Processes

In general and as detailed in Tien (2006), customers would like to be empowered as individuals; that is, as individuals, they should be able to (a) contribute in a collaborative situation, (b) receive customized or personalized attention, (c) access an integrated system or process, and (d) obtain adaptive real-time or just-in-time input . Thus, services—and service innovations—should be about how best to (a) collaborate (especially in regard to self-serving, contributing, communicating, standardizing, and globalizing issues), (b) customize (especially in regard to profiling and personalizing issues), (c) integrate (especially in regard to supply chaining, demand chaining, data warehousing, and systematizing issues), and (d) adapt (especially in regard to real-timing, automating, organizing and motivating issues). In sum, collaboration, customization, integration, and adaptation may be considered to be the four major service processes.

Collaboration—especially intercompany collaboration—is perhaps the most surprising service process. After all, patents were established to protect intellectual property, long enough for the inventors to recoup a good return on their creative investment. However, since services are, by necessity, cocreated or coproduced, collaboration is essential. As noted by Palmisano (2004), innovation in services is becoming too complex; it requires collaboration across disciplines, specialties, organizations, and cultures. Govindarajan and Trimble (2005) recommend that past assumptions, mindsets, and biases must be forgotten (especially in regard to collaboration), and Sanford and Taylor (2005) further underscore this point by suggesting that companies must "let go to grow." Indeed, individuals are collaborating—for free—to enhance an open source software (e.g., Apache), to network, including meeting new friends, sharing photos, or emailing internally (e.g., MySpace), to play a global game with guilds and imaginary gold (e.g., World of Warcraft), or even to live in a virtual world with individualized avatars and Linden dollars (e.g., Second Life), all to satisfy their creative, if not altruistic, and competitive needs. There is much to learn about collaboration, organization, and other real-time business applications from the always-on virtual world. In fact, the real world is using Second Life type of environments to introduce and test new ads, to train new hires, and to design and market new products. Of course, the two worlds are becoming more intertwined when virtual land development companies act in a very realistic manner and when 300 Linden dollars can be exchanged for \$1.

Customization is a critical by-product of collaboration. Tien et al. (2004) have identified several levels of customization. Partial customization occurs in an assemble-to-order environment; that is, upon the arrival of a customer order, the stocked components are assembled into a finished product. As examples, in addition to computer assemblers like Dell and Gateway, Nike offers a program called NikeiD that allows customers to choose the color, material, cushioning, and other attributes of their athletic shoe order, and Procter & Gamble allows women to create and order custom personal-care products such as cosmetics, fragrances, and shampoos. Mass customization occurs when the customer market is partitioned into a very large number of segments, with each segment being a single individual. Customization of clothing, car seats, and other body-fitted products is being advanced through laser-based, 3D body scanners that not only capture a "point cloud" of the targeted body surface (e.g., some 150,000 points are required to create a digital skin of the entire body) but also the software algorithms that integrate the points and extract the needed size measurements. For example, European shoe makers recently initiated a project called EUROShoE, in which an individual's feet are laser scanned and the data are forwarded to a CAD/CAM computer that controls the manufacturing process. Real-time mass customization occurs when the needs of an individualized customer market are met on a real-time basis (e.g., a tailor who laser scans an individual's upper torso and then delivers a uniquely fitted jacket within a reasonable period, while the individual is waiting). Tien et al. (2004) also suggest that goods and services will become indistinguishable when real-time mass customization becomes a reality.

Integration in regard to services allows for "one-stop shopping," a highly desirable situation for a consumer or customer. Integration of financial services has resulted in giant banks (e.g., Citigroup); integration of home building goods and services has resulted in super stores (e.g., Home Depot); and integration of software services has resulted complex software packages (e.g., Microsoft Office). Integration also enhances system efficiency, if not its effectiveness. For example, the radio frequency identification (RFID) tag—or computer chip with a transmitter—serves to integrate the supply chain. The tags are being placed on pallets or individual items passing through the supply chain. In essence, RFID serves to make the supply chains more visible in real time, and as the price of tags decreases, RFID will become ever more popular and critical to the efficient functioning of any supply chain, including the distribution and shipping of goods.

Finally, adaptation is yet another highly desirable service process, especially in regard to electronic services. As detailed in Section 13.3, decision informatics facilitates adaptation in real time. For example, Tien (2005) demonstrates why adaptive decision making is critical when confronting urban disruptions. While terrorist acts are the most insidious and onerous of all urban disruptions, it is obvious that there are many similarities in the way one should deal with these willful acts and those caused by natural and accidental incidents that have also resulted in adverse and severe consequences. However, there is one major and critical difference between terrorist acts and the other types of disruptions: the terrorist acts are willful and therefore also adaptive. One must counter these acts with the same, if not more sophisticated, willful, and adaptive and informed approach. The right decisions must be made at the right time and for the right reason, including those concerned with the preparation for a major disruption, the prediction of such a disruption, the prevention or mitigation of the disruption, the detection of the disruption, the response to the disruption, and the recovery steps that are necessary to adequately, if not fully, recuperate from the disruption. As a consequence, one must trade off between productivity and security; between just-in-time interdependencies and just-in-case inventories; and between high-probability, low-risk life-as-usual situations and low-probability, high-risk catastrophes.

13.2.3 Products

In regard to services, one can group products into two categories. First, there are those physical products or goods (e.g., autos, aircrafts, satellites, computers.) that, as indicated in Section 13.1, enable the delivery of effective and high-quality services (e.g., road travel, air travel, global positioning, electronic services.). Second, there are those more virtual products or goods that can themselves be considered services. From a business perspective, there are, of course, three reasons to act—to create a new service, to solve a particular problem, or to compete in a specific area. These three actions or foci can best be described by relating them to the aforementioned four service processes. As identified in Fig. 13.7, one can create a new service through collaboration (e.g., social networks, either on the web or physically), customization (e.g., express delivery, by air or land), integration (e.g., snowboarding, which combine skiing and surfing), or adaptation (e.g., configuring mutual funds, which focus on new market segments). Alternatively, one can solve a problem through collaboration (e.g., designing minivans, which meet the needs of "soccer moms"), customization (e.g., developing degree programs, which meet the different professional needs of society), integration (e.g., establishing Big Box stores, which meet the shopping needs of customers), or adaptation (e.g., producing home videos, which meet the documenting needs of individual consumers). Additionally, one can compete through collaboration (e.g., HD-DVD formats, either by Blu-Ray Disc or by Advanced Optical Disc), customization (e.g., PC configurations, based on different module combinations), integration (e.g., cell

Service processes	Business foci		
	Creation focused	Solution focused	Competition focused
Collaboration	Social networks	Minivans	HD-DVD formats
Customization	Express delivery	Degree programs	PC configurations
Integration	Snowboarding	Big box stores	Cell phones
Adaptation	Mutual funds	Home videos	Discount retail

FIGURE 13.7 Example services: service processes and business Foci

phones, with different feature and price combinations), or adaptation (e.g., discount retail, focused on revenue enhancement).

An urban center's infrastructures can be considered service products that are critical to its very existence. Historically, these infrastructures (e.g., emergency services, travel services, financial services.) have been physically and logically separate, with little interdependence. However, as a result of advances in information technology and the necessity for improved efficiency and effectiveness, these infrastructures have become increasingly automated and interlinked or interdependent. In fact, because the information technology revolution has changed the way business is transacted, government is operated, and national defense is conducted, the U.S. President (2001) singled it out as the most critical infrastructure to protect in the wake of 9/11. Thus, while the United States is considered a superpower because of its military strength and economic prowess, nontraditional attacks on its interdependent and cyber-supported infrastructures could significantly harm both the nation's military power and economy. Clearly, infrastructures, especially the information infrastructure, are among the nation's weakest links; they are vulnerable to willful acts of sabotage. Recent technological advancements on imbuing infrastructures with "intelligence" make it increasingly feasible to address the safety and security issues, allowing for the continuous monitoring and real-time control of critical infrastructures.

Sadly, the same advances that have enhanced interconnectedness have created new vulnerabilities, especially related to equipment failure, human error, weather and other natural causes, and physical and cyber attacks. Thus, electronic viruses, biological agents, and other toxic materials can turn a nation's "lifelines" into "deathlines" (Beroggi and Wallace, 1995), in that they can be used to facilitate the spread of these materials—whether by accident or by willful act. Even the Internet—with almost a billion users—has become a terrorist tool; jihad websites are recruiting members, soliciting funds, and promoting violence (e.g., by showing the beheading of hostages). Thus, the tools or technologies that underpin a modern society can likewise serve as weapons for undermining, if not destroying, society. Biological, chemical, and nuclear breakthroughs can be employed as weapons of mass destruction; the highly effective Internet can be a medium for cyber viruses, hackers, and spammers; and airplanes can be employed as missiles against people, infrastructures, and commerce.

13.3   SYSTEM INTEGRATION

Inasmuch as a service system is an integrated system, it is, in essence, a system of systems (Hipel et al., 2007). As identified in Fig. 13.8, the objectives of system integration are to enhance its efficiency (leading to greater interdependency), effectiveness (leading to greater usefulness), and adaptiveness (leading to greater responsiveness). Figure 13.8 also identifies the integrative methods in regard to a component's design, interface, and interdependency; a decision's strategic, tactical, and operational orientation; and an organization's data, modelling, and cybernetic considerations .

Methods	Objectives		
	Efficient	Effective	Adaptive
Component integration: • Design: creativity, computer-aided, holistic • Interface: human factors, cognition, visualization • Interdependency: complexity, multidisciplinary	 ✓ ✓ ✓	 ✓ ✓ ✓	 ✓ ✓ ✓
Decision integration: • Strategic: values, simulation, optimization • Tactical: risk/decision analysis, game theory • Operational: decision informatics, improvisation	 ✓ ✓ ✓	 ✓ ✓ ✓	 ✓ ✓ ✓
Management integration: • Data: fusion, mining, visualization • Modeling: architecting, adaptive learning • Cybernetic: monitoring, real-time control	 ✓ ✓ ✓	 ✓ ✓ ✓	 ✓ ✓ ✓

FIGURE 13.8   System of systems: integrative methods and objectives

13.3.1   Component Integration

Systems design, interface, and interdependency constitute the foundation upon which systems research methods can be supported and broadened in scope. Design, or creative problem solving, constitutes the philosophical foundation upon which all engineering disciplines can flourish and mature. The design process permits humans to employ the imaginative or "right brain" component of their intelligence in concert with their analytical or "left brain" capabilities to creatively solve, often in an iterative manner, tough problems, ranging from designing intelligent transportation systems to developing effective government policies. The information technology revolution has permitted the analysis part of design to be largely replaced by computers. For example, a human can tentatively imagine the main features of an advanced transportation vehicle having certain capabilities for satisfying transportation objectives, which can then be rigorously analyzed and viewed graphically using a Computer Aided Design (CAD)/Computer Aided Manufacturing (CAM) program. Based on this analytical and visual feedback, the vehicle can be redesigned and analyzed again in an iterative manner until a satisfactory design is achieved that meets specified performance (i.e., human interface, environmental, fuel efficiency) criteria.

System interface could include the interactions between and among software agents, humans, machines, subsystems, and systems of systems. Human factors constitute a discipline that deals with many of these interactions. However, another critical interface concerns how humans interact with information. In developing appropriate human–information interfaces, one must pay careful attention to a number of factors. First, human–information interfaces are actually a part of any decision support model; they structure the manner in which the model output or information is provided to the decision maker. Cognition represents the point of interface between the human and the information presented. The presentation must enhance the cognitive process of mental visualization, capable of creating images from complex multidimensional data, including structured and unstructured text documents, measurements, images, and video. Second, constructing and communicating a mental image common to a team of, say, emergency responders facilitate collaboration and leads to more effective decision making at all levels, from operational to tactical to strategic. Nevertheless, cognitive facilitation is especially necessary in operational settings that are under high stress. Third, cognitive modeling and decision making must combine machine learning technology with a priori knowledge in a probabilistic data mining framework to develop models of an individual's tasks, goals, interests, and intent. These user-behavior models must be designed to adapt to the individual decision maker so as to promote better understanding of the needs and actions of the individual, including adversarial behaviors and intents.

System interdependency refers to the progressive linking and testing of system components to merge their functional and technical characteristics into a comprehensive, interoperable system of systems (SoS). For example, in a fully integrated SoS, each system can communicate and interact with the entire SoS, without any compatibility issues. For this purpose, an SoS needs a common language, without which the SoS components cannot be fully functional and the SoS cannot be adaptive in the sense that new system components cannot be appropriately integrated into the SoS without a major effort. The concept of an SoS arises from the need to more effectively and efficiently implement and analyze large, complex, and heterogeneous systems working in a cooperative and interdependent manner. The driving force behind the desire to view these systems as an SoS is to achieve higher capabilities and performance than would be possible with the components as stand-alone systems.

13.3.2 Decision Integration

The people- or human-related methods included in this subsection are focused on integrated decision making—at the strategic, tactical, and operational levels. For example, in the context of an urban disruption (Tien, 2005), at the strategic level (which includes the preparation and recovery stages of a disruption), decisions must be made in terms of months, if not weeks; at the tactical level (which includes the prediction and prevention stages of a disruption), decisions must be made in terms of days, if not hours; and at the operational level (which includes the detection and response stages of a disruption), decisions must be made in real time. Alternatively, one could consider the different decision-making levels in terms of a data, information, knowledge, and wisdom continuum (Tien, 2003). As depicted in Fig. 13.9, data represent basic transactions captured during operations, while information represents processed data (e.g., derivations, groupings, patterns.). Clearly, except for simple operational decisions, decision making at the tactical or higher levels requires, at a minimum, appropriate information or processed data. Figure 13.9 also identifies knowledge as processed information (together with experiences, beliefs, values, cultures, etc.) and wisdom as processed knowledge (together with insights, theories, etc.). Thus, strategic decisions require knowledge, while systemic decisions require wisdom.

Strategic decision making is usually distinguished from tactical and operational decision making by the organizational and financial impact of the decisions (i.e., the impact of a strategic decision being significantly greater than those at the tactical and

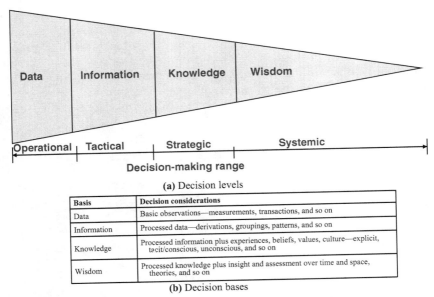

(a) Decision levels

Basis	Decision considerations
Data	Basic observations—measurements, transactions, and so on
Information	Processed data—derivations, groupings, patterns, and so on
Knowledge	Processed information plus experiences, beliefs, values, culture—explicit, tacit/conscious, unconscious, and so on
Wisdom	Processed knowledge plus insight and assessment over time and space, theories, and so on

(b) Decision bases

FIGURE 13.9 Decision-making framework

operational levels); by the "clock speed" (i.e., major strategic decisions usually do not arise as often as tactical and operational decisions and the amount of time available for strategic decision making is usually greater than that for tactical and operational decision making, sometimes significantly so); and by the complexity or scope of the decisions (i.e., strategic decisions—in contrast to tactical and operational decisions – must also take into consideration political, legal, social, and ethical issues).

Tactical decision making is concerned with tackling more medium-term problems and associated objectives. Appropriate decision-making techniques developed in systems engineering and operations research can be effectively utilized in regard to tactical decisions. More specifically, systems engineering focuses on all levels of decision making, including the strategic and tactical levels; on unstructured and complex problems; on qualitative and quantitative data; on soft and hard systems; on the integration of technical, institutional, cultural, financial and other inputs; on multiple conflicting objectives; and, quite appropriately, on a system of systems perspective. As one progresses from the operational to the tactical to the strategic level of decision making, one tends to employ more societal system models and fewer physical system models. Moreover, as indicated in the earlier design subsection, many of these system models recognize the multiple-participant multiple-objective characteristics of real-world problems.

At the operational level, decision making is not only about making the right decisions but also about making timely—and therefore adaptive—decisions. This is especially true at the operational level, where humans must react in seconds and software programs must react in milliseconds. As an example, real-time, information-based decision making—which Tien (2003) calls decision informatics—is needed

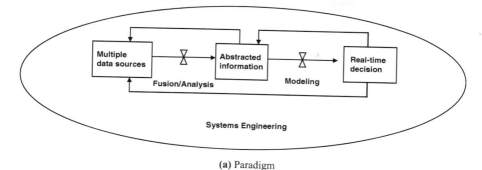

(a) Paradigm

Disciplinary core	Related methods
Decision foci (for services innovation)	• Collaboration: self-serving, contributing, communicating, networking, standardizing, globalizing • Customization: profiling, personalizing • Integration: supply chaining, demand chaining, data warehousing, systematizing • Adaptation: real-timing, automating, organizing, motivating
Data mining (fusion and analysis)	• Statistics: nonhomogeneous data fusion, fuzzy logic, neural networks, biometrics • Mathematics: probability, classification, clustering, association, sequencing • Management science: expectation management, yield management • Cognitive science: visualization, cognition
Decision modeling	• Operations research: optimization, simulation, prediction • Decision science: game theory, risk analysis, dynamic pricing, Bayesian networks • Computer science: service-oriented architecture (SOA), XML, genetic algorithms • Industrial engineering: project management, scheduling,
Systems engineering	• Electrical engineering: cybernetics, networks, pattern recognition • Human machine systems: human factors, cognitive ergonomics, ethnography • System performance: life cycle, value chain • System biology: predictive medicine, preventive medicine, personalized medicine

(b) Methods

FIGURE 13.10   Operational development: a decision informatics paradigm

for enhancing the production and delivery of services, especially emerging e-services. As shown in Fig. 13.10, the nature of the required real-time decision (e.g., regarding the production and/or delivery of a service) determines, where appropriate and from a systems engineering perspective, the data to be collected (possibly, from multiple, nonhomogeneous sources) and the real-time fusion/analysis to be undertaken to obtain the needed information for input to the modeling effort which, in turn, provides the knowledge to support the required decision in a timely and informed manner. The feedback loops in Fig. 13.10 are within the context of systems engineering; they serve to refine the analysis and modeling steps.

Thus, operational decision making or decision informatics is supported by two sets of technologies (i.e., information and decision technologies) and underpinned by three disciplines: data fusion/analysis, decision modeling, and systems engineering. Data fusion/analysis methods include data mining, visualization, data management, probability, statistics, quality, reliability, fuzzy logic, multivariable testing, and pattern analysis; however, real-time data fusion/analysis is more complex and requires additional research. Decision-modeling methods include discrete simulation, finite element analysis, stochastic methods, neural networks, genetic algorithms, optimization, and so on; however, real-time decision modeling, like real-time data fusion/analysis, also requires additional research, especially since all steady-state models become irrelevant in a real-time environment. Systems engineering includes cybernetics or feedback and control; it integrates people, processes and products from a holistic perspective, especially human-centered systems that are computationally-intensive and intelligence-oriented. Similarly, undertaking systems engineering within a real-time environment requires additional thought and research.

Finally, it should be noted that the decision informatics paradigm depicted in Fig. 13.10 is, as a framework, generic and applicable to most, if not all, decision problems. In fact, since any data analysis or modeling effort should only be undertaken in support of some kind of a decision (including the design of a product or a service), all analyses and modeling activities should be able to be viewed within the decision informatics framework. Thus, the framework can be very appropriately applied to critical issues in regard to a particular service, product, and infrastructure or transportation system. Additionally, the adaptive nature of decision informatics is very much akin to the evidence-based—or, more appropriately, risk-based—medicine that is becoming increasingly employed in health care.

13.3.3   Organizational Integration

Organizational or management integration relies on data, models, and cybernetic control. Data are acquired by sensors, which could be in the form of humans, robotic networks, aerial images, radio frequency signals, and other measures and signatures. In regard to tsunamis, for example, seismographs, deep ocean detection devices with buoy transmitters, and/or tide gauges can all sense a potential tsunami. More recently, data warehouses are proliferating and data mining techniques are gaining in popularity. No matter how large a data warehouse and how sophisticated a data mining technique, problems can, of course, occur if the data do not possess the desirable attributes of measurability, availability, consistency, validity, reliability, stability, accuracy, independence, robustness, and completeness. Nevertheless, through the careful analysis or mining of the data, Davenport and Harris (2007) describe how high-performing companies are developing their competitive strategies around data-driven insights based on sophisticated statistical analysis and predictive modeling. Companies as diverse as Capital One, Procter & Gamble, the Boston Red Sox, Best Buy, and Amazon have made better decisions by identifying profitable customers, accelerating innovation, optimizing supply chains and pricing, and discovering new measures of performance. However, in most situations, data alone are useless unless access to and analysis of the data are in real time.

In developing real-time, adaptive data processors, one must consider several critical issues. First, as depicted in Fig. 13.10, these data processors must be able to combine (i.e., fuse and analyze) streaming data from sensors and appropriate input from knowledge bases (including output from tactical and strategic databases) to generate information that could serve as input to operational decision support models and/or provide the basis for making informed decisions. Second, as also shown in Fig. 13.10, the type of data to collect and how to process it depend on what decision is to be made; these dependencies highlight the difficulty of developing effective and adaptive data processors or data miners. Further, once a decision is made, it may constrain subsequent decisions that, in turn, may change future data requirements and information needs. Third, inasmuch as the data processors must function in real time and be adaptable to an ongoing stream of data, genetic algorithms, which equations can mutate repeatedly in an evolutionary manner until a solution emerges that best fit the observed data, are becoming the tools of choice in this area.

In regard to models at the strategic or policy level, there are a number of appropriate models that can support decisions. As examples, Kaplan et al. (2002) have developed a set of complex models to demonstrate that the best prevention strategy to a smallpox attack would be to undertake immediate and widespread vaccination. Unfortunately, models, including simulations, dealing with multiple systems are still relatively immature and must be the focus of additional research and development. Such system of systems models are quite complex and will require a multidisciplinary approach. At the tactical level and as Larson (2005) details, there is a range of decision models for, say, response planning. Indeed, response is about allocating or reallocating resources, which is the essence of operations research—a science that helped the United States minimize shipping losses during World War II, brought efficiencies in production, and developed optimal scheduling of personnel. Another set of critical emergency response models includes those that can simulate, as examples, the impact of an airliner hitting a chemical plant, the dispersion of radioactive material following the explosion of a dirty bomb, and the spread of illness due to a contaminated water supply.

At the operational level, there is a need for real-time decision support models. In such a situation, it is not just about speeding up steady-state models and their solution algorithms; indeed, steady-state models become irrelevant in real-time environments. In essence, it concerns reasoning under both uncertainty and severe time constraints. The development of operational decision support models must recognize several critical issues. First, in addition to defining what data to collect and how they should be fused and analyzed, decisions also drive what kind of models or simulations are needed. These operational models are, in turn, based on abstracted information and output from tactical and strategic decision support models. The models must capture changing behaviors and conditions and adaptively—usually, by employing Bayesian networks—be appropriately responsive within the changing environment. Second, most adaptive models are closely aligned with evolutionary models, also known as genetic algorithms; thus, they function in a manner similar to biological evolution or natural selection. Today, computationally intensive evolutionary algorithms have been employed to coordinate airport operations, to enhance autonomous operations in unmanned aircrafts, and to determine sniper locations while on patrol in Iraq. Third, computational improvisation is another operational modeling approach that can be employed when one cannot predict and plan for every possible contingency. (Indeed, much of what happened on 9/11 was improvised, based on the ingenuity of the responders.) Improvisation involves reexamining and reorganizing past knowledge in time to meet the requirements of an unexpected situation; it may be conceptualized as a search and assembly problem, influenced by such factors as time available for planning, prevailing risk, and constraints imposed by prior decisions (Mendonca and Wallace, 2004).

At all levels of decision making, there are a range of possible simulation models that can be utilized, ranging from those that can be regarded as supporting war games to those with avatars and virtual environments. The World Wide Web will soon be part of a World Wide Simulation where an immersive, three-dimensional environment may well combine elements of such virtual worlds as Second Life with such mapping tools as Google Earth. Moreover, sight and sound will be complemented with virtual-touch technology or haptics, which can give users the sensation that they are feeling solid objects through tactile interfaces and physical resistance.

Cybernetics is derived from the Greek word "kybernetics," which refers to a steersman or governor. Within a system and as indicated earlier, cybernetics is about feedback (through evaluation of performance relative to stated objectives) and control (through communication, self-regulation, adaptation, optimization, and/or management). In regard to system control, it is perhaps the most critical challenge facing SoS designers. Due to the difficulty or impossibility of developing a comprehensive SoS model, either analytically or through simulation, SoS control remains an open problem and is, of course, different for each application domain. Moreover, real-time control—which is required in almost all application domains—of interdependent systems poses an especially difficult problem. The cooperative control of an SoS assumes that it can be characterized by a set of interconnected systems or agents with a common goal. Classical techniques of control design, optimisation, and estimation could be used to create parallel architectures for, as an example, coordinating numerous sensors. However, many issues dealing with real-time cooperative control have not been addressed, even in non-SoS structures. A critical issue concerns controlling an SoS in the presence of communication delays to and among the SoS systems.

Autonomous control of an SoS assumes that it can be characterized by a set of "intelligent systems" that can be implicitly or autonomously controlled. Although the concept of autonomous or intelligent control was first introduced three decades ago by Gupta et al. (1977), the control community has only recently paid substantial attention to such an approach, especially in regard to a variety of industrial applications (e.g., cameras, dishwashers, automobiles.). Most of these applications are due to Zadeh (1996) and involve fuzzy logic, neural networks, evolutionary algorithms, and soft computing; the strength of these methods is in its ability to cope with imprecision, uncertainties, and partial truth. Moreover, the methods can be used to process information, adapt to changing environmental conditions, and learn from the environment; thus, they are adaptive and, to a large extent, responsive to real-time input. However, additional research is required before autonomous control can make full

Dimensions of Integration		Levels of Integration:Examples (Component Systems)		
Attribute	Definition	Low	Medium	High
Physical	Degree of systems co-location	Marketing (consumers, data mining, algorithms)	Virtual experiences (customers, algorithms)	Security (users, biometrics, identification, verification, algorithms)
Temporal	Degree of systems co-timing	eBay (sellers, buyers, algorithms)	Travelocity (travelers, suppliers, algorithms)	Global positioning (satellites, electronics, maps, users, algorithms)
Organizational	Degree of systems co-management	e-training (trainees, curricula, algorithms)	Government services (users, civil servants, procedures/algorithms)	Health/wellness (customers, one-stop facilities, procedures)
Operational	Degree of systems co-dependency	credit ratings (consumers, providers, data bases, algorithms)	e-health (imaging, monitoring, triaging, algorithms)	Standards (interfaces, interoperability, procedures/algorithms)
Functional	Degree of systems co-functioning	My space (customers, networking, algorithms)	Amazon.com (ordering, shipping, algorithms)	FedEx (packaging, shipping, routing, delivering, procedures/algorithms)

FIGURE 13.11   Service system of systems: dimensions and levels of integration

use of a continuous data stream, including taking into consideration the possible future state of a system of systems.

13.4   CONCLUDING INSIGHTS

A number of insights can be made concerning a system of systems view of services. First, it is obvious that such a view can be taken from several perspectives. In addition to the system components and integration perspective detailed herein, Fig. 13.11 considers a service system of systems in terms of its dimensions and levels of integration. The systems that underpin a service SoS can be integrated in a physical (i.e., degree of systems colocation), temporal (i.e., degree of systems cotiming), organizational (i.e., degree of systems comanagement), operational (i.e., degree of systems codependency), and/or functional (i.e., degree of systems cofunctioning) dimension. Along each dimension or attribute, the level of integration can be low, medium, or high. Within each attribute, Fig. 13.11 identifies a low, medium, and high level of service integration example, including their respective underlying systems.

Second, it is obvious that systems – especially service systems – are becoming increasingly more complex; indeed, each system can be regarded as a system of systems, together with all the attendant life cycle design, human interface, and system integration issues. For example, central to the mission of transforming the current U.S. Army into a leaner, more technologically advanced fighting force is a vast computerized network—called the Future Combat Systems (FCS) – that would link humans (i.e., soldiers, commanders.) to a panoply of sensors, satellites, robots, drones, and armored vehicles. Initiated in 2002, FCS has a projected price tag of $230 billion through 2030; it would require the writing of over 60 million lines of computer code, the most ever for any system. The Boeing Company and its main subcontractor, Science Applications International Corporation (SAIC), are managing this very complex effort that, if successful, could change the nature of warfare and lift the proverbial fog of war from the battlefield. The decision informatics paradigm (Tien, 2003) must, of course, be at the heart of any FCS that requires adaptive, real-time decision making.

Third, as real-time decisions must be made in an accelerated and coproduced manner, the human service provider will increasingly become a bottleneck; he/she must make way for a smart robot or software agent. For example, everyone could use a smart alter ego or agent that could analyze, and perhaps fuse, all the existing and incoming e-mails, phone calls, Web pages, news clips, and stock quotes, and assigns every item a priority based on the individual's preferences and observed behaviors. It should be able to perform a semantic analysis of a message text, judge the sender-recipient relationships by examining an organizational chart and recall the urgency of the recipient's responses to previous messages from the same sender. To this, it might add information gathered by watching the user via a video camera or by scrutinizing his/her calendar. Most probably, such an agent would be based on a Bayesian statistical model—capable of evaluating hundreds of user-related factors linked by probabilities, causes and effects in a vast web of contingent outcomes—that infers the likelihood that a given decision on the software's part would lead to the user's desired outcome. The ultimate goal is to judge when the user can safely be interrupted, with what kind of message, and via which device. Perhaps the same agent could serve as a travel agent by searching the Internet and gathering all the relevant information about airline schedules and hotel prices, and, with the user's consent, returning with the electronic tickets and reservations. Clearly, such agents must be a part of any complex service system.

Finally, as a critical aspect of complexity, modern systems of systems are also becoming increasingly more human centered, if not human focused; thus, products and services are becoming more personalized or customized. Certainly, services coproduction implies the existence of a human customer, if not a human service provider. The implication is profound: a multidisciplinary approach must be employed—it must also include techniques from the social sciences (i.e., sociology, psychology, and philosophy) and management (i.e., organization, economics, and entrepreneurship). As a consequence, researchers must expand their systems (i.e., holistic-oriented), man (i.e., decision-oriented), and cybernetic (i.e., adaptive-oriented) methods to include and be integrated with those techniques that are beyond science and engineering. For example, higher customer satisfaction can be achieved not only by improving service quality but also by lowering customer expectation. In essence, as stated by Hipel et al. (2007), systems, man, and cybernetics constitute an integrative, adaptive, and multidisciplinary approach to creative problem solving, which takes into account stakeholders' value systems and satisfies important societal, environmental, economic, and other criteria to enhance the decision-making process when designing, implementing, operating, and maintaining a system or system of systems to meet societal needs in a fair, ethical, and sustainable manner throughout the system's life cycle.

ACKNOWLEDGMENT

The author would like to express his appreciation to Dr. Anuj Goel, a recent graduate of Department of Decision Sciences & Engineering Systems at Rensselaer Polytechnic Institute. He has contributed to the resultant Chapter.

REFERENCES

Beroggi, B.E.G., Wallace, W.A., 1995, Real-time decision support for emergency management: an integration of advanced computer and communications technology, *Journal of Contingencies and Crisis Management*, 3(1): 18–26.

Davenport, T.H., Harris, J.G., 2007, *Competing on Analytics: The New Science of Winning*, Harvard Business School Press, Boston, MA.

Friedman, T.L., 2005, *The World is Flat: a Brief History of the Twenty-First Century*, Farrar, Strauss & Giroux, New York, NY.

Govindarajan, V., Trimble, C., 2005, *Ten Rules for Strategic Innovators: from Idea to Execution*, Harvard Business School Press, Boston, MA.

Gupta, M., Saridis, G., Gaines, B., 1977, *Fuzzy Automata and Decision Processes*, North-Holland, New York.

Hipel, K.W., Jamshidi, M.M., Tien, J.M., White, C.C., 2007, The future of systems, man and cybernetics: application domains and research methods, *IEEE Transactions on Systems, Man, and Cybernetics Part C*, 30(2): 213–218.

Kaplan, E.H., Craft, D.L., Wein, L.M., 2002, Emergency response to a smallpox attack: the case for mass vaccination, *Proceedings of the National Academy of Sciences*, 99(16): 10935–10940.

Larson, R.C., 2005, Decision models for emergency response planning, In Kamien, D. (Ed.), *Handbook of Homeland Security*, McGraw-Hill, New York, NY.

Mendonca, D., Wallace, W.A., 2004, Studying organizationally-situated improvisation in response to extreme events, *International Journal of Mass Emergencies and Disasters*, 22(2): 5–29.

National Academies, 2006, *Engineering the Health Care System*, National Academies Press, Washington, DC.

Palmisano, S.J., 2004, *Global Innovation Outlook*, IBM, Armonk, NY.

Pine B.J., II, Gilmore, J.H., 1999, *The Experience Economy*, Harvard Business School Press, Boston, MA.

Quinn, J.B., Baruch, J.J., Paquette, P.C., 1987, Technology in services, *Scientific American*, 257(6):50–58.

Sanford, L.S., Taylor, D., 2005, *Let Go to Grow: Escaping the Commodity Trap*, Upper Saddle River, Pearson Education, NJ.

Tien, J.M., 2000, Individual-centered education: an any one, any time, any where approach to engineering education, *IEEE Transactions on Systems, Man, and Cybernetics Part C, Special issue on Systems Engineering Education*, 30(2): 213–218.

Tien, J.M., 2003, Towards a decision informatics paradigm: a real-time, information-based approach to decision making, *IEEE Transactions on Systems, Man, and Cybernetics, Special issue, Part C*, 33(1): 102–113.

Tien, J.M., 2005, Viewing urban disruptions from a decision informatics perspective, *Journal of Systems Science and Systems Engineering*, 14(3): 257–288.

Tien, J.M., 2006, Services innovation: decision attributes, innovation enablers, and innovation drivers, Chapter 2, In Cheng Hsu,(Ed.), *Service Enterprise Integration: an Enterprise System Engineering Perspective*, Springer Science, Norwell, MA.

Tien, J.M., Berg, D., 1995, Systems engineering in the growing service economy, *IEEE Transactions on Systems, Man, and Cybernetics*, 25(5): 321–326.

Tien, J.M., Berg, D., 2003, A case for service systems engineering, *Journal of Systems Science and Systems Engineering*, 12(1): 13–38.

Tien, J.M., Berg, D., 2006, On services research and education, *Journal of Systems Science and Systems Engineering*, 15(3): 257–283.

Tien, J.M., Berg, D., 2007, A calculus for services innovation, *Journal of Systems Science and Systems Engineering*, 16(2): 129–165.

Tien, J.M., Cahn, M.F., 1981, *An Evaluation of the Wilmington Management of Demand Program*, National Institute of Justice, Washington, DC.

Tien, J.M., Krishnamurthy, A., Yasar, A., 2004, Towards Real-Time Customized Management of Supply and Demand Chains, *Journal of Systems Science and Systems Engineering*, 13(3): 257–278.

U.S. President 2001, Executive Order on Critical Infrastructure Protection, October 16, The White House, Washington, DC.

Yasar, A. 2005, Real-time and simultaneous management of supply and demand chains, Ph.D. Thesis, Rensselaer Polytechnic Institute, Troy, NY.

Zadeh, L.A., 1996, The evolution of systems analysis and control: a personal perspective, *IEEE Control Systems Magazine*, 16(3): 95–98.

Chapter **14**

System of Systems Engineering in Space Exploration

STEVEN D. JOLLY[1] and BRIAN K. MUIRHEAD[2]

[1]Lockheed Martin Space Systems, Littleton, CO, USA
[2]NASA Jet Propulsion Laboratory, Pasadena, CA, USA

14.1 KEY ISSUES IN SPACE EXPLORATION SYSTEM OF SYSTEMS

As demonstrated in the International Council on Systems Engineering (INCOSE) systems engineering handbook (INCOSE, 2006), the application of SoSE principles does not require that the system be very large in physical size or employ hundreds or thousands of people for many, many years. This is also true in the space exploration world. Even the most austere deep space mission consists of an enormous complexity of other systems to achieve success (e.g., flight system, ground data system, Deep Space Network System, etc.) Regardless of the terms, current and future missions are requiring greater and greater interdependency between physically separate elements, in the limit acting as a single system of systems (SoS). This not only increases the likelihood and consequence of traditional design and development issues (like critical path management) but also introduces new problems that can threaten mission success (e.g., geographical dispersion). Since the beginning of the space age, systems engineers have had to find ways to exploit the technology of their day to perform extraordinary missions. While the physics governing space exploration has not changed, the technology is in constant upward flux in capability, and yet in some ways has become more vulnerable to human error and more susceptible to certain aspects of the space environment (like radiation). This section highlights 10 important issues that every space exploration SoS system engineer (SoSE) must understand, ranging from technical issues like software and computation to seemingly esoteric and bothersome issues like International Traffic in Arms

System of Systems Engineering: Innovations for the 21st Century, Edited by Mo Jamshidi
Copyright © 2009 John Wiley & Sons, Inc., Publication

Regulation (ITAR) and proprietary information. These were not chosen at random. For the SoSE, these are important topics of our time (and the foreseeable future). These are issues that have demonstrably defined the thin margin between mission success or mission loss.

14.1.1 What Software has Become

Space systems today have fundamentally changed from systems in the 1980s and even the early 1990s, probably nowhere more than in the area of software and computation, both flight and ground. Spacecraft and their associated ground stations in the first 30–40 years of spaceflight were fundamentally limited by the speed of processors and the limited availability of data and code storage. Thus, early systems like the Gemini flight computer or later with the Apollo Guidance Computer (AGC) had 13–36K words of storage for the flight code (Tomayko, 1988). So by necessity the software was focused on the key disciplines of guidance and control, and some mission critical sequencing that could not be done reliably in hardware. The flight software (FSW) code was heavily optimized in assembly language to fit in the available memory.

Since the FSW size and functionality were limited by hardware performance, these early space systems relied heavily on ground commanding for sequencing and timing (for unmanned missions) or astronauts (Fig. 14.1), or hardware—executed in largely analog electrical designs. This limited the number of interfaces that FSW had with hardware, and FSW could be more easily viewed as a bounded subsystem of the larger spacecraft system.

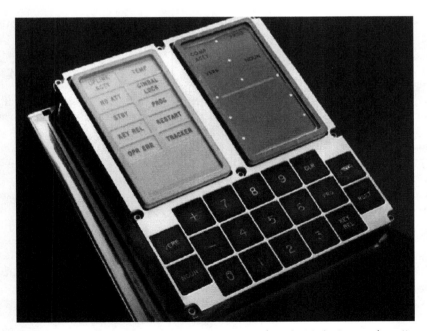

FIGURE 14.1 Apollo guidance computer user interface (NASA). *Source*: http://www.nasa.gov/centers/dryden/images/content/86985main_TF-2001-02_popup2.jpg

As Moore's law would dictate, processors and memory devices (as well as the entire area of integrated digital circuits) progressed in an exponential fashion and forever changed the way spacecraft are designed. The 13 K words of assembly language instructions running on early processors and ferrous rope memory of the Gemini era grew into 40,000 lines of code (LOC)—high-level C code running on reduced instruction set commands (RISC)-based processors and integrated circuit dynamic random access memory—for Mars Surveyor missions of the late 1990s, and now 166,000 LOC for NASA's recent Mars Reconnaissance Orbiter (MRO). The human exploration program was even more profound as it grew from an Apollo era assembly language instruction set of 36 K words executing on the AGC to Orion, which will have an estimated 1–1.5 million LOC running on advanced avionics employing self-checking processor pairs.

Why did this happen? It was because the revolution in the processor and electronics worlds (themselves linked) overtook and drew the space program along. Today's space programs exploit commercial electronics development not just because of the increased performance but also to avoid obsolescence and reduce life cycle costs. Hardware-based timing and sequencing as known in the early days was gone, replaced by large FSW loads running on ever-faster processors, application-specific integrated circuits (ASIC), and field-programmable gate arrays (FPGA).

What does all this mean? The answer is best illustrated by thinking about the technique of the N-squared diagram. This diagram is used by systems engineers to understand which elements, subsystems, and components of the end-to-end system have an interface with one another (Lalli et al. 1997). The greater the number of interfaces between pieces of the system, the greater the complexity and the risk (development and flight) to the space mission. For software, space and ground, the N-squared diagrams have exponentially progressed from lightly populated diagrams to diagrams that show software has an interface to virtually everything (Fig. 14.2). One result is that software cannot readily be developed and managed as a subsystem anymore—but is now clearly the domain of the SoSE.

The implication of this on SoS engineering for space exploration is absolutely profound. First, new approaches are needed to promote and control horizontal integration (across all those interfaces), and simply attempting to employ traditional interface techniques such as interface control drawings (ICDs) will bring the program to ruin due to the overhead cost to produce and maintain them. Similarly, vertical integration (including top-down requirements derivation) cannot be relegated to the software subsystem or integrated product team (IPT) but must be the purview of the systems engineering and integration team (SEIT). In addition to the software and firmware, the SoSE faces an equally difficult challenge in dealing with the parameterization of software. There are literally hundreds of thousands of parameters (numerical data) that are uploaded and ingested by software, parameters that can subtly or radically change the behavior of software and result in the difference between mission success and failure. For reference, MRO has approximately 120 K parameters of which 25 K are assumed to be mission critical. If there is an error in any one of the critical parameters, it could lead to failure. Therefore, with this difficulty and complexity of validating SW systems, today it is necessary to formulate a

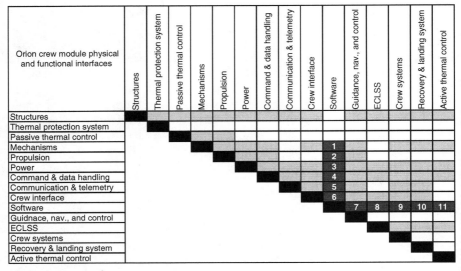

FIGURE 14.2 N^2 Analysis of Orion's crew module: FSW interfaces with 11 of 14 possible subsystems, only 2 less than structure

comprehensive plan of integrated testing. Testing for off-nominal and stressing cases, using high-fidelity simulations of the end-to-end system in a space environment, must be rigorous and thorough in order to prove that parameters, both individually and collectively, are valid for the mission.

Design and testing of FSW are also affected by the level of sophistication of the system architecture for failure detection, isolation, and response (FDIR). This critical and complex area of systems engineering requires early understanding of the relationships between hardware and software elements, their failure modes, and how to recognize, compartmentalize, and limit problems/failures. A significant part of the FSW test program must be dedicated to the integrated testing of the HW, SW, and FDIR system.

14.1.2 Requirements Complexity and Growth

SoS space missions face unprecedented challenges in requirements development, management, and verification, both because of the shear number of them and because of the preponderance of major driving requirements, including many requirements that, result in large consequences to system design, execution, and performance with small changes. This problem is compounded by a growing tendency to overspecify requirements (i.e., too detailed and too prescriptive) to control outcomes and behavior unnecessary to the functioning of the design. Thus, it is paramount that validation of early requirements be accomplished. Normally, projects talk of requirements verification and validation (V&V) being formally bought off in the Phase C/D time frame, when the actual hardware and software are tested at the acceptance level and then integrated for system-level and end-to-end testing. This testing is said to accomplish verification (did we build it right) and validation (did we build the right thing) in a

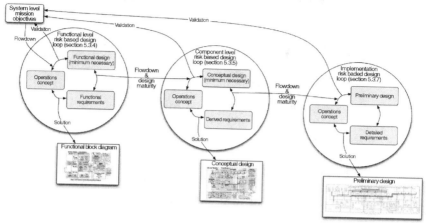

FIGURE 14.3   Example of the CONOP's role in SoS development: Davis time phase state diagram (NASA). *Source*: Coauthor Muirhead

mostly simultaneous manner. But how can a project know that it has the necessary and sufficient requirements that lead to the right system being designed? How can it know sufficiently early in the life cycle so as to make course corrections without huge cost, schedule, or performance penalties? The answer lies in a more specific pursuit of validation (as opposed to verification), and how it can be achieved regarding requirements in the early phases.

The process begins with the discipline of developing a technical concept of operations (CONOP). This is not a one-time, high-level mission description that is put on a shelf and forgotten (Fig. 14.3). This is a living document that develops throughout the project's life cycle, a document that increases in fidelity at every major milestone. The CONOP provides the context in which requirements have their meaning, and without sufficiently detailed CONOPs or design reference missions (DRMs), the project cannot be certain that its top-level requirements (much less the other derived specifications) are valid. The CONOP should be time-domain oriented and become sufficiently detailed so that it provides the story of how this SoS space mission will unfold and perform (eventually becoming the basis of mission operation procedures).

With the CONOP/DRM as the backdrop, early requirements validation methodologies are developed using combinations of analysis, simulations, test beds, and tests/demonstrations that completely cover the operational timeline. For in addition to the CONOP, the project needs a physical and functional concept of how requirements would be satisfied in a conceptual design. The design concept or particular aspects of it are represented by these tools. Together with the CONOP (the story) and these tools (a possible instantiation of a design), one can explore the validity of the requirements set. It is important not to attempt to validate the entirety of the requirements all at once, but to start with the critical-to-customer subset, and then incrementally approach validation on the larger set as the life cycle unfolds (targeting completion between PDR and CDR of this early effort). Proportional to the complexity of the project will be the levels of hardware/software demonstration employed, which should be built-in to the integrated master plan and integrated master schedule (IMP/IMS).

In SoS-class missions, the system under development will almost always require a staged or block implementation (sometimes called preplanned product improvement) to achieve the final state. So in designing an element of an SoS like Constellation's Orion (the Crew Exploration Vehicle), which over its life cycle first visits the International Space Station (ISS) and later on experiences progressively more difficult Lunar missions (shorter "sorties" followed by 6 months long "outpost" missions), needs special consideration in the requirements process to not overconstrain the system. In this example, the reliability requirements defined for initial capability missions to ISS, currently 1 in 200 for loss of mission, are unrealistic for the much more complex lunar missions (current requirement is 1 in 20 for loss of mission), and so Orion is being designed to have the capability to evolve or upgrade from the ISS to lunar expedition missions.

Another key challenge facing the SoS space missions is the general area of requirements creep. It is true that requirements growth, whether creep or avalanche, leads to cost and schedule increases, as well as potentially threatening mission success (Rustan, 2005). There are many causes, both external and internal to the project. The shear size of SoS projects means that both the probability and the consequences of creep are greatly increased. A chief issue with new requirements is being able to correctly estimate their affect on cost and schedule. In the early phases of development, like the conceptual design phase, the design is more flexible and the team is able to more readily balance out the effects of mutually exclusive requirements and constraints to size a system that can close. As the system design matures forward through SRR to PDR, there is inherently less flexibility to adapt the architecture and rebalance to new or changing requirements. By the time the project gets to PDR, the only way to accommodate changing and new requirements is to force fit them into the existing design. If allowed to proceed, performance margins begin to erode and the many requirements that were previously thought to be independent collectively conspire to break the system. There is also the reality that not all requirements are created equal, so small changes in a driving requirement can produce the same effect. Bottom line: undisciplined requirements management will lead to cost and schedule growth, which can lead to program cancellation (GAO, 2006).

14.1.3 Interface Complexity and Growth

The design and execution of SoS Space Exploration missions is basically an extreme exercise of interface definition and control. These systems are so large and complex that they must be parsed into manageable elements, and then likewise the elements into subsystems and so forth. But the real challenge for the SoSE is recombining the parts into a functioning whole. It is the nature of such missions that they involve many government and industrial partners, teammates, contractors, and subcontractors—with many diverse institutions and cultures. With such intertwining and interleaving of responsibility and the resulting contractual complexity, it is imperative that a rigorous interface control strategy be invoked or the critical path will be at risk with enormous potential for schedule delays and cost growth. The consequences of poor interface control are not limited to development—they can lead to mission failures that can be devastating not only to individual teams but to a whole industry! Though there are many interfaces to be controlled in an SOS space exploration mission other than strictly technical, we will focus here on technical interface control of the overall system.

An interface in an SOS is simply the boundary between two or more entities, software, hardware, and/or a user. The interface defines how an element interacts with the outside world and as such must provide the translation of an element's internal functionality to its broader or higher system function. Interfaces range from very simple, low levels, such as 0 or 1 discrete commands between software and hardware, to complex human–machine interactions. Many of the most difficult problems and greatest potential for mistakes show up at interfaces. The most difficult to define, understand, and manage/control are between different disciplines, different subsystems, and/or between people and organizations. A good rule of thumb is always to minimize the number of interfaces, however, for SoS a large number of complex interfaces cannot be avoided.

The number of interfaces between elements that are point-to-point interconnected (e.g., must communicate/interact with each other in some way) is equal to $n(n-1)/2$, where n is the number of elements. So, for example, a flight system within a single project, with eight subsystems plus a system engineering organization and a management office could have 45 technical and management interfaces to define and manage. Although interfaces may be documented through various types of interface requirements/control documents (mechanical, electrical, informational, and organizational), they also need a mechanism for effective communication between the elements to work problems and be sure that the interfaces are understood. Obviously, for a large SoS the combinatorics can result in a large number of possible interfaces. This forces much greater responsibility onto the SoS engineering team to define, develop, and manage increasingly clever techniques for effective interface definition and control.

Some of the most complex and important interfaces are the user interface for safety-critical or mission-critical systems, where the user (ground or crew) has time limitations within which to make correct decisions. User interfaces for these types of systems must be well designed, easily understood, easily used, and capable of thorough validation with humans in the loop.

Among options for more efficient techniques to help understand and manage interfaces are use of model-based representations of elements (i.e., functional software models as opposed to discrete parameters and requirements) and standards (e.g., standard communication protocols) (Watson, 2003). Other options include use of interface description languages for bridging between different software languages and state-of-the-art electronic information management systems that ensure control, accuracy, and rapid access to the most current data.

Obviously, the validation of interfaces is critical to the success of any SoS. How an interface is to be tested and validated must be fully thought through at the time in interface definition. Given that complexity of the interfaces in an SoS, the validation must be done early and repeated at higher and higher levels of integration. The validation strategy must include clear responsibility for the validation process, specific measurable parameters, and specific milestones in the development cycle for conducting thorough testing (with consideration given to time to fix the problems that will inevitably appear). Use of software-based functional models can greatly facilitate the validation process for electrical and informational interfaces.

For each SoS space mission the major and submajor interfaces could be different and must be judged on their own merit. Unfortunately, projects often do not spend enough time at the architectural level playing out the risks associated with these decisions, and sometimes they are necessarily made on the basis of partner core capabilities or organization structures rather than system interface optimization. Nevertheless, it is incumbent on project management and systems engineering to recognize this non-optimality and step up the rigor by which the interfaces will be managed.

Although doing this in practice is a very complex issue, there are three basic tenets of interface development that are essential to mission success:

1. Thorough requirements discovery and validation;
2. Early interface validation; and
3. Rigorous use of interface control documentation.

Interface driving requirements must be understood in early conceptual design of the SoS. The methodologies that can be used to discover them are varied but include the CONOP, simulations, Monte Carlo analysis, mission phase fault-tree analysis, system compatibility analysis, and critical signals analysis—to name a few. Increasingly more popular in Department of Defense (DoD) missions is the use of architecture views, such as the DoD Architecture Framework (DoDAF). Whichever set of tools a project applies, it is critical that the "day in the life" approach to analysis be used. This involves thinking operationally of the time–domain interaction of the SoS space mission elements, subsystems, and components. The best practice approach is to divide the mission into phases and subphases and describe the events step by step and the interactions of the parts of the system as time proceeds forward through the phase. On all of NASA's recent Mars Lander missions, Mars Pathfinder, Mars Exploration Rovers, Phoenix Mars Scout, and the Mars Science Laboratory (MSL), the teams have defined entry, descent, and landing (EDL) as a critical phase and have created engineering products (at appropriate fidelities corresponding to the program life cycle) that accomplish both time domain and functional domain analysis (Fig. 14.4). This facilitated the discovery of critical requirements that led to the development of whole new systems needed to assure safe landing (rocket-assisted deceleration, horizontal velocity control, etc.).

Early interface validation is the key for proving compatibility across the interface and allows the project to provide course corrections to engineering design should problems arise. Because it is early, the overall effect on the critical path is mitigated and actions can be taken. Exchange of breadboards, engineering models, interface simulations, and so on are the most common ways of achieving this validation.

The third tenet surrounds the use of the ICD for major interfaces. The ICD is usually derived from a parent requirements document like an interface requirements

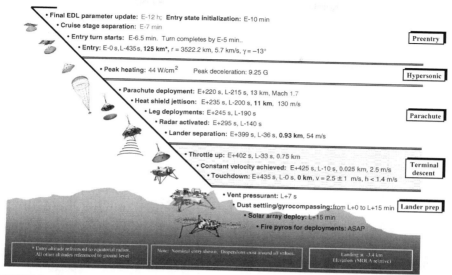

- **Final EDL parameter update:** E-12 h; **Entry state initialization:** E-10 min
  - **Cruise stage separation:** E-7 min
    - **Entry turn starts:** E-6.5 min. Turn completes by E-5 min..
      - **Entry:** E-0 s, L-435 s, **125 km***, r = 3522.2 km, 5.7 km/s, γ = −13° | **Preentry** |

- **Peak heating:** 44 W/cm²   Peak deceleration: 9.25 G | **Hypersonic** |

- **Parachute deployment:** E+220 s, L-215 s, 13 km, Mach 1.7
  - **Heat shield jettison:** E+235 s, L-200 s, **11 km**, 130 m/s
    - **Leg deployments:** E+245 s, L-190 s | **Parachute** |
      - **Radar activated:** E+295 s, L-140 s
        - **Lander separation:** E+399 s, L-36 s, **0.93 km**, 54 m/s

- **Throttle up:** E+402 s, L-33 s, 0.75 km
  - **Constant velocity achieved:** E+425 s, L-10 s, 0.025 km, 2.5 m/s | **Terminal descent** |
    - **Touchdown:** E+435 s, L-0 s, **0 km**, v = 2.5 ± 1 m/s, h < 1.4 m/s

- **Vent pressurant:** L+7 s
  - **Dust settling/gyrocompassing:** from L+0 to L+15 min | **Lander prep** |
    - **Solar array deploy:** L+15 min
      - **Fire pyros for deployments:** ASAP

* Entry altitude referenced to equatorial radius.
All other altitudes referenced to ground level   Note: Nominal entry shown. Dispersions exist around all values.   Landing at -3.4 km Elevation (MOLA relative)

FIGURE 14.4 Example of time–domain CONOP/DRM, Phoenix Mars Scout entry, descent, and landing (NASA)

document (IRD) or interface requirements specification (IRS). It can be thought of as a legal instrument that captures all necessary information to achieve mission success across the interface (Lalli et al., 1997). This is much more than the requirements, as the ICD captures various details of the design of the interface (e.g., timing, command and data interface, mechanical and thermal interface design, software). It is produced by one side of the interface but signed by both, and signed by the project management (typically the government or the prime contractor). So in fact it is a triangular arrangement, wherein both sides of the technical interface have agreement, and the project office (which has both contractual and technical responsibility) has also agreed. It is critical that early ICDs be established, even in the conceptual design phase, and that they are used as a living document—but under configuration control— throughout the development life cycle. Best practice is to perform formal verification to the ICDs; they are not merely another location for information but are authoritative. Subsequently, ICDs can be very expensive to produce and maintain and are generally employed on only major and submajor interfaces, as applying them to too many interfaces can crush a project financially, becoming a false economy.

14.1.4 Technical Performance Measures and Margin Management

The coin of the realm for systems engineers on SoS space missions are technical margins. Technical margins are managed for such quantities as mass, volume, power/ energy, data rate, bus bandwidth, CPU utilization, and memory (to name just a few). A margin is simply the difference between two numbers. Typically, these two numbers are a current estimate and an allocation or requirement. The sign and magnitude of the difference, over time, can make all the difference in the success or failure of a system

System resource/mission phase	SDR	PDR	CDR	ATLO start	Launch
Mass	25%	20%	15%	10%	3%
Energy/power	30%	20%	15%	10%	10%
Power switches	35%	30%	20%	15%	10%
CPU utilization	75%	60%	50%	30%	20%
Memory					
SSR (bulk storage)	30%	20%	20%	15%	10%
DRAM	75%	60%	50%	30%	20%
NVM (flash)	75%	60%	50%	40%	30%
SFC EEPROM	75%	60%	50%	40%	30%
Avionics					
Serial port assignments	3	3	2	2	2
Bus slot assignments	3	2	2	1	1
Discrete I/O	30%	20%	15%	12.50%	10%
Analog I/O	30%	20%	15%	12.50%	10%
Earth to S/C link(C)	3 db	3 db	3 db	3 db	3 db
Link margin bit error rate (3 sigma)	1.00E-06	1.00E-05	1.00E-05	1.00E-05	1.00E-05
Bus bandwidth	60%	60%	55%	55%	50%
Mission data volume	20%	20%	15%	10%	10%
ASIC/FPGA gates remaining	40%	30%	20%	15%	10%
Crew IVA time	40%	30%	20%	10%	10%

FIGURE 14.5 Example margins table from constellation (NASA)

development. For SoS space missions resources and need for margins can cut across systems and therefore require a hierarchy of allocations and management techniques consistent with the systems engineering structure of the effort.

Required margins, throughout the development life cycle, are a strong function of requirements stability, technology readiness, and design maturity, generally in that order. Margins-use schedules, that are typically historically based, need to be adjusted for other factors including amount and readiness levels of new technology and complexity of interfaces.

It is up to the SoSE to specify the technical parameters that need to be defined, understood, tracked, and managed. The following table (Fig. 14.5) is representative of technical margins a spaceflight mission manages over time.

At the top of the technical margins for space exploration list is almost always mass. This is certainly the case for SoS space missions, at least in the initial development. Mass at launch, mass at injection to a specific orbit or trajectory, mass at entry, mass at landing all need to be tracked and managed. Mass is broken down into various ways, but of particular interest is dry mass and usable propellant (wet mass). Mass management often starts out as a bookkeeping process but sooner or later (generally sooner, when the first mass margin crisis hits) becomes a major systems engineering focus and challenge that lasts the life of the program.

The art of margins management is metering them out at the right time to handle problems and reduce risks while preserving the right amount for future needs. This includes brokering margin across major elements or subsystems to balance risk. It is the role of systems engineering to define the margins tracking metrics and review them on a regular basis. Systems engineering also needs to have clear criteria for knowing when the margins are inadequate and leading the charge to reconstitute them. This may include descoping capability, changing operational approaches, and/or spending money to change/improve designs. Obviously, later in the life cycle one runs out of margin the more expensive it gets. On an interplanetary robotic mission, a rate of >$200K/kg might be the cost of reconstituting mass late in the development cycle. For human missions, the cost can easily be 10× greater.

Active use of a comprehensive threats and opportunities (T/Os) list is essential to manage margin risk. Identification and management of T/Os need to be layered consistent with the SoS engineering structure and covered by management reserves at these levels. Monte Carlo techniques are being developed and utilized to bring more rigor and ease of use to effectively quantify margin risk. A fraction of management reserves also need to be unencumbered to handle completely unanticipated problems (commonly referred to as unknown-unknowns).

In multistaged SoS (e.g., launch vehicles, lunar/planetary Landers), the mass, gear ratios associated with the physics of the rocket equation need to be considered (NASA, 2005). This effect concerns margin compounding, where mass margin on a downstream system, for example, a Lander, has additional margin applied to every element that carries it. This results in margin being piled on top of margin, generally resulting in excessive margin requirements. Here again development and use of Monte Carlo techniques can help optimize the allocation and management of mass margin.

14.1.5   Electronic, Electrical, and Electromechanical (EEE) Parts and Common-Cause Failure

In the distant past, the space mission design and the push for advancement in the military and aerospace sectors heavily influenced the development of EEE parts to be used in extreme environments. But with the advent of the electronics "explosion" in commercial markets, space missions are the users, rather than the developers of this technology and more and more missions are developed and flown with commercial hardware, some of it radiation hardened, some of it up-screened, and some of it just as it is. There are at least three major implications for the designers of SoS space missions that need to be considered:

1. Feature size
2. Common-cause failure
3. Obsolescence

With the smaller feature size of the each new generation of EEE digital devices (now in the sub-100 nm range), the aerospace industry is experiencing decreased robustness in the area of radiation tolerance (e.g., single event effects). In fact, certain phenomenological effects associated with even low-dose radiation are causing some newer devices to fail, a susceptibility only relatively recently experienced (Sarsfield, 1998). Systems must not only be designed with combinations of shielding and redundancy, but extensive testing and characterization of the parts must also be accomplished. In addition, denser device structures enabled by reduced feature size can result in potential thermal runaway problems in the severe environments of space. The implications on reliability and fault detection, isolation, and response can be daunting (Seale, 2003). With the potential for more upsets on these parts, space assets must have both graceful degradation modes and what are being termed "absolute safety nets" using not only dissimilar technology but also "harder" technology.

This leads directly to the issue of common-cause failure in EEE devices and the implications on system design and operations. There is the probability that all the devices of the same type can fail in a certain period and cause complete loss of some functionality onboard a spacecraft—or on the ground. From a parts reliability perspective, the root cause may be a flaw in the device common to all of them or a certain lot. From a component or subsystem perspective the root cause may the implementation (how the part is being used in the design). But from a systems perspective, one can postulate and observe other root causes that cause the same effect—causes external to the device or the subsystem. A good example was the simultaneous loss of four of six Russian module computers on ISS in 2007 due to water condensation in a zero-g environment (Oberg, 2007), or the loss of attitude and all computer control onboard the Stardust mission in 2004 for an extended period while the vehicle was bombarded by solar protons and the computers and star cameras were unusable. The increased susceptibility of modern EEE parts combined with the increased reliance on such devices (like ASICs and FPGAs) for all mission critical events only underscores the challenge for the space mission designer. For SoS programs like constellation, the system engineers are developing concepts like emergency return mode to battle this susceptibility. Rather that adding additional strings of equipment—which themselves do not escape these common mode failures—NASA and their contractors are considering adding the capability for the astronauts to intercede and control propulsion actuators, pyrotechnic events, and environmental functions with very simple, high reliability devices such as circuit breakers, hand controllers, and valves.

Finally, for big systems that must operate over multiple decades (or even one), parts obsolescence is a major challenge. It is less and less common to rapidly produce all the space vehicles in a short period, put them in storage, and then launch them to replace ailing assets. The Space Shuttle is still sporting technology that is 30 years old. Today's ongoing revolution in EEE parts and the near-total reliance that modern space missions have on the commercial sector implies that the build-and-store design model would require buying large quantities of these parts before they are discontinued and hope for technical support from the manufacturer. This is not a good model. Thankfully, more typical of today's SoS missions is the notion that the space vehicles will have preplanned product improvements (P3I) and therefore through evolution or block upgrade to both capability and obsolescence can be addressed (Naegeli et al., 2004). Of course, this increases the risk of the unwanted phenomena of requirements creep and cost growth.

14.1.6   Integrated Risk Management

Risk management has existed in many forms in the history of space missions, although it is only in the past 15 years or so that it has become a formal part of program management with separate terminologies and processes. In some ways, what was previously intrinsic in engineering practice has now become so formal that the overhead associated with labeling something a risk and maintaining it (tracking, mitigating/handling) self-limits the numbers and types of risk that can have program-level attention. This has resulted in mission losses. This is not to say that only those risks having program level attention get handled correctly, but rather risk management needs to be integrated into the broad SoS engineering culture at all levels—and not just be seen as a separate discipline. Tom Young, former president of Martin Marietta, once said "We're in a one strike and you're out business." (NASA, 2004) Therefore, every engineer needs to think like a systems engineer when it comes to risks on space missions.

For SoS space missions the numbers of risks that need to be identified and mitigated at any one time can range from hundreds to thousands and span all the way from component development through integration of the systems into flight operations. Labeling and handling all those risks through a single database at the program level would be cumbersome, costly, and ineffective. In contrast, world-class SoSE teams integrate risk thinking into all the engineering disciplines, and although there will always be the more formal organizational and process evidence of risk management, the average engineer on the program would plan and execute his or he day-to-day technical activities around a risk paradigm. Literally, every-thing that an engineer does, especially in the early program phases, can be directly mapped to reducing risk, technical and programmatic. Moreover, it is the job of the SoSE team and management to achieve the right balance between risk and performance, and cost and schedule.

A critical factor in integrated risk management for SoS space missions is the nonlinear affect of accumulated risk. This is best illustrated by considering the losses of the Mars Climate Orbiter and the Mars Polar Lander in 1999 (NASA, 1999). The missions were very aggressive in nature, both in the technical aspects of capturing or landing at Mars and in the low costs for the missions. Every week the development team (codeveloping both missions simultaneously) had to make risk decisions at the engineering review board level, approving or disapproving designs and design changes. Each was considered on its own merit and although the mission context was always used in making decisions, the decision to approve or disapprove was essentially atomic in nature. Yes, there were changes and designs that clearly had dependent risks, and those were considered together, but those that were considered independent were judged on their own merit. In retrospect, following the losses of the missions and the resulting failure review boards, the projects had accumulated risk to a level that was not understood (Euler et al., 2001). In fact, many independent risks somehow combined to raise the overall level in a nonlinear fashion—it is not the mere addition of risk after risk. Best practice techniques of mission phase fault-tree analysis (Beutelschies, 2002), mission stress (including off-nominal conditions) testing, and test-like-you-fly processes (White and Wright, 2005) measure, expose, and control accumulated risk for today's SoS space missions.

Other techniques should also be employed such as "pause and learn" events (used by NASA, adapted from the U.S. Army's "after action review" AAR (Rogers and Milam, 2005)) and the use of retired engineers/managers to help identify risks. Similarly, use of probabilistic risk assessment (PRA) techniques can be helpful in the risk assessment of SoS missions (NASA, 2007) and is often required, especially in understanding the relative improvement in reliability to a system between different architectures (e.g., single-and two-fault tolerance). But it should always be used by the SoSE in conjunction with many other sources of data for risk mitigation (including engineering intuition, resulting system complexity, and interactions that cannot be modeled well within such techniques).

14.1.7 Critical Path Execution and Consequences of Failure

As systems are added to systems to create the large complex infrastructure of an overall space exploration SoS (like NASA's Constellation Program), the number of major interfaces between these systems results in multiple critical or near-critical paths and potentially enormous penalties in cost and schedule as execution problems inevitably arise. Due to the large numbers of personnel working on such a project, even a delay of only weeks or a few months can result in many millions of dollars cost increase due to the marching army effect. As dramatic cost increases occur, SoS projects can breach the recognized cost constraint (for example, the Nunn-McCurdy 25% growth limit, which if breached requires Congressional recertification to proceed), raising the possibility of project cancellation or major restructuring. Add to this situation planetary launch windows that have typically short durations (20 days), or the need to get critical assets launched on time to replace aging or damaged spacecraft, or even time-critical deployment of ground systems to meet the country's needs for protection of life, property and defense—the result of not making critical path has staggering consequences.

There are basic tenets of systems engineering that work in smaller systems and can be effectively applied to the SoS to avoid or reduce the consequences of failures in critical path execution. An example is early interface validation between the major systems of the SoS through deliberate usage of the integrated master plan and integrated master schedule. In SoS engineering, this is more difficult than smaller systems because each system or segment of the SoS is a major project in its own right and has different teams performing the development, different prime contractors, and different government customers. This results in much more challenging (and expensive) collaboration between leaders and engineers across the interface in order to create early, effective interface validation demonstrations and tests. It is not unusual for SoS projects to employ a separate system integrator (SI) to achieve integration of the various interfaces at the SoS level. Whether this approach is used or not, the major interfaces should be of ultimate concern for the SoS systems engineer and must be addressed with explicit planning and control through the IMP/IMS.

Another example of critical path mitigation that is crucial to SoSE is new technology management. This includes early identification of new technology and the creation of technology "on ramps," fallback technologies, or even competing technology developments. These plans must be laid in the early stages of project development, not reactively later in the project when difficulties occur.

14.1.8 ITAR and Proprietary Issues Across Vast Teams

International Traffic in Arms Regulations has become a major issue for all space system developers. In SoS projects, the use of non-U.S. contractors/subcontractors and technology poses special challenges to the successful deployment of the system. It is imperative that projects and individuals comply with ITAR (and other similar trade regulations), and therefore the system engineers and architects must carefully reflect on the execution and mission success impacts of having, in particular, major technical interfaces that are international. This is not to discourage such interfaces but due to the limits of ITAR, it is necessary to make the technical interface clean and simple. The project needs to very early on develop and get approval of the technical assistance agreement (TAA) and ensure that interface is specified sufficiently to achieve mission success, within the scope of the TAA.

Protecting the proprietary rights of the various contractors, teammates, and partners in SoS projects is greatly complicated due to the very large size of these projects and their many interfaces, both major and minor. The actual system architecture and parsing of functionality across the systems and the subsystems can have an effect on the difficulty in protecting these rights while simultaneously meeting performance, cost, and schedule. Although the system engineer would rather relegate this topic to the program management and contracting officials, it must be realized that poor system architectures can manifest themselves as impediments to mission success during development or operations when critical information cannot be shared in a timely fashion to resolve a situation or anomaly. It is one thing to experience it before deployment of a space exploration SoS, it is quite another in the post-launch operations arena where time is of the essence. This is true of both proprietary issues and ITAR issues.

14.1.9 Geographical Distribution

As would be expected, the SoSE faces enormous challenges in the area of communications and the development, control, and distribution of requirements, design specifications, data, processes, notes, technical memos, analyses, and myriads of other products. Trying to actually perform work and not find oneself in continuous telecons, videocons, net meetings—and airplanes—is most frustrating for engineers and leaders alike. These projects are not, and in the future will not, be colocated. Geographical distribution of SoS projects is a reality and with international participation, it becomes even more challenging.

Even as Internet technology and processing power and storage continue to climb exponentially and indeed offer tremendous advantages for technical collaboration, the engineer faces an inevitable data overload situation and control of the technical baseline is at risk due to too many servers, too many desktops, and too many engineering review boards. The successful space exploration SoS must not only establish business rhythms and processes that create time for collaboration (and work), but again, the very system architecture itself must also be constructed to promote execution. Architectures with unclear and overly complex segmentation and interfaces will simply amplify the disadvantages of geographical distribution, and the project will struggle. This can map directly into critical path problems and even mission success failures as illustrated with the loss of the Mars Climate Orbiter in 1999 due in part to geographical dispersion of the project team and misunderstandings regarding the design and navigation of the orbiter (Euler et al., 2001). Project Constellation faces challenges as it is not only distributed across NASA and its contractors, but also across all the NASA centers themselves (NASA, 2006) as depicted in Fig. 14.6. This requires special attention in setting up collaborative working groups, system integration groups (SIGs), integrated product teams, and the IT systems that capture, store, and share data, while adjusting management infrastructures to accomplish both horizontal and vertical integration to overcome both geographical and time zone differences, while preserving centers of excellence .

14.1.10 Key Qualities, Training, and Practices of SoS Engineers

Systems engineering is not cookbook engineering, how much more so for tomorrow's SoSE. It is all about analysis, synthesis, and optimization of solutions to complex problems in a highly resource-constrained environment. While there is a common thread of the systems thinking (Davidz, 2007) that so characterizes all systems engineer, within space exploration it can be seen that there are at least two primary vocations for SEs: system design and integration leaders and system engineers/ analysts, and the necessary training and personality traits for each are different.

The system design and integration leaders can be described as engineers endowed with strong leadership/people skills and have a work experience background that could be described as a so-called "T profile."[1] The vertical trunk of the T is indicative of their deep technical knowledge in one or more spacecraft subsystems/disciplines— particularly of having really delivered hardware and/or software and experienced flight operations. The horizontal member that caps the trunk of the T represents their broad experience across many disciplines and their ability to understand, probe, and comprehend the overall system of system's function. This is their mental model of how the entire system functions. Their people skills combined with these technical skills equip them to be the technical leaders of SoS development, and they often represent the future program managers and chief engineers.

System engineers/analysts are the highly skilled engineers that perform a variety of critical functions within the discipline of SE including mathematical simulations, compatibility and specialty engineering, power, mass, and pointing budgets, technical performance measures, requirements verification and validation, and so on. While multidisciplined within SE, it is not necessary that these personnel have spent time in

---

[1]A concept popularized by George W. Morgenthaler, formerly of Martin Marietta Corporation and the University of Colorado.

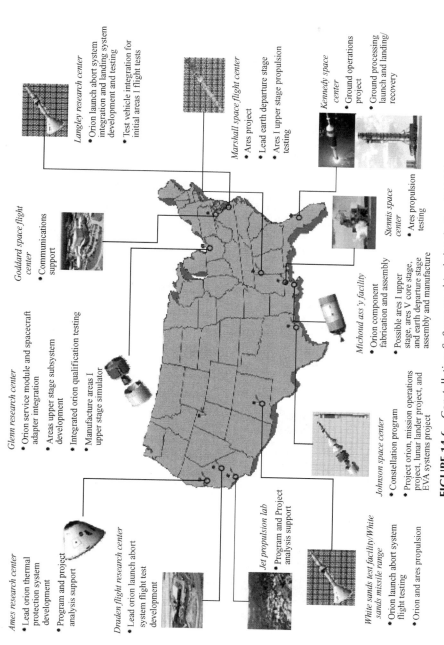

FIGURE 14.6 Constellation SoS geographical distribution (NASA)

Langley research center
• Orion launch abort system integration and landing system development and testing
• Test vehicle integration for initial areas I flight tests

Marshall space flight center
• Ares project
• Lead earth departure stage
• Ares I upper stage propulsion testing

Kennedy space center
• Ground operations project
• Ground processing launch and landing/ recovery

Stennis space center
• Ares propulsion testing

Goddard space flight center
• Communications support

Glenn research center
• Orion service module and spacecraft adapter integration
• Areas upper stage subsystem development
• Integrated orion qualification testing
• Manufacture areas I upper stage simulator

Michoud ass'y facility
• Orion component fabrication and assembly
• Possible ares I upper stage, ares V core stage, and earth departure stage assembly and manufacture

Ames research center
• Lead orion thermal protection system development
• Program and project analysis support

Dryden flight research center
• Lead orion launch abort system flight test development

Jet propulsion lab
• Program and Project analysis support

White sands test facility/White sands missile range
• Orion launch abort system flight testing
• Orion and ares propulsion

Johnson space center
• Constellation program
• Project orion, mission operations project, lunar lander project, and EVA systems project

333

Developing system of systems engineers		
General characteristics	Elements of training	Seen in practice
Generalist, architect, firefighter	On the job training, how work gets done, mentoring	Know what they know and what they do not know
Intellectually curiosity, self-confident, energetic	Hands-on experience, end-to-end ownership develops judgment	Big picture, end to end, concept to operations, the *systems view*
Big picture oriented, end-to-end and concept- to-operations thinker	Working across subsystems and with new technologies	Tracks and knows state of key technical /program resources and their margin
Comfortable with change and uncertainty	Classes for fundamentals, familiarity with tools, lessons learned	Understands difference between requirements & capabilities
Good communicator and listener	Learn processes as useful tools	Knows processes are tools and not an end to themselves
Healthy paranoia	Multiple job and project experience	Builds in robustness, overlapping capability
Team player, works well as part of a diverse team	Test and tune decision-making skills and judgment	Conducts objective trade studies, balances technical and programmatic

FIGURE 14.7   Training system of system engineers for space exploration

subsystem disciplines and many will have been hired into SE early in their career. This is in contrast to the system design and integration leaders that have had to spend their early career distinguishing themselves within a subsystem. Similarly, while superior people skills are very helpful for this category of SE, it is really not necessary—for their strengths lie in their analytical skill set.

In both cases, SoS projects face the daunting challenge of identifying, recruiting, and shaping the development of these system engineers—especially in the face of an aging technical workforce. Preferably, the development of these systems engineers will start at the university where they could experience many elements of complex project development in the form of senior or graduate design projects, or better yet, real spacecraft development experience on university projects with sponsorship from NASA or other government agencies. Combined with internships with the government and its contractors, entry level professionals can then be more effectively guided into SoS systems engineering career development. Figure 14.7 summarizes the overall characteristics, training, and practice of tomorrow's SoS systems engineer.

14.2   PROGRESS IN SPACE EXPLORATION SoSE

This section outlines three projects that are currently engaged in various stages of the SoSE challenge. ISS is in the final stages of space-based assembly that involves international partners and many, many years of design and development. MSL has passed the critical design review phase (as of this writing) and is in final development, headed for launch in 2009. And finally, constellation is approaching the preliminary design maturity for Orion (the crew exploration vehicle, CEV) and Ares I, the launch system for Orion. Similarly, Altair (the Constellation Program Lunar Lander) design is underway and will reach PDR maturity somewhere in the next several years.

14.2.1 International Space Station

One of the longest ongoing missions in NASA's history, the SoS International Space Station is nearing its completion of assembly at the end of this decade. It has a long and interesting development life cycle that stretches back into the late 1980s. Consisting of many pressurized modules, load carrying truss structures, and enormous 78 m (256 feet) solar arrays, this the largest and heaviest (over 218,000 kg, 482,000 lbs) orbiting man-made device in the history of the space age that can be seen easily with the human eye at night. ISS construction began with the launch of the Russian-built Zarya Control Module on a Proton Rocket in 1998 and evolved through a long series of assembly flights of the Space Shuttle, 23 at the time of this writing. In every aspect the design, development, assembly, and operation of the ISS represent the greatest SoS achievement to date for space exploration, rivaled only perhaps by the Apollo program. This is especially true when one considers how the Space Shuttle and Soyuz missions are integral to the ISS mission success, both from an assembly and sustaining operations perspective.

Notwithstanding the ISS's stated purpose of providing a scientific laboratory in space for research and development, from an engineering perspective, it is a fabulous pathfinder for in-space assembly, repair, replacement, and very complex mission operations with humans and machines in the loop. One of the most complex ISS missions in the entire construction sequence STS-120 became even more challenging when a solar array was damaged during unfurling after being relocated from one position in the station to another. At the same time, the crew also discovered debris from one of the large solar array gimbals indicating internal damage, unrelated to the other incident. In a dazzling display of what only humans in space can do, EVAs were planned and executed to affect repairs on the solar array (Fig. 14.8)—the suited astronaut attached at the end of the very limit of the reach of the robotic arm carefully using makeshift electrically insulated tools around the high-voltage array (Schwartz, 2007). This adaptation to the unknown-unknowns has been the hallmark of space exploration from the very beginning for both robotic and human missions, and it is magnified in complexity and risk when dealing with an SoS like ISS. The numerous lessons learned from ISS are influencing the approach for human missions to the Moon and Mars.

Clearly, one of the most challenging aspects for the ISS has been managing the enormous number of technical interfaces in the system, an extraordinarily large number of which are international (NASA, 2007). System architecture defined most of these interfaces at the major assembly level. In addition to the Zarya module discussed above, examples are the Russian Zvezda Service Module, the European Columbus Module, the Italian-built Multipurpose Logistics Module, the Russian Progress and Soyuz spacecraft, and the Japanese Kibo Experiment Module. Modular system architecture promoted cleaner interfaces in which ITAR and proprietary constraints

FIGURE 14.8   Astronaut Scott Parazynski repairs an ISS solar array (NASA)

could be more easily accommodated. Similarly, ISS's number of major interfaces and foreign partners, global mission controls, and the variety of docked vehicles (Shuttle, Soyuz, Progress, Autonomous Transfer Vehicle, H-II Transfer Vehicle, and potentially commercial systems) underscore the vast complexity of this system. Add to that the multiple life cycles of major elements and subsystems, such that when new modules are being added, other systems are beginning to fade and fail, ISS has no analog in space exploration history!

14.2.2   Mars Science Laboratory

The Mars Science Laboratory is the most ambitious Mars mission yet, and it involves a revolutionary terminal descent system utilizing an overhead powered descent propulsion configuration and bridle assembly to gently land the radioisotope thermoelectric generator (RTG)-powered rover directly on its wheels. This avoids the complexity of having to egress off a landing platform via a ramp (as the MER and Pathfinder rovers did) or lowering mechanism. This will be the first Lander mission to ever fly such a configuration, and the system architecture of this "descent stage" concept essentially separates the touchdown gear (whether wheels, legs, air bags, or pallets) from the terminal descent propulsion. For MSL, only the first example of how this approach can be applied, this enables mass and volume resources sufficient to place a rover the size of a Mini Cooper directly on the surface.

The brilliance of this concept is found in how the overall system was architected and how the interfaces were defined and drawn using solid SoS engineering approaches to interface definition and functional allocation. By returning to first principles of EDL and analyzing the dynamics, dispersions, and handoffs between the various subphases, it was discovered that the touchdown method was the determining factor in the final landing system configuration. Thus, powered descent soft Landers (such as Viking, MPL, and Phoenix) all place the propulsion below the payload deck, but attempt to be as squat as possible so that the payload is close to the surface (to interact with it after touchdown). But in the wildly successful Pathfinder mission of 1996, it was demonstrated that there was another method of touchdown using a two-body system of overhead configured vertical descent retros and a Lander encased in 4π steradian elastic air bags connected via a bridle that is cut at 10 m above the surface (NASA-JPL, 2008). In contrast to missions with propulsion underneath the payload, Pathfinder's technique places the payload in very close proximity to the surface. This method was successfully repeated with MER (spirit and opportunity)—upgraded to add additional retros for horizontal velocity control.

Due to the limitations of ballistic entry mass technology that Viking pioneered, we have reached the limits of an MER-like system. But not having the inherent robustness of elastic air bags means that the velocity box (vertical and horizontal) must be greatly reduced to meet the capabilities of touchdown gears (legs or pallets), which have very limited ability to absorb energy.

Through a series of trades, it became clear that the Pathfinder experience was demonstrating the principle that a simpler physical interface between the propulsive terminal descent and the actual landed payload would have huge advantages for MSL-class rovers. In fact, when one considers the design of a launch vehicle upper stage, its function is to get the spacecraft and its payload to a certain velocity trajectory (velocity) and then separate. So to, a "descent stage" concept proved to be for a mission like MSL, and the so-called Sky Crane was born (Fig. 14.9). Instead of using an overhead system to only reach a certain altitude and cut the payload loose, MSL employs an overhead throttled liquid system and a bridle that reels down the payload to create a soft landing in which the payload encounters the surface with a residual velocity of less than 2 m/s vertical and 1 m/s horizontal (compared to 100 km/hr for

FIGURE 14.9   Depiction of MSL's Sky Crane in terminal descent (NASA-JPL)

air bag systems) on its highly compliant landing gear (as needed for mobility) and completely avoids the complexities of detecting touchdown and controlling engine shutdown at precisely the right moment. After safely delivering the Lander, the Sky Crane flies away to crash-land safely away from the landing site. This approach holds promise for future Landers and can be exploited to provide precision soft landing of very large and fragile payloads in hazardous regions.

14.2.3 Constellation Program

As an SoS it does not get much more complex than constellation (Muirhead, 2008). NASA's Constellation Program (CxP) is responsible for the design, development, test, and operations of the flight, ground, and mission operations elements needed for humans to explore the Moon and ultimately to explore Mars (Rhatigan et al., 2007). The program is on a "go as you pay" plan that will develop an initial capability (IC), whose missions will be to the ISS (and possibly other targets) and Lunar Capability (LC), whose missions include lunar sorties to anywhere on the Moon and establishment and operation of a permanent lunar outpost. The major elements of the CxP are shown in Fig. 14.10.

The baseline CxP architecture for Lunar Capability is based on what is called the 1.5 strategy. This means that for a lunar mission we have a heavy-lift vehicle that carries the lunar Lander and the translunar injection (TLI) stage to LEO; this is the 1.0 part. The crew launch to LEO on the smaller Ares I vehicle (the same used for the initial capability mission to the ISS) and rendezvous with the Lander and TLI stage in LEO; this is the 0.5 part. Apollo was a 1.0 strategy with injection stage, Lander, and crew on one vehicle. The need for the 1.5 strategy is driven by the size of the systems needed for establishing a significant outpost capability on the Moon.

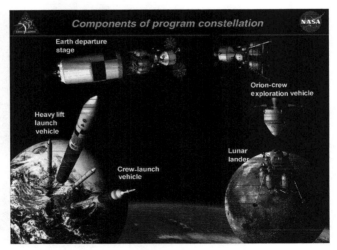

FIGURE 14.10 SoS approach for human space exploration (NASA)

The SoS engineering structure of CxP starts with a large systems engineering and integration (SE&I) organization at the program level. This organization is mirrored on a smaller scale in each of the four current major projects. There are SIGs responsible for developing requirements and doing system analysis and trades across major discipline areas including flight performance, loads/structures/mechanisms, software, avionics, human systems, thermal, environments, and ground/mission operations. There are also significant horizontal and vertical integration activities associated with schedule, risk, resources, architecture, analysis, and process management. The challenge within a program of this scale is to establish and maintain clear communications, and there is a significant effort to do so through face-to-face interactions as well as every form of electronic communication.

The current culture of the aerospace industry has become driven by and obsessed with requirements. The tendency of management at all levels is to use requirements to control outcomes and behaviors, with the generally unintended consequence of stifling creativity in the design process. Of course, in a large program and with numerous industrial partners requirements are needed. However, the tendency is to overspecify requirements, often resulting in overly constrained design conditions for which there is no reasonable design solution that works within the overall mission constraints. One of the roles of the SoSE is to identify where overly constraining requirements are driving potentially bad design decisions (e.g., the flight loads environments) and advocating/leading reassessment/elimination of the low probability design cases. The basic strategy is then, to work to a reasonable design capability as the designs progress, reassess the various failure modes or contingency scenarios for increases in likelihood of occurrence and, if necessary, either incorporate design changes or make other changes to "guarantee" that such conditions will not occur in flight.

Based on Shuttle and ISS heritage, the initial CxP approach was to design the existing human rating requirement to be two-failure tolerant (i.e., at least three strings of everything), fail-operational, then fail-safe. However, the mass limitations of travel beyond Earth orbit can allow very limited, if any, application of a blind cookbook adherence to a two-failure-tolerant requirement. NASA has recognized this and modified the governing document for human rating requirements, NPR 8705.2B. The current wording does not explicitly specify a level of fault tolerance but instead states that "The space system shall provide failure tolerance to catastrophic events, with the specific level of failure tolerance (1, 2 or more) and implementation (similar or dissimilar redundancy) derived from an integrated design and safety analysis." This requirement puts the responsibility for meeting it squarely in the hands of the SoSE, which must work with design engineers, technical authorities (engineering, safety, and medical), and program/project management to meet the intent of the requirement.

History has shown that exploration is one of the most exciting and dangerous of human endeavors. In addition, the exploration of space by machines or humans continues in that rich and risky tradition. The CxP is on a course to explore the Moon and Mars while facing many of the same challenges that Apollo faced. At this point in time, the greatest challenge is establishing preliminary designs for crew and launch vehicles that are reliable, robust, and can be developed as quickly as possible on a "go as you pay" funding plan. CxP will be a highly visible laboratory for SoS engineering.

14.3 FUTURE CHALLENGES IN SPACE EXPLORATION SoSE

In the near future, we should see multivehicle formation flying become a reality, enabling fantastic new space-based observatories that can resolve the cosmos at a level never achieved before. And in this same time frame (10–15 years?), astrobiologists and geologists are anxiously awaiting the next step in Mars exploration, the return of actual samples from the surface. But as we look further, we can see in the distance the towering but daunting crown jewel of space exploration—the human exploration of Mars! These three example SoS future challenges, from remote sensing, sample return, and human exploration, are briefly analyzed in this final section. Each represents a revolutionary increase in complexity compared to today's missions, and each in its time will be the state of the art for SoSE in space exploration.

14.3.1 Formation Flying

Missions involving precision formations of a few to many spacecraft have been proposed for many years and demonstration missions of the technology required have already been conducted. Formation flying (FF) concepts like NASA's Terrestrial Planet Finder (Aung et al., 2004) employ various sparse/synthetic aperture and/or interferometeric techniques with physically separate spacecraft (Fig. 14.11). This promises space-based astronomical missions that greatly exceed the performance of any of the current great observatories such as Hubble, Spitzer, or the upcoming Kepler and James Webb Space Telescope missions.

In addition to the greatly increased navigation and pointing control, and position and attitude knowledge requirements to which FF missions must be designed, the

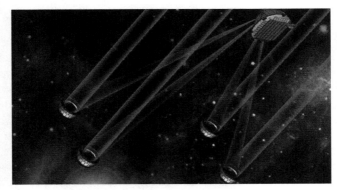

FIGURE 14.11 A TPF mission concept, formation flying of five spacecraft (NASA-JPL)

near-earth missions will likely use global positioning knowledge from GPS and other systems. So in addition to having many spacecraft in this SoS application, FF missions will have a critical interface to yet another constellation of spacecraft, another SoS. FF spacecraft will undoubtedly require high individual autonomy for orbital position and fault protection, and the constellation will need cross-links between the spacecraft to execute formation control. In addition, the flight and ground systems would incorporate increased autonomy beyond that used for current missions to control and limit the operations costs while flying many spacecraft simultaneously.

A key challenge for such systems will be system availability. Even with the increased reliability of modern spacecraft, by the time the project with a large constellation has launched the constellation (assuming that a large fraction cannot be launched on one EELV), some mission life of the first assets will have been consumed. Nevertheless, even if the constellation can be launched at one time, the science phase of such missions will depend upon some threshold number of the constellation being in place and meeting individual performance so that the system as a whole is observing at its science performance floor. Will formation flying missions be able to meet the performance floor even when one or more elements of the constellation are not available? Perhaps, depending on the nature of the mission, but it may cost prohibitive. So it is likely that the science campaign will be intense and system availability will have to be very high, approaching the nearly 100% numbers that the next generation of NOAA geostationary satellites (GOES-R) require. Either way constellations of formation flying missions represent a major challenge to the SoSE.

14.3.2 Mars Sample Return

Preceding the first human missions to Mars, Mars sample return (MSR) is very high on the robotic exploration mission priority list (NASA-JPL, 2000). In addition to exciting science, such a mission would provide needed information on basic properties of Martian soils and atmosphere needed for human safety. The mission has been studied for decades but was first started in earnest by NASA in the late 1990s and then cancelled in 2000 due to a variety of factors affecting the Mars Exploration Program (including the loss of the Mars Surveyor Program spacecraft MCO and MPL). The remnants of the program became known as the Large Lander study that concentrated on generating design alternatives for delivering a mini-pickup sized RTG-powered rover to the surface but set aside other aspects of the sample return chain such as the Mars Ascent Vehicle (MAV), sample canister, and the earth return vehicle. Concentrating on entry, descent, and landing techniques, this study paved the way for the current MSL mission slated for launch in 2009.

Notwithstanding the difficulty of EDL, one of the daunting aspects of MSR is the number of spacecraft assets required and the complicated sample transfer chain—beginning from collection and encapsulation of the sample and proceeding back up out of the Mars' gravity well, to rendezvous with the Earth return vehicle and into an Earth bound transfer orbit (Fig. 14.12). To just bring a modest sample of a few kilograms back to earth is an enormous SoS undertaking and represents

NASA-JPL

FIGURE 14.12 Mars ascent vehicle fires to carry sample canister into Mars orbit for rendezvous with Earth return vehicle (NASA-JPL)

the most complex deep space mission ever conceived to date. The physics involved and the nature of the rocket equation demand that the sample be placed in low Mars orbit and then an Earth return vehicle rendezvous and capture the sample canister before initiating the Earth-return burn. There are other significant challenges associated with forward and back contamination of biological organisms that add significant complications to the mission and hardware design and affect the way that the space hardware is processed and how the sample is contained and possibly sterilized for Earth return.

For all deep space missions (including constellation), the most serious aspect of SoSE design is that of the complex CONOP/DRM. These missions undergo many different critical events, far more than near-Earth nonreturn missions, including planetary capture with combinations of propulsive deceleration and aerodynamic techniques like aerobraking/aerocapture and hypersonic entry (where atmospheres are present). Similarly, these missions involve time-domain changes in spacecraft physical and nonphysical configuration that are radical. Envision a Lander packaged inside an aeroshell with multiple staging events undergoing EDL, or an orbiter repositioning appendages to achieve aerostability as it enters the outer atmosphere to scrub orbital velocity.

MSR has all of these aspects but multiplied by 5 or 10. It is essential that time and functional domain aspects of the mission be architected in detail to ensure mission success. The interfaces on this type of mission cannot be easily simplified due to the operational complexity, and the international partnerships anticipated will only underscore the need for a sophisticated CONOP/DRM that provides for both derived requirements and the context in which they are understood.

14.3.3 Human Exploration of Mars

Probably the penultimate goal in space exploration is the human exploration of Mars, dreamed for many years by writers and engineers alike (Aldridge, 2004). Considering Apollo, ISS, and even the lunar outpost missions anticipated in the next 15 years, human missions to Mars are the greatest technological and logistical challenge that humans will have ever faced in exploration (Fig. 14.13). MSR's time-domain and functional domain complexity represents only a fraction of that required to safely transport humans to and from the surface of Mars (NASA, 2005). However, the issue facing SoS engineers is not really the individual technologies required to enable such exploration, for they largely exist today (with a few notable exceptions). The issue is the SoSE aspect of designing and developing cost-effective, reliable system architectures within the limitations of the next generation of launch vehicles. As will be demonstrated with the upcoming Orion missions, a human mission to Mars will require that major elements be launched into low Earth orbit where they rendezvous, perform proximity operations and then dock (RPOD), and then perform a burn as a larger system for the Earth departure into a Mars trajectory. A major challenge will be the timing of the Earth departure burns for Mars (Fig. 14.14), which will be undoubtedly bounded by similar launch windows by which current robotic missions are constrained (optimally every 26 months). Thus, we again see an SoSE design that is governed by serious time–domain requirements, all the necessary launches and RPOD operations must be completed and the burn initiated within a time constraint probably measured in weeks, but not months (given conventional propulsion technology or even nuclear thermal).

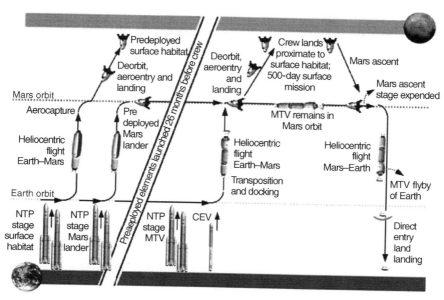

FIGURE 14.13 SoS challenge of the human exploration of Mars (NASA)

FIGURE 14.14   earth departure stage fires engines to leave Earth's orbit (NASA/John Frassanito and Associates)

There are numerous decisions that need to be made in this architecture, and many trade studies that need completion. The duration of the stays at Mars is a direct function of interplanetary trajectories and the choice of propulsion (e.g., conventional versus nuclear thermal) that, ultimately governed by the rocket equation, limits the mass that can be delivered. How will entry descent and landing be accomplished? Propulsive capture or aerocapture? Or perhaps direct entry like the Mars Exploration Rovers (MER) and Phoenix? Does the architecture exploit *in-situ* resource utilization (ISRU) such as the manufacture of propellants and breathing air from ice or water below the surface? These are just the beginning of many architecture questions for this SoS, introducing many permutations of the system design and underscoring the immense complexity of this mission.

As with MSR, exploitation of the practice of detailed CONOP/DRM derivation and the functional architecture of the overall system will be imperative for mission success. Similarly, when planetary launch windows are involved (not true for lunar missions where constellation can design for access almost anytime), then critical path issues can bring about consequences that even more money cannot solve. With windows that constrain schedule, a poorly architected and executed human exploration SoS will only have cost and risk to trade—an unfortunate and unacceptable situation to face. In contrast, upfront SoSE, using hard-won and proven system engineering techniques, can achieve world-class performance and a launch-on-time record, ultimately rewarded by mission success!

14.4   CONCLUSIONS

Space exploration has long captured the imagination of the public and serves an important role in allowing people to interact and learn about our solar system and the universe beyond. Perhaps more importantly, it encourages young people to pursue high technology careers. Advanced technology and the people who develop and apply it are what enable both human and robotic missions—as well as contribute significantly to the health and welfare of our country in areas of defense and commerce. Even as the International Space Station nears completion and the Space Shuttle retires after a noble career of 30 years, amazing pictures and science data are streaming in from dozens of robotic missions across our solar system over the last 40 years—including more data from one mission at Mars than all previous Mars missions combined! And now, poised on the frontier of the largest, most complex human mission ever undertaken, the Orion and Ares I projects, being joined by the Altar Lunar lander and heavy-lift Ares V, NASA and its contractors are building hardware to take humans back to the Moon and beyond. At the same time, robotic missions are increasing in performance, size, and complexity, as in the revolutionary development of the MSL's "Sky Crane" approach to landing a RTG-powered rover on the surface of Mars.

From each of these SoS missions, engineers and scientists are experiencing challenges not only proportional to the increased complexity, but many that are nonlinear. To achieve mission success, tomorrow's space exploration SoSE teams need to be greatly strengthened in many areas ranging from understanding how software is a system in its own right (i.e., no longer a subsystem) to dealing with complex interface management and requirements validation. And they must tackle head-on, difficult problems presented by ITAR and geographical dispersion. Even as the much needed tools and processes are developed for these various challenges, at the root of the practice of SoSE will always remain the system engineer. Training and grooming of world-class systems engineers that have the "systems view" must begin in the university with real project experience, followed by thoughtful, customized career path guidance from mentors and senior management.

Space exploration has long been the crucible in which the practice of systems engineering is tried and tested in a very visible and accountable fashion. From the most stunning successes of the past, such as Apollo and Viking, to those of the present, such as the Space Shuttle, ISS, MER, Deep Impact, Stardust, and MRO (naming only a few), these successes were all hard won. They result in great part from the incorporation of lessons learned from near misses and, unfortunately, tragic mission losses—some of which cost much more than hardware. Space exploration is as exhilarating, dangerous, and potentially rewarding as the early exploration of our planet. As space exploration projects increase in their size and complexity, the practice of SoSE will remain the crucial discipline to ensure mission success.

ACKNOWLEDGMENT

The authors would like to thank Lisa Guerra of NASA-HQ, Rob Manning of NASA-JPL, and Greg Bollendonk of Lockheed Martin Space Systems for their inputs and guidance in preparing this chapter.

REFERENCES

Aldridge, E., 2004, Report of the President's Commission on Implementation of United States Space Exploration Policy, June 2004.

Aung, M., Ahmed, A., Wette, M., Scharf, D., Tien, J., Purcell, G., Regehr, M., Landin, B., 2004, An overview of formation flying technology development for the terrestrial planet finder mission, IEEE, Aerospace Conference, Vol. 4, pp. 2667–2679.

Beutelschies, G., 2002, That one's gotta work' Mars Odyssey's use of a Fault Tree Driven Risk Assessment Process, *IEEE Aerospace Conference Proceedings*, Vol. 2, pp. 651–671.

Davidz, H., 2007, Developing the Next Generation of Systems Engineers, Crosslink, The Aerospace Corporation, Spring 2007.

Euler, E., Jolly, S., Curtis, H., 2001, Failures of the Mars Climate Orbiter and Mars Polar Lander: a Perspective from the People Involved, AAS 01-074, *American Astronautical Society 24th Annual Guidance and Control Conference*, Breckenridge 2001.

GAO, 2006, Space Acquisitions Improvements Needed in Space Systems Acquisitions and Keys to Achieving Them, United States Government Accountability Office, GAO-06-626T, 2006.

INCOSE, 2006, *Systems Engineering Handbook* Version 3, p. 2.4.

Lalli, V., Kastner, R., Hartt, H., 1997, *Training Manual for Elements of Interface Definition and Control*, NASA Reference Publication 1370.

Muirhead, B., Constellation Major Technical Challenges of 2007, *IEEE Aerospace Conference*, 2008.

Naegeli, C., Sjoberg, B., Schneider, S., Lee, P., Fara, D., Adkins, D., Andreoli, L., 2004, National Polar-orbiting Operational Environmental Satellite System (NPOESS) Potential Pre-Planned Product Improvement (P3I) Status, AIAA 2004-6058, Space 2004 Conference and Exhibit, San Diego.

NASA, 1999, Mars Climate Orbiter Mishap Investigation Board Phase I Report, November 10, 1999.

NASA, 2004, Lessons Learned Study Final Report for the Exploration Systems Mission Directorate.

NASA, 2005, Exploration Systems Architecture Study Final Report, NASA-TM-2005-214062.

NASA, 2006, NASA Announces Distribution of Constellation Work, NASA Press Release 06-233, June 5, 2006.

NASA, 2007, Final Report of the International Space Station Independent Safety Task Force, February 2007.

NASA-JPL, 2000, NASA Outlines Mars Exploration Program for Next Two Decades, NASA-JPL Press Release 2000-104, October 26, 2000.

NASA-JPL, 2008, Mars Exploration Rovers web page. http://marsrovers.jpl.nasa.gov/mission/tl_entry1.html.

Oberg, J., 2007, Space Station: Internal NASA Report Explains Origins of June Computer Crisis, IEEE Spectrum, October 2007.

Rhatigan, J., Hanley, J., Geyer, M., 2007, Formulation of NASA's Constellation Program, IAC-07-B3.1.06, NASA-JSC.

Rogers, E., Milam, J., 2005, Pausing for learning: applying the after action review process at the NASA Goddard Space Flight Center, *IEEE Aerospace Conference Proceedings*, pp. 4383–4388.

Rustan, P., 2005, Testimony of Dr. Pedro "Pete" L. Rustan, Director Advanced Systems and Technology National Reconnaissance Office, United States House of Representatives House Armed Services Committee Hearing Space Acquisition.

Sarsfield, L., 1998, The Cosmos on a Shoestring, RAND, p. 143, Santa Monica, CA.

Schwartz, J., 2007, Space Station is repaired in spacewalk, *The New York Times*, November 4, 2007.

Seale, E., 2003, The Evolution of SPIDER fault protection, incremental development, and the Mars Reconnaissance Orbiter Mission, *IEEE Aerospace Conference Proceedings*, Vol. 5, pp. 2493–2499.

Tomayko, J., 1988, Computers in Spaceflight The NASA Experience, NASA Contractor Report 182505, NASA History Office.

Watson, D., 2003, Model-based autonomy in deep space missions, *IEEE Intelligent Systems*, 18(3):8–11.

White, J., Wright, C., 2005, Presentation: Test Like You Fly: A Risk Management Approach, Space Systems Engineering & Risk Management Symposium, The Aerospace Corporation.

Chapter **15**

Communication and Navigation Networks in Space System of Systems

KUL B. BHASIN and JEFFREY L. HAYDEN

NASA Glenn Research Center, Cleveland, OH, USA

15.1 HISTORICAL PERSPECTIVE

Since the development of rocketry, the space system of systems (space SoS) definition has been emerging. Space systems architecturally consist of three main segments: (1) space, (2) launch, and (3) ground. The space segment contains satellites, the International Space Station (ISS), the Space Shuttle, and other spacecraft as its elements. The launch segment consists of launch vehicles, launch pads, and platforms, as well as the supporting infrastructure. The ground segment includes large and small antennas and communication stations, ground data distribution infrastructure, and operations control centers. As we begin to follow the Vision for Space Exploration (The White House, 2004) and explore the Moon and Mars surfaces (NASA, 2006), the space systems can be defined within frameworks based on the same three segments. That is, the Moon has a space segment with orbiting elements, a launch segment for leaving the surface of the Moon, and a Lunar surface segment for interfacing with the space segment and for handling local communications. A space system, for the purpose of this chapter, is defined as space system elements, their supporting infrastructure, and managing and supporting organizations.

In the beginning, space systems directly connected to Earth ground systems for telemetry, command and control, and data delivery. As a result, existing communications, navigation, and networking for space systems have grown in an independent fashion with experts in each field solving the problem just for that field. Radio

System of Systems Engineering: Innovations for the 21st Century, Edited by Mo Jamshidi
Copyright © 2009 John Wiley & Sons, Inc., Publication

engineers designed the payloads for today's "bent pipe" communication satellites. The Global Positioning Satellite (GPS) system design for providing precise Earth location determination is an extrapolation of the LOng RAnge Navigation (LORAN) technique of the 1950s. In this technique, precise reception of time markers relative to reference emitter positions are correlated to precise position on the Earth. Other space navigation techniques use artifacts in the radio frequency (RF) communication path and time transfer techniques to determine the location and velocity of a spacecraft within the solar system. Networking in space today is point to point among ground terminals and spacecraft, requiring most communication paths to/from space to be scheduled such that communication is available only on an operational plan and is not easily adapted to handle multidirectional communications under dynamic conditions.

15.1.1 Early Communications Satellites

After the Russian launch of Sputnik I on October 4, 1957, the U.S. government got serious about exploring and using space. The Congress and President Eisenhower created the National Aeronautics and Space Administration (NASA) on October 1, 1958 "to provide for research into the problems of flight within and outside the Earth's atmosphere and for other purposes." NASA launched the first American communication satellite, the passive 100- foot diameter, RF reflector, Echo balloon, into an elliptical, low-altitude Earth orbit (LEO) on August 12, 1960 (Pierce, 1990). Two-way voice links were accomplished between Bell Laboratories in Holmdel, NJ, and the NASA station at Goldstone, CA, by bouncing RF transmissions off the surface of Echo as shown in Fig. 15.1.

In 1961, NASA awarded a contract to RCA to build RELAY, a medium Earth orbit (MEO — 4000 miles high) active communication satellite (Whelan, 2007). RELAY 1 was launched on December 13, 1962, and RELAY 2 was launched on January 21, 1964. The RELAY satellites were small and simply transmitted a received signal from one ground station to another as shown in Fig. 15.2. During that same period, NASA assisted AT&T by launching TELSTAR 1 on July 10, 1962 and TELSTAR 2 on May 7, 1963. These were medium orbit prototype relay satellites that AT&T intended to be the forerunners of a 50-satellite constellation spaced around the Earth to provide global commercial telecom services. However, when the Kennedy administration decided to

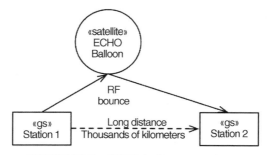

FIGURE 15.1 ECHO balloon experiment

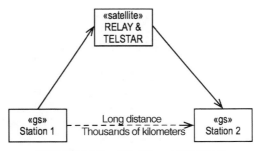

FIGURE 15.2   RELAY and TELSTAR

give the monopoly on satellite communications to the Communication Satellite Corporation (COMSAT), AT&T's TELSTAR satellite project was halted.

15.1.2   First Geosynchronous Communication Satellites

On July 26, 1963, NASA launched the first communication satellite into synchronous orbit, the SYNCOM 2 (the SYNCOM 1 launch was a failure) (Whelan, 2007). While the orbit was synchronous, it was also inclined by 33° due to being launched from Cape Canaveral. The inclination caused the satellite position to rise 33° above the equator to 33° below the equator each day, so the ground antennas had to track this motion to maintain the communication link. SYNCOM 3 was launched on August 19, 1964, into a true geostationary Earth orbit (GEO) that had an inclination of less than 1° with respect to the equator. A GEO-located satellite enabled the ground station antenna to be simplified by relieving it from having to provide major attitude changes to track the satellite. As shown in Fig. 15.3, communications were normally between two ground stations, each of which extended communication service to user nodes by early surface wired networks.

15.1.3   Early Commercial Communications Satellites

In 1960 AT&T filed with the Federal Communications Commission (FCC) for permission to launch an experimental communications satellite with a view to rapidly

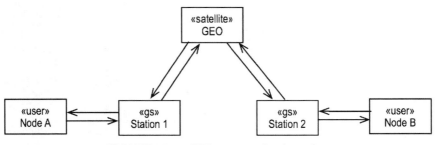

FIGURE 15.3   GEO communication relay

implementing an operational system (Whelan, 2007). The U.S. government had no policy in place at the time to enable AT&T to go forward with its proposal. In 1961, NASA awarded a competitive contract to RCA to build the RELAY satellite mentioned previously. Concurrently, AT&T built its own medium-orbit satellite (TELSTAR), which NASA then launched on a cost-reimbursable basis. Additionally, NASA had awarded a sole-source contract to Hughes Aircraft Company to build a 24-h (20,000 miles high) satellite (SYNCOM). The military program, ADVENT, was canceled a year later due to complexity of the spacecraft, delay in launcher availability, and cost overruns.

By 1964, two TELSTARs, two RELAYs, and two SYNCOMs had operated successfully in space. This timing was fortunate because COMSAT, which was formed as a result of the Communications Satellite Act of 1962, was in the process of contracting for their first satellite. COMSAT's initial capital of 200 million dollars was considered sufficient to build a system of dozens of medium orbit satellites. For a variety of reasons, including costs, COMSAT ultimately chose to reject the joint AT&T/RCA offer of a medium-orbit satellite incorporating the best of TELSTAR and RELAY. Instead, they chose the 24-h-orbit (geosynchronous) satellite offered by Hughes Aircraft Company for their first two systems and a TRW geosynchronous satellite for their third system as shown in Fig. 15.4.

The Hughes Aircraft Company, later Hughes Space and Communications Company and now Boeing Satellite Systems, built Early Bird, the first commercial communications satellite, for COMSAT. The satellite was launched into synchronous orbit on April 6, 1965. It was placed in commercial service on June 28. Early Bird's design stemmed from the SYNCOM satellites Hughes had built for NASA to demonstrate the feasibility of communications from synchronous orbit. On station in orbit 22,300 miles above the equator, Early Bird provided line-of-sight communications between Europe and North America. As a communications repeater, Early Bird handled communications that were representative of all types of common carrier network traffic, including telephone, television, telegraph, and facsimile transmissions.

NASA initiated the development of the advanced communications technology satellite (ACTS) in the late 1980s (Ivancic et al., 1999). The purpose of the ACTS was to test the use of Internet technologies over satellites placed in GEO. The ACTS was capable of in-space switching of received data streams to multiple transmit streams, thereby enabling a user to obtain access to any location under the ACTS footprint.

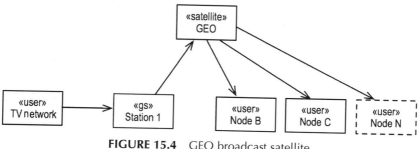

FIGURE 15.4   GEO broadcast satellite

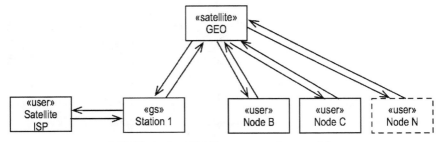

FIGURE 15.5   GEO internet service provider

Communications via the ACTS was bidirectional among the ground users geographically spread out as shown in Fig. 15.5. The successful ACTS experiments contributed directly to the development of the Internet Protocols (IP)-compliant HughesNet and Spaceway networks.

15.1.4   Department of Defense (DoD) Communication Satellites

The Signal Communications Orbit Relay Equipment (SCORE) was the world's first active communications satellite (Whelan, 2007). On December 18, 1958, the Advanced Research Projects Agency (ARPA) of the Department of Defense (DoD) launched SCORE. SCORE was an experimental, store and forward (Fig. 15.6), communications satellite in an elliptical LEO that stored received messages from ground stations on a tape recorder and transmitted them to other ground stations later when the satellite passed overhead. The first message sent to the world as SCORE orbited the Earth was a 1958 Christmas message from President Eisenhower.

In 1960 ARPA attempted to develop Advent, the first geosynchronous Earth orbiting (GEO) communication satellite. Advent was very sophisticated for the time — it was a large, three-axis stabilized satellite bus that was to be launched on an early Atlas Centaur rocket. The program had many technical difficulties and the Centaur proved not to be ready until 1968, so Advent was cancelled in 1962.

Upon the cancellation of the Advent program, the first operational (not experimental) Initial Defense Communication Satellite Program (IDCSP) was begun (Kucherman et al., 1966). The satellite was a simple design with each IDCSP payload

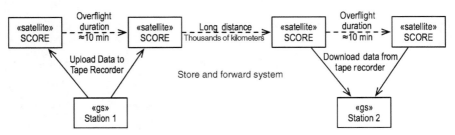

FIGURE 15.6   Store and forward satellite

having a single repeater with a capacity of 1 megabit per second of data or 10 voice circuits when communicating with large terminals on Earth. By 1968, a total of 28 IDCSP satellites were spaced around the world in an orbit just under the GEO radius. That same year, the constellation was declared operational and renamed to the Initial Defense Satellite Communication System.

DoD operated three types of communication satellite systems in the 1980s: the Defense Satellite Communications System (DSCS) and the US Navy's Fleet Satellite Communications System (FLTSATCOM), supplemented by gapfiller satellites were used. To provide global unscheduled access and world connectivity to all types of military platforms, the Milstar satellite system was defined in 1985 and the satellites were launched from 1994 to 2003. The five Milstar satellites in GEO, capable of cross-linking and communicating to each other to pass messages around the globe, make up the constellation aided by mission control and ground terminals (Martin et al., 2007).

15.1.5 The Tracking and Data Relay Satellite System (TDRSS)

Early NASA space systems provided data communications directly to ground systems. As NASA began to plan the space shuttle and further Earth science missions in LEO in the late 1970s, it needed nearly continuous communication around the Earth (Elwell et al., 1992). This was a major transition point for how space systems would communicate with the users. Now a global relay satellite communication infrastructure was designed for global coverage. To achieve near- continuous coverage, NASA decided to place a constellation of Tracking and Data Relay Satellites (TDRS) into GEO along with a ground station system to control it at White Sand Complex in New Mexico. An additional ground station was later placed on Guam in the Pacific to control a TDRS over the Indian Ocean. The geometric coverage of the Earth by the TDRSS constellation is shown in Fig. 15.7.

TRW manufactured the first set of seven TDRSs (with one lost) that were launched in the 1983–1995 time frame. The TDRSS infrastructure provides communication services at low and high data rates in the S-band (2–4 GHz) and Ku-band (12–18 GHz). Later a second set of three TDRSs that are capable of communicating in the Ka-band (18–31 GHz) were built by Boeing and launched in the early 2000s. The nine remaining TDRSs in GEO constitute the present TDRSS constellation with six active satellites placed near the longitudes shown in Fig. 15.7, two stored in parking longitudes, and one swinging out of plane but servicing the South Pole.

15.1.6 The Deep Space Network

The Jet Propulsion Laboratory began construction of the first dish antenna at what later became the U.S. site of the Deep Space Network (DSN) at Goldstone, CA (NASA, 2005). This first antenna was erected to communicate with the 1958 flights of Pioneer missions 3 and 4 to the Moon. Pioneer 3 did not make it due to a launch error, but Pioneer 4 flew beyond the Moon and the antenna received its signals out to 655,000 km, a record distance for the time. The first overseas DSN site was established at Woomera, Australia, in 1960 with the installation of a 26-m dish antenna of the same design as the

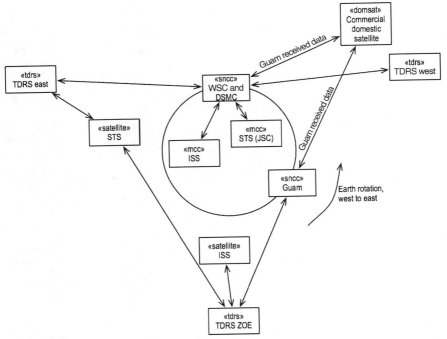

FIGURE 15.7 Nominal orbital geometry of the space network (SN) tracking and data relay satellite system (TDRSS)

Pioneer antenna. A third Pioneer style antenna was erected near Hartebeesthoek, South Africa, in 1961. These early antennas supported the Ranger 1 mission to the Moon. As the number of robotic missions increased in the early 1960s (Ranger, Surveyor, and Lunar Orbiter to the Moon and Mariner to the solar system), a new style of 26-m antennas was added at Goldstone, Tidbinbilla, Australia (near Canberra), and Robledo, Spain (near Madrid). These three sites became the permanent DSN sites in use today. Today, each complex consists of at least four deep space stations equipped with ultrasensitive receiving systems and large parabolic dish antennas, which are as follows:

- One 34-m (111-foot) diameter high-efficiency antenna
- One 34-m beam-waveguide antenna (three at the Goldstone complex and two in Madrid)
- One 26-m (85-foot) antenna
- One 70-m (230-foot) antenna

The DSN is able to provide communication services to several deep space missions simultaneously. Figure 15.8 illustrates how the DSN handles multiple missions and handoffs between DSN sites as the Earth rotates from west to east.

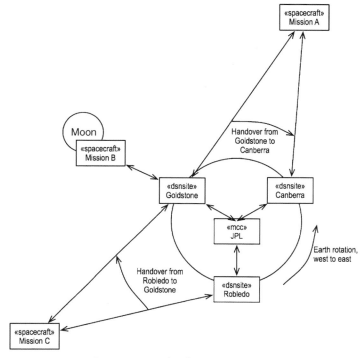

FIGURE 15.8   The deep space network

15.1.7   Early System of Systems Communications and Navigation for Apollo

The Apollo capsule's signal was picked up by the three new stations that had been built for the Apollo mission near the JPL DSN sites, one each in California, Australia, and Spain (Kimberlin, 2004). These were backed up by the DSN stations (then called "wing stations"). All of the signals were routed to NASA's communications center (now the Goddard Space Flight Center) in Greenbelt, Maryland. Figure 15.9 shows the nodes involved for Apollo during the Moon missions.

For Project Apollo, NASA decided to combine all communications between the spacecraft and the Earth into a single multiplexed feed called "The Unified S-Band System," including audio communications, television images, crew medical telemetry, and the spacecraft systems telemetry. NASA also reallocated transoceanic transmissions circuits it had with Intelsat to the Apollo Program.

NASA had installed a "real-time computer complex" (RTCC) for instantaneous information on and control of manned space missions at Goddard during the Mercury missions and the early part of Gemini. The complex linked 17 ground stations around the globe and permitted observers to monitor manned flights on virtually a continuous basis. After lengthy technical and administrative arguments, NASA moved the computer complex to Houston to form an "integrated mission control center." The center was used for processing global signals for display to flight controllers,

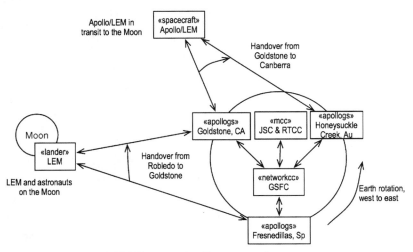

FIGURE 15.9   Apollo system of systems

computing and sending antenna-aiming directions to the global tracking stations, providing navigation information to the spacecraft, and simulating all mission data for personnel training and equipment checkout.

15.2   SPACE SYSTEMS OF SYSTEMS

The questions now arise: What comprises a space system of systems (Space SoS)? What are the current status and trends? What role do communication and navigation networks play in a Space SoS? Although the definition of Space SoS is emerging, it can be defined as a network of assets on Earth, in orbit around Earth, in orbit around solar system bodies, and on the surface of solar system bodies that are interconnected and/or interoperated to perform a mission, and/or provide services that cannot be performed by monolithic space systems alone. Space SoSs are being operated, modified, developed, and planned mainly in civil, commercial, and military sectors. Initially, space legacy systems are being integrated and/or interconnected to newer systems. Complex systems of systems for disparate missions are currently in the planning or system design stages. The following trends are emerging in the civil, commercial, and military space system of systems.

Civil—NASA was the first civil organization to deploy Space SoS through its Apollo Mission to the Moon. Subsequently, it has created small Space SoSs by integrating terrestrial, Internet, robotic systems, and Mars surface and space communication network infrastructure (TDRSS) into its missions. NASA has embarked on space exploration missions to the Moon and Mars, which have increasing complexity; it will integrate these advanced systems to create an exploration system of systems. NOAA is also developing an Earth observation system called GEOSS as a complex SoS.

Commercial—The commercial sector has deployed large space systems for communication applications such as the Iridium satellite constellation. Iridium mainly provides voice services. By integrating terrestrial phone networks with a space network of interconnected satellites, it becomes an SoS.

Military—Military space organizations are using space to communicate, navigate, command, control, gather intelligence, obtain weather conditions and weather forecasts, and execute strategic and tactical plans for defense. With the advent of space and terrestrial communication networks, computers, and software, complex Space SoS are emerging.

15.2.1 Communication and Navigation Network of Networks

Communications and navigation networks are architected, designed, developed, and deployed based on the type of Space SoS services needed. Communications and navigation networks of the future will be architected as an integrated set of new assets and a federation of upgraded legacy systems capable of routing IP traffic from any node to any other node on the network (Hayden, 2000). Figure 15.10 depicts a complex yet generic set of system of system node types that are expected to be involved in missions of exploration.

The bottom of the figure indicates the Earth ground systems involved in support of exploration missions. Legacy systems shown include the DSN and its control center, the TDRSS and its controlling ground stations, and the navigation support center. The communication user control centers are shown for exploration missions, launch, and ISS. A new Distant IP Network Control Center and its ground stations around the Earth are included in the diagram. The ground stations of this network addition may be collocated at the DSN sites; however, they will likely require new system hardware and antennas to handle high communication data rates at long distances, new navigational services, and modern networking technologies. This new or highly modified legacy equipment must be dedicated full- time to the exploration missions. These ground stations will handle direct communications to and control of communication relay satellites in orbit around the solar system body, as well as direct communications to any visible lunar or planetary surface communication terminal.

As shown at the lower right in the figure, ground network sites handle capturing data from the crew launch vehicle before and during launch. Range safety is also on hand to send a flight termination signal in the event the launch goes astray. Additionally, a contingency voice system is available in the event normal communication is lost to the flight vehicle.

The next level up, the Earth orbit systems, displays the communications needed at that level. Communication support needs are provided by the legacy TDRSS to support crew vehicle rendezvous with the ISS as the crew vehicle replaces the Shuttle in replenishing and recrewing the ISS. The TDRSS can also support the launch vehicle and crew vehicle during launch and rendezvous of the crew vehicle with the lander and Earth departure rocket to get ready for the transfer to the Moon, Mars, or elsewhere. The Space network can also support communications to the crew return capsule during descent and landing.

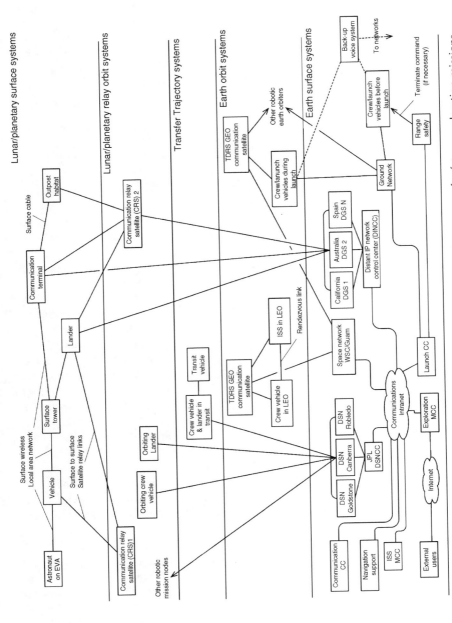

FIGURE 15.10 General diagram of communication and navigation systems for future exploration missions

The third level (transfer trajectory systems) shows a crew vehicle and lander in transit to the solar system object. The legacy DSN system can be modified to provide communications to support this phase and also the in-orbit crew vehicle and lander, shown before landing on the solar system object at level 4 (lunar/planetary relay orbit systems).

The fourth level also includes communication relay satellites in orbit around the Moon or Mars. These satellites support communication networking among surface nodes that are spread widely apart, such as astronauts, surface vehicles, and fixed habitats. They also provide high-rate trunks to route communications to and from Earth. These satellites may include cross-links among them to route networks from the backside to the front side of the body being explored (Bhasin and Hayden, 2001).

The fifth level shows communications with the relay satellites, the Earth's surface, and over the surface of the solar system body. The network over the surface will be of short range and limited by line of sight. However, the surface local area networks can be extended with towers similar to the architecture used in cell phone networks on Earth (Bhasin et al., 2005).

It is imperative to implement a system of systems at the lowest cost of operations in an infrastructure such as this one for space exploration. Automation of ground systems, modernization of legacy assets, highly capable new assets, autonomy of operations of in-space communications assets, and fully IP-compliant networking technologies will be necessary for maintaining a low cost of operations (Bhasin and Hayden, 2002). Cost of development will not be insignificant but in the very extended long-range plans being considered, the cost of operations will dwarf the cost of improvement in technologies. Much of the cost of the new technologies has already been carried by the commercial world. IP networking solutions are largely available for direct use. Work in this area does need to be done to handle IP addressing and networking among moving nodes.

15.3 COMMUNICATIONS AND NAVIGATION NETWORK ARCHITECTURES

15.3.1 Space Communication Architecture Types

Communication and navigation architecture has evolved from those shown in Fig. 15.1 through Fig. 15.9 and summarized in Fig. 15.11. The most complex of these architectures are the SN and today's DSN that are controlled by a combination of antenna scheduling computers and human operators. The communication processing in the TDRSS low data rate multiple access (MA) equipment at WSC has recently been modified to provide IP-compliant communication service autonomously on demand of the user. It is not possible to obtain service on demand from the high data rate TDRSS or the DSN systems as these services are under preprogrammed automatic and human operator control and must be scheduled in advance, or coordinated with the network's operations when needed, usually to support an emergency.

Progression of complexity in communication and navigation network architecture

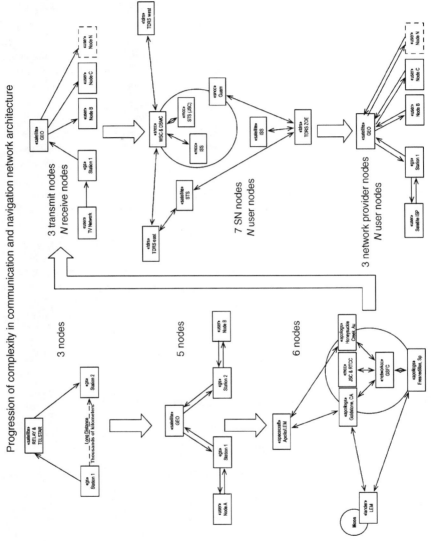

FIGURE 15.11 Progression of complexity in communication and navigation network

A much more complex system of systems will be required in an architecture that provides space communications and navigation services for space exploration. Some services will be provided by legacy systems such as the DSN (Fig. 15.8) and the SN (Fig. 15.7) that will have to be modified to handle modern networking technologies. Other services, particularly the very high data rate communications, new navigational aids, and modern networking capabilities, will be provided when they are needed by new systems designed to provide them. This federation of systems shown in Fig. 15.10 will work as an integrated Space SoS for supporting future space exploration missions. Communication systems that must interoperate are included in the crew vehicle, Earth relay satellites such as TDRSS, communication relay satellites, and surface communication infrastructures such as might be in orbit around (Sand et al., 2002) and on the surface of the Moon or Mars. The complexity of architecture for a Space SoS that provides communication and navigation services to astronauts and systems on the Moon becomes apparent in Fig. 15.12 (Bhasin et al., 2006). This figure illustrates several different types of communication links, including the long-range high data rate trunk lines reaching from the distant IP network's (DIN) ground stations (DGS) to the communication relay satellites (CRS) (Soloff et al., 2005) in lunar orbit (LCRS) and directly to the communication unit on the lunar surface (if that unit is in view of the Earth); the shorter range high data rate links from a LCRS to the surface communication unit or human occupied vehicles; the low data rate links from a LCRSs' multiple access subsystems to rovers, robots, science instruments, fuel extraction equipment (*in-situ* resource unit, ISRU), and astronauts performing extra-vehicular activities (EVAs) on the surface; long range surface wireless wide area networks (WANs) that interconnect the surface communication unit, human occupied vehicles, rovers, robots, and network repeaters; and short range surface wireless local area networks (LANs) that interconnect astronauts on EVA, in vehicles, and in habitat, with each other and with rovers, robots, and scientific instruments in close proximity. The acronyms DIN, DGS, CRS, LCRS, and their terms were fabricated solely to describe architecture in this chapter separately from the existing legacy architecture and are not terms in general use.

The LANs and WANs enable astronauts to collaborate on their activities by conversing, sharing their data, and sharing their vision by passing video among them while performing construction duties or exploring. The low data rate links to relay satellites enable communications among astronauts, vehicles, and other surface nodes that are widely spread apart beyond the line of sight of a WAN. The high data rate links enable the Earth operations centers to send voice, video, directives, commands, software, and data to surface nodes on the Moon. Likewise, they enable the astronauts and other surface nodes to send voice, video, and science data from exploration activities; health data; navigational location coordinates; and engineering data to interested parties on the Earth.

The complexity of communication on the Moon is further illustrated in Fig. 15.12 by showing the types of networks that may reside within the communication relay and surface nodes. In the figure, the onboard communication subsystems and the control subsystems in a communication relay satellite are interconnected with an IP-compliant network. Each subsystem has its own IP address dispensed by the central

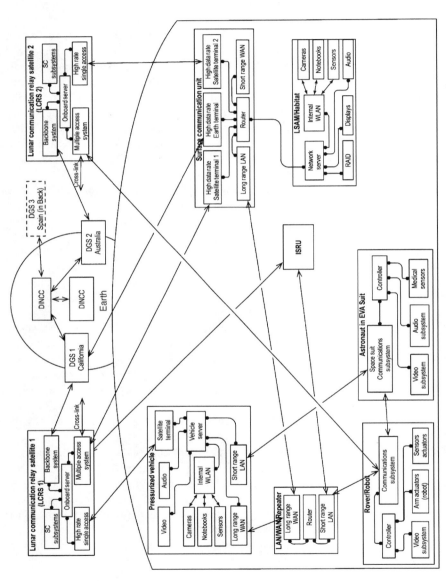

FIGURE 15.12 Illustration of the highly complex networking needed for exploration missions

computer or server for configuring, operating, and monitoring each of the sub-systems on the satellite. The satellites provide IP-compliant communication services on demand of the using nodes and, along with a high degree of autonomy, reduce the need for continuous human operations for operating the satellites and for scheduling the services they provide. The other surface nodes in the figure have their own controlling computers/servers and internal IP-compliant networks to interconnect all the subsystems within each node. The IP-compliant interfaces and routers, prevalent throughout the networks and user nodes, can route data throughout the infrastructure over the tortuous path between any two (or more) nodes on the Moon and/or the Earth.

The architecture must describe the joining of these disparate systems so they can intercommunicate and hand off communication links among each other as an exploration mission vehicle passes from one system's domain to another's. Further-more, the communication services must join so that an exploration customer node may communicate with another node across system domains. The development process for this architecture must lay out the methods, operations, and interfaces needed to join together a federation of multiple, heterogeneous, legacy, and new systems that incorporate networks at multiple levels and in multiple domains. The architecture must address the resolution of the problems and incompatibilities associated with incorporating those disparate systems into a federation of communication and navigation services for space exploration.

All of the existing and future infrastructure must interoperate to achieve the high capacity, high performance Space SoS attributes needed to provide superior com-munication, navigation, timing, and information services. These capabilities, driven by emerging exploration requirements, will enable future astronauts to conduct space exploration, communicate with Earth-based scientists, excite future genera-tions of explorers with high definition video to enable a virtual-presence participa-tion in the astronauts' exploration activities, and return safely to Earth at the end of the mission.

15.4 COMMUNICATIONS AND NAVIGATION INFRASTRUCTURE-BASED APPROACH

15.4.1 Architecture Decomposition Process

The process for describing the needed communications, navigation, and networking entities as a Space SoS construct is accomplished by using standard system engineer-ing methods for defining and gathering capability requirements and identifying a generalized initial architecture (Maier, 1998). Additional steps include decomposing the architecture and using the Department of Defense Architecture Framework (DoDAF) operational, system, and technical view diagrams (Bhasin et al., 2007), and then refining the architecture as actual operational activities and functional requirements are identified during the architecture development process (DoDAF, 2003; DoDAF, 2007).

The space communication and navigation networks must interface with all exploration missions from prelaunch testing through crew recovery at the end of each mission. The networks will support each mission at launch, in LEO, in early docking, in transit to and orbit around the Moon or Mars, during lunar or Mars landing, on the lunar or Mars surface, and during the trip back to Earth. The space networks will be compatible with space exploration customers to provide seamless communication, tracking, and timing services throughout their missions.

The system development of the space networks will take place over multiple phases, each spanning multiple years during which the political, management, and contractual arrangement can change. Further complexity is added by the fact that the post-phase 1 system has to provide communication and tracking services to space and planetary surface assets based on multilayered architectural information systems while maintaining services for legacy missions.

The architecture views will convey the space communication and navigation network architecture to all the stakeholders (astronauts, exploration mission customers, system engineers, implementers, test engineers, international partners, potential equipment vendors, etc.) to assist them in verifying that the Space SoS will address their concerns or to help them understand the Space SoS and how they may take part in implementing it.

The network requires the integration of architectural practices across multiple disciplines. Such practices are based on the views of stakeholders of varying disciplines and managerial levels in the architecture; model-based analyses of elements of the architecture by subject matter experts; and collaboration among the stakeholders, the architect, and the experts during the evolution of the architecture (as captured by view development) to final consensus.

The DoDAF methods of developing the architectural views were chosen for describing the space communication and navigation networks Space SoS because they are particularly suited to developing and defining complex communication architecture and have been vetted by use in defining "as-is" and "to-be" communication architectures for major DoD communication projects. A hierarchical document and diagramming structure is used to present the complex communication and navigation network architecture in an efficient manner to a variety of users with differing needs. Thus, the amount of detail presented increases with each successive level, as shown in Fig. 15.13. The network architecture is described in terms of its operational, system, and technical attributes so that the user can gain insight into how it fulfills its mission objectives. The network architecture is decomposed into segments and elements. The purpose of the descriptions for each level is to show how segments and elements support the operational, system, and technical aspects of the architecture and thus to relate the parts of the network architecture to its whole.

15.4.2   Defining Exploration Mission Systems and Their Interfaces

A number of NASA exploration architecture teams and study groups have defined the systems to be deployed during the exploration missions. The crew vehicle is a critical

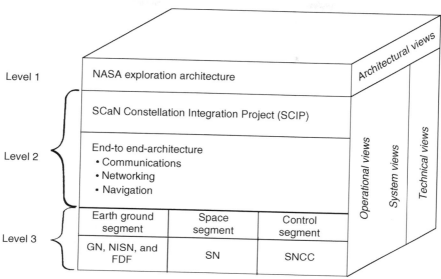

FIGURE 15.13 Architecture decomposition

system that will carry the astronauts to Earth orbit, beyond, and back to Earth. Several additional systems come into play to carry the crew vehicle into space and then to the Moon. An outpost developed on the surface of the Moon will require another set of systems to support the lunar exploration missions that follow. In Fig. 15.10, the notional communication and navigation systems shown provide the interfaces among the explorations systems on the Earth surface, in Earth orbit, in translunar space, in lunar orbit, and on the lunar surface. Many of the systems and interfaces are dynamic in nature, which adds to the complexity that must be addressed during the development of the architecture.

15.5 END-TO-END COMMUNICATION ARCHITECTING PROCESS

The communication and navigation network architecture must pursue an aggressive adoption of common communications links and protocols, drawing from standards-based IP network technology and effective use of layering to isolate functions, aggregate common functions to maximize interoperability, provide autonomy in data handling operations, and enable the overall architecture to evolve through frequent system upgrades at minimum cost. The space communication and navigation network architecture must be flexible and evolvable enough to adapt to technology, missions, and user needs that will inevitably change dramatically over the next 20 years. With the short Earth-based Internet generational life cycles and communication-rich application training of future astronauts, the space-based command, control, communication, and information (C3I) capabilities will need to be raised to meet the expectations of future users.

15.5.1   IP-Centric Networks Architecture

The communication and navigation network architecture must provide end-to-end connectivity services at low data rates in the S-band and at high data rates in the Ka-band to the crew vehicle and end-to-end S-band services to the crew launch vehicle during all mission phases. The end-to-end S-band and Ka-band services will support communications, radiometric tracking, and timing synchronization activities.

The communication and navigation network will consist of the SN, GN, and DSN legacy system elements, the new communication Intranet, and the network elements, previously identified by the fabricated acronyms DIN, DGS, and the CRS in lunar orbit (LCRS) and later Mars orbit (MCRS), which will be added as space exploration extends into distant and deep space. Some system elements, such as the DSN and the mission control facilities at Johnson Space Center (JSC), have over 25 years of legacy hardware, software, and policies. These networks are administered by different control center software systems and operational policies. To put together, an IP-compliant end-to-end management system with human-rated real-time responsiveness requires stitching together numerous different administrative systems and domains with different policies, priorities, and management systems to form an integrated system that can plan, allocate, control, deploy, coordinate, and monitor the resources of the network to support space exploration. Many of the legacy systems supporting existing space missions also have to evolve into the future architecture to meet the space exploration mission requirements. Currently, GN, SN, and DSN do not provide a network interface to missions; however, they will need to extend network layer capabilities to space. Quality of service (QoS) over different domains poses another challenge in mapping and preserving QoS integrity over multiple domains. With human lives at stake on Mars or the Moon, automation, automatic error recovery, and local distributed control will have to be pervasive in the architecture.

Later in the text after Fig. 15.22, an example in the section for applying DoDAF views is discussed that describes the data on the needlines that need to be passed over the communication path for an astronaut controlling an instrument while on EVA and passing on instrument data to a collaborating scientist on Earth. The application and capabilities of IP layer-based network architecture is illustrated in Fig. 15.14 by building on the example described in that text. Figure 15.14 shows the naming of the network layers and the type of data carried over them in the legend at the upper right of the diagram. Starting at the lower left is a scientific instrument. The instrument has its own simple internal network that begins at the top of a five-layer stack, the instrument controlling application layer. The diagram is a network diagram concerned with protocol processing, protocol interfaces, and transport mediums. While the protocols reside in system blocks, the internal subsystems are not discussed in detail. The point of this diagram is to follow the communication of scientific, voice, and video data through protocol layers along a set of communication paths that provides communication and networking service between the Moon and the Earth. In the example, the astronaut or scientist would initiate a voice and video connection with the other. Then

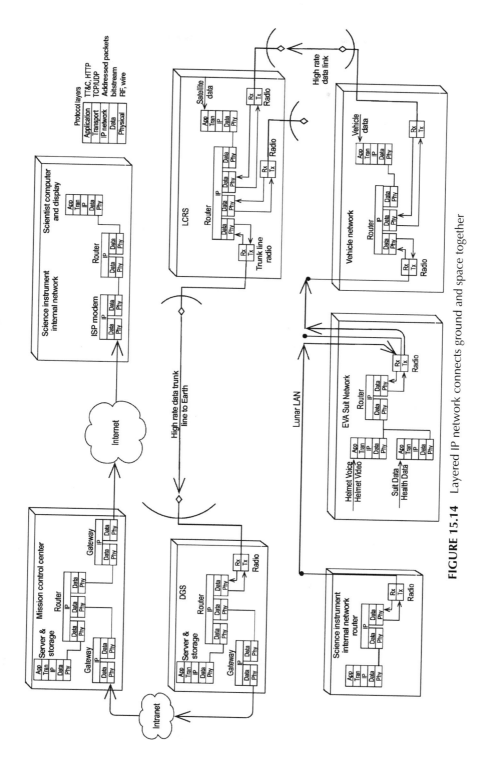

FIGURE 15.14 Layered IP network connects ground and space together

the astronaut can initiate the transport of the instrument data by pressing buttons on the instrument or by verbally commanding it to release its data to the scientist's address on Earth. The diagram indicates the paths through the protocol stacks, routers, and gateways the instrument data must take to reach the scientist and then stream the data to him or her. These same network paths also handle the voice and video from the astronaut to the scientist in a separate parallel stream. The scientist, in turn, converses with the astronaut on a return path through the same network nodes.

15.5.2 Propagating Navigation Architecture to Space

As the network is deployed in space from Earth-based to lunar and Mars network support, a system of coarse and fine-grained navigation references will be integrated within the network to enable future space explorers and applications to access navigation and position services and to have autonomy in choosing paths of navigation. In other words, astronauts must be able to "fly" crew vehicles and landers independently from an Earth-based mission control center simply because a mission to the Moon or Mars could result in a landing occurring out of sight of Earth. Furthermore, a landing on Mars will occur at a great distance from Earth, so far that real-time communication from Earth to Mars is impossible. The astronauts will be in control of their safe landing on Mars, and the networks must provide the communication and navigation services to enable them to act autonomously.

15.6 APPLYING DoDAF VIEWS

The DoDAF methodologies and a few typical products are described below to illustrate the architecture. Other crosscutting views of the multitude of "as-is" legacy systems, "to-be" future systems, and future IP network decomposition views of the network layers and network-centric architecture may also be useful in describing the architecture. The architecture diagramming methods used range from the defined DoDAF graphical views to custom diagrams that use pictorial icons to represent nodal entities.

The DoDAF Deskbook for Version 1.0 recommends six general steps that should be followed for developing the architecture and resultant products. Please refer to the Deskbook for a more detailed description of the process steps illustrated in Fig. 15.15 (DoDAF, 2003).

Three general view categories are defined within DoDAF for an architectural description, namely, operational views (OVs), system views (SVs), and technical standard views (TVs). A fourth all view (AV) category is used to provide a written summary description of the architecture and to maintain a dictionary of the terms used in the architecture. Figure 15.16 shows a sequence for generating architectural view that is recommended in the DoDAF Version 1.0, Deskbook instructions (DoDAF, 2003). While not all diagrams are displayed, the example diagrams given below, OV-1, OV-4, OV-5, OV-2, and SV-1, were developed in the order indicated in Fig. 15.16.

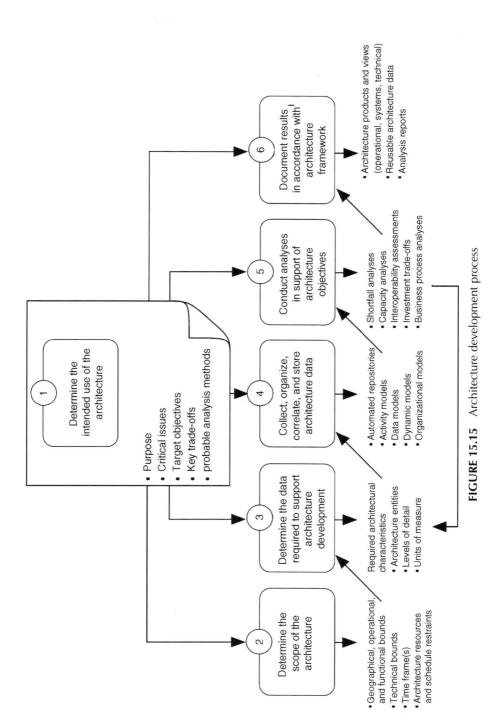

FIGURE 15.15 Architecture development process

1. Determine the intended use of the architecture
- Purpose
- Critical issues
- Target objectives
- Key trade-offs
- probable analysis methods

2. Determine the scope of the architecture
- Geographical, operational, and functional bounds
- Technical bounds
- Time frame(s)
- Architecture resources and schedule restraints

3. Determine the data required to support architecture development
- Required architectural characteristics
- Architecture entities
- Levels of detail
- Units of measure

4. Collect, organize, correlate, and store architecture data
- Automated repositories
- Activity models
- Data models
- Dynamic models
- Organizational models

5. Conduct analyses in support of architecture objectives
- Shortfall analyses
- Capacity analyses
- Interoperability assessments
- Investment trade-offs
- Business process analyses

6. Document results in accordance with architecture framework
- Architecture products and views (operational, systems, technical)
- Reusable architecture data
- Analysis reports

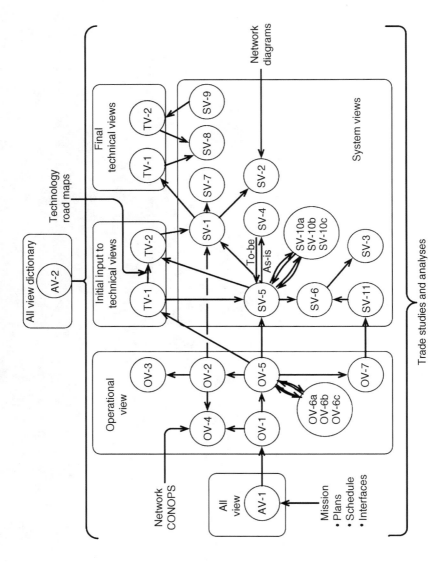

FIGURE 15.16 Architecture build sequence

The OV and SV products shown below are used to describe architecture that provides communication and navigation services to nodes on the Moon and the Earth. Note that the communication and navigation systems and operational activities will evolve over different mission stages. For example, testing communications to crew and launch vehicles prior to launch, versus supporting communications and location tracking of the human and robotic nodes on the surface of the Moon, involve different systems and operations. The graphical product examples below are used to capture the architectural characteristics needed to support communication and navigation activities of nodes on the surface of the Moon.

15.6.1 Example Diagrams

The first operational view, OV-1, describes the overall operational concept for the architecture being described. Typically, the OV-1 is drawn with icons that represent the various system and operational nodes of that architecture. Here, we use a simpler diagramming technique of using labeled boxes to indicate the nodes. Figure 15.17, a simplified OV-1, shows the operational concept for interaction among the DIN control center (DINCC), the DGSs, the exploration MCC, the LCRSs, and the lunar surface nodes. Essentially, the figure indicates that any node may communicate with any other node in its vicinity. What is also implied but not clearly obvious is that the use of IP in the networks enables any node to communicate with any other node on the networks by routing the data through any of the intermediate nodes shown in the figure.

Figure 15.18 is an OV-4 diagram that shows the relationships among the organizations involved with providing the communications and navigation services to the using nodes. The customer for the services is represented by the exploration mission control center (Exploration MCC) shown at the top. The customer interacts with the communication and navigation control center and management by requesting services to support its operations. The communication and navigation control center, in turn, requests network services from, and passes mission customer schedule information to, the distant IP network control center and to the communication Intranet management center. The communication and navigation control center coordinates and schedules the network services with the DINCC that controls the DGS remote sites around the Earth and the LCRSs in orbit around the Moon. The DINCC also monitors the operation and health of the DGS site antennas and the LCRSs. Ideally, the operation of the DGS antennas and the LCRSs would be fully autonomous with the DGS antennas maintaining continuous connectivity with the LCRSs through regular handoffs between stations as they come into and pass out of view of the Moon, as the Earth rotates, and as the LCRSs may become obscured by the Moon. Likewise, the LCRSs would maintain connection with Earth through the handoffs and would provide lunar surface users with service both scheduled and on demand of the user. Make and break of a communication link with a surface node would be an autonomous operation. The right side of the figure shows that the networks provide a virtual communications path between the Exploration MCC and any/all of the lunar surface nodes. Of course, the real path goes through multiple routing nodes in the network. The bottom of the figure shows that the lunar surface nodes, including the lunar surface terminal, can

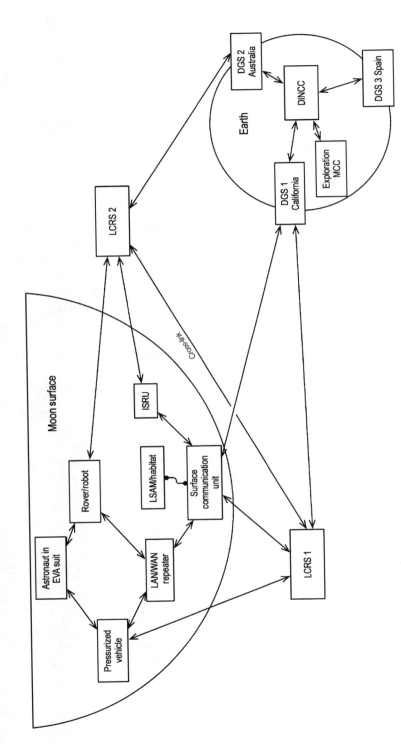

FIGURE 15.17 OV-1, operational concept for lunar exploration

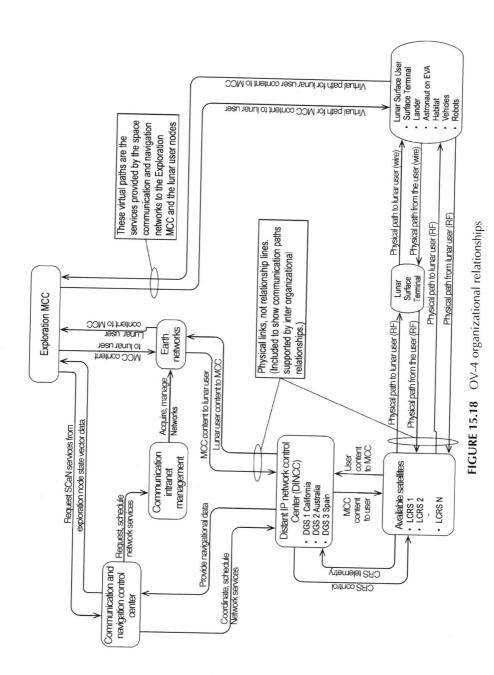

FIGURE 15.18 OV-4 organizational relationships

communicate directly with the LCRSs, for communicating either with Earth or with other surface nodes. This figure does not show the surface network details because it is expected that that network is fully autonomous and needs no external operations, except for maintenance that can be administered over the network or by astronauts on the surface.

The activity diagrams, OV-5, can be drawn in several ways. Here, the architecture hierarchy diagram shows the relationship among the activities supported by the services and infrastructure. Figure 15.19 starts at the top level 0 with activity A0 providing the services and infrastructure for the entire exploration program for the Moon and the Mars. Several level 1 activities are indicated; activity A3 for the lunar missions is an example. Below A3 at level 2 is activity A3.9, the services and infrastructure needed to support human operations on the Moon. Below this at level 3 are three of the activities that require the ability to intercommunicate on the surface: A3.9.1, communications with the Earth; A3.9.2, communications and navigation among nodes spaced up to 10 km apart; and A3.9.3, communications and navigation among nodes that are far apart and need a LCRS to complete the connection.

The activities are then decomposed to finer detail to show the sequence of activities that the services and infrastructure need to support. Figure 15.20 shows the use case and activity sequence diagrams in universal modeling language (UML) icons for the A3.9.1 activity in which a surface user communicates with Earth. The use of common modeling tools and standards are encouraged for diagramming DoDAF architecture. The UML is an accepted diagramming technique for modeling software, and the DoDAF encourages its use as one of the standards for modeling architecture (DoDAF, 2007). These diagrams indicate the kind of activities the communication and navigation services and infrastructure will need to support the lunar surface user.

Figure 15.21 shows just the use case diagrams for A3.9.2 and A3.9.3. Sequence diagrams similar to those in Fig. 15.20 are normally drawn for each of these activities, but are not shown here.

Figure 15.22 is an OV-2 diagram that describes the operational node connectivity for the mission architecture being described. It is recommended that this diagram be drawn after the organization is defined (OV-4) and after the activities (OV-5s) are identified. Then the information that each organization needs to do its work can be identified along with the source of the information. The rectangular boxes in the diagram indicate the operational nodes. The connections between the nodes are called needlines, and the rounded rectangles along the needlines display the information that must be passed from one node to another to enable each node to accomplish its operational tasks. Each data item is numbered for tracking and cross-referencing throughout the architecture development. For example, an astronaut who is communicating with a scientist on Earth may wish to show the scientist interesting mineral deposits. Such a communication path might be as shown in the Fig. 15.22, where the astronaut communicates with voice, microscope video of the minerals, and data readout of a portable analysis device, such as an X-ray fluorescence instrument that is passed from the astronaut's helmet radio to the vehicle shown. The vehicle aggregates

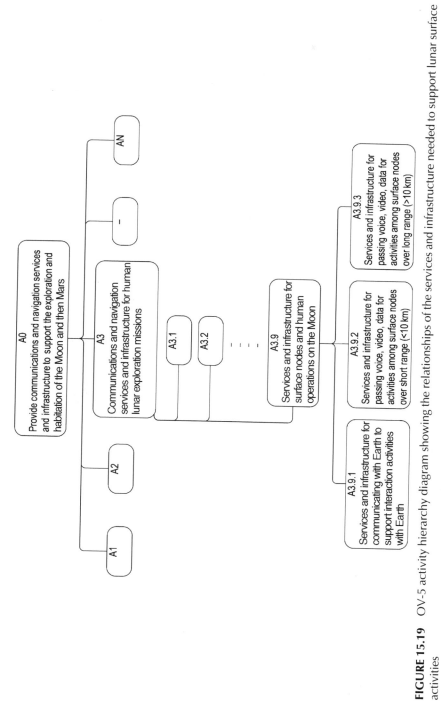

FIGURE 15.19 OV-5 activity hierarchy diagram showing the relationships of the services and infrastructure needed to support lunar surface activities

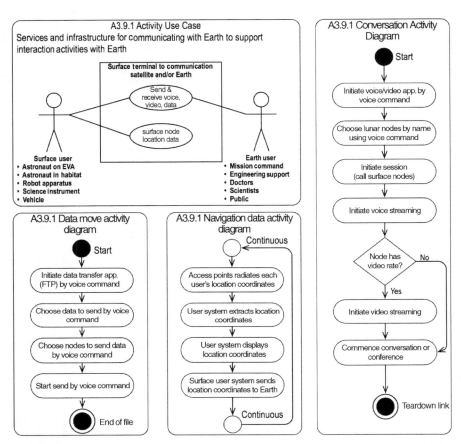

FIGURE 15.20 OV-5, use case and activity sequence diagrams for A.3.9.1, in which a lunar surface user communicates with Earth

all the data in its vicinity and passes those to an overhead LCRS. The LCRS forms a larger aggregate of all the data passed to it and sends the data that are addressed to Earth on to an Earth ground station. The ground station extracts the IP data and sends it on to the Exploration MCC that, in turn, routes the voice, video, and data on to the scientist, who may be accessing his or her end of the connection over the Internet. Communication from the scientist is returned on the same path. The many possible communication paths among any and all nodes can be constructed from figures similar to the one shown.

Figure 15.23 shows an example of the interface diagram (SV-1) for the communication and navigation systems and services needed to support the activities of nodes on the surface of the Moon. This diagram uses UML component (rectangle with a small component artifact at the upper right corner) and system (block-shaped) icons to describe the Space SoS layout. The little rectangles indicate ports into and out of a system block. In the diagram, the ground stations and DINCC are shown as

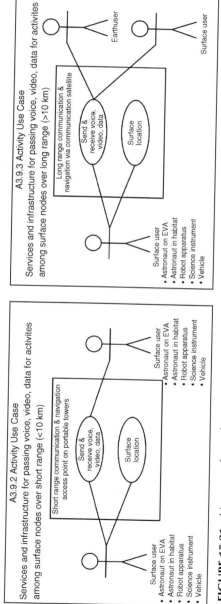

FIGURE 15.21 Use cases for surface user to surface user communications and navigation over short and long distances

A3.9.2 Activity Use Case

Services and infrastructure for passing voice, video, data for activites among surface nodes over short range (<10 km)

Short range communication & navigation access point on portable towers

Send & receive voice, video, data

Surface location

Surface user
• Astronaut on EVA
• Astronaut in habitat
• Robot apparatus
• Science instrument
• Vehicle

Surface user
• Astronaut on EVA
• Astronaut in habitat
• Robot apparatus
• Science instrument
• Vehicle

A3.9.3 Activity Use Case

Services and infrastructure for passing voice, video, data for activites among surface nodes over long range (>10 km)

Long range communication & navigation via communication satellite

Send & receive voice, video, data

Surface location

Earthuser

Surface user

Surface user
• Astronaut on EVA
• Astronaut in habitat
• Robot apparatus
• Science instrument
• Vehicle

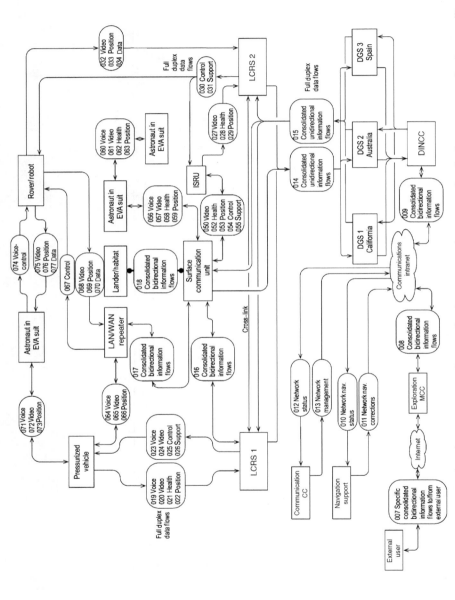

FIGURE 15.22 OV-2, operational node connectivity diagram for the Moon surface networks, the lunar communication satellites, and the Earth-based ground networks

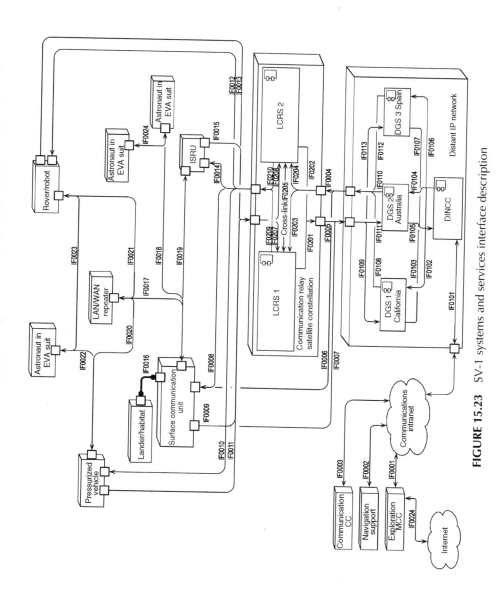

FIGURE 15.23 SV-1 systems and services interface description

components of the larger distant IP network system. Likewise, the LCRSs are aggregated as components of a larger LCRS Constellation. It is possible to make this simplification if all the satellites in the constellation are identical. If not, they should be treated as unique individuals in the diagrams. The purpose of SV-1 is to identify all the interfaces among all the system nodes; here, they are numbered. If a software tool were to be used to build and track the evolution of the architecture, the interfaces would also be numbered as they were entered into the tool's database. SV-1 leads to SV-2 (not shown), which is a similar diagram that shows the type of communication that occurs on each interface line, such as radio frequency band or network type. Ultimately, a system data exchange matrix (SV-6) is constructed that contains all the pertinent information that can be identified or decided during architecture development that passes over each interface. Included for each interface in the matrix is the needline and operational node information (obtained from the OV-2) that passes over the interface. It is from this matrix that the system engineers can extract the requirements for the interfaces.

15.7 MODELING, SIMULATION, AND SYSTEM ENGINEERING OF COMMUNICATION AND NAVIGATION NETWORKS

The roles of architecture, system and network modeling, simulation, and emulation for communication and navigation networks for Space SoS are becoming prominent prior to their implementation (Endres, et al., 2004). In most of systems of systems, the communication and navigation network consists of legacy, terrestrial Internet, and modified networks. To interoperate these networks with the various systems, extensive modeling and simulation are needed to save cost and increase reliability. The modeling and simulation processes are rapidly evolving for the communications and navigation networks needed for Space SoS based on methodologies and tools that are available for space and terrestrial communication and navigation network currently deployed.

A top-level view of the architecture to modeling to simulation process for the communications and navigation networks for Space SoS is shown in Fig. 15.24. The use of an architecture framework for the communications and navigation networks allows the creation of the architecture definition documents that contain diagrams and tabular data sets needed as input for the architecture modeling and analysis. A number of modeling tools are available to perform the design trade-off and performance analysis. At this stage, data traffic models and analysis are needed for the data flow among the systems of systems for optimization and to meet the constraints of the communications and navigation network, especially for the legacy space communications and navigation networks. IP network centric communications and navigation networks are based on layered network architecture adding another level of complexity. The network simulation and emulation tools based on space and terrestrial protocols are being developed and used. Using the current Internet technologies, distributed simulation test beds are being developed for the Space SoS during the formulation design phases.

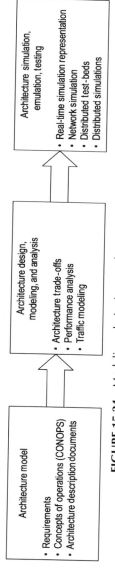

FIGURE 15.24 Modeling, designing, analyzing, and testing the architecture

Architecture model

- Requirements
- Concepts of operations (CONOPS)
- Architecture description documents

Architecture design, modeling, and analysis

- Architecture trade-offs
- Performance analysis
- Traffic modeling

Architecture simulation, emulation, testing

- Real-time simulation representation
- Network simulation
- Distributed test-beds
- Distributed simulations

15.8   SUMMARY

Communication and navigation networks are essential elements of Space SoS. Their level of involvement can range from a few space links to a complex infrastructure or system of systems. The challenge to the communication and navigation networks is to make them highly reliable, interoperable, easy to use, operationally simple, and easy to manage. Additional challenges are presented below.

15.8.1   System Engineering Challenges

Success factors for communication and navigation networks for Space SoS will require advanced planning and extensive formulation. System of system engineering methodologies and tools will be equally applicable to development of communication and navigation networks. During the architecture and requirements definition phases of the Space SoS program, advanced architecture frameworks are used to study architecture trade-offs for making key architectural decisions based on identified driving requirements, validation of requirements feasibility, trade studies, performance optimization, and risk reduction by simulating the operational scenarios. System engineering tools to manage requirements and architecture definition are key in the automation of system engineering functions, such as requirement flowdown, test planning, risk management, and functional and operational flow decompositions.

The next step is to pursue an aggressive plan to develop and use a rich suite of integrated modeling and simulation processes and tools, a suite of system engineering and integration tools, and a distributed network of testbeds in multiple system integration laboratories. With effective integration of the development environment and tool fidelities, the modeling and simulation tools can be useful in all phases of implementation, including architecture definition, requirement decomposition, design, test definition, test validation, training, and resulting operations. During the test phase, modeling and simulation tools integrated with an emulation test bed environment can cost-effectively reduce program risk by verifying and validating operational test scenarios, identifying early integration and validation risk areas, performing metric collection, and refining details for concept of operations (CONOPS) and operational procedures.

15.8.2   Development and Operational Challenges

Integration of legacy space communication networks with IP networks, in addition to design of new networks, will require management and operational reorganization. The transformation of network management from scheduling, monitoring, and control to an access on demand paradigm will require new techniques and approaches.

15.8.3   Technology Challenges

The technology challenges include the introduction of autonomous operations among the nodes of the new Space SoS and the implementation of inexpensive, standards-compliant network interface radios and networks devices within those nodes.

REFERENCES

Bhasin, K., Hayden, J., 2001, Inter-Spacecraft Communication Architectures and Technologies for Coordinated Spacecraft Missions. *AIAA 2001 Space Conference and Exposition,* August 28–30, 2001 Albuquerque, NM. Available from the AIAA 2001-4709.

Bhasin, K., Hayden, J., 2002, Space internet architectures and technologies for NASA enterprises, *International Journal of Satellite Communications,* 20(5): 311–332 (DOI:10. 2002 sat. 727).

Bhasin, K., Linsky T., Hayden J., Tseng, S., 2005, Surface Communication Network Architectures for Exploration Missions AIAA Space 2005, Long Beach. Available from the AIAA-2005-6695.

Bhasin, K., Hackenberg, Anthony W., Slywczak, R., Bose, P., Bergamo, M., Hayden, J., 2006, Lunar relay satellite network for space exploration: architecture, technologies and challenges, *Proceedings 24nd AIAA International Communications Satellite Systems Conference,* June 12–14, 2006, San Diego, CA. Available from the AIAA-2006-5363.

Bhasin K., Putt C., Hayden J., Tseng S., Biswas A., Kennedy B., Jennings E., Miller R., Hudiburg J., Miller D., Jeffries A., Sartwell T., 2007, Architecting the communication and navigation networks for NASA's space exploration systems, *2007 IEEE International Conference on System of Systems Engineering,* April 16–18, 2007, San Antonio, TX. Available from the 2007 IEEE 1-4244-1160-2/07.

DoDAF V. 1.0, 2003, DoD Architecture Framework Version 1.0, the Deskbook and Volumes I & II, Department of Defense Architecture Framework Working Group. Superceded by DoDAF V. 1.5

DoDAF V. 1.5, 2007, DoD Architecture Framework Version 1.5, Volume I, II and III, Department of Defense Architecture Framework Working Group. Available at https:// dars1.army.mil/IER/index.jsp. Accessed Dec 1 2007.

Elwell D.W., Levine A.J., Harris D.W., 1992, The tracking and data relay satellite system: an historical perspective, *AIAA 14th International Communication Satellite Systems Conference.,* pp. 92–1882.

Endres, S., Griffith, M., Malakooti B., Bhasin, K., Holtz A., 2004, Spaced-Based internet network emulation for deep space mission applications, *Proceedings 22nd AIAA International Communications Satellite Systems Conference,* May 9–12, 2004, Monterey, CA. Available from the AIAA-2004-3210.

Hayden, J., 2000, Spacecraft/ground architectures using internet protocols to facilitate autonomous mission operations. *Proceedings of the 2000 IEEE Aerospace Conference, March, 2000.* Available from the IEEE 0-7803-5846-5/00.

Ivancic W., Zernic M., Hoder D., Brooks D., Beering D., Welch A., 1999, ACTS 118X final report high-speed tcp interoperability testing, *5th Ka-Band Utilization Conference.*

Kimberlin D., 2004, Camelot on the Moon, *Radio Guide,* 7(12): 4, 6, 23.

Kucherman H.B., Pritcher W.L., Wall, V.W., 1966, The initial defense communication satellite program, Paper 67–267, AIAA Communication Satellite Systems Conference.

Maier, M., 1998, Architecting principles for systems-of-systems, *System Engineering,* 1(4): 267–284.

Martin D.H., Anderson P.R., Bartamian L., 2007, *Communications Satellites,* 5th Edition, Aerospace Press, California.

NASA, 2005, History of the Deep Space Network. Available at http://deepspace.jpl.nasa.gov/dsn/history/index.html. Accessed Dec 1, 2007.

NASA, 2006, AIAA 2nd Space Exploration Conference, Dec 4–6, 2006, Houston, TX. Proceedings are available at http://www.nasa.gov/mission_pages/exploration/main/2nd_exploration_conf.html. Accessed Dec 1, 2007.

Pierce J. 1990, ECHO — America's First Communications Satellite, SMEC Vintage Electrics Volume 2 #1.

Sand O.S., Bhasin K., Hayden J., 2002, Relay Station Based Architectures and Technology for Space Missions to the Outer Planets, May 12–15, 2002, AIAA, Montreal, Canada., Available at the AIAA-2002-2066.

Soloff J., Israel D., Deutsch L., 2005, A sustained proximity network for multi-mission lunar exploration, *AIAA 1st Space Exploration Conference,* Jan 30–Feb 1, 2005, Orlando, FL, Available from the AIAA-2005-2505.

The White House, 2004, President Bush Announces New Vision for Space Exploration Program, Available at http://www.whitehouse.gov/news/releases/2004/01/20040114-1.html. Accessed Dec 1, 2007.

Whelan, D., 2007, Communications Satellites: Making the Global Village Possible. Available at http://www.hq.nasa.gov/office/pao/History/satcomhistory.html. Accessed Dec1, 2007.

Chapter 16

Operation and Control of Electrical Power Systems

PETR KORBA[1] and IAN A. HISKENS[2]

[1]ABB Corporate Research Ltd, Switzerland
[2]University of Wisconsin-Madison, Madison, WI, USA

16.1 BACKGROUND

Modern society is dependent upon a highly reliable supply of electricity. Power system operations must therefore seek to minimize interruptions to consumers, even when the system is subjected to large disturbances. Operation and control strategies are not straightforward though, because electrical energy cannot be stored on a large scale. Generation must always closely match load, even though loads undergo large variations throughout a daily cycle. Many important systems of systems have been developed over the years to ensure the desired level of reliability is achieved.

Power systems began as local, independent entities. Vertically integrated utilities owned the generation facilities, and the transmission and distribution networks, necessary to serve their customers. Over time it became apparent that there were benefits to interconnecting with neighboring utilities. If a generator tripped off-line, for example, the utility could "lean" on its neighbors for the additional power needed to achieve generation–load balance. As a result, power systems have become more and more interconnected.

Operating and control strategies are continually evolving to meet the needs of larger, more tightly interconnected systems and to exploit improvements in computers, communications, and algorithms. A summary of current approaches to system control is provided in Section 16.2. Power systems are facing numerous changes and challenges though. Interconnected systems are growing ever larger, and market

System of Systems Engineering: Innovations for the 21st Century, Edited by Mo Jamshidi
Copyright © 2009 John Wiley & Sons, Inc., Publication

philosophies are altering traditional generating and power flow patterns. Environmental concerns are providing the impetus for much greater utilization of renewable-based generation. Meanwhile, the underlying physical infrastructure is aging. These factors are considered in Section 16.3.

In the face of these changes, power systems must move to more sophisticated monitoring and control strategies. Section 16.4 provides an overview of wide area monitoring and control (WAMC) schemes, Section 16.5 discusses flexible AC transmission system (FACTS) devices, and Section 16.6 reviews trends in the application of various control techniques to power systems. Section 16.7 brings together monitoring and control, to present a number of examples of advanced strategies. Conclusions are provided in Section 16.8, and future challenges are discussed in Section 16.9.

16.2   CURRENT PRACTICES IN OPERATION AND CONTROL OF ELECTRICAL POWER SYSTEMS

Power systems experience many different forms of disturbances, from lightning-induced flashovers to large load fluctuations. To ensure robust response to disturbances, power systems employ a hierarchical control structure.

At the component level, protection schemes provide very fast, local response to faults. Fault detection typically involves fairly simple measurement manipulation and threshold comparisons. The resulting control action is discrete; the protected device is either tripped or remains in service. The system-wide consequences of the control action cannot be taken into account at this level. Without coordination at a higher level, though, the global effects of these local decisions can be significant (Andersson et al., 2005).

At a higher level, local feedback loops are responsible for regulating the active and reactive power delivered by generating sources and for controlling voltages at strategic nodes within the transmission/distribution network, including generator terminal buses and load buses. Generator governors measure frequency deviation and adjust the turbine mechanical power according to a droop characteristic of the form shown in Fig. 16.1. System dynamics ensure that this process provides appropriate regulation of the electrical power delivered to the network.

Bus voltages are regulated by numerous different types of devices, including generators, transformers, and static var compensators (SVCs). In all cases, the controller seeks to minimize the error between the measured voltage and its set-point value. Often droop characteristics, similar to Fig. 16.1, are used to ensure an equitable sharing of the reactive support role among sources.[1]

Oversight of power system operations is achieved through regional control centers and "independent system operators" (ISO). Supervisory control and data acquisition (SCADA) systems acquire measurements of key system quantities, such as generator outputs, bus voltages, transmission line flows, and distribution substation loads.

[1]In this case, though, the droop characteristic describes the relationship between reactive power along the horizontal axis and voltage set oint on the vertical axis.

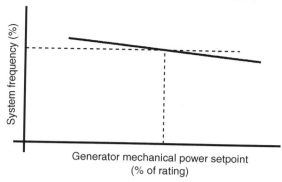

FIGURE 16.1   Governor droop characteristic

The SCADA system provides a two-way flow of signals. It allows operators to monitor system performance and also communicates operator actions back to appropriate control devices within the system. Typical operator actions could range from opening/closing circuit breakers to remote control of hydropower plants.

SCADA systems are usually linked with an energy management system (EMS), allowing operators to assess system security and economic performance. A state estimator filters the SCADA measurements to provide a consistent estimate of all system variables (Monticelli, 1999; Abur and Gómez Expósito, 2004). In most cases, the filtering process is formulated as a weighted nonlinear least squares problem and solved using a Gauss–Newton algorithm.

The state estimator output forms the basis for a power flow model of the system (Monticelli, 1999). An automated process evaluates the likely system response for a range of credible contingencies, such as line or generator outage scenarios. It is also common for the state estimator model to be linked with load and generation forecasts, to predict response to contingencies in the near future. If this contingency evaluation process identifies system vulnerabilities, operators must decide whether to accept the risk or improve system security by adjusting appropriate set points.

An increasing number of EMSs also incorporate an optimal power flow (OPF) (Wood and Wollenberg, 1996), an optimization formulation that embeds the power flow model within the constraint set. The objective of an OPF may take different forms, with numerous implementations minimizing overall system losses, while others minimize the reactive power output of sources such as generators or SVCs. The OPF determines optimal values for operational parameters, such as set points for voltage regulating devices. These optimal values may be telemetered directly to the actual devices (Van Cutsem and Vournas, 1998) or may be used by operators as guidance in tuning the system.

Load frequency control (LFC) has played a vital role in power system operations for many years (Wood and Wollenberg, 1996). LFC is a decentralized control strategy that relies upon decomposition of the power system into control areas. The control areas often have historical significance, being closely aligned with the boundaries of the original vertically integrated utilities. Within each control area, an area control error

(ACE) signal is calculated from the frequency deviation and the net unscheduled power interchange with neighboring areas. The ACE signal for each area drives the adjustment, via integral control, of power set points for generators within the area. LFC therefore restores frequency and tie-line power flows to their scheduled values, with the control effort distributed across all participating generators. LFC is often combined with a generation dispatch process[2] to form automatic generation control (AGC) (Wood and Wollenberg, 1996).

16.3   THE CHANGING NATURE OF ELECTRICAL POWER SYSTEMS

Many power systems were originally designed for local power supply needs, but are now expanding beyond state borders. Furthermore, as a result of market deregulation, they are operated in a very competitive environment. The deregulated market has changed business philosophy and transmission network operating policies considerably. In order to maximize profit, wheeling of electrical power from distant generators is becoming a common practice for many power utilities. Competition and deregulation have led to situations where multiple energy producers share the same transmission network, which often lacks the necessary capacity and/or safety margins. This leads frequently to transmission bottlenecks and overloads along the transmission path (AutoSoft, 2006).

A reliable electricity grid is one of the main infrastructures supporting a developed economy and one of the key premises for an economy to become developed. Reliable delivery of electricity at an acceptable price depends on the electrical transmission and distribution networks that link power generation with consumers. Electric power supply has become so important for the entire society that every effort must be made to prevent power systems from collapse scenarios that may lead to power system black-outs. A number of developments are driving changes in electrical power system grids. The following list provides an indication of some of the changes that should be taken into account in operation and control of modern power systems (Korba et al., 2005).

Interconnecting electrical grids is a common trend, with the European connection to eastern grids providing a topical example. Such interconnections can lead to synchronization difficulties that may have a negative effect on network angular stability. Undesired, or even unstable, power oscillations can directly affect system security.

Deregulated power markets are leading to heavily loaded transmission lines, as large amounts of power are transported over long distances. Electricity does not even have to be produced in the country where it is consumed. It gives rise to anomalous situations where weak lines built in the past become major corridors. As a result, undesirable power flows and oscillations occur.

Growing concern for *environmental issues* is reflected by the opposition of the general public to overhead lines, the pressing CO_2 problems, and phasing out of

---

[2]Traditionally, economic dispatch was employed. The move toward electricity markets is bringing about changes though, with a variety of approaches emerging.

nuclear power in some countries. This leads to more distributed, renewable-based power generation. As a consequence, a greater proportion of the generation mix has lower power ratings and is connected at lower voltage levels.

Distributed generation is seeing growth in two important areas: (1) backup generation embedded within existing distribution networks and (2) renewable generation. In the latter case, power production is typically some distance from the major loads. Hydro, solar, or wind power generation plants may be built in locations with weak connections to the transmission or distribution grids. Power production depends on weather conditions, and so may be highly variable. As a consequence, some form of compensation (energy storage) may be necessary to secure the power supply to consumers. This leads to further changes in power generation and flow patterns in electrical power system grids.

Load control has been available in various forms, such as demand-side management and underfrequency load shedding (UFLS), for many years. Advances in communications allow much more precise control though. In fact, it is possible to target individual loads that can be temporarily switched without causing any disruption to consumers. Examples include air-conditioning compressors and numerous home appliances (California Energy Commission, 2003a,b). Consolidation of these many small loads yields a nondisruptive load control resource that can effectively influence overall system behavior (Hiskens and Gong, 2005).

Aging grid infrastructure is an issue that is highlighted by deregulation of the power market. Even if new power system components are used to replace older ones, they are typically based on different technologies, for example power electronics, with different impacts on the physical characteristics of the electrical grids.

As a result of these changes, power systems tend to be operating closer to their limits. Dynamics and nonlinearities are having a greater influence on system behavior, and consequently, systems are exhibiting increased sensitivity to disturbances. Therefore, the need to manage grid security and power flows, without adequate grid expansion, will likely lead to greater use of high-impact controllers. Unless carefully tuned, interactions could substantially degrade overall performance. The original electrical grids were planned and built without taking the above considerations into account. Future expansion strategies will require rethinking of power system operations, as well as development of new technologies that are able to cope with the changing conditions. The basic objectives remain to control the flow of power from producer to consumer and maintain a stable grid. This refers to the need for maintaining an appropriate voltage profile (voltage stability), frequency within the nominal range (frequency stability), and synchronism between all generators connected to the same AC network (angle stability).

16.4 WIDE AREA MONITORING AND CONTROL

A promising and challenging approach to ensuring system security and reliability is to provide system-wide protection and control that complements conventional local equipment and SCADA/EMS. While it is virtually impossible to predict and prevent all contingencies that may lead to power system collapses, a wide area monitoring and control system can provide reliable security predictions and optimized coordinated actions to mitigate or prevent large area disturbances. The main tasks that are currently being addressed include early recognition of instabilities, increased power system availability, operation closer to limits, increased power transmission capability with no reduction in security, better access to low-cost generation, fewer load shedding events, and minimization of the amount of load shedding. The main disadvantages of common monitoring and control systems lie in their inadequate view of the system state and in uncoordinated local actions of decentralized protection and control systems. The solution is to complement SCADA/EMS with a dynamic measurement system using synchronized Phasor Measurement Units (PMU) (Phadke et al., 1986). Together with stability assessment and stabilization algorithms, such a system is called a WAMC system.

PMUs allow measurement of voltage and current phasors, together with a satellite triggered time stamp, in time intervals of a few milliseconds. This is illustrated in Fig. 16.2. Recently developed WAMC systems that address diverse phenomena will be discussed in Section 16.7. The core idea of WAMC systems is the centralized processing of data collected from various locations across a power system and evaluation of the actual power system operating conditions with respect to stability limits. Although the particular application range of WAMC is quite wide, the fundamental structure remains the same. From the hardware point of view, there are

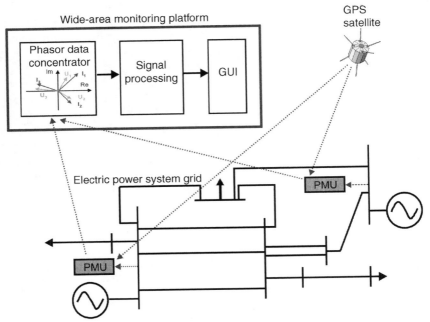

FIGURE 16.2 Wide area monitoring of electrical power systems

three stages of data handling: data acquisition, data delivery, and data processing, with the latter stage incorporating algorithms for monitoring, protection, and control.

To properly observe overall system dynamics, the required measurements can be characterized as follows: they must be taken from various network locations, with sufficiently high sampling rates and synchronized sampling. Today, the only feasible way of ensuring synchronization of many devices that are geographically distant is to use Global Positioning System (GPS) time synchronization signals. This is the concept employed in PMUs. Their outputs are collected centrally in a phasor data concentrating platform (PDC), without need for communication between the units themselves. For this reason, they are ideal basic components of the centralized WAMC structure. In addition, they have the capability of providing various types of data, and of preprocessing that data.

The quantities to be measured by PMUs, and the locations for installation of the PMUs, are dependent on many factors. The most significant criteria are the instability phenomena that should be mitigated, followed by practical economic aspects such as utilization of existing communication infrastructures. Transmission of the measured data, and of control commands in the opposite direction, may require dedicated communications channels, particularly if WAMC command execution is time critical. Otherwise, the use of available TCP/IP networks in substations is recommended (Zima et al., 2005).

16.5 FLEXIBLE AC TRANSMISSION SYSTEMS

A fundamental characteristic of alternating current (AC) electrical power systems is the impossibility of controlling power flow along a specific "contract path" from a generator to a (possibly geographically distant) load. The flow of electricity over each path is determined by the physical characteristics of the transmission lines. Therefore, "loop flows" may take power far away from the most direct route between a generator and the load center and affect numerous interconnected utilities, even though they have no interest in the intended transaction.

For any transmission line, three key parameters determine the power flow: terminal bus voltages, line impedance, and the phase angle between the sending and receiving ends of the line. Figure 16.3 shows a simplified two-bus transmission line. The bus voltage magnitudes at the line ends are V_1 and V_2, and the angle across the line is $\delta = \delta_1 - \delta_2$. The line is assumed to have inductive impedance X_{12}, and the line's resistance and capacitance are ignored (Kundur, 1994). In this case, the expression for active power flow can be written as.

$$P_{12} = \frac{V_1 V_2}{X_{12}} \sin\delta$$

Figure 16.3 provides useful mathematical insights into the techniques available for network-based power system control. Voltage magnitudes cannot be varied significantly, since they must be kept within regulated limits, normally ±5–10%,

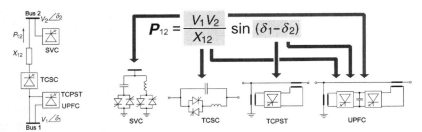

FIGURE 16.3 Principles of flexible AC transmission systems—examples of FACTS devices and their influence on physical parameters of electrical power systems (SVC—static var compensation, TCSC—thyristor controlled series compensation, TCPST—thyristor controlled phase shifting transformer, UPFC—unified power flow controller)

providing very limited scope for power flow control. The line reactance and the voltage angle difference are not bound by such restrictions though, and so provide the only practical alternatives for power flow control. Present approaches to modifying these parameters use a variety of mechanically switched devices, such as phase shifting transformers or series capacitors. They have limited control capabilities and high maintenance costs, however, and do not respond quickly to changing operational conditions. Fortunately, recent developments in power electronics and semiconductor technologies have enabled a new family of devices, which have acquired the common name FACTS. These devices allow control of power flows over desired paths, allowing vastly improved utilization of transmission line thermal capacity.

The basic idea of flexible AC transmission systems is depicted in Fig. 16.3. There are two general groups of FACTS devices, categorized according to whether they are series or shunt connected. Figure 16.3 illustrates how different FACTS devices influence transmission line quantities, and hence affect power flow. Installation of multiple FACTS devices offers unparalleled opportunities for system-wide power flow control and congestion management. If poorly tuned, however, their control actions may potentially have negative effects, which could reduce system security. These issues have been the focus of considerable research and development activity in the past decade. Theoretical analysis and practical investigations have demonstrated the feasibility of a range of FACTS applications, including damping critical interarea oscillations and optimally regulating power flows in electrical grids. These investigations have shown that coordinated control of multiple FACTS devices is effective in managing system congestion in a continuously changing deregulated environment. More details about recent achievements can be found, for example, in Oudalov et al. (1999), Noroozian et al. (2004), AutoSoft (2006), Chaudhuri et al. (2006), Majumder et al. (2006), and Oudalov et al. (2006).

16.6 TRENDS IN CONTROL OF ELECTRICAL POWER SYSTEMS

In response to large disturbances, heavily stressed systems tend to exhibit quite complicated behavior. Operation of power systems closer to limits will therefore

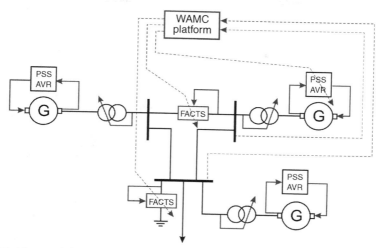

FIGURE 16.4 While conventional controls in power systems are based on single independent local feedbacks, future trends go toward processing of multivariable remote feedback signals

require increasingly sophisticated control schemes. This will entail a transition from the current reliance on predominantly local-measurement-based control to schemes that make much greater use of remote measurements, see Fig. 16.4. Coordination of conventional controllers, such as automatic voltage regulators (AVRs) and power system stabilizers (PSSs), with FACTS devices will be vital for such wide area control. A range of control techniques offer possibilities for addressing the control needs of future power systems.

Multivariable control underlies the development of controllers that use multiple, possibly remote, measurements, such as voltages, currents, and line flows. It provides a framework for assessing the relative benefits of multiple measurements (Sadikovic et al., 2006).

Hierarchical control refers to multilayered control structures that involve distributed decision making, see Fig. 16.4. Higher levels are typified by access to widespread information and decisions that coordinate lower levels. Lower levels access local information and implement the commands that are received from higher levels.

Multiobjective control provides a framework for coordinated design of control laws that reflect conflicting objectives. For example, improvements in power flow control may raise the possibility of poorly damped oscillations.

Adaptive control provides a systematic mechanism for controller retuning as system conditions evolve with time.

Model predictive control relies on an approximate model of the system to predict dynamic behavior and to formulate an optimal open-loop control sequence. The initial instance of this sequence is applied to the system. The resulting system response is subsequently measured, and the prediction/control-application process is repeated.

This repetition provides a feedback mechanism that compensates for model error (Camacho and Bordons, 2004; Venkat et al., Forthcoming). Predictive control offers strategies for maneuvering systems through highly disturbed conditions (Hiskens and Gong, 2004).

16.7 NEW APPROACHES AND OPPORTUNITIES

A prerequisite for the realization of the described monitoring and control algorithms is the existence of WAMC platforms. Such platforms preprocess and concentrate all the necessary signals that are typically obtained from PMUs that are distributed throughout the entire power system. Recent pioneering work has led to the production of the first WAMC platforms (Zima et al., 2005; Leirbukt et al., 2006). Current WAMC experience relates predominantly to wide area monitoring (Larsson et al., 2007). Practical demonstrations have shown that applying advanced algorithms to remote measurements enables monitoring of phenomena such as the frequency and damping of electromechanical oscillations, line thermal overloads, and voltage instability (Larsson et al., 2002; Zima et al., 2005; Leirbukt et al., 2006; Korba, 2007). This was not possible prior to WAMC technology. Higher quality information is now available to operators. That assists in decision making, but ultimately the decisions are still the responsibility of human operators. The next logical step is to route this additional information directly to fast-acting power system controllers. Initial research has demonstrated the benefits of making full use of wide area information. Powerful new approaches to damping electromechanical oscillations, and to voltage and power flow control, have been developed and evaluated in Oudalov et al. (1999), Rehtanz et al. (2002), Werner et al. (2003), Zima et al. (2003), Noroozian et al. (2004), Sadikovic et al. (2004), Oudalov and Korba (2005), Chaudhuri et al. (2006), Majumder et al. (2006), Oudalov et al. (2006) Sadikovic et al. (2006) and Korba et al. (2007).

16.7.1 Examples

The nature of instability phenomena and the power system topology properties can be used to categorize the WAMC algorithms, which have been developed so far and published, into two groups: (a) *algorithms requiring full observability* of the network (e.g., frequency instability assessment or voltage instability assessment of meshed networks) and (b) *algorithms not requiring entire network observation* (e.g., assessment of power oscillation, voltage instability assessment of transmission corridors, or line temperature monitoring). Wide area control of FACTS can make use of all monitoring algorithms. The first group of monitoring algorithms needs, as a prerequisite, almost the complete network observation, which can be expressed in terms of the following data availability/knowledge: network topology, all voltage magnitudes and angles, all branch current magnitudes and angles, and current magnitudes and angles of selected loads and generators (Zima et al., 2005). This is achieved by the application of multistage linear state estimation using the measured PMU data.

The basic principles are described in Phadke et al. (1986) and Rehtanz et al. (2002). Other algorithms are briefly explained in the following subsections.

A part of a real power system is shown in Fig. 16.5. It will be used here to describe some wide area algorithms and methodologies. Note that in the considered power system, the northern (upper) part represents the area with the dominant generation,

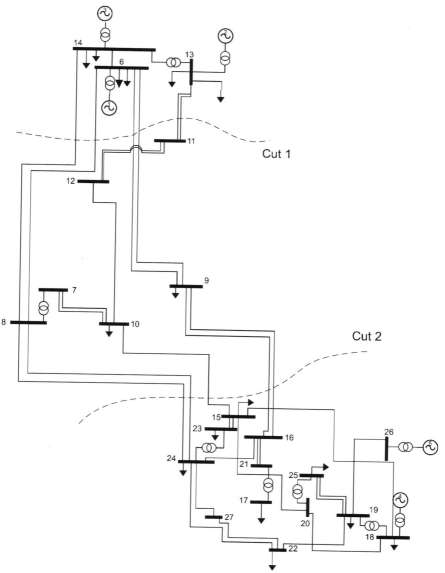

FIGURE 16.5 Example showing a part of an existing power network with a transmission corridor enclosed by transfer cuts

whereas the southern (bottom) part has dominant consumption. For this reason, transmission lines connecting these two areas are subjected to a heavy power flow loading. This leads to a high sensitivity of this power system to any outage and the occurring problems include almost all types of known instabilities often having severe consequences in the form of partial or even system wide blackouts. The installation of a WAMC system could significantly reduce this risk as demonstrated in examples described in the next sections.

16.7.2 Frequency Stability Assessment

Frequency stability is a major concern in the operation of power systems. Following severe disturbances, such as an outage of a large generation station or an interconnection to a neighboring system, the average frequency will drop in the system. Unless the frequency drop is arrested before the frequency reaches 47–48 Hz in a 50 Hz system (or 57–58 Hz in a 60 Hz system), thermal units will be tripped since their turbines might be damaged from prolonged underfrequency operation. A blackout is then highly likely. It can therefore be necessary to disconnect loads to preserve system integrity. UFLS is the most widely used protection against frequency instability. Typically, load is shed based on a local frequency measurement in several steps of 5–20% each. Typical threshold values are 57–58.5 Hz for 60 Hz systems or 48–48.5 Hz for a 50 Hz system. Usually, there is also a time delay intended for a noise rejection. The drawback of these types of relays is their slow response since they have to wait until the frequency is already low before ordering load shedding and since overshedding is commonly ordered by these relays.

It is well known that the power mismatch ΔP following the outage of a generator or a tie line can be estimated from the initial rate of change of the frequency $d\omega/dt$ and the system inertia constant H_{system} according to the following formula

$$\Delta P = 2H_{\text{system}} \frac{d\omega}{dt}$$

This value can be used as an indication of the amount of load that has to be shed to arrest further frequency decline. Several papers have proposed UFLS relays based on this measurement. For example, in Anderson and Mirheydar (1990, 1992) and Huang and Huang (2000) the so-called adaptive UFLS has been presented. The approaches use only local measurements and can avoid the slow response associated with the conventional approach. The disadvantage of these approaches is a lack of coordination resulting in the fact that it is difficult to tune them in such a way that overshedding is avoided. A further drawback of both types of relays is that their tuning depends on off-line assumptions of load response to frequency deviations and of the remaining system inertia. Since these vary from disturbance to disturbance, the relays must be tuned for the worst-case scenario and may take excessive control action resulting in overfrequency as well.

The proposed approach is based on a predictive method for frequency stability and wide area phasor measurements and avoids the drawbacks of conventional relays.

The control actions are based on online measurements instead of conservative off-line assumptions. A single-machine equivalent model is formed online and used for the monitoring and control of frequency stability. This model is then used for monitoring frequency stability and for calculating the amount of load or generation shedding that is required to maintain the frequency within acceptable bounds. This method is also applicable for generation-rich islands, where overfrequency might be a problem. In this case, the control action reduces generation instead of load. The following differential equations determine the frequency dynamics:

$$\frac{d\omega_{avg}}{dt} = \frac{1}{2H_{system}}(P_m - P_e)$$

where

$$H_{system} = \sum_{i=1...N} H_i, \quad P_m = \sum_{i=1...N} P_{m,i}, \quad P_e = P_{loss} + \sum_{i=1...M} P_{l,i}$$

for N generators, M load buses, and active power losses P_{loss}. The average frequency of generators is ω_{avg}. P_m and P_e are the total mechanical power delivered to the generator shafts by the turbines and the total electrical load (the sum of all active power losses and loads P_l on M busses) on the generators, respectively. The linearized dynamics of a single-machine model can be described using state–space equations as

$$\frac{d\Delta x}{dt} = A_{ode}\Delta x + B_{ode}\Delta k + E_{ode}\Delta d$$

$$\Delta y = C_{ode}\Delta x$$

where Δx is the dynamic state vector (containing the average frequency ω_{avg}), Δk is an external input modeling load shedding, and Δd is a disturbance input modeling, for example, generator tripping. With the matrices A_{ode}, B_{ode}, C_{ode}, and E_{ode} containing the corresponding sensitivity coefficients of the underlying ordinary differential equation (ODE), the power mismatch in the system can now be calculated as below:

$$\Delta d = 2H_{system}\frac{d\omega_{avg}}{dt}$$

This power mismatch will change due to load frequency and voltage sensitivity as well as applied control actions. The quantity $d\omega_{avg}/dt$ can be obtained from measurements and H_{system} is a parameter to be estimated by the wide area monitoring system (Scholtz et al., 2007). The calculated power mismatch thus corresponds to the amount of generation and load lost at the instant of disconnection. Using this power mismatch, the predicted steady-state output (for $s = 0$) can be calculated as in the following:

$$\Delta y^* = C_{ode}(sI - A_{ode})^{-1}(B_{ode}\Delta k + E_{ode}\Delta d)$$

Subsequently, the predicted steady-state frequency y^* is obtained as $y^* = y_0 + \Delta y^*$. This value of steady-state frequency is used as a frequency stability indicator—if the steady-state frequency is unacceptable, corrective controls such as load shedding must be applied.

16.7.3   Voltage Stability Assessment in Transmission Corridors

This method extends ideas used for undervoltage load shedding (UVLS) relays that detect voltage instability typically using only local measurements. If voltage becomes abnormally low (below a threshold value), voltage instability is assumed to be present. In response, load is disconnected (typically using a certain time delay) until voltages return above their preset threshold values. This technique is referred to as UVLS and is the most widely used technique for preventing voltage instabilities in high-voltage networks. The drawback of this approach is that the voltage threshold must be set in advance, and thus the relay cannot adapt to changing operating conditions in the power system. Note that voltage can be fairly close to its nominal value even though the system is approaching its maximum loadability. Therefore, voltage level alone cannot be used as a reliable indicator of proximity to the maximum loadability. The main idea of the approach discussed here is to use remote measurements from both ends of the transmission corridor and to split the assessment into two stages. First, the parameters of a T-equivalent of the transmission corridor are determined through a direct calculation (and therefore without the time delay of local approaches such as Balanathan et al. (1998). Second, the Thevenin equivalent of the feeding generators is computed. The considered equivalents are shown in Fig. 16.6. Its main advantage over the traditional approaches is that parameters of the equivalent network can be directly computed from phasor measurements, and thus very precise and without any time delay.

When a WAMC system installation with complete observability is not available but the power system network has clearly defined corridors (in practice, it is often the case as shown in the realistic example in Fig. 16.5), the described voltage instability assessment can still be applied. As an example, the network considered in Fig. 16.5 has a longitudinal shape with a clearly defined generation area in the northern end above cut 1. Generation in that area is about 2440 MW and the local load is about 578 MW. The area also contains shunt reactors rated at a total of 275 MVAr. At the southern end is a load area with about 2159 MW of load and 700 MW of local generation. In between the cuts is a transmission corridor with five parallel transmission lines, but also about 300 MW of load. To apply the transmission corridor voltage instability assessment to this network, the five parallel lines are

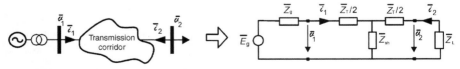

FIGURE 16.6   T- and Thevenin equivalents of transmission corridor and generation

modeled as a single corridor equivalent. Corridor boundaries are defined as the buses neighboring the two cuts on either side of the corridor. After identifying the boundaries given by the two transfer cuts, one can define two virtual buses; one for each end of the transmission corridor. Buses 6, 13, and 14 of the original system are grouped into virtual bus 1, and buses 24, 15, and 16 into virtual bus 2. The part of the system between cuts 1 and 2 becomes the virtual transmission corridor. At least one voltage in the area of each virtual bus and the currents on each line crossing the cut must be measured. The currents at either end of the virtual transmission corridor can be calculated as

$$\bar{i}_i = \left(\frac{p_{\text{cut}-i} + jq_{\text{cut}-i}}{v_i} \right)^*, \quad i \in 1,2$$

For example, here $p_{\text{cut}-i}$ and $q_{\text{cut}-i}$ refer to the sum of the power transfers through cut i and v_i to the average of the voltages of the buses included in virtual bus i.

Figure 16.7 shows the simulation results where the two lines between buses 9 and 16 are tripped at 20 and 30 s, respectively. The latter makes the system unstable and eventually the system collapses at about 82 s. The top-left plot shows the calculated voltages of the two virtual buses, and the top-right plot illustrates the so-called *PV*- curve, where the maximum power transfer is given by.

$$p_{\text{Lmax}} = \Re \left[\bar{Z}_{\text{th}} \left| \frac{\bar{E}_{\text{th}}}{2\bar{Z}_{\text{th}}} \right|^2 \right]$$

where

$$\bar{E}_{\text{th}} = \bar{v}_2 \frac{\bar{Z}_{\text{th}} + \bar{Z}_{\text{L}}}{\bar{Z}_{\text{L}}} \quad \text{and} \quad \bar{Z}_{\text{th}} = \frac{\bar{Z}_{\text{T}}}{2} + \frac{1}{\frac{1}{\bar{Z}_{\text{sh}}} + \frac{1}{\bar{Z}_{\text{T}}/2 + \bar{Z}_{\text{g}}}}.$$

The rightmost point (nose) of this curve corresponds to the stability boundary where $k = k_{\text{crit}}$ (or equivalently, $\bar{Z}_{\text{th}} = \bar{Z}_{\text{L}}$).

In this example, the stability boundary is crossed at about 60 s, and the collapse of the system progresses rapidly after this point. The two lower plots show the estimated Thevenin impedance $\bar{Z}_{\text{th}}$ and voltage $\bar{E}_{\text{th}}$ as well as the estimated load impedance $\bar{Z}_{\text{L}}$. In the bottom-left plot, the stability boundary is given by the intersection of the curves for the Thevenin and load impedances. Using directly only the measured phasors $\bar{v}_1, \bar{i}_1$ and $\bar{v}_2, \bar{i}_2$, the required impedances $\bar{Z}_{\text{T}}, \bar{Z}_{\text{sh}}$ and $\bar{Z}_{\text{L}}$ are given by

$$\bar{Z}_{\text{T}} = 2 \frac{\bar{v}_1 - \bar{v}_2}{\bar{i}_1 - \bar{i}_2}$$

$$\bar{Z}_{\text{sh}} = \frac{\bar{v}_2 \bar{i}_1 - \bar{v}_1 \bar{i}_2}{\bar{i}_1{}^2 - \bar{i}_2{}^2}$$

$$\bar{Z}_{L} = \frac{\bar{v}_2}{-\bar{i}_2}$$

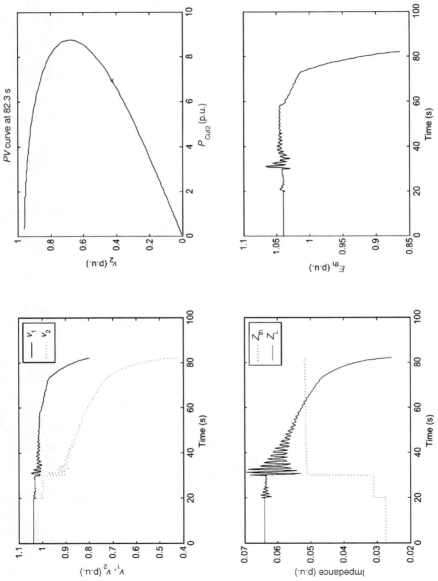

FIGURE 16.7 Simulation results for tripping of the two lines between buses 9 and 16 at 20 and 30 s, respectively

However, the complex voltage $\bar{E}_g$ and the impedance of the equivalent voltage source $\bar{Z}_g$ cannot be calculated in the same straightforward manner. To avoid a time delay in the estimation procedure, one of them is assumed to be known. If the generators have voltage controllers operating within their limits, $\bar{E}_g$ can be assumed constant and $\bar{Z}_g$ is then calculated as

$$\bar{Z}_g = \frac{\bar{E}_g - \bar{v}_1}{\bar{i}_1}$$

In most practical cases, it is more realistic to assume that $\bar{Z}_g$ is known, since it typically comprises the step-up transformers and short transmission line to the beginning of the transmission corridor. It is therefore possible to calculate the equivalent complex voltage of the generators as $\bar{E}_g = \bar{v}_1 + \bar{Z}_g \bar{i}_1$. In networks where transmission corridors cannot be easily identified (or where generation and load are more evenly distributed), yet another method can be applied (Zima et al., 2005). Hence, in a WAMC environment, load shedding can be based on evaluation of the actual loading point on a PV curve, and its proximity to the point of maximum loadability (distance to the "nose" of the PV curve), or on a sudden drop in the estimated Thevenin voltage $\bar{E}_{th}$ instead of doing an undervoltage load shedding.

16.7.4   Power Oscillation Assessment

Electromechanical oscillations occur in power systems due to a lack of damping torque at the generator rotors. Oscillations of generator rotors induce oscillations of other quantities, such as bus voltages, system frequency, and active and reactive power flows on transmission lines. Depending on the number of oscillating generators and the size of the power network, power system oscillations have been reported in the range of 0.1–2 Hz (Rogers, 2000; Korba et al., 2003; Leirbukt et al., 2006).

In general, power system oscillations are ever present, poorly damped and not dangerous as long as they do not become unstable. The objective is to develop an algorithm for real-time monitoring of oscillations from on line measured signals, that is, to estimate the parameters that characterize the oscillations in terms of frequency, damping, and amplitude. These parameters typically vary with time and can hardly be obtained by watching the measured signals displayed on the screen of an operator station. The algorithm presented here is model-based and utilizes an autoregressive (AR) model

$$y(k) = \sum_{i=1}^{n} a_i(k)y(k-i) + \varepsilon(k)$$

with ε given by $\varepsilon(k) = \hat{y}(k|k-1) - y(k)$.

Figure 16.8 schematically shows the approach developed and productized recently (Leirbukt et al., 2006; Korba, 2007). The power system is excited by a sequence of disturbances represented by the noise e and modeled by a linear AR model with

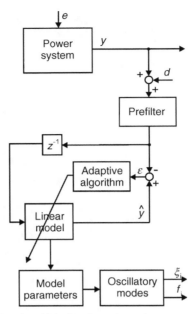

FIGURE 16.8   An algorithm for detection of power system oscillations

adjustable (time-varying) coefficients. The appropriate signal y (the measurement provided by a PMU) is selected either based on results of modal analysis of the power system model or by consulting an experienced operator, in order to ensure high observability of the modes of interest, since the measured signal y may also contain some measurement noise d. An adaptive algorithm recursively optimizes a criterion and yields the AR model parameters that give the best match between the model output $\hat{y}$ and measurement y. The overall goal is to estimate the frequency f_i and the damping ξ_i of the detected oscillations. They are obtained repeatedly, once per given refresh time.

The first step is to estimate recursively the coefficients $a_i(k)$ that minimize the sum of squared prediction errors:

$$J = \min_{a_i(k)} \sum \varepsilon^T \varepsilon = \min_{a_i(k)} \sum (\hat{y}(k|k-1) - y(k))^2$$

where $\hat{y}(k|k-1)$ denotes the predicted value of $y(k)$ based on the measurements available only up to the time instant $(k-1)$. This model captures the information about the time-varying system dynamics, which depends on the operating point of the power system. The poles of the resulting transfer function can be calculated by solving the following characteristic equation for a set of $a_i(k)$ frozen at time k:

$$z^n - a_1(k)z^{n-1} \cdots - a_{n-1}(k)z - a_n(k) = 0$$

It is assumed that the power system remains at an operating point long enough for the estimated coefficients to converge. Indeed, this is not a constraint in practice. For the algorithm presented here, the estimated model parameters converge quickly compared to the dynamics of the power system. The estimated model is converted from the discrete-time domain into the continuous-time domain to obtain the required poles $s_i = \alpha_i + i\omega_i$. For the known sampling period T_s, the relationship between the discrete-time operator z and the continuous-time operator s is given by (bilinear Tustin's approximation) $z = (1 + sT_s/2)/(1 - sT_s/2)$. Complex poles of the continuous-time model characterize the oscillations; the real part α_i describes the decay rate, and the imaginary part $\omega_i = 2\pi f_i$ gives the modal frequency f_i in Hertz. The relative damping, given by

$$\xi_i = \frac{-\alpha_i}{\sqrt{\alpha_i^2 + \omega_i^2}} \, 100(\%)$$

provides a practical measure of system dynamic performance. When a power system is stable, the relative damping lies in the range $0\% < \xi_i < 100\%$ for all values of i.

The optimal model parameters $a_i(k)$ are updated recursively once per sampling period T_s with each new measurement $y(k)$. However, in order to economize computational power, the solution of the characteristic equation, which gives the estimates of modal frequency and damping, can be calculated and displayed for the operator only once per refresh time T_r, where $T_r \gg T_s$.

Kalman filtering techniques (Haykin, 2002) have been employed to identify the optimal model parameters online. The standard set of equations to be solved recursively is

$$g(k) = K(k-1)u(k)[u^T(k)K(k-1)u(k)+Q_m]^{-1}$$
$$\hat{y}(k) = u^T(k)p(k-1)$$
$$\varepsilon(k) = \hat{y}(k) - y(k)$$
$$p(k) = p(k-1) + \varepsilon(k)g(k)$$
$$K(k) = K(k-1) - g(k)u^T K(k-1) + Q_p$$
$$K(k) = \frac{1}{2}\left[K(k) + K^T(k)\right]$$

where k is the actual time (iteration), n is the number of estimated parameters, $u(k)$ are the buffered measurements $u(k) = [y(k-1), \ldots, y(k-n)]$, $y(k)$, $\hat{y}(k)$ are the measured and predicted signals, $p(k)$ is the vector of the estimated model parameters $a_i(k) = [1, -p_1(k), \ldots, -p_n(k)]$, $K(k)$ is the correlation matrix of estimation error $\varepsilon(k)$, $g(k)$ is the Kalman gain, and Q_p, Q_m is the correlation matrix of process and measurement noise.

The model order n is the most important tuning parameter for AR models. Selection of a model order that is too low will result in the spectrum being highly smoothed. However, if the order is too high, fake low-level peaks in the spectrum will be introduced. An effective and systematic method to find the appropriate model order n is to use criteria proposed by Akaike (Proakis and Manolakis, 1987) followed by a trial-and-error procedure. The measured signal is filtered through a digital band-pass filter. If, over a short period of time, the band-pass filtered signal consists predominately of noise (low

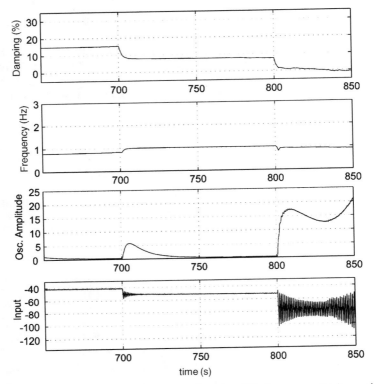

FIGURE 16.9   Estimated dominant relative damping (%), frequency (Hz), amplitude of oscillation, and the measured input signal (angle of current measured with PMU in line connecting buses 19 and 22) subject to the following scenario: normal operation until 700 s, 700–800 s disconnecting one line between buses 9 and 16, and disconnecting the second line between buses 9 and 16 at 800 s

signal power) rather than of realistic data, it is practical to freeze the estimation during this time. Details about tuning and practical experience with this approach can be found in Korba (2007), for example. Besides the estimation of frequency and damping, a simple algorithm has been developed to calculate the running mean value and the actual amplitude of oscillations. The results of this monitoring algorithm, obtained for the test power system depicted in Fig. 16.5 are shown in Fig. 16.9.

From the control point of view, a challenging step is to close the loop. The presented algorithm (or similar) could be used by fast-acting controllers of suitable actuators (such as power system stabilizers, FACTS, or high-voltage DC converters) to improve damping of the detected oscillations (Korba et al., 2007).

16.7.5   Thermal Monitoring

The maximum power flow on a transmission line is often restricted by thermal limits that take account of heating due to high current flow through the conductor resistance.

These limit values are often established off-line and are based on worst-case assumptions, such as high ambient temperature and no wind, that is, no cooling factors. But this is a very rare situation and the limit is often conservative. An online method for monitoring the actual line temperature has been proposed, implemented, and productized (Zima et al., 2005). The principle is very simple and is based on measuring the voltage and current phasors at both ends of the line. Parameters of a model of the transmission line are calculated directly, as in the case of the voltage stability assessment algorithm presented earlier. Using these measurements, variations in line resistance can be determined. When the material properties of the transmission line are known, in particular the thermoresistivity[3], the average temperature of the line can be calculated. This information can be used in a wide area control scheme as an up-to-date upper limit for power flow control, thereby ensuring that the power system is always operated within its true thermal limits.

16.8   CONCLUDING INSIGHTS

A number of insights can be made concerning a system of systems view of electrical power systems. They have attained high reliability through the use of a hierarchical structure. At the lowest level, protection systems use local informa-tion to achieve very fast responses. However, operational decisions at the highest level are based on centralized processing of remote data that are acquired relatively slowly from across the power system that can be entirely spread over a continent. Power systems are, however, experiencing numerous changes that are challenging current control capabilities. These changes include deregulation, ever-larger interconnections, environmental issues, distributed generation, and aging infrastructure.

Wide area monitoring and control offers a promising approach to ensuring system security. The monitoring features of WAMC are reliant upon PMUs for providing fast, synchronized, system-wide data. FACTS technology provides enhanced network-based control capabilities that integrate well with WAMC concepts. Coordination of FACTS controllers with existing feedback controls is vital though.

A number of WAMC applications have been discussed and demonstrated. First installations of WAMC have shown in practice that conventional SCADA/EMS systems can be successfully extended to incorporate dynamic phasor measurements. Power system security has been enhanced through provision of real-time monitoring of key quantities, and by enabling emergency controls. While the first wide area monitoring platforms have already been commercialized, closing the control loops using powerful FACTS devices and remote feedback signals is the subject of ongoing research.

---

[3]The thermoresistivity describes the thermal dependence of the material resistance.

16.9   FUTURE CHALLENGES IN OPERATION AND CONTROL OF ELECTRICAL POWER SYSTEMS

As mentioned previously, numerous changes are impacting power system operation and control. Electricity market deregulation is resulting in less predictable generation and power flow patterns. Accordingly, off-line studies of postulated system conditions are of limited use, placing greater reliance on measurements and online analysis. A further consequence of the altered flow patterns, and of reluctance to construct new transmission assets, is that power systems tend to be operating closer to security limits. Maintaining high reliability will require increasingly sophisticated control strategies, including closing feedback loops that are currently the domain of human operators. WAMC systems that incorporate FACTS devices offer benefits for enhancing system controllability. Such schemes must be carefully integrated though, to prevent destabilizing interactions with existing controls.

Numerous challenges are arising with growth in the use of distributed generation. Coordinated scheduling of many small sources is much more difficult than the traditional dispatching of large generators. This is particularly so for renewable generation, where primary energy source availability is often not well correlated with demand. (For example, wind tends to be most plentiful over night, when load is lowest.) Large-scale energy storage would be extremely useful, but it is not currently technically feasible. Plug hybrid electric vehicles (PHEVs), and other forms of nondisruptive load control, offer some possibilities, but investigations of the required hierarchical control structures have only just begun. Furthermore, standard power system analysis tools are not suited to such a proliferation of controllable devices. New analysis techniques are required for assessing the impacts of many small devices on overall system performance.

REFERENCES

Abur, A., Gómez Expósito, A., 2004, *Power System State Estimation*, Marcel Dekker, New York, NY.

Anderson, P.M., Mirheydar, M., 1990, A low-order system frequency response model, *IEEE Transactions on Power Systems*, 5(3): 720–729.

Anderson, P.M., Mirheydar, M., 1992, An adaptive method for setting under-frequency load shedding relays, *IEEE Transactions on Power Systems*, 7(2) 647–655.

Andersson, G., Donalek, P., Farmer, R., Hatziargyriou, N., Kamwa, I., Kundur, P., Martins, N., Paserba, J., Pourbeik, P., Sanchez-Gasca, J., Schulz, R., Stankovic, A., Taylor, C., Vittal, V., 2005, Causes of the 2003 major grid blackouts in North America and Europe, and recommended means to improve system dynamic performance, *IEEE Transactions on Power Systems*, 20(4): 1922–1928.

AutoSoft: International Journal on Automation and Soft Computing, Special Issue on Intelligent Automation of Power Systems, 12(1) January 2006.

Balanathan, R., Pahalawatha, N.C., Annakkage, U.D., Sharp, P.W., 1998, Under-voltage load shedding to avoid voltage instability, *IEE Proceedings of the Generation, Transmission & Distribution*, 145(2): 175–181.

California Energy Commission, 2003a, Final Report Compilation for Smart Load Control and Grid-Friendly Appliances, Technical Report P-500-03-096-A10, October 2003.

California Energy Commission, 2003b, Final Report Compilation for Aggregated Load Shedding, Technical Report P-500-03-096-A12, October 2003.

Camacho, E., Bordons, C., 2004, *Model Predictive Control*, 2nd edition, Springer, Berlin.

Chaudhuri, B., Korba, P., Pal, B.C., 2006, Design of damping controllers through simultaneous stabilization technique, *International Journal on Automation and Soft Computing*, 12(1): 41–50.

Haykin, S., 2002, *Adaptive Filter Theory*, 4th edition, Prentice Hall.

Hiskens, I.A., Gong, B., 2004, Voltage stability enhancement via model predictive control of load, *Proceedings of the Symposium on Bulk Power System Dynamics and Control VI*, August 2004, Cortina d'Ampezzo, Italy.

Hiskens, I.A., Gong, B., 2005, MPC-based load shedding for voltage stability enhancement, *Proceedings of the 44th IEEE Conference on Decision and Control*, December 2005, Seville, Spain, pp. 4463–4468.

Huang, S.-J., Huang, C.-C., 2000, An adaptive load shedding method with time-based design for isolated power systems, *Electrical Power and Energy Systems*, 22: 51–58.

Korba, P., 2007, Real-time monitoring of electromechanical oscillations in power systems, *IET Generation Transmission & Distribution*, 1: 80–88.

Korba, P., Larsson, M., Oudalov, A., Preiss, O., 2005, Towards the future of power system control, *ABB Review Journal*, (2/05): 35–38.

Korba, P., Larsson, M., Rehtanz, C., 2003, Detection of oscillations in power systems using kalman filtering techniques, *Proceedings of the IEEE Conference on Control Applications*, Vol. 1, June 2003, Istanbul, Turkey, pp. 183–188.

Korba, P., Larsson, M., Sadikovic, R., Andersson, G., Chaudhuri, B., Pal, B., Majumder, R., 2007, Towards real-time implementation of adaptive controllers for FACTS devices, *Proceedings of the IEEE Power Engineering Society General Meeting*, June 28, 2007, Tampa, FL.

Kundur, P., 1994, *Power System Stability and Control*, McGraw-Hill, NY.

Larsson, M., Korba, P., Zima, M., 2007, Implementation and applications of wide-area monitoring systems, *Proceeding of IEEE Power Engineering Society General Meeting*, June 28, 2007, Tampa, FL.

Larsson, M., Rehtanz, C., Bertsch, J., 2002, Real-time voltage stability assessment of transmission corridors, *IFAC Symposium on Power Plants and Power Systems Control*, Seoul, Korea.

Leirbukt, A.B., Korba, P., Gjerde, J.O., Uhlen, K., Vormedal, L.K., Warland, L., 2006, Wide area monitoring experiences in Norway, *Power Systems Conference & Exposition* (PCSE), October–November 1, 2006, Atlanta.

Majumder, R., Pal, B.C., Dufour, C., Korba, P., 2006, Design and real-time implementation of robust FACTS controller for damping inter-area oscillation, *IEEE Transactions on Power Systems*, 21(2): 809–816.

Monticelli, A., 1999, *State Estimation in Electric Power Systems*, Kluwer Academic Publishers, Norwell, MA.

Noroozian, M., Halvarson, P., Rehtanz, C., Korba, P., Preiss, O., Pal, B., Chaudhuri B., Green, T., 2004, Incorporating new development in control theory and wide-area measurement in

SVC function for improving power system performance, *Proceedings of 15th Conference on Electric Power Supply Industry* (CEPSI 2004), October 2004, Shanghai, China.

Oudalov, A., Korba, P., 2005, Coordinated power flow control using FACTS devices, *Proceedings of the 16th IFAC World Congress,* July 4–8, 2005, Prague.

Oudalov, A., Cherkaoui, R., Germond, A., 1999, Coordinated control of multiple FACTS devices, *Proceedings of 11th Power Systems. Automation and Control Conference*, Bled, Slovenia, pp. 71–77.

Oudalov, A., Korba, P., Cherkaoui, R., Germond, A.J., 2006, Fuzzy gain scheduling technique for power flow control, *International Journal of Computer Applications in Technology*, 27 (2/3): 119–132.

Phadke, A., Thorpe, J.S., Karimi, J., 1986, State estimation with phasor measurements, *IEEE Transsactions on Power Systems*, 1(1): 233–241.

Proakis, J., Manolakis, D.G., 1987, *Introduction to Signal Processing*, Macmillan.

Rehtanz, C., Larsson, M., Zima, M., Kaba, M., Bertsch, J., 2002, System for wide area protection, control and optimization based on phasor measurements, *Power Systems and Communication Systems Infrastructures for the Future*, September 23–27, 2002, Beijing.

Rogers, G., 2000, *Power System Oscillations*, Kluwer Academic Publishers.

Sadikovic, R., Andersson, G., Korba, P., 2004, Power flow control strategy for FACTS device, *Proceedings of WAC* 2004, Vol. 15, June 28–July 1 2004, World Automation Congress, Seville, Spain pp. 31–36.

Sadikovic, R., Korba, P., Andersson, G., 2006, Optimal location of FACTS devices for multiple control objectives, *Proceedings on Symposium of Specialist in Electric Operational and Expansion Planning* (X SEPOPE), May 21–25, 2006, Florianopolis, Brazil.

Scholtz, E., Larsson, M., Korba, P., 2007, Real-time parameter estimation of dynamic power system models using multiple observers, *Proceeding of IEEE PowerTech,* July 1–5, 2007, Lausanne, Switzerland, Paper 533.

Van Cutsem, T., Vournas, C., 1998, *Voltage Stability of Electric Power Systems*, Kluwer Academic Publishers, Norwell, MA.

Venkat, A.N., Hiskens, I.A., Rawlings, J.B., Wright, S.J., Distributed MPC strategies with application to power system automatic generation control, *IEEE Transactions on Control Systems Technology*. Forthcoming.

Werner, H., Korba, P., Yang, T.S., 2003, Robust tuning of power system stabilizers using LMI-techniques, *IEEE Transactions on Control Systems Technology*, 11(1): 147–152.

Wood, A.J., Wollenberg, B.F., 1996, *Power Generation Operation and Control*, John Wiley and Sons, New York, NY.

Zima, M., Korba, P., Andersson, G., 2003, Power systems emergency control approach using trajectory sensitivities, *Proceedings of the IEEE Conference on Control Applications,* Vol. 1, June 2003, Istanbul, Turkey, pp. 189–194.

Zima, M., Larsson, M., Korba, P., Rehtanz, C., Andersson, G., 2005, Design aspects for wide area monitoring and control systems, *Proceedings of the IEEE*, 93(5): 980–996.

Chapter **17**

Future Transportation Fuel System of Systems

MICHAEL DUFFY, BOBI GARRETT, CYNTHIA RILEY, and DEBRA SANDOR

National Renewable Energy Laboratory, Golden, CO, USA

17.1 INTRODUCTION

Ready access to affordable oil is the cornerstone of the U.S. economy. In 2004, the United States consumed almost 21 million barrels of crude oil and refined products per day (EIA, 2006)— one-quarter of the world's total crude oil consumption of 84 million barrels per day. Approximately, 60 percent of the U.S. demand was met by imports (EIA, 2006). The transportation sector, which receives nearly all of its energy from petroleum products, accounts for two-thirds of U.S. petroleum use. As President Bush aptly noted in his 2006 State of the Union Address, "America is addicted to oil."

The rest of the world is rapidly following suit. Global demand for petroleum and other fuels is projected to grow from 84 million barrels of oil equivalent per day in 2004 to 97 million barrels per day in 2015 and 118 million barrels per day in 2030 (EIA, 2007). Most of this growth will occur in developing countries, driven both by population growth and by realization of economic development aspirations. But U.S. demand also continues to grow, with petroleum imports expected to top 26 million barrels per day by 2025 (EIA, 2006). In the light of increasing worldwide oil demand, our increased reliance on imported sources of energy threatens our national security, economy, and future competitiveness.

How this growing demand for energy is met poses one of the most complex and challenging issues of our time. The current national energy dialogue reflects the challenge in simultaneously considering the social, political, economic, and environmental issues as the energy system is defined, necessary technological capabilities are established, and programs and investments are implemented to meet those

System of Systems Engineering: Innovations for the 21st Century, Edited by Mo Jamshidi
Copyright © 2009 John Wiley & Sons, Inc., Publication

capabilities. Efforts to change energy infrastructure will require public and private sector commitment and investment over a multidecade period.

17.1.1   First Steps: Federal Goals

In his 2003 State of the Union Address, President Bush announced a $1.2 billion Hydrogen Fuel Initiative to reverse America's growing dependence on foreign oil by developing the technology needed for commercially viable, hydrogen-powered fuel cells—a way to power cars, trucks, homes, and businesses—that produces no pollution and no greenhouse gases. Through partnerships with the private sector, the President's Hydrogen Fuel Initiative seeks to develop hydrogen, fuel cell, and infrastructure technologies needed to make it practical and cost-effective for a large number of Americans to choose to use fuel cell vehicles by 2020 (Whitehouse, 2003).

In 2006, President Bush unveiled the Advanced Energy Initiative, which outlined significant new investments and policies to (1) accelerate deployment of efficient hybrid and clean diesel vehicles in the near-term, (2) develop domestic renewable alternatives to gasoline and diesel fuels in the mid-term, and (3) invest in the advanced battery and hydrogen fuel-cell technologies needed for substantial reductions in future oil demand. Together, these efforts will help the United States reach the President's long-term goal of replacing more than 75 percent of U.S. oil imports from the Middle East by 2025 (Bush, 2006).

As a follow-on supporting initiative, in the 2007 State of the Union address, the President announced his "Twenty in Ten" goal to cut U.S. gasoline consumption by 20 percent over the next 10 years. His proposal focuses on two key elements (Whitehouse, 2007):

- *Fuel Switching*: Increase mandatory fuel standards to require the equivalent of 35 billion gallons of renewable and alternative fuels in 2017 (15 percent reduction).
- *Increased Vehicle Efficiency*: Tighten Corporate Average Fuel Economy (CAFE) standards to reduce projected annual gasoline use by 8.5 billion gallons in 2017 (5 percent reduction).

A variety of renewable and alternative fuels—ethanol, natural gas, propane, coal to liquids, and electricity—will be required to meet the 35 billion gallon fuel target. Deploying significant volumes of these types of fuels in this timeframe will require significant technology advancements in concert with effective government policies to motivate transformation of the current petroleum-based transportation fuel market. Tasked with advancing reliable, clean, and affordable energy technologies, the U.S. Department of Energy (DOE) is playing a key role in developing and implementing strategies to bring renewable and alternative fuels to commercialization.

The federal initiatives provide a starting point for moving from virtual total reliance on petroleum to other options in the nation's transportation sector. Ultimately, the timeframe in which a significant change in oil consumption might be seen highly depends on when technology options will be available for market introduction, construction lead times for fuel production facilities, whether existing fuel delivery and fueling infrastructure can be used or new infrastructure must be developed, and how quickly on-the-road vehicles are replaced with new vehicles.

17.1.2 Purpose and Scope

The transition from our current petroleum-based transportation fuel economy to a future fuel economy that incorporates significant amounts of alternative and renewable transportation fuels can be characterized and addressed as a system of systems (SoS) problem. This chapter describes the current efforts of the DOE Biomass and Hydrogen Programs from an SoS perspective. It provides the working definition of SoS used to frame the discussion; describes the current transportation fuel SoS; provides a vision for a future transportation fuel SoS, with a focus on the intermediate transition to biofuels in the near- to mid-term and to hydrogen in the long-term; and identifies modeling tools and approaches that are used by the DOE to understand and track progress toward the future transportation fuel SoS.

17.2 SYSTEM OF SYSTEMS OVERVIEW

System of systems is a capability or enterprise-based approach to solving complex problems, in which multiple complex systems interact both independently and as an integrated whole. While SoS has its roots in the established systems engineering discipline, as outlined in Fig. 17.1, addressing SoS goes beyond traditional systems engineering in a number of ways.

	System engineering perspective	System of systems engineering perspective
Scope	Project/product	Enterprise/capability
	Autonomous/well-bounded	Interdependent
Objective	Enable fulfillment of requirements	Enable evolving capability
	Structured project process	Guide integrated portfolio
Time Frame	System lifecycle	Multiple, interacting system life cycles
	Discrete beginning and end	Amorphous beginning Important history and precursors
Organization	Unified and authoritative	Collaborative network
Development	Design follows requirements	Design is likely legacy-constrained
Verification	System in network context	Ensemble as a whole
	One time, final event	Continuous, iterative

FIGURE 17.1 Differences between traditional systems engineering and system of systems engineering (SOSECE, 2007)

The field of SoS engineering is still emerging and the SoS community has not yet come to agreement upon a single commonly accepted definition of SoS. As a starting point, the International Council on Systems Engineering (INCOSE) defines SoS as follows:

> System of systems applies to a system of interest whose system elements are themselves systems; typically these entail large-scale inter-disciplinary problems with multiple, heterogeneous, and distributed systems" (INCOSE, 2006).

A key aspect of SoS that is not called out in this definition is the importance of context in developing a desired physical capability. According to the System of Systems Center of Excellence, "SoS engineering addresses a complex system in terms of relationships, politics, operations, logistics, stakeholders, patterns, policies, training and doctrine, context, environment, conceptual frame, geography and boundaries" (SOSECE, 2007). This broader definition is needed to characterize and transform the transportation fuel SoS.

17.3 TRANSPORTATION FUEL SYSTEM AS A SYSTEM OF SYSTEMS

Conceptually, the transportation fuel SoS can be represented as shown in Fig. 17.2 and described in terms of capability and context.

17.3.1 Capability

The physical systems and infrastructure included in the transportation SoS can be organized around five interdependent systems that comprise the feedstock-to-fuel supply chain. The primary objective of each system is described in the context of the existing transportation SoS, which moves crude oil from its source to the final processed fuel used by consumers, as illustrated in Fig. 17.3.

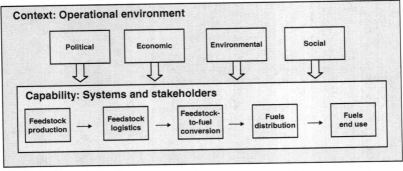

FIGURE 17.2 Transportation fuel system of systems

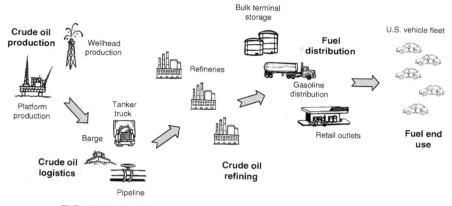

FIGURE 17.3   The existing petroleum-based transportation SoS

- *Feedstock Production System*: The objective of the feedstock production system is to produce large quantities of high-quality, raw feedstock cost-effectively, efficiently, and in compliance with applicable safety and environmental regulations. In the case of the current petroleum-based fuel economy, this is all of the exploration and production infrastructure (e.g., drilling rigs, production platforms) required to extract crude oil from reserves around the globe.

- *Feedstock Logistics System*: The objective of the feedstock logistics system is to collect, store, and transport raw feedstock from the production point to the fuel production facility cost-effectively, efficiently, and in compliance with all applicable safety and environmental regulations. In the case of the current petroleum-based system, this includes all of the infrastructure required to move crude oil from the field to the refinery. The nation's extensive network of petroleum transmission pipelines are the primary means of moving crude oil from oil fields on land and offshore to refineries where the oil is turned into fuels and other products. There are approximately 55,000 miles of crude oil trunk lines in the United States that connect regional markets (Pipeline 101, 2007a). The crude oil logistics system also includes ports, storage tanks, barges, and tankers depending on where the oil supply originates.

- *Feedstock-to-Fuel Conversion System*: The objective of the feedstock-to-fuel conversion system is to process raw feedstock into specification-compliant transportation fuel cost-effectively, efficiently, and in compliance with all applicable safety and environmental regulations. In the case of the current petroleum-based system, this includes all of the infrastructure (e.g., reactors, distillation columns, etc.) required to operate a refinery. U.S. refining capacity stands at approximately 17 million barrels per day (EIA, 2004b). Refineries process the crude oil feedstock into gasoline, diesel fuel, heating oil, jet fuel, liquefied petroleum gases, and other petroleum-based products. Gasoline, which represents nearly 45 percent of the domestic production of all refined products, is the petroleum product most demanded by U.S. consumers (EIA, 2005).

- *Fuels Distribution System*: The objective of the fuels distribution system is to move transportation fuel from the refinery to the consumer point-of-use (i.e., the consumer's vehicle) cost-effectively, efficiently, and in compliance with all applicable safety and environmental regulations. In the case of the current petroleum-based system, this includes all of the infrastructure (e.g., pipelines, storage tanks, and fuel dispensers) required to transport, store, and dispense transportation fuel. There are approximately 95,000 miles nationwide of refined products pipelines, which move gasoline, diesel fuel, and other petroleum products to consumer markets (API, 2006). Refined products pipelines are found in almost every state in the United States (Pipeline, 2007b). The majority of gasoline is shipped by pipeline to bulk storage terminals near consuming areas. At these terminals, the gasoline is loaded into tanker trucks and then delivered to one of the approximately 167,000 retail outlets in the United States, where the gasoline is unloaded into the underground tanks at the gas station (EIA, 2005).

- *Fuels End-Use System (Vehicle)*: The objective of the fuels end-use system is to provide high-performance, reliable, affordable, and safe vehicles to consumers cost-effectively, efficiently, and in compliance with all applicable safety and environmental regulations. In the case of the current petroleum-based transportation fuel system, this is all of the infrastructure (e.g., auto industry and supporting industries—rubber, computer chips, steel, etc.) required to manufacture and distribute vehicles to consumers. In 2005, almost 66 million cars and commercial vehicles were produced worldwide (12 million of these vehicles were produced in the United States) (DOT, 2006a). In 2005, there were almost 250 million highway vehicles registered in the United States (DOT, 2006b).

17.3.2 Context

The transportation fuel SoS must operate within the context of political, economic, social, and environmental conditions that influence its physical domain. Together, these perspectives serve to define "the interrelated conditions which exemplify a system's state of being and which describe its purpose, scope, and meaning for services it may offer" (Polzer et al., 2007). A brief description of each perspective with respect to the transportation fuel SoS follows.

- *Political Context*: Government policies, incentives, laws, and regulations have affected the transportation fuel SoS for many decades. Global politics is, and will continue to be, a key consideration in the operating environment, given the fact that over two-thirds of the world's remaining global oil reserves lie in the Middle East (Rifkin, 2002). Significant government incentives that directly support the U.S. petroleum-based industry have been in place for years. For example, between 1968 and 2000, the petroleum industry received over $150 billion dollars in tax breaks—for exploring and for producing petroleum within the

United States through percentage depletion deductions, expensing of exploration, development and production of nonconventional fuels (GAO, 2000). Since the 1970s, laws and regulations related to the transportation fuel SoS have been primarily driven by increasing concerns for the environment, safety, and energy efficiency (e.g., mandated vehicle emissions requirements and fleet average fuel economy standards).

- *Economic Context*: The transportation fuel SoS operates within a global marketplace. "Oil is the world economy's most important source of energy and is, therefore, critical to economic growth (API, 2006)." With an estimated total value of between $2 trillion and $5 trillion, the petrochemical industry is the largest business in the world (Rifkin, 2002). The automobile industry is also a major contributor to global and U.S. economies. In 2005, the total sales of automobiles were about 4% of the nation's gross domestic product, equivalent to around $500 billion dollars in sales. It is estimated that for every autoworker, 7.5 jobs are created in other industries (Ford Motor Company, 2007). On the downside, U.S. reliance on imported oil has significant negative impacts on the U.S. economy. The last three major oil price shocks, which were driven by political events in the Middle East, pushed the United States into an economic recession. In addition, the United States spends an estimated $200,000 per minute on foreign oil, accounting for about one-fourth of its annual trade deficit (UCS, 2002). In the future, petroleum prices are expected to rise as the worldwide oil demand continues to increase and oil supplies begin to wane, further increasing cash flow out of the U.S. economy.

- *Environmental Context*: The current petroleum-based transportation fuel SoS has had significant negative impacts on the environment. "Oil extraction, refining and transportation are responsible for land destruction and toxic contamination at the extraction point, oil spills in oceans around the world, and toxic air and water emissions from oil refining operations"(CWAC, 2007). Millions of acres of farmland and wildlife habitat have been lost to roads and highways across the United States, and vehicle emissions are major contributors to air and water pollution, as well as global climate change. Today, the transportation sector accounts for about a third of total U.S. emissions of carbon dioxide (an important greenhouse gas) (DOE Biomass Program, 2007). Not surprisingly, the transportation fuel SoS operates under a multitude of environmental protection laws and regulations regarding oil production, transport, and use. Pressure from environmental advocacy groups continues to motivate government action to mitigate and minimize the environmental impacts of our current transportation fuel SoS through sustainable, energy-efficient, and clean alternatives.

- *Social Context*: Affordable transportation fuel and personal mobility are virtual "rights" in the United States today. Even as traffic congestion and air pollution plague our cities, oil and the automobile are the foundation of our twenty-first century commercial and social lives. Today, the average American consumes about 25 barrels of oil per year; for perspective, the average person in China uses less than 2 barrels of oil per year (Nationmaster, 2007). Modern food production and distribution almost exclusively depend on oil and natural gas—from producing the fertilizer used to grow crops to fueling harvesting equipment and refrigerated trucks that deliver food to consumers—and today, agriculture is one of the world's most energy-intensive industries (Worldwatch Institute, 2007). Recognition that this oil-dependent lifestyle cannot be sustained into the future is slowly building—driven in part by the pinch consumers feel as the costs of food, gasoline, and consumer products rise in response to higher oil prices, as well as increasing concern for the environment. For example, since 2004, the popularity of sport utility vehicles (SUVs) has waned and consumer demand for more fuel efficient vehicles has risen (Worldwatch Institute, 2007). Nonetheless, consumers maintain their high expectations regarding performance, comfort, safety, reliability, cost, and size of the vehicles they purchase and drive. This illustrates the primary challenge from a social perspective— overcoming our natural resistance to change and managing expectations as alternative transportation fuels enter the market.

17.4 VISION FOR FUTURE TRANSPORTATION FUEL SoS

A future energy system will need to include a greater diversity of supply, provide a more flexible and agile infrastructure for moving energy from source to use, and use energy much more efficiently. The energy system will need to minimize its environmental footprint at both global and local levels, moving toward carbon-neutral technologies, minimizing their use of water, and minimizing and mitigating other emissions that affect local air, water, or land resources. The transportation fuel SoS will continually evolve as needs change and technologies become available. It is envisioned that the transportation fuel SoS will evolve to integrate the biomass-to-biofuels SoS in the near- to mid-term (2012–2030) and then expand and transform more dramatically to accommodate the hydrogen SoS in the longer term (beyond 2030).

17.4.1 Near- to Mid-Term Vision (2007–2030): Biomass-Based Fuels

Biomass is the only domestic, sustainable, and renewable primary energy resource that can replace liquid transportation fuels currently produced from petroleum sources. Biomass resources include crops such as corn and soybeans, agricultural and forest residues, consisting of corn stover (stalks and leaves that remain after the corn grain is harvested), wheat straw, and forest thinnings. Also included are nonedible perennial crops such as switchgrass and hybrid poplar, which can be grown as energy crops. It is estimated that the United States has the potential to produce over 1.3 billion tons of biomass annually (Perlack, et al., 2005)—more than enough to replace about 30 percent of current U.S. gasoline consumption—sustainably, without impacting food, feed, and fiber uses.

The vision of the DOE Biomass Program is "A viable, sustainable domestic biomass industry that produces renewable biofuels, bioproducts and biopower, enhances U.S. energy security, reduces our dependence on oil, provides environmental benefits including reduced greenhouse gas emissions, and creates economic

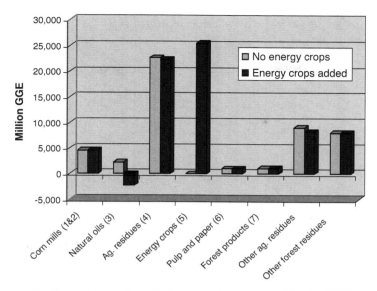

FIGURE 17.4   Potential gasoline displacement by pathway (Total ~70 billion GGE)

opportunities across the nation" (DOE, 2007). In 2030, a robust biomass-based energy industry and supporting infrastructure will be in place, fully operational and capable of producing 60 billion gallons of cellulosic ethanol annually.

- *Biomass Feedstocks Production System*: In 2030, a variety of sustainable, cost-effective, regionally-available, lignocellulosic feedstocks will be integrated into the current agricultural and forestry industries and available for biofuels production. Agricultural resources (corn stover, straw, and switchgrass) and forest resources (forest thinnings, logging residues, and urban wood residues) dedicated to biofuels production will total 600 million dry tons. The feedstocks with the greatest ultimate ethanol production potential include agricultural residues, perennial energy crops, and forest residues, as shown in Fig. 17.4.

- *Biomass Feedstocks Logistics System*: In 2030, mature technologies for collection, storage, and preprocessing both wet and dry biomass feedstocks will be integrated into the agricultural and forestry industries and in use in all regions of the country. Approximately, 30 billion ton-miles of transport will be in place to deliver 600 million dry tons of "reactor-throat ready" feedstock to biofuels production facilities.

- *Biofuels Production System*: In 2030, an estimated 300 commercial cellulosic biofuels production plants will be producing 48 billion gge (gallons of gasoline equivalent)[1] of biofuels per year. These plants will integrate advanced biochemical and thermochemical conversion technologies to maximize production of biofuels

---

[1] A unit of measure used to express the energy content of alternative fuels that have different energy densities on a common volumetric basis.

from biomass. Coproducts will include heat and power, and, in some cases, materials/chemicals/products that improve the overall plant economics.

- *Biofuels Distribution System*: In 2030, shipping biofuels through pipelines will be standard practice as an integrated component of the nation's petroleum distribution system. In total 48 billion gge of biofuels will be shipped from production facilities to bulk terminals either through dedicated lines or through common carrier pipelines. At the terminal, biofuels will be loaded into tanker trucks and then delivered to one of the 100,000 plus dedicated biofuel pumps in retail outlets across the United States.

- *Biofuels End-Use System*: In 2030, 230 million biofuels-compatible vehicles will be on the road, consuming 48 billion gge of biofuels annually. These vehicles will run on biofuel/gasoline blends and biofuel/hybrid platforms and will have the same or better performance than today's conventional fuel vehicles.

17.4.2   Long-Term Vision (2030 and Beyond): Hydrogen-Based Economy

An alternative pathway to a future transportation fuel SoS is focused on a hydrogen-based economy. Hydrogen is the most abundant element in the universe but hydrogen does not exist alone in nature; it must be isolated from hydrogen-containing substances or feedstocks—water (H_2O), natural gas (CH_4), biomass (cellulose, hemicellulose, or lignin), and hydrocarbons such as coal. The United States has enough domestic energy resources—wind, solar, biomass, hydroelectric, coal, and nuclear—to meet all of its energy needs, not just the transportation needs. These resources are widely distributed, which allows the use of decentralized hydrogen production facilities from coast to coast. The ultimate vision of an operational hydrogen economy is based on the concept of a renewable hydrogen cycle. In this cycle, renewable energy sources such as the sun and wind are converted to electricity, which is then used to produce hydrogen from water. The stored hydrogen is converted back to electricity cleanly and efficiently in a fuel cell that powers the end-use vehicles. The entire cycle is sustainable and non-polluting.

The vision of the DOE Hydrogen Program is "Hydrogen is America's clean energy choice. Hydrogen is flexible, affordable, safe, domestically produced, used in all sectors of the economy, and in all regions of the country" (DOE, 2002). In 2040, a robust hydrogen-based energy industry and supporting infrastructure will be in place, fully operational and capable of producing 64 million metric tons of hydrogen to fuel 300 million fuel cell vehicles annually.

- *Hydrogen Feedstock Production System*: In 2040, a variety of domestic feedstocks will be available for hydrogen production including biomass, coal (with carbon sequestration), water in combination with electricity-generating renewables such as wind and solar power, and nuclear power. Feedstock selection will be based on the resources and processes that are most economical or specifically selected by particular states or regions. In 2040, the amount of each resource required to produce 13 million metric tons of hydrogen (20% of the total projected hydrogen demand of 64 million metric tons) is summarized in Fig. 17.5.

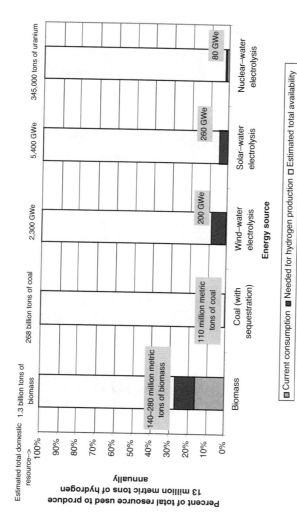

FIGURE 17.5 Potential contribution of domestic resources to hydrogen production (DOE and DOT, 22006) (resource required to produce 13 million metric tons of hydrogen annually in gray)

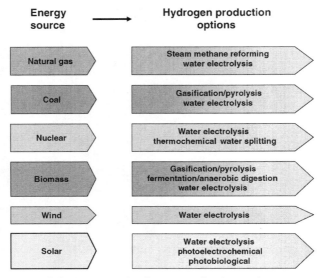

FIGURE 17.6   Hydrogen production options

- *Hydrogen Feedstock Logistics System*: In 2040, the infrastructure for moving hydrogen feedstocks—biomass, coal and water—to hydrogen production facilities will be well established.[2] For example, biomass feedstock logistics systems will be fully mature, with the capability to deliver over 600 million dry tons of feedstock per year to production facilities. The current coal delivery infrastructure will be expanded by over 50 percent (DOE and DOT, 2006).
- *Hydrogen Production System*: In 2040, 64 million metric tons of hydrogen will be produced in centralized facilities in remote locations, in power parks and fueling stations in our communities, in distributed facilities in rural areas, and at customers' homes and businesses. As detailed in Fig. 17.6, thermal and electrochemical processes will use fossil fuels, biomass, or water as feedstocks and release little or no carbon dioxide into the atmosphere. Water-splitting microorganisms and biomass fermentation will also become viable sources for renewable hydrogen (DOE and DOT, 2006).
- *Hydrogen Distribution System*: In 2040, a national supply network will evolve from the existing fossil fuel-based infrastructure to accommodate distribution of 64 million metric tons of hydrogen produced in both centralized and decentralized production facilities. Pipelines will distribute hydrogen to high-demand areas, and trucks and rail will distribute hydrogen to rural and lower demand areas. On-site hydrogen production and distribution facilities will be built where demand is high enough to sustain maintenance of the technologies.

---

[2]Natural gas is not considered to be a long-term feedstock for hydrogen production because it would result in increased imports of natural gas.

- *Hydrogen End-Use System*: In 2040, the on-board hydrogen storage problem will be solved with lightweight, low-cost, and compact storage devices. One specific solution will be the use of high-tech solid materials, such as metal hydrides and carbon nanotubes. Fuel cells will be a mature, cost-competitive technology in commercial production. Hydrogen will become the dominant fuel for vehicle fleets across the country, with a projected 300 million fuel cell vehicles on the road.

17.5   CHALLENGES

To meet the growing demand for energy without compromising security, environmental quality, or the ability to sustain economies, a significant transformation in the global energy enterprise is required over the coming decades. The United States has undergone two energy transitions over the last two centuries—from wood to coal during the late 1800s and from coal to oil and gas during the early 1900s—driven by the availability of abundant, more productive, and less costly energy sources. In contrast, the energy transition the nation is now facing requires alternative energy sources that, while abundant, are more expensive than the petroleum fuels in the market today. This significantly increases the difficulty of integrating new fuels into the transportation fuel SoS. The challenges associated with the envisioned transition to alternative and renewable fuels are summarized from the perspective of the six SoS characteristics outlined in Fig. 17.1.

- *Scope*: The essence of this endeavor is to transform the transportation fuels sector of the U.S. economy by developing the capability to displace significant quantities of petroleum-based transportation fuels with renewable alternatives in combination with more energy efficient technologies and systems.
- *Objective*: No single solution is envisioned for the future and the "best" solution (an optimum mix of technologies and systems) will likely change over time. A phased approach will allow new energy systems to be integrated into the existing transportation fuel infrastructure as technologies and systems advance to the point of commercial readiness. For example, today, ethanol produced from grain is already established in the fuel marketplace, both as a blending agent and as a fuel for flexible fuel vehicles (FFVs), and energy-efficient gasoline-electric hybrid vehicles are steadily increasing market share. Real-world experience gained with these technologies and systems will be used to guide the RD&D needed to enable the next generation of alternative and renewable fuels (e.g., biomass and hydrogen). This is the overarching strategy for transformation of the transportation fuel SoS over the next several decades. Figure 17.7 illustrates the envisioned evolution of vehicle technologies over time.
- *Timeframe*: Numerous initiatives to reduce our dependence on petroleum have been proposed and implemented (with varying levels of funding and degrees of success) since the oil shocks of the early 1970s. Development of energy efficient and renewable technologies has typically been an integral part of these initiatives.

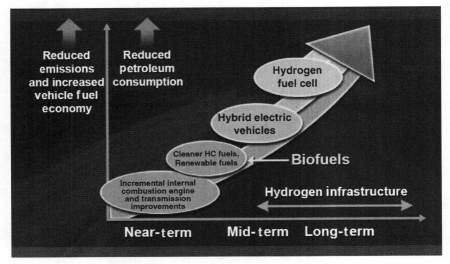

FIGURE 17.7 A future vision for transportation

Progress made to date will be leveraged as this work continues in the coming years. New advanced fuel technologies will be under development (early stages of systems life cycle), at the same time commercially ready technologies are being demonstrated and deployed (middle stages of systems life cycle). For example, fundamental hydrogen storage R&D is being conducted to enable a future hydrogen economy that is decades away, at the same time full-scale biomass-based ethanol production plants are being constructed in support of a biomass-to-biofuels SoS that appears to be on the verge of commercial viability. In fact, some of the elements of the biomass-to-biofuels SoS are already available to customers (e.g., to date, an estimated 5 million flexible fuel vehicles that can run on ethanol (E85) or gasoline have been sold in the United States).

- *Organization*: The transformation from a petroleum to a nonpetroleum-based transportation fuel SoS will require the focused, coordinated, and collaborative action of a multitude of diverse stakeholders. For example, the existing transportation fuel SoS comprises a global network of producers, refiners, marketers, brokers, traders, automobile manufacturers, and consumers. In the future, the stakeholders involved will expand as each new type of fuel comes into play. The many stakeholders with key roles in the development of the future biofuels and hydrogen fuel systems are shown in Fig. 17.8.

- *Development*: The existing petroleum-based, feedstock-to-fuel supply chain is a decades-old, fully mature, highly networked, and relatively efficient SoS. As such, it contains great inertia and several persistent trends—resulting from sunk costs in the current infrastructure and consumer expectations regarding performance, cost, and safety—that will influence the energy economy well into the future. Initially, new technologies must be designed to operate within the

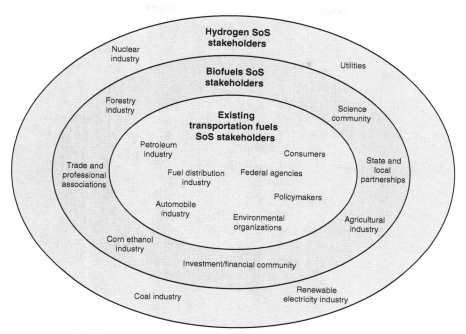

FIGURE 17.8   Stakeholders in transportation SoS transition

existing SoS and compete effectively with the existing system to enable a transition to new/alternative fuels to begin to occur.

- *Verification*: The ultimate success of this endeavor will be measured by the overall reduction in our nation's petroleum consumption. Consequently, at any point in time, the transportation fuel SoS can only be verified as a unified whole, based on the effectiveness of the interrelated systems working together. The new SoS will be evaluated on an ongoing basis as technologies and systems are developed, improved, and implemented within the SoS over time.

17.6   NEAR-TERM EVOLUTION OF SoS: TRANSITION TO BIOMASS TRANSPORTATION FUEL SoS

17.6.1   Where are we now? Existing Ethanol Industry

Ethanol and biodiesel are the only biofuels in commercial production today. The existing ethanol industry is the focus of this discussion. Ethanol is used throughout the United States as a blend component of gasoline to reduce vehicle emissions and improve octane. Consequently, the supply chain from the farmer's field to the customer's vehicle is mature and integrated into the existing fuel supply infrastructure.

- *Ethanol Feedstock Production System*: Starch crops such as corn and sorghum are the predominant ethanol feedstocks today. In 2005, 1.43 billion bushels of corn was used for ethanol production, representing nearly 13% of the U.S. corn crop. With an average yield of 147.9 bushels of corn per acre, about 10 million acres of the total 81.8 million acres of corn planted in 2005 was used to produce fuel ethanol (NGCA, 2007). Ethanol represents the third largest market for U.S. corn, behind only livestock feed and exports. Ethanol production also consumed 15% of the nation's grain sorghum crop.

- *Ethanol Feedstock Logistics System*: The equipment and systems used to harvest, collect, transport, and store corn have been optimized for grain harvest over many years. The basic field equipment includes combines (which cut, gather, thresh, separate, and clean the corn), grain carts (which shuttle corn grain from combines to grain trucks or grain receiving facilities), and grain trucks (which haul the grain to ethanol production facilities). Corn is typically stored in bins or buildings to preserve the quality of the grain during storage.

- *Ethanol Production Systems*: Two commercial production processes are used to convert the starch in grains to ethanol: dry milling and wet milling. In dry milling, the entire corn kernel or other starchy grain is ground into meal and processed without separating out the various component of the grain. The major coproduct from the dry milling process is distillers dried grains (DDG). In wet milling, the grain is soaked in a dilute sulfurous acid solution for 24 to 48 hs, which facilitates the separation of the grain into its many components. The coproducts of wet milling include corn oil, corn germ, corn gluten feed, and high-fructose corn syrup. Wet mills are typically much larger, producing hundreds of millions of gallons of ethanol per year, than dry mills, which produce tens of millions of gallons of ethanol per year. According to the Renewable Fuels Association, in 2005, 4 billion gallons of ethanol was produced in 95 wet and dry mill facilities (production capacity of 79% from wet and 21% from dry) in 19 states across the country—mostly from corn (RFA, 2006). Tax incentives provided for ethanol have amounted to approximately $11 billion since 1979 (GAO, 2000).

- *Ethanol Distribution System*: Today, most fuel ethanol is used as gasoline blending stock to increase fuel octane and help meet gasoline oxygenate requirements in urban ozone nonattainment areas. Typical reformulated gasoline (E10) contains 5–10% ethanol and accounts for about 2% by volume of all gasoline sold in the United States. Ethanol is also blended with gasoline to create E85, a blend of 85% ethanol and 15% gasoline. Global Trade Consulting estimated that almost 13 million gallons of E85 fuel were consumed in the United States in 2004 (Maren, 2004). There are currently more than 1,100 E85 fueling stations (out of 170,000 total fueling stations) across the United States, predominantly located in the Midwest (see Fig. 17.9).Today, approximately 75% of ethanol is moved by rail and the remaining 25% by truck, with barge and ship movements representing transfers of rail or truck shipments (RFA, 2006).

- *Ethanol End-Use System*: Since the early 1990s, more than 5 million flexible fuel vehicles (FFVs) have been sold in the United States (RFA, 2006). An

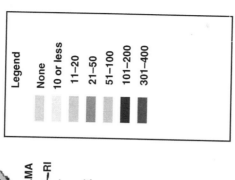

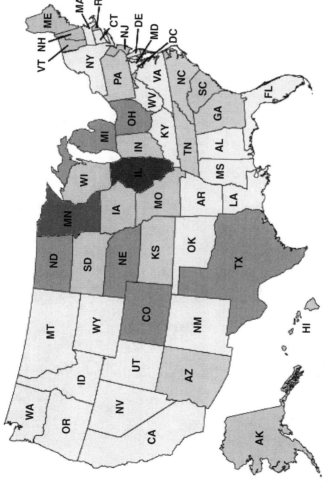

FIGURE 17.9 E85 fueling station locations (AFDC, 2007)

estimated 147,000 E85 FFVs were registered in the United States in 2004 (EIA, 2004b). In 2006, Ford Motor Co. and General Motors Corp. announced an aggressive push to manufacture a record number of E85 FFVs—a total of 650,000 new vehicles (General Motors Corporation, 2006). GM also launched its "Live Green, Go Yellow" marketing campaign drawing attention to E85 and its E85-ready cars and trucks (General Motors Corporation, 2007).

17.6.2   How Do We Get There? Transition to Biofuels SoS

The transition to a biofuels SoS will require initially integration with and ultimately transformation of industries of the current transportation fuel SoS supply chain—that is, petroleum fuel production and distribution industries and the automobile industry. A future biofuels SoS will also require transformation of the well-established industries that will provide biomass feedstock (e.g., agricultural, forestry, and waste management industries) and produce biofuels (corn ethanol industry).

Because of the wide diversity of biomass feedstocks, conversion technologies, integration options, and potential products, a multitude of options are possible. Seven conceptual technology pathways have been developed to guide the research efforts and identify the key interfaces that will enable the establishment of commercially viable biofuel production facilities and supporting infrastructure. Each pathway represents a generic set of potential feedstock production, feedstock logistics, and biofuel production system configurations for a specific biomass resource base, as shown in Fig. 17.10.

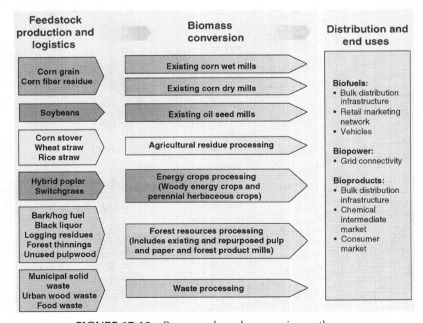

FIGURE 17.10   Resource-based conversion pathways

Within each pathway, there are a number of viable alternative routes to biofuel production.

The feedstocks with the greatest ultimate biofuel production potential include agricultural residues, perennial energy crops, and forest residues (see Fig. 17.3). Agricultural and forest residues such as corn stover, wheat straw, and forest thinnings are considered the best near-term lignocellulosic feedstocks because of their availability and low cost. Perennial energy crops such as switchgrass and hybrid poplar are longer term options because they require reallocation of existing land use as well as a number of years to mature before they are ready to harvest. The initial transition strategy (2007–2017) is focused on developing, demonstrating, and deploying cellulosic ethanol (over other biofuels). In the longer term (2017–2030), additional biofuels (e.g., biobutanol, renewable diesel, and Fischer Tropsch biofuels) will likely be incorporated as technologies are developed and the transportation fuel SoS evolves.

The overall transition strategy is based on three key elements:

- Build on and leverage established commercial processes and systems;
- Conduct R&D to continually advance technology; and
- Incrementally establish a new biomass-based industry as technologies become commercially viable.

For example, the transition to a lignocellulosic ethanol SoS is envisioned as a three-phase effort as illustrated in Fig. 17.11.

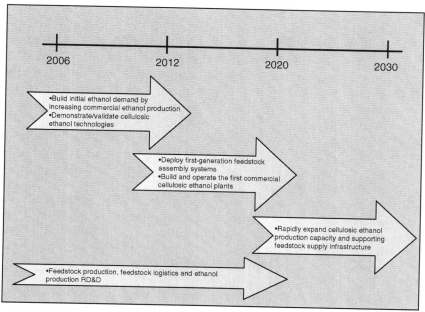

FIGURE 17.11   Phased transition to a lignocellulosic ethanol SoS

Phases I and II are focused on developing and demonstrating technologies. Phase III is focused on rapid deployment and market penetration.

- *Phase I (2007–2012)*: In phase I, the primary goals are to build initial ethanol demand in the marketplace and demonstrate and validate first-generation cellullosic ethanol technologies using agricultural residues such as corn stover. Expansion of the current starch ethanol industry production capacity will provide the foundation for future cellulosic ethanol demand by motivating installation of additional E85 fueling stations and increased FFV production. In addition, DOE-sponsored regional feedstock partnerships will begin to motivate the identification, evaluation, and establishment of viable, regionally based feedstock resources for future biorefineries. Simultaneously, R&D efforts to develop and prove agricultural feedstock assembly technologies and systems and biomass conversion technologies in integrated processes will support the deployment objectives of phase II. Government policies mandating the use of alternative and renewable fuels and funding of precompetitive biomass R&D are essential to the successful completion of phase I.

- *Phase II (2013–2017)*: In phase II, the primary goals are to maintain the 2012 starch ethanol industry production capacity, establish commercial-scale agricultural residue collection systems, and build and operate the first commercial cellulosic ethanol plants. The cellulosic ethanol facilities will build on existing bioindustry segments to tap into biomass feedstocks that are readily available, such as agricultural residues (e.g., corn stover) and existing bioindustry process feedstocks and intermediates (e.g., residual corn fiber, wood chips used in pulp and paper mills, and waste materials).[3] To support the coming demand for high-quality cellulosic biomass, the first energy crop plantations will be established. R&D efforts will focus on the development and validation of feedstocks tailored specifically for biofuel production, sustainable feedstock practices, and second-generation biomass conversion and biofuel end-use technologies that are more efficient and cost-effective. Jointly funded government–industry partnerships to enable construction of the first commercial biomass conversion plants and motivate energy crop production, in addition to continued funding of R&D to advance biofuel technologies and systems, are key to the success of phase II.

- *Phase III (2017–2030)*: In phase III, the primary goals are to rapidly expand the cellulosic fuel ethanol industry—by reallocating land to dedicated energy crop production, building new commercial-scale ethanol production facilities and supporting ethanol-dedicated infrastructure, and continuing to add to FFV fleet. As technologies mature, advanced second-generation systems will be integrated into the growing cellulosic ethanol infrastructure. In this timeframe, the emerging biomass-to-biofuels industry will become self-sustaining and government policies and financial incentives will no longer be required.

---

[3]EPAct 2005 Section 932(d) requires DOE to let a solicitation to design, build, construct, and operate a commercial-scale integrated cellulosic biorefinery; 40% federal cost share of project; $56 million planned for FY 07; $160 million in total.

17.7 LONG-TERM EVOLUTION OF SoS: TRANSITION TO HYDROGEN TRANSPORTATION FUEL SoS

17.7.1 Where are We Now? Existing Hydrogen Economy

Outside the space industry, hydrogen is not used as a fuel today. Most hydrogen is used to convert heavy petroleum sources into lighter fractions suitable for use as fuels and to produce ammonia for fertilizer production. The existing hydrogen production through end-use supply chain for these chemical applications is mature but not sufficient to support significant transportation applications.

- *Hydrogen Feedstock Production System*: Today, the primary feedstock for hydrogen production is natural gas. Coal is also used as a hydrogen feedstock, and ethane, propane, butane, and naphtha are also occasionally used as feedstocks for hydrogen production in refineries. Grid electricity, produced primarily from coal and natural gas, in combination with water, is also used as hydrogen feedstock.

- *Hydrogen Feedstock Logistics System*: An extensive natural gas distribution network is in place in the United States today. This network includes about 20,000 miles of natural gas gathering lines, which move natural gas to large cross-country transmission pipelines, and 278,000 miles of onshore and offshore natural gas transmission lines (Pipeline 101, 2007c). Most bulk coal transportation is by rail, with trucks used for local transport (Rifkin, 2002).

- *Hydrogen Production System*: Approximately 9 million tons of hydrogen is produced annually for the industrial sector, largely by steam methane reforming of natural gas. The steam reforming process produces a mixture of hydrogen and carbon monoxide, or synthesis gas (syngas). The carbon monoxide in the syngas is converted to carbon dioxide and additional hydrogen via the water–gas shift reaction. In industrial practice, steam reforming of hydrocarbons takes place at high pressure over a nickel catalyst, usually supported on alumina, at 650–950 °C (SRI Consulting, 2004). To a lesser degree, hydrogen is produced via electrolysis by passing electricity through two electrodes in water. The water molecule is split and produces oxygen at the anode and hydrogen at the cathode. Electrolyzers are used primarily to produce small volumes of relatively high-purity hydrogen and oxygen for specialized applications.

- *Hydrogen Distribution System*: Hydrogen gas is difficult to store and transport because it is very light. To make it easier to handle, the energy density of hydrogen gas is increased in one of the two ways. Hydrogen gas can be compressed to a high pressure (up to 7,000 psi) or cooled to a very low temperature (below 423°F) so that the hydrogen gas becomes a liquid. In either case, the hydrogen is stored in especially designed tanks. The largest volumes of gaseous hydrogen are delivered by pipeline. The most extensive pipeline networks are located in the Gulf Coast, where large quantities of hydrogen are consumed at refineries and in the manufacture of chemicals (SRI Consulting, 2004). Hydrogen is also transported by road in compressed hydrogen cylinders and cryogenic liquid tankers.

Today, there are 70 hydrogen refueling stations in operation in the United States and Canada; of these, 15 are publicly accessible. (NHA, 2007) Figure 17.12 shows the distribution of these stations across the United States.

- *Hydrogen End-Use System*: Today, hydrogen is primarily used as a chemical, rather than a fuel, to make cleaner gasoline, fertilizer, and food products. High-purity hydrogen is used for electronics manufacturing. Hydrogen's primary use as a fuel is in the U.S. Space Program for both the main engine of the space shuttle and the fuel cells that provide the shuttle's electric power.

In the race for the potential fuel cell vehicle market, GM is dispatching 100 fuel cell Chevy Equinox crossovers to drivers in New York, Los Angeles, and Washington, D.C.; Ford has a handful of fuel cell Ford Focus vehicles being tested throughout the world; DaimlerChrysler has more than 100 fuel cell vehicles on the road, including 60 passenger cars and 30 buses; Toyota is testing 20 fuel cell Highlanders; and Honda recently unveiled its next-generation fuel cell vehicle, the FCX (Detroit News, 2007).

17.7.2   How Do We Get There? Transition to a Hydrogen Future

The transition to a hydrogen SoS will require initially integration with and ultimately transformation of industries of the current transportation fuel SoS (i.e., petroleum fuel production and distribution industries and the automobile industry), the emerging biofuel SoS, the established chemical-based hydrogen production industry, the natural gas industry, and utilities that provide electricity for the grid.

The transition to a hydrogen SoS will occur incrementally over many decades as technologies advance, market acceptance increases, and large investments in infrastructure that can support the expanded production, delivery, storage and use of hydrogen are made. An overview of the step-wise transition is presented in Fig. 17.13.

The transition to a hydrogen SoS can be envisioned to occur in two major phases.

- *Near-to-Mid Term (2007–2020)*: Similar to the biofuels SoS transition, the first steps toward a clean hydrogen future will build on the commercial processes and systems in use today. To build initial hydrogen fuel demand, hydrogen production capacity via steam reforming natural gas and water electrolysis using conventional electricity-generating sources (coal and natural gas) will be expanded. Distributed hydrogen generation will begin to play a role. Existing hydrogen distribution systems will be expanded dramatically. Fleets will transition to hydrogen fuel vehicles, and the first hydrogen fuel cell vehicles will be introduced to consumer markets. Government-supported R&D will advance hydrogen production, storage, and end-use technologies, and improve system performance and economics. Government policies and financial incentives will be implemented to stimulate industry investment and encourage consumer adoption of hydrogen technologies and systems.
- *Mid-to-Long Term (2020–2040)*: As hydrogen markets grow, costs across the hydrogen transportation fuel supply chain will decrease through economies of

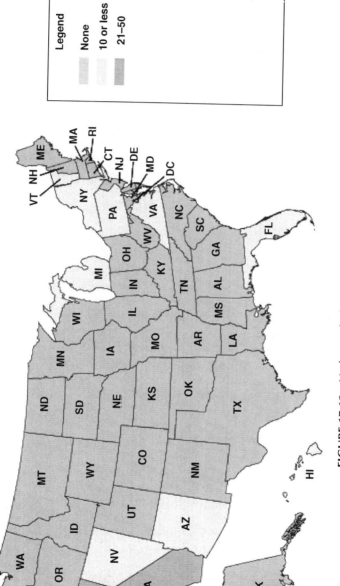

FIGURE 17.12 Hydrogen fueling station locations (AFDC, 2007)

Legend

None

10 or less

21–50

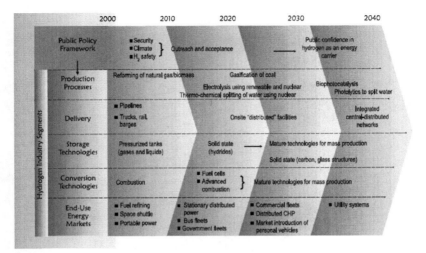

FIGURE 17.13   Overview of the transition to a hydrogen economy (DOE and DOT, 22006)

scale and dramatic technological advances. Natural gas and coal will be replaced by biomass and water as primary hydrogen feedstocks. Distributed hydrogen generation capability will expand significantly to enable hydrogen to be used where it is produced. Hydrogen distribution systems will also be expanded to all regions of the United States as hydrogen-fueled vehicles become widely available to consumers. Government-supported R&D will continue to advance hydrogen production, storage, and end-use technologies for next-generation applications. In this timeframe, government financial support for commercial ventures and consumer acceptance will no longer be needed as the hydrogen fuel SoS becomes firmly established.

17.8   MANAGING THE TRANSITION USING SE TOOLS

As illustrated in Fig. 17.14, the SoS challenge is how to design and implement a smooth transition from the existing legacy of a gasoline-fueled transportation economy to a future biomass-fueled or hydrogen-fueled transportation economy.

A variety of system of systems engineering (SoSE) tools are available to model, document, and track the transition from our current "legacy" transportation system to a future sustainable transportation system. Two of the tools currently used by DOE include

- STELLA, a system dynamics tool, to model and understand the key factors influencing the transition to biofuels,
- CORE®, a model-based systems engineering tool, to define and track the progress toward a future hydrogen economy.

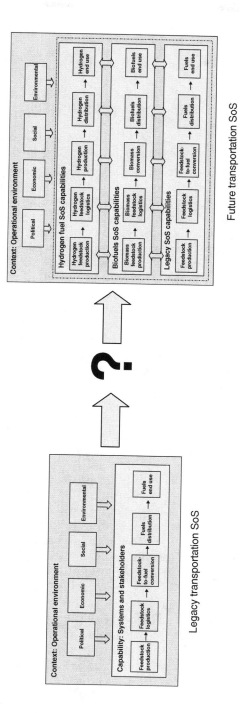

FIGURE 17.14 The system of systems engineering challenge

433

17.8.1   Using System Dynamics to Understand/Accelerate the Transition to Biofuels

System dynamics is a methodology for analyzing the behavior of complex feedback systems over time. It originated at the Massachusetts Institute of Technology in the 1950s as a modeling technique to improve the understanding of industrial processes. Since then, the span of applications has grown to include strategic energy planning and policy analysis, among others.

Energy policy analysis is well suited for a system dynamics approach because designing and implementing national energy policies requires knowledge of the complexities of our nation's energy sector, along with an understanding of the cause and effect relationships of national policies on economic growth, technology development and deployment, national security, international trade, environmental sustainability, and global climate change. System dynamics helps the decision maker sort through the complexity of these connections by providing a framework and set of tools to investigate how and why complex real-world systems behave the way they do over time. The goal is to leverage this added understanding to design and implement more efficient and effective policies.

The Biomass Scenario Model (BSM) was developed (as a prototype) to serve as a vehicle for investigating the dynamics associated with the potential evolutionary trajectories of a biofuel industry in the United States. The model uses a system dynamics framework, built on the STELLA software platform (ISEE Systems, 2007), to represent the dynamic interactions of the five major systems that comprise the biomass-to-biofuels supply chain. Figure 17.15 provides a conceptual overview of the model structure and points to the key features and components of the dynamic framework—supply chain, supply infrastructure, R&D, investment decision, policy space, and external economy.

In the current BSM, the dynamics of the growth across the supply chain is determined by the timing of the buildup of the infrastructure associated with each system. From a system of systems perspective, this buildup of the supply chain infrastructure is the *capability* that must be deployed to realize a future biofuel industry.

This capability is developed in the *context* of the competing oil market, vehicle demand for biofuel, and various government policies over time. In the model, the buildup of the infrastructure is determined by the dynamics of investor decisions. Investor response is driven by the performance and cost competitiveness of the fuels (which are driven by technology advancements due to R&D progress) along with the potential demand for them in the marketplace. Specific government policies and external economic factors (e.g., interest rates, price of gasoline) are evaluated as to their impact on the relative attractiveness of investing in new biofuel technology.

The logical framework for the model is based on input from stakeholders with expertise across the biomass-to-biofuels supply chain—from farming, forestry, financing, and process engineering to transportation and automobile manufacturing. Data for the model is derived from more detailed and narrowly focused studies of individual systems of the supply chain including feedstock logistics analyses,

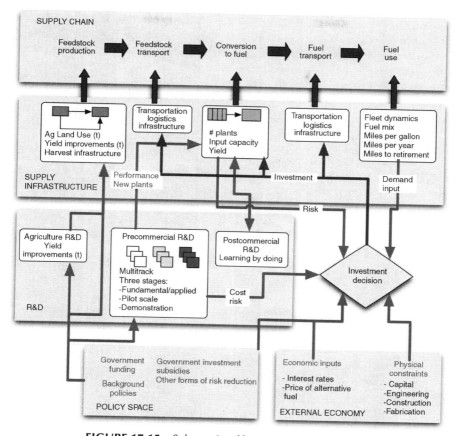

FIGURE 17.15   Schematic of biomass scenario model

agricultural economics analyses, process models life cycle assessment, economic models, and market analyses.

Incorporating additional aspects of the "real-world" operational context into the model will lead to a more robust characterization and improved understanding of the full impacts of the transition to a biomass-based transportation fuel SoS. These include the following:

- *Environmental Context*: Key environmental issues associated with the biomass-to-biofuels system include soil sustainability, water impacts of feedstock production and conversion processes, and greenhouse gas emissions.
- *Social Context*: The success of the transition to a biofuel-based SoS is linked to society's willingness to change—e.g., farmers must switch from cash crops to energy crops and consumers must select alternative vehicles over conventional vehicles.

Scenarios evaluated in the BSM help policy makers to envision, in broad terms, the emergence of a sustainable biomass-based industry and plausible future conditions. The power of the model is not primarily in the specific numbers it generates, but rather in the insight it provides into how specific actions facilitate or impede the takeoff of a biomass-based transportation fuel industry.

17.8.2 Using Model-based Systems of Systems Engineering to Define the Hydrogen Economy

Model-Based Systems Engineering (MBSE) is a methodology for developing and analyzing system requirements using graphical representations of the underlying functions, requirements, architecture, relationships, and interfaces that define the system. The MBSE approach helps to improve understanding and communication of complex system and system-of-systems designs. Beginning with the end in mind, that is, the vision of a future hydrogen economy, MSBE can help define and document what it will take, in terms of *capability*, to get from here to there.

The systems engineering fundamentals underlying the MBSE approach can be readily expanded to SoS applications. Specifically, the capabilities required for a future hydrogen fuel SoS can be defined and managed in terms of the functions, requirements, architecture, and tests (FRAT) framework (Mar and Morais, 2002) as follows:

1. Describe each system within the SoS in terms of four views:
   - What the system does (functions)?
   - How well the system performs its functions (requirements)?
   - What the system actually is (architecture)?
   - Verification and validation activities that provide the proof that the actual system satisfies the intended functions and requirements (test).
2. Define and understand the three interacting systems:
   - *Physical System*: The required infrastructure for an operational hydrogen economy.
   - *Program System*: The RD&D program that brings the desired infrastructure into being.
   - *External System*: Everything else that interacts with the physical and program systems.

The desired SoS infrastructure and the RD&D program for establishing the desired infrastructure are derived from the vision and mission of the DOE Hydrogen Program, and captured in an integrated baseline, as illustrated in Fig. 17.16. The purpose of the integrated baseline is to establish and control the documented, traceable relationships between requirements, functions, strategies, technologies, and so on of the SoS. The integrated baseline consists of two components. The technical baseline, which focuses on defining the technological capabilities in terms of configuration, performance, and characteristics of the SoS, ensures that we will be *doing the right things*. The technical baseline defines where the program is at any point in time and where it ultimately must

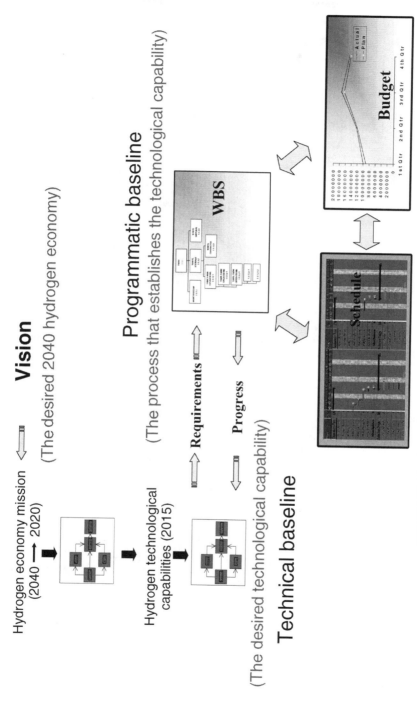

FIGURE 17.16 A requirements-driven, mission-oriented transition program

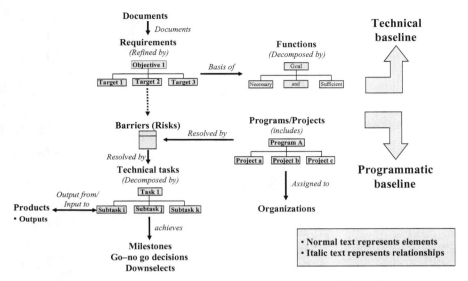

FIGURE 17.17   Implementing the SoSE process within CORE

be (from a technology development standpoint). The programmatic baseline, which focuses on the scope of work, schedule, and costs of activities required to bring the technological capabilities into being, ensures that we will *do things right*. The goal is an integrated baseline for a requirements-driven, mission-oriented program and, hence, a firm foundation for designing and implementing the SoS necessary to transition to an operational hydrogen economy.

The hydrogen integrated baseline is built on the CORE® MBSE software (Vitech Corporation, 2007) platform, a database-driven integrated systems engineering and program management support tool, designed to document and manage each step of the FRAT process in a centralized electronic repository. In addition to providing a graphical representation of the system, CORE® allows program organization and efforts to be viewed from a variety of perspectives through automatic generation of standard reports, tables, and diagrams. Figure 17.17 illustrates how CORE® can be used to organize, coordinate, and document both technical and programmatic baselines necessary to efficiently achieve the desired operational hydrogen economy. This diagram highlights the traceability of the information back to its source document and shows the key relationships between the model components (or "elements").

The technical information for the baseline—requirements (in the form of performance and cost targets), barriers (risks), and technical tasks and milestones—is drawn from high-level guiding documents for the DOE Hydrogen Program.[4]

---

[4]Guiding documents include DOE and EERE Strategic Plans, DOE Hydrogen Program Posture Plan and Multi-Year Program Plans. Available from DOE Hydrogen Program Web site. http://www.hydrogen. energy.gov/.

The programmatic information for the baseline—integrated portfolio of projects (which form the basis of the program's work breakdown structure), along with the associated budgets, schedules, and resource requirements—is developed in conjunction with the DOE Hydrogen Program and Project management teams. The integrated baseline is updated regularly as technologies advance, requirements evolve, and budgets expand and contract.

The integrated baseline as implemented in the CORE® MBSE software provides a robust tool to (1) document and track the current state and progress of the hydrogen SoS and (2) establish a defensible basis for budget estimates and technical decisions as the hydrogen fuel SoS moves from concept to commercial reality.

17.9 SUMMARY

A significant shift in the transportation energy sector is on the horizon as oil supplies tighten and the growth in worldwide energy demand accelerates. Increased pressure for environmentally and economically sustainable technologies and concerns about national energy security will also shape our future transportation systems. There is no single "best" solution to the transportation fuel conundrum the world is facing today. A variety of renewable and alternative fuels—biofuels, coal-to-liquids, and renewable electricity—will play a role in our transportation future. Deploying significant volumes of alternative transportation fuels will require major technology advancements along with government policies to spur the transformation of the current petroleum-based market and infrastructure.

A system of systems framework can effectively guide this complex evolutionary process, in which advanced energy technologies and systems will be continually integrated into the existing transportation fuel infrastructure as they become commercially available. The system of systems approach addresses the physical systems of the transportation fuel supply chain, as independent systems and as an integrated whole, in the context of ever-changing political, economic, social, and environmental conditions. The system of systems framework also aids in aligning and coordinating the efforts of the many diverse stakeholders that will be required to develop and implement future transportation fuel systems.

The success of today's transportation fuel transformation efforts will be determined many decades from now by the effectiveness of new fuels, technologies, and systems working together in a unified transportation fuel system of systems, as measured by the overall reduction in our nation's petroleum consumption.

REFERENCES

AFDC (DOE Alternative Fuels Data Center), 2007, DOE Alternative Fueling Station Locator. Available at http://www.eere.energy.gov/afdc/stations/find_station.php. Accessed October 6, 2007.

API (American Petroleum Institute), 2006, Understanding Today's Crude Oil and Product Markets. Available at http://www.api.org/aboutoilgas/sectors/pipeline/index.cfm; http://www.factsonfuel.org/gasoline/OilPrimer.pdf.

Bush, George W., 2006, State of the Union Address, January 31, 2006. Available at http://www.whitehouse.gov/stateoftheunion/2006/.

CWAC (Clean Water Action Council), 2007, Energy Efficiency and Conservation. Available at http://www.cwac.net/energy/index.html. Accessed October 2, 2007.

DOE (U.S. Department of Energy), 2002, *A National Vision of America's Transition to a Hydrogen Economy—to 2030 and Beyond*, U.S. Department of Energy. Available at http://www1.eere.energy.gov/hydrogenandfuelcells/pdfs/vision_doc.pdf

DOE (U.S. Department of Energy), 2007, *Biomass Multiyear Program Plan*, Office of the Biomass Program, Energy Efficiency and Renewable Energy, U.S. Department of Energy. Available at http://www1.eere.energy.gov/biomass/pdfs/biomass_program_mypp.pdf

DOE Biomass Program, 2007, Environmental Benefits. Available at http://www1.eere.energy.gov/biomass/environmental.html. Accessed October 2, 2007.

DOE and DOT, 2006, Hydrogen Posture Plan: An Integrated Research, Development and Demonstration Plan, U.S. Department of Energy (DOE) and U.S. Department of Transportation (DOT). Available at http://www.hydrogen.energy.gov/pdfs/hydrogen_posture_ plan_dec06.pdf.

DOT (U.S. Department of Transportation), 2006a, World Motor Vehicle Production, Bureau of Transportation Statistics. Available at http://www.bts.gov/publications/national_ transportation_statistics/html/table_01_22.html.

DOT (U.S. Department of Transportation), 2006b, World Motor Vehicle Production, Bureau of Transportation Statistics. (Highway, total registered vehicles includes passenger cars, motorcycles, other two-axle four-tire vehicles, single-unit two-axle six-tire or more trucks, combination trucks, and buses.) Available at http://www.bts.gov/publications/national_ transportation_statistics/excel/table_01_11.xls.

EIA (Energy Information Administration), 2004a, Estimated Number of Alternative Fuel Vehicles, U.S. Department of Energy. Available at http://www.eia.doe.gov/cneaf/alternate/page/datatables/aft1-13_03.html.

EIA (Energy Information Administration), 2004b, *Petroleum Supply Annual 2004*, Vol. I (Tables 36 and 40), Office of Oil and Gas, U.S. Department of Energy, Washington, D.C.

EIA (Energy Information Administration), 2005, Where Does My Gasoline Come From? U.S. Energy Information Administration. Available at http://www.eia.doe.gov/bookshelf/brochures/gasoline/printer_friendly.pdf

EIA (Energy Information Adminstration), 2006, Annual Energy Outlook. Available at http://www.eia.doe.gov/oiaf/archive/aeo06/gas.html.

EIA (Energy Information Administration), 2007, International Energy Outlook (IEO) Highlights. Available at http://www.eia.doe.gov/oiaf/ieo/highlights.html.

Ford Motor Company, 2007, Economic Impact of the Automotive Industry. Available at http://www.ford.com/aboutford/microsites/sustainability-report-2006-07/finCaseEconomic.htm. Accessed October 2, 2007.

GAO (U.S. General Accounting Office), 2000, Letter to Tom Harkin, U.S. Senate. (September 25), Resources, Community, and Economic Development Division, U.S. General Accounting Office. Available at http://www.gao.gov/new.items/rc00301r.pdf.

General Motors Corporation, 2006, GM Launches Ethanol Image Campaign, Press Release, January 25. Available at http://media.gm.com/servlet/GatewayServlet?target=http://image.emerald.gm.com/gmnews/viewmonthlyreleasedetail.do?domain=74&docid=22343.

General Motors Corporation, 2007, Live Green, Go Yellow. Available at http://www.gm.com/company/onlygm/livegreengoyellow/. Accessed October 6, 2007.

INCOSE (International Council on Systems Engineering), 2006, *Systems Engineering Handbook: A Guide for System Life Cycle Processes and Activities*, INCOSE-TP-2003-002-03.

ISEE Systems, 2007, STELLA Systems Thinking Software. Available at http://www.iseesystems.com/.

Mar, B.W., Bernard, G.M., 2002, FRAT-A Basic Framework For Systems Engineering, INCOSE 2002 International Symposium, July 28–August 1, Las Vegas, Nevada. Available at http://www.eskimo.com/~jjs-sbw/524_bmar.PDF.

Maren, G., 2004, The Rise of E85, *Motor Age*. Available at http://www.search-autoparts.com/searchautoparts/article/articleDetail.jsp;jsessionid=HXwBdskk1CKj9wTjmncSBYGTZhvP2DK3Lfz1GCGcgyP2Q84BvqPg!-1212263997?id=141080. Accessed October 17, 2007.

Nationmaster, 2007, Energy Statistics. Available at http://www.nationmaster.com/graph/ene_oil_con_percap-energy-oil-consumption-per-capita. Accessed October 2, 2007.

NGCA (National Corn Growers Association), 2007, World of Corn. Available at http://www.ncga.com/WorldOfCorn/main/production1.asp. Accessed October 2, 2007.

NHA (National Hydrogen Association), 2007, Hydrogen Fueling Stations. Available at http://www.hydrogenassociation.org/general/fuelingSearch.asp. Accessed October 3, 2007.

Perlack, R.D., Wright, L.L., Turhollow, A.F., Graham, R.L., Stokes, B.J., Erbach, D.C., 2005, *Biomass as a Feedstock for a Bioenergy and Bioproducts Industry: The Technical Feasibility of a Billion-Ton Annual Supply*, USDA/DOE, DOE/GO-102005-213. Available at http://www1.eere.energy.gov/biomass/pdfs/final_billionton_vision_report2.pdf

Pipeline 101, 2007a, Crude Oil Pipelines. Available at http://www.pipeline101.com/Overview/crude-pl.html. Accessed October 2, 2007.

Pipeline 101, 2007b, Refined Products Pipelines. Available at http://www.pipeline101.com/Overview/products-pl.html. Accessed October 2, 2007.

Pipeline 101, 2007c, Natural Gas Pipelines. Available at http://www.pipeline101.com/Overview/natgas-pl.html. Accessed October 2, 2007.

Polzer, H., DeLaurentis, D., Fry, D., 2007, *Multiplicity of Perspectives, Context Scope, and Context Shifting Events*. Available at http://colab.cim3.net/file/work/SICoP/2007-08-09/HPlozer03192007.doc.

RFA (Renewable Fuels Association), 2006, From Niche to Nation: Ethanol Industry Outlook 2006. Available at http://www.ethanolrfa.org/objects/pdf/outlook/outlook_2006.pdf. Accessed October 2, 2007.

Rifkin, J., 2002, *The Hydrogen Economy*, Penguin Group (USA).

SRI Consulting, 2004, *CEH Marketing Research Report: Hydrogen*, August 2004, *Chemical Economics Handbook*.

SOSECE (System of Systems Engineering Center of Excellence), 2007, SE vs. SoS Engineering. Available at http://www.sosece.org/index.cfm?fuseaction=79B5908B-802C-E84F-655D66D5E5B166F0. Accessed September 10, 2007.

The Detroit News, 2007, GM Aims to Market Fuel Cell Vehicles, June 15. Available at http://www.detnews.com/apps/pbcs.dll/article?AID=/20070615/AUTO01/706150354/1001/biz. Accessed October 2, 2007.

UCS (Union of Concerned Scientists), 2002, *Energy* Security: Solutions to Protect America's Power Supply and Reduce America's Oil Dependence. Available at www.ucsusa.org.

Vitech Corporation, 2007, Available at http://www.vitechcorp.com/.

Whitehouse, 2003, Hydrogen Fuel: A Clean and Secure Energy Future, Press Release January 30. Available at http://www.whitehouse.gov/news/releases/2003/01/20030130-20.html.

Whitehouse, 2007, Twenty in Ten: Strengthening America's Energy Security. Available at http://www.whitehouse.gov/stateoftheunion/2007/initiatives/energy.html.

Worldwatch Institute, 2007, *Biofuels for Transport: Global Potential and Implications for Sustainable Energy and Agriculture*, EarthScan, London, p. 117.

Chapter 18

Sustainable Environmental Management from a System of Systems Engineering Perspective

KEITH W. HIPEL,[1] AMER OBEIDI,[1] LIPING FANG,[2] and D. MARC KILGOUR[3]

[1]University of Waterloo, Ontario, Canada
[2]Ryerson University, Ontario, Canada
[3]Wilfrid Laurier University, Ontario, Canada

18.1 ENVIRONMENTAL SYSTEM OF SYSTEMS

18.1.1 A System and System of Systems

It was not the strongest hurricane in the history of the United States, but it was the deadliest and costliest (Knabb et al., 2006). At least 2541 persons were reported dead or missing as a result of hurricane Katrina and subsequent floods that hit along much of the north-central Gulf Cost of the United States from August 25 to 29, 2005. The U.S. Insurance Information Institute estimates the total incurred loss in property damage caused by the hurricane to be around $40.6 billion. The most devastation occurred in New Orleans, Louisiana, which accounted for 85% of the total deaths and 62% of insured losses. Damages from Katrina also extended to other areas in the states of Mississippi, Alabama, Florida, Tennessee, and Georgia, but to a lesser extent than New Orleans. Almost every levee as part of the federal flood protection system in New Orleans was breached, causing 80% of the city to be inundated.

Hurricane Katrina constitutes one of the most striking and bleakest examples of how human-induced interventions, which are purposefully designed to control rather than work in harmony with natural forces, can wreak catastrophic consequences on societal systems (Van Heerden and Bryan, 2006). In an impressive study about what caused the complete annihilation of some ancient civilizations such as the people of

System of Systems Engineering: Innovations for the 21st Century, Edited by Mo Jamshidi
Copyright © 2009 John Wiley & Sons, Inc., Publication

Easter Island, the Maya in Mexico, and the Norse people of Greenland, Diamond (2005) convincingly attributes societal collapse to a combination of cultural and population factors, a disregard of the environment, and the inability or unwillingness to take corrective actions to mitigate evolving ecological damages. In spite of advances in technology and knowledge of current complex societal systems, he professes that future societies are likely to face the same fate as old civilizations and will collapse because of environmental and population pressures, exacerbated by failures in strategic judgment and group decision-making processes. Diamond (2005) outlines a road map of features of group decision making that most likely would contribute to harming the environment, which are: failure to anticipate a problem before it occurs, failure to perceive a problem despite the fact that it has already happened, failure to solve the problem, and carrying out unsuccessful solutions to the problem. Situations involving multiple participants having multiple and conflicting objectives, interests, priorities, value systems, and beliefs are more highly susceptible to following the fate of ancient civilizations that totally vanished.

The disastrous effects of hurricane Katrina in New Orleans and the Mississippi Gulf Coast were not solely caused by nature, but rather they were created by unintended and systematic man-made mistakes in the planning, designing, and executing of a 100-year flood protection system put into practice by the construction of a system of levees around New Orleans, wetlands, and swamps to protect societal systems (Van Heerden and Bryan, 2006). Grunwald (2007) argues that a combination of unsubstantiated and faulty engineering designed solutions and misaligned or conflicting priorities of different stakeholders, such as the U.S. Army Corps of Engineers, politicians, and other commercial-interests lobby groups in the Oil and Shipping industry, exacerbated the severity of the ecological topography of the area and constituted the main reasons for the catastrophic effects on societal, energy, industry, agriculture, and other systems and systems of systems. As it will be argued in upcoming sections, currently, multiple participant-multiple objective decision making is a key characteristic of all systems of systems (Hipel and Fang, 2005).

Glaciers, coral reefs, boreal forests, wetlands, and polar and alpine life zones are examples of systems that are parts of a larger system called the natural system of systems. An important example of a class of societal system is a service system that provides intangible added value assets (Quinn et al., 1987) to be utilized by people, but cannot be owned, such as health care, amusement, garbage management, and defense and security. Infrastructure systems are used for the physical movement of people and goods and enabled by physical components such as roads, bridges, seaports, airports, and pipelines (Hipel et al., 2007). On the other hand, an electrical power supply system is an example of a complex adaptive system of systems, created entirely by human beings, and is part of the energy, services and infrastructure, industrial, and agricultural systems. To be more specific, this system consists of many different types of entities or agents that are actively interacting together for electricity supply and consumption within and among societal systems of systems. Some of these agents, for instance, are privately run supply and generating plants, government-owned transmission and distribution companies, and consumers, be they individual house-holds, large service and industrial firms, or agricultural complexes. Some of the agents possess the properties of being intelligent, can adapt to changing situations, and can learn from experience. In this sense, these agents can be regarded as systems, and the combination of all agents constitutes a system of systems.

A brief literature review reveals that there are many definitions for both a system and a system of systems depending on the domain that one is adopting for analysis. Senge (1994, p. 90), for instance, defines a system as "a perceived whole whose elements hang together because they continually affect each other over time and operate toward a common purpose." Rouse (2003, p. 154) views a system as "a group or combination of interrelated, interdependent, or interacting elements that form a collective entity. Elements may include physical, behavioral, or symbolic entities. Elements may interact physically, mathematically, and/or by exchange of information. Systems tend to have purposes, although in some cases the observer ascribes such purposes." In this sense, the composite of entities in a system that provides a capability to accomplish specific functions and objectives must be *interrelated*, *interdependent*, and *mutually interacting*.

A system consists of a large number of mutually interrelated entities. Interrelatedness in property or behavior is important for the system to exist and to maintain an order of compatibility for this existence. Also, the property and behavior of every entity affect and depend on the property and behavior of at least one other entity in the composition that makes the system (Ackoff, 1973; Blanchard and Fabrycky, 1990). Ackoff (1973, p. 664) argues that "A system is more than the sum of its parts; it is an indivisible whole. It loses its essential properties when it is taken apart." Therefore, interdependency among entities means that fragmentation of the system into its constituent entities or subcompositions cannot be done without affecting the integrity of the system as a whole. In other words, the system would lose some of its fundamental features. Mutual interactions among entities of a system lead to synergy in the performance of the system, which otherwise would not be possible to attain when considering actions and behaviors of individual entities (Ackoff, 1973).

Interrelatedness, interdependency, and mutual interactions of entities in a system define the main characteristics of the system. Sophisticated architectural composites of entities give rise to the system's characteristics of being complex, emergent, and adaptive. A complex system is one that is laterally structured, difficult to understand using a reductionism approach for the lack of simple cause-and-effect relationships within entities, sensitive to small perturbations, and interactions among entities are unpredictable (Senge, 1994; Sage, 1999; Calvano and John, 2004; Miller and Page, 2007). A system has the property of emergence when the behavior of the whole cannot be predicted from the behavior of the composite entities (Sage, 1999). The system in this case is dynamically evolving over time to a higher order system behavior. A system, which is complex and emergent, is also adaptive, so that it has the capacity to continuously respond to variations in the environment as a result of learning from experience (Miller and Page, 2007). A system can also be formed by the integration of subsystems. In this case, the system is considered to be a metasubsystem.

Many large and complex real-life phenomena, however, consist of interdependent, heterogeneous systems, working cooperatively to achieve a common objective. At an early date, Ackoff (1973, p. 664) recognized these systems and argued that "[t]he elements of a system may themselves be systems, and every system may be part of a larger system." A system, in this case, is in reality a system of systems, or a metasystem, which represents a new class of systems created by the integration of multiple harmonious systems operating in a synergistic manner that otherwise would not be possible to realize (Sage and Cuppan, 2001; Hipel et al., 2007). Environmental, societal, intelligent, and integrated systems can all be described as complex systems of systems (see Casti, 1997; Bar-Yam, 2005; Hipel and Fang, 2005; Braha et al., 2006; Hipel et al., 2007; Rouse, 2007). For instance, at the basic level of a catchment or river basin, water may be studied as an ecological system, so a holistic approach to managing it requires that surface water and groundwater issues be integrated with land use. However, water can also be part of a higher order, large and complex system consisting of water, land, atmosphere, and biosphere. To ensure the integrity and sustainability of ecosystems, both terrestrial and aquatic, the focus must be on managing the relationships among the various subsystems of this complex system of systems.

Sage and Biemer (2007, p. 7) define a system of systems as "a large-scale, complex system, involving a combination of technologies, humans, and organizations, and consisting of components which are systems themselves, achieving a unique end-state by providing synergistic capability from its component systems". Hence, a system of systems involves coupling, combination, or integration of multiple, autonomous, heterogeneous, and trans-domain systems, which collaboratively as a whole perform functions not possible by any of the individual systems (Luskasik, 1998; Maier, 1999; Carlock and Fenton, 2001; Keating et al., 2003). A system of systems is by default a complex system, where each individual constituent system is autonomous because it may be evolving independently from other systems in the whole. Sage and Biemer (2007), drawing on earlier research (see Maier, 1999; Sage and Cuppan, 2001), argue that a system of systems has a majority of the following characteristics: (1) operational independence of the individual systems; (2) managerial independence of the individual systems; (3) geographical distribution of the individual systems; (4) emergent behavior, where the system performs functions not possible by any of the individual stand-alone systems; (5) evolutionary development caused by continuous interoperability relationships between constituent systems; (6) self-organization; and (7) adaptation.

18.1.2 Types of Systems of Systems

A universal characteristic of virtually all systems or systems of systems is that they involve multiple participants with multiple objectives (Hipel and Fang, 2005). In systems of systems, multiple participants pervade because during any process of planning, designing, construction, operation, or managements of these systems there is often a hierarchy of various stakeholders distributed across different and interconnected systems, which range from power brokers at the level of government, and corporate and private investors, to the level of those mostly affected, be that human or biological life. Multiple objectives are often present in the form of a hierarchy of

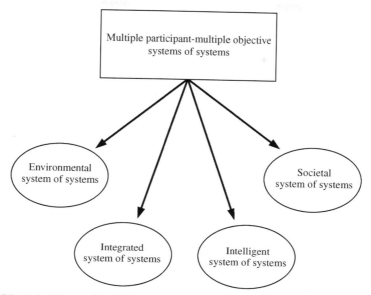

FIGURE 18.1   Types of multiple participant-multiple objective systems of systems

objectives, subobjectives, and sub-subobjectives, because of the cross-disciplinary and cross-domain and interoperability of these systems. In many systems, these multiple objectives are incommensurate, conflicting, or competing. Figure 18.1 displays some higher level multiple participant-multiple objective systems of systems conceptualized as four main types of systems: environmental, societal, integrated, and intelligent (Hipel and Fang, 2005).

Environmental systems of systems (see Fig. 18.3 in Section 18.2.1) refer to the natural incubators that provide, permit, and encourage the formation, the development, and the evolution of all systems of systems on planet Earth. The environment constitutes a set of complex systems encompassing and interacting with all biological systems, consisting of microbiological systems (proteins and nucleic acids), zoological systems (animal life), and botanical systems (plant life). Clearly, the environment supports the life of all societal, integrated, and intelligent systems. Examples of complex environmental systems include hydrological systems, natural systems, ecological systems, and geographical systems. For instance, hydrological systems represent a number of other systems such as atmosphere, glaciers and ice caps, underground soil, surface of land and rock, oceans, rivers and lakes, and oceans, in which the common throughput of these systems is water in its different states (Hipel and McLeod, 1994; Hipel and Ben-Haim, 1999). A biological system interacting with an environmental system at a given location creates an ecological system.

Societal systems broadly comprise a range of man-made activities and artifacts that have been created to sustain human life, to serve individuals' interests, and to provide recreation to groups of people. These systems represent the realms that have the most influence on all other systems, especially detrimental effects on the environment.

Examples of societal systems include economical, political, government, financial, agricultural, industrial, urban, and energy systems. Human ingenuity and technological innovations in recent years have led to rapid advancements in the emergent fields of robotics, mechatronics, information, and computerized systems. This has created a new type of system of systems that may be grouped under the category of intelligent systems (Hipel and Fang, 2005; Miller and Page, 2007). This category of systems comprises computerized-worlds inhabited by agents, which are programmed to interact with one another, cooperatively or competitively, to achieve certain prescribed objectives. Software agents commonly used to assist in online bidding strategies on eBay, such as eBay Bid Assistant and eSnipe, are examples of an intelligent system of agents that employ competitive strategies against other agents. These agents place a bid on behalf of a bidder for an item listed on eBay just a few seconds before auction ends in an attempt to win an item, saving the bidder time, money, and aggravation. A mechatronics entity, such as a robot, is another example of an intelligent system that behaves cooperatively with other robots working on an assembly line of a car manufacturing facility. Whatever the case, an intelligent system is entirely designed by humans to act interdependently and it contains certain decision-based strategies and protocols inscribed into its coding to achieve certain goals (Rosenschein and Zlotkin, 1994; Ho and Kamel, 1998). As emphasized by Hipel and Fang (2005), because intelligent systems are completely designed, operated, and maintained by humans, professionals, including systems engineers and computer scientists, have an important responsibility to create ethical systems that prioritize environmental stewardship and societal well-being over bottom-line profits.

In situations in which a software agent interacts with people or a mix of people and agents, the combination of societal and intelligent systems produces an integrated system. Figure 18.2 depicts that combination within the context of environmental systems. Notice that some artificial intelligent systems are not shown to be part of societal systems because some of these created systems could operate for indefinite time periods independent of the societal systems; but when societal and intelligent systems overlap this gives rise to integrated systems. In the eBay auction example, if a software agent, representing a client, employs a sniping strategy against other bidders, the combination comprises an integrated system of systems. Another example of an integrated system is the landing control of many modern jumbo jet airplanes. These aircrafts are equipped with sophisticated software agents capable of making a fully

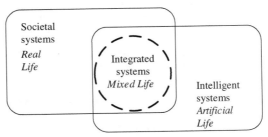

FIGURE 18.2   World systems of systems

automatic takeoff and landing, and can be safely flown entirely on their own, without any intervention by pilots. However, often pilots exercise overall control of the aircraft within an integrated systems framework. In fact, one can think of the pilot as a "safety device" who can takeover when an unforeseen emergency emerges as a result of emergent properties of the complex aircraft system of systems or some unexpected outside interference such as a sudden downdraft.

18.2 ENVIRONMENTAL ISSUES

18.2.1 Natural, Societal, and Environmental Systems of Systems (SoS)

Using the jargon of geological timescales, the last 8000 years is called the *Holocene Epoch*, which is part of the *Quarternary Period*, which in turn is a subset of the *Cainozoic Era*. What is particularly remarkable about the *Holocene Epoch* is that the climate around the globe has been relatively stable and warm—hence this epoch is often labeled as "the long summer." Even though human beings came into existence about 150,000 years ago in Africa, from where they migrated around the globe, it was only during "the long summer" that humans had access to climatic conditions within which they were able to domesticate plants and animals to form farming communities, upon which villages, towns, cities, and societies were constructed.

Biological life on planet Earth can be classified according to a hierarchy of living organisms. At the basic level, there are microorganisms (microbiological species such as proteins, nucleic acids, and cells), then, plant life (botanical species), and finally a higher life form, animal life (zoological species). The part of Earth upon which all biological species survive, evolve, and thrive is called the biosphere. As shown in the right-hand side of Fig. 18.3, the three crucial natural systems supporting the biosphere are the lithosphere (soil, rocks), hydrosphere (water), and atmosphere (air). When a biological life form functions with a natural system, the resulting system is often referred to as an ecological system, or ecosystem. Hence, a natural system of systems exists on planet Earth as a result of the highly balanced and complex interrelatedness, interdependence, and mutual interactions among natural systems and ecosystems. Notice also that systems composing a natural system of systems are autonomous (evolving independently) and heterogeneous, which as a whole are interdependent and homogenous and function synergistically to achieve an ultimate objective (Luskasik, 1998; Maier, 1999; Carlock and Fenton, 2001; Keating et al., 2003). However, a disruption of any of these systems has high implications on other systems. An example of a complex system of systems is the intersection of water and biological systems. The bulk transfer of water from its natural drainage basin can have negative consequential effects on native habitats and exotic species, as well as social, economic, and environmental impacts.

A healthy and sustainable natural system of systems is vital for life support and prosperity of human beings. The left-hand side of Fig. 18.3 depicts some of the major systems that compose a large and complex societal (or human) system of systems: energy, services and infrastructure, industrial, and agricultural systems. All societal

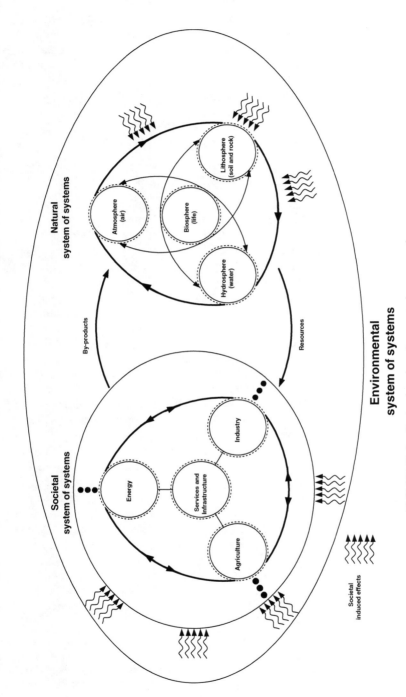

FIGURE 18.3 Societal and environmental systems of systems

systems are designed by human beings with the ultimate purpose of serving their individual or collective needs and desires. Notice that a societal system of systems consists of a combination of autonomous and heterogeneous technologies, humans, and organizational systems, which are operationally independent, managerially independent, and geographically distributed. Moreover, the system of systems as a whole is emergent, evolving, self-organized, and adaptive, as a result of interoperability among constituent systems (Sage and Biemer, 2007). An energy system, for instance, is a system of systems and consists of renewable and nonrenewable energy systems (see Fig. 18.4 and Section 18.2.2). A disruption of the energy system can have dire consequences on other industrial, services and infrastructure, and agricultural systems as exemplified by the great North American power shutdown of August 14, 2003,

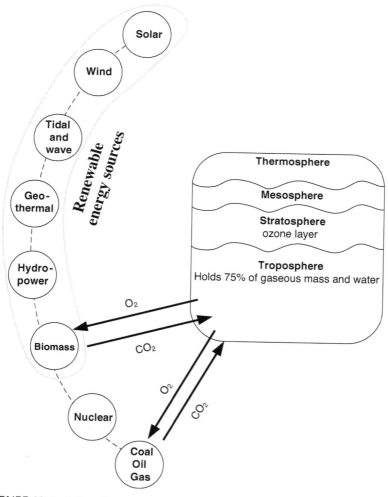

FIGURE 18.4   Interactions among energy and atmospheric systems of systems

which is described in the beginning of Section 18.3.1. Hence, systems within the societal systems of systems are highly dependent and completely interrelated with other systems, be they in the societal or natural systems domains.

Arrows connecting societal and natural systems of systems signify a higher order of interrelatedness and mutual interaction, which gives rise to a large-scale complex environmental system of systems. The arrow at the bottom of Fig. 18.3 represents natural resources that are mined, drawn, excavated, and exploited for serving the societal systems. For instance, water is drawn from the hydrosphere (rivers, lakes, and groundwater); hydrocarbon fossil oil is drilled and mined from the lithosphere. On the contrary, unwanted by-products from the societal systems are dumped back into the natural systems of systems in the form of wastes and contaminating gases. For example, the defense industry in nations having nuclear weapons can produce large quantities of nuclear wastes that can ruin huge tracts of land as witnessed by the infamous U.S. Department of Energy Hanford Site in the American state of Washington. Also, highly populated cities discharge huge quantities of sewage and industrial wastes into rivers and lakes causing an alarming contamination of both marine and freshwater ecosystems.

Overall, the extraction and exploitation of natural resources accompanied by dumping of by-products into natural system of systems create *societal induced effects*, which impact both societal and natural systems and place stress on the environment. In some cases, societal induced effects are merely human actions that inadvertently cause the destruction of the environment. For instance, poor logging practices in mountainous areas with heavy rainfall are susceptible to excessive deforestation and habitat destruction because of massive landslides that suddenly take place. Also, human population growth, especially in developing countries, means that there is more demand for valuable natural resources such as water, minerals, metals, and agricultural land that are necessary for sustaining societal systems. This massive appetite for resources is accompanied by an increase in the amount and complexity of by-products being dumped back into the natural system of systems.

But some of the societal-induced effects are ecological damages inflicted intentionally by humans such as the buildup of toxic chemicals in the environment and human-caused greenhouse gas emissions and associated climate change that result from massive energy consumption. For instance, extreme climatic conditions are a key characteristic of climate change and are already well underway as evidenced by severe droughts in the Sahel region of Africa and in Australia, or massive flood damages as a result of hurricanes (e.g., hurricane Katrina burst the dikes protecting the American city of New Orleans). Scientists now believe that the intensity and frequency of hurricanes are brought about by the heating of the oceans. As a result of global warming, the Gulf Stream could slow down and thereby cause Europe to "freeze over." Melting of the polar ice caps and glaciers will cause sea levels to rise and coastal cities in low-lying areas, such as Amsterdam and Singapore, to be flooded. Atoll nations in the Pacific Ocean, including Kiribats, Maldives, Marshall Islands, Tokelan, and Tuvala, will simply disappear under the water due to higher sea levels and bleaching of the coral reefs. The foregoing and many other tragic impacts of global warming will bring about famine, mass migrations of people, poverty, pronounced conflict, and perhaps even the collapse of large civilizations, such as China and India, which can barely feed their enormous populations. Whatever the situation, one known urgent problem in sustainable relationships, which must be wisely addressed in the short, medium, and long terms, is the global warming crisis, which requires systems thinking to ensure that this most pressing of sustainable relationships is effectively solved.

As a dramatic illustration of the great urgency for the immediate development of effective systems-thinking analytical approaches for developing environmental policies that can adapt and react to problems in real time based on information collected and processed virtually instantaneously on a global scale, consider a global disaster which nearly occurred and was just narrowly avoided simply by good providence. The Ozone Hole crisis caused by the release of ozone-depleting substances, principally chlorofluorocarbons or CFCs, is being rectified through the 1987 Montreal Protocol and strengthened versions thereof. Based upon the scientific work of their predecessors, Paul Crutzen, F. Sherwood Rowland, and Mario Molina established that ozone depletion was real and the culprit was man-made chemicals, especially CFCs. However, as pointed out by Flannery (2005, Ch. 23), humanity was very fortunate that chemists used chlorine rather than bromine to make CFCs. Because bromine is 45 times more lethal in destroying ozone than chlorine, if BFCs had been made in place of CFCs, most of the ozone in the stratosphere shown on the right in Fig. 18.4 could have been destroyed before humankind ever realized what was happening. Hence, biological and ecological systems shown on the right in Fig. 18.3 could have failed on a large scale as a result of intense ultra-violet radiation, thereby causing the societal systems on the left to collapse.

18.2.2   Energy Systems and the Environment

To illustrate why systems thinking approaches are urgently needed to tackle serious environmental issues now being played out as a result of societal induced effects, consider the interaction between energy systems from the societal realm and atmospheric systems in the natural domain, as illustrated in Fig. 18.4. To achieve economic growth and development in industrialized countries and to satisfy the basic needs and demands of human beings, a range of technologies for producing energy has been developed. Traditionally, there has been extensive extraction and utilization of hydrocarbon fossil fuels such as coal, oil, and gas for use in agriculture, industry, various services, and infrastructure systems. When fossil fuel is burned, greenhouse gases, mainly carbon dioxide (CO_2) along with other chemical compounds such as methane and nitrous oxide, are released into the atmosphere. These greenhouse gases act as a blanket and trap heat in the lower troposphere, causing global warming and severe climate change. In fact, carbon dioxide released as by-products from societal systems is the biggest cause of global warming—80% of global warming is attributed to emissions of carbon dioxide. However, other synthetic chemicals such as chlorofluorocarbons or CFCs are destroying the ozone layer in the stratosphere, causing serious health problems to people.

At the present time there are 380 ppm (parts per million) of CO_2 in the atmosphere, which represents about a 20% increase in CO_2 levels since 1958 (315 ppm), and this number will rapidly increase, unless, of course, rapid and decisive mitigating actions are taken by society to bring this frightening situation under control (Hipel et al., 2007; IPCC, 2007; Stern, 2007). In fact, when all greenhouse gases are taken into account, the CO_2 equivalent is currently about 430 ppm. It is alarming to know that 450 ppm of CO_2 in the atmosphere is a level that is unparalleled to any level in the last 650,000 years, and that the accompanying temperatures would also be the highest. Moreover, even if appropriate remedial actions were taken now, there would still be negative consequences, but at least they would be bearable and manageable. In reality, the annual costs of achieving stabilization of greenhouse gases in terms of CO_2 equivalent units at 500–550 ppm is only about 1% of global gross domestic product (GDP) (Stern, 2007).

To counter the effects of greenhouse gases that cause climate change, societal systems must reduce all CO_2 emissions by 70% below the 1990 levels by 2050 (Flannery, 2005), which would cause CO_2 levels to stabilize at 450 ppm. Hence, energy systems must be properly designed and managed to meet the vision of 450 ppm of CO_2 in the atmosphere. As suggested by Jaccard (2005), the ultimate mission is to have "a low impact and low risk energy system that can meet expanded human energy needs indefinitely and do this as inexpensively as possible, without succumbing to cataclysmic force (like excessive global warming) at some future time." The left-hand side in Fig. 18.4 displays the main categories of energy production systems that are available for utilization. The top-six-categories consisting of solar, wind, tidal and wave, geothermal, hydropower, and biomass are considered renewable energy sources, among which only biomass burning releases CO_2. Nuclear power generation is a CO_2 emission free source of energy, although one must ultimately dispense of the nuclear residual wastes, and, of course, have rigorous safety, regulation, and inspection systems to prevent nuclear accidents and the development of nuclear weapons. New generations of nuclear reactors, which are safer and more fuel efficient, are already under development.

As argued by Gore (2006), political will is required to control greenhouse gases, which translates into the need for governments at all levels to develop and implement sound policy and governance in cooperation with all stakeholders. Gore further states that the control of greenhouse gases is essentially a moral issue. One should not imperil the very survival of the human species because of economic greed. For his timely and valuable contributions to creating a global awareness of the seriousness of global warming and political approaches for solving it, former Vice-President Gore was announced as a corecipient of the 2007 Nobel Peace Prize on October 7, 2007, along with the many members of the Intergovernmental Panel on Climate Change (IPCC) who provided the rigorous scientific studies. The "atmospheric commons" must not be allowed to succumb to the "Tragedy of the Commons" in which a resource collapses as a direct result of individual greed by a myriad of interest groups. Hipel and Fang (2005) maintain that technological, societal, and hybrid systems must directly account for stakeholders' value systems and should be designed, operated, and maintained according to sound ethical principles. When designing a physical system, such as a commercial aircraft, systems engineers carefully employ a vast array of physical systems methods to interactively arrive at a final design that is reliable, safe, and economical. Likewise, this deliberate type of logical and creative thinking is required in the purposeful design and analysis of policies. As explained in Section 18.3, societal systems models that could prove to be useful in policy analysis include conflict resolution to handle strategic aspects of policy analysis and multiple objective decision analysis to account for the multidimensional characteristics of stakeholders' value systems. Undoubtedly, the expertise and creativity of systems thinkers are sorely needed in the design of policies, in general, and for the design of effective treaties for reducing greenhouse gases, in particular.

Governments, scientific bodies, professional societies, and other organizations confirm that global warming is real and suggest courses of action to follow. On October 30, 2006, the Government of the United Kingdom published the Stern Report (Stern, 2007) in which it states "The scientific evidence is now overwhelming: climate change presents very serious global risks, and it demands an urgent global response." The Review investigates in detail the economic costs of the impacts of climate change, as well as the costs and benefits of actions to reduce the emissions of greenhouse gases. The simple conclusion of the report is clear: "The benefits of strong, early action considerably outweigh the costs." In reality, climatic change would cost 20 times less than doing nothing, which is considered by many to be highly immoral. Furthermore, as pointed out in the conclusions, "There is still time to avoid the worst impacts of climate change, if we take strong action now." Hence, a sustainable relationship must be purposefully maintained between the societal and environmental systems in Fig. 18.3, as well as between the more specific systems of systems depicted in Fig. 18.4.

18.3 SYSTEM OF SYSTEMS DECISION METHODOLOGIES

18.3.1 SoS Thinking for Policy Development and Governance of Resources

On August 14, 2003, a sudden and unexpected electrical blackout covered most of the province of Ontario in Canada and several states in the northeastern part of the United States. Almost 50 million people were affected by the complete loss of electricity that caused a dramatic effect on economic and finance, service and infrastructure, security, and other societal systems. Although electricity was restored within hours to many customers, in some areas in the United States the loss of power lasted several days and many large industrial facilities in the province of Ontario experienced rolling blackouts due to generation capacity shortages. The estimated economic impacts alone for the loss of electricity is in the range of US$4–10 billion in the United States, while in Canadian the gross domestic product fell by 0.7% in the month of August and the province lost approximately 18.9 million work hours and Can$ 2.3 billion (NAERC, 2004). The immense ramification of the collapse of this complex infrastructure system is a clear example of the intricate dependency of societal, natural, and environmental systems or systems of systems. However, the conclusions of the task force (U.S.—Canada Power System Outage Task Force) that was created to investigate the incident, galvanized the importance of adopting an integrative approach for managing this complex electrical generation and distribution system of systems, and highlighted the need for effective and adaptive multiple participant-multiple objective decision-making approaches that provide strategic and tactical recommendations in a timely fashion.

As explained in Section 18.2, modern societal developments are stressing planet Earth's terrestrial, aquatic, and atmospheric ecosystems as a result of global warming and climate change (IPCC, 2007). Hence, there is an imperative need for holistic and adaptive approaches that deal with prevalent socioeconomic and sociocultural developments and challenges. Gharajedaghi (2006) argues that complexity and chaos encountered in many systems are not intrinsic characteristics of these systems—rather they are the consequence of our perceptions, understanding, and approaches to managing these systems. A systems thinking approach is required to mitigate the tremendous pressure caused by population explosion, deepening poverty, and prevailing economic and industrial activities on the environment. Such an approach for the management of environmental and societal systems of systems has to embrace both a holistic and adaptive analytical framework as a viable paradigm for achieving the objectives of sustainable development and protecting vital ecosystem integrity (Hipel and Fang, 2005; Gharajedaghi, 2006; Hipel et al., 2007). As an example of an effective management system, a holistic approach to managing water resources requires that surface and ground water issues such as quantity, quality, and upstream and downstream interests be integrated with land use at the level of the catchment and river basin.

For an integrated system of systems approach to be complete, realistic, and viable, the physical design of the system has to reflect the value systems, interests, and objectives, of all stakeholders, and has to be directly linked with proper governance systems to ensure goals are met in an integrative manner. In other words, an integrated approach to management of the environment and other societal systems must take into account the behavior of various constituencies within and between the natural and societal dimensions to guarantee maintaining a balanced and sustainable relationship between these systems of systems.

A careful integration of various consistencies of systems of systems is important but not entirely sufficient for guaranteeing success. Another point of import for effective systems thinking is adaptability. The complexity and scope of many environmental and societal systems of systems requires flexible and timely responsive theoretical and practical methodologies for dealing with the unpredictable, variable, and uncertain nature of these systems. For example, current studies on global warming and climate change indicate that as the atmospheric systems remain under abuse from technological and man-made stresses, future mitigation measures required to stop and reverse deteriorations in these systems have to be integrative and highly adaptable (IPCC, 2007; Stern, 2007). Hence, an adaptive systems approach to decision making has to be closely coupled with the integrative framework to embrace complexity and high uncertainty. The approach is adaptive because it acknowledges the dynamic nature of systems, requires continues monitoring, and flourishes on active learning with an understanding of human decision-making processes (Gunderson et al., 1995; Walters, 1997; Gunderson, 1999).

The process of adaptive and integrative systems approach for policy governance and management of systems of systems begins by ascertaining systems' assumptions and boundaries, and then decomposing the systems into subsystems. All relevant constituencies must be identified and all stakeholders' perspectives concerning the behavior and state of a system have to be considered before planning, implementing, enforcing, and evaluating different policies. Then, one must embrace all of the characteristics of system of systems thinking in the process, as articulated by Hipel et al. (2007). Specifically, several characteristics must be contemplated when considering an adaptive and integrative analytical framework. First, most real-world problems require a holistic viewpoint in the design and analysis of systems of systems in which various system constituencies are part of the analytical model. Second, systems thinking is a multidisciplinary approach that requires the interdisciplinary collaboration of many experts including people working in economics, social science, engineering, and other fields in science. Third, multiple participants and their value systems, beliefs, priorities, objectives, and perceptions should be entertained when analyzing complex systems of systems. Finally, since multiple participant-multiple objective decision making is a main feature of any system of systems (Hipel and Fang, 2005), ethical designs and fairness principles (Young, 1994; Syme et al., 1999; Wang et al., 2007b) should be considered as part of an adaptive and integrated approach.

To address the complexity of systems, both art and science have to be entertained in the process of systems thinking. Innovative design, modeling procedures, and analytical tools can be employed for developing, managing, and governing systems of systems, but human judgment and decision-making processes play important roles for understanding these complex systems (Newbern and Nolte, 1999; Hipel et al., 2007). The art of systems thinking in decision making, especially in situations characterized by the presence of multiple participants, each with multiple objectives, comes from the continuous and iterative revisions of systems' assumptions and boundaries as information and insights from other systems analyses are garnered to procure a better understanding of the problem. Hence, this qualitative part of the systems thinking approach leads to enlightened decision making and provides effective cooperation among decision makers.

18.3.2 Systems Decision Tools for Integrative and Adaptive Management

Formal scientific approaches have been traditionally used for modeling and analyzing real-world problems. Most notably, the established field of operations research (OR) is based on a set of rigorous procedures and methods, such as mathematical programming, decision analysis, heuristics, discrete simulation, stochastic methods, neural networks, and genetic algorithms, which have been intensively applied to diverse practical problems in industry, business, military operations, and many other domains. In fact, the Institute for Operations Research and the Management Sciences (INFORMS) and the International Federation of Operational Research Societies (IFORS) view OR as the application of quantitative modeling processes to complex systems that helps management determine its policy and actions scientifically.

Hipel et al. (2008) explain how one matches the capabilities of OR techniques to the characteristics of the problem being studied to decide upon which OR techniques to employ. Historically, OR techniques were mainly developed for tackling highly structured problems that are expressed quantitatively and have well-defined objectives, constraints, and boundaries, for which the ultimate purpose is utilization at the tactical or operational level of decision making.

However, as discussed in Section 18.2, unprecedented energy resource consumptions, depletion of natural resources (land and marine) caused by globalization and the adoption of an exclusive market-driven economy, rapid population growth, and many other technological advancement issues are causing environmental degradation, climatic changes, and threatening planet Earth's ecosystem viability. Hence, while the use of traditional analytical tools such as operations research are often used for modeling and analyzing many real-world phenomena, and perhaps some complex systems, there is an urgent need for new ways of systems thinking that encourages a holistic approach to managing chaos and understanding complexity (Gharajedaghi, 2006). Some innovative methodological and theoretical modeling procedures that incorporate language necessary for addressing the complexity of large-scale systems and system of systems and embrace all the characteristics of system of systems thinking (mentioned in Section 18.3.1— holistic viewpoint, interdisciplinary, multiple participants with multiple objectives, and ethics and fairness) are purposefully designed to be employed in assisting all critical stakeholders in a system. This implies that the field of systems engineering (Sage, 1992) must be radically expanded to handle the forgoing and many other complex systems of systems issues (Hipel et al., 2007).

Figure 18.5 depicts aspects of a system of systems decision-making framework, based on the suggestions of Hipel (1992), for considering strategic decision making, where problems are less well-defined, and require both quantitative and qualitative modeling. For example, consider the situation in which a developed country wishes to evaluate whether or not to adopt the Kyoto Protocol for the reduction of greenhouse gas emissions and the extent of that reduction for, say, the next 20 years. The left column in Fig. 18.5 contains some of the main factors that must be considered in implementing a policy at a scale of that significance. A sound physical design, in this case, includes designing and developing several climate models, and choosing the ones that provide realistic descriptions of the current state of affairs and verifiable spatial and temporal predictions over time. A climate model is a highly complex computer model that represents the atmospheric system with many feedback loops and interactions, and includes factors such as precipitations, interactive clouds, oceans, land surface and aerosols, carbon dioxide concentration and cycle, and radiation levels of various surfaces on Earth. A challenging task among scientists in this field is reaching a consensus on the most appropriate climate model to adopt. However, such conclusions must also be evaluated with respect to other environmental, economic, financial, political, and societal systems of systems considerations. Some of these, such as those having political and social impacts, are nonquantitative in nature, while others, such as those having economic, risk, and certain environmental factors, may be measurable and quantitative in nature.

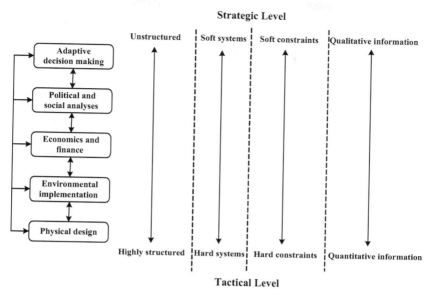

FIGURE 18.5   System of systems decision-making framework

A main feature of the system of systems shown in Fig. 18.3 is openness, where a full understanding of the behavior of any system requires understanding of the context in which the system interacts and interoperates with other systems (Ackoff and Emery, 1972; Gharajedaghi, 2006). A system's performance is controlled by its physical structure and its environment that include, among other things, all stakeholders that influence the system (Hipel and Fang, 2005). Hence, a holistic systems thinking of stakeholders' choices in a system of systems requires a repertoire of analytical tools that are capable of addressing tactical (or local level) and strategic (embodies system of systems level) decisions. The right side in Fig. 18.5 shows the four main characteristics of this decision-making hierarchical framework (Hipel, 1992; Hipel et al., 2007). A value focused thinking approach can be used to help in determining both short- and long-term objectives needed for sound decision making (Keeney, 1992). Depending on the depth of the analysis, one moves from analyses that require tactical level decision making, where systems are highly structured with quantitative constraints and quantitative information to unstructured systems with qualitative constraints and qualitative information. As indicated on the top-left cell in Fig. 18.5, the output from all of the systems analyses must be taken into account to reach an overall informed, adaptive, and integrative strategic system of systems decision, which has long-lasting and generally agreeable terms for all critical stakeholders. Hence, one must choose an appropriate set of systems tools to study and analyze the relevant aspects of the system or system of systems being studied (Hipel et al., 2007).

As can be appreciated from the foregoing, interactions among key stakeholders at the strategic level of a system or system of systems are likely to cause conflict. In fact, as one moves in the hierarchy from the tactical to the strategic level more decision makers are expected to be involved, which naturally embody further differences in value systems, beliefs, criteria, priorities, and incommensurate objectives, some of which are conflicting. Because of the universal nature of conflict whenever human beings interact with one another, research on multiple participant-multiple objective decision making and conflict resolution has taken place in a wide range of disciplines, including OR and systems engineering. Figure 18.6 shows the genealogy of formal multiple participant system of systems tools founded upon various mathematical assumptions. At the top is game theory, which was firmly established in 1944 by Von Neumann and Morgenstern (1953) with the publication of their seminal book entitled *Theory of Games and Economic Behavior*, for dealing with interactive decision situations.

Over the years many game theoretic models have been developed for describing decision situations that range from purely competitive, or noncooperative conflicts, to those having a high level of cooperation. For instance, approaches listed in the left column of Fig. 18.6 are generally used in situations where each decision maker has independent control of its specific set of actions or options, but has limited control over the outcome of the conflict as it interacts with other decision makers. This interdependency in decisions is what adds to the complexity of system of systems shown in Fig. 18.3. In cooperative game theory, shown in the right column in Fig. 18.6, two or more decision makers form a coalition and adopt cooperative behavior toward members of their own coalition and a competitive behavior toward those decision makers outside the coalition. In another situation, decision makers may choose to share in the use of a resource, but the size of allocated pieces must be decided by employing a rule for fair division or resource allocation.

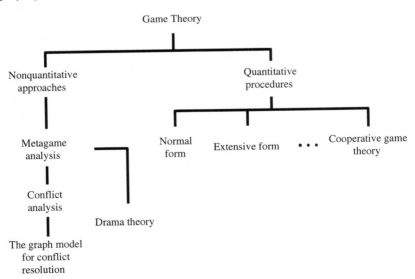

FIGURE 18.6 Genealogy of formal multiple participant SoS tools

Another way to classify a game theoretic method is according to the type of decision makers' preference information. A decision maker's preferences reflect the decision maker's beliefs, value system, priority, and objectives. Hence, to calibrate a conflict model, one must elicit each decision maker's preference over available outcomes. The techniques listed in the left column in Fig. 18.6 are usually categorized as being nonquantitative approaches as they only assume relative preference information. Accordingly, it is not necessary to know exactly the degree to which one prefers something over another. However, techniques falling under the right column in Fig. 18.6 generally require cardinal preference information expressed by Von Neumann and Morgenstern utility functions. Therefore, these techniques are usually labeled as being quantitative. It should be emphasized again that both procedures in the left and right columns in Fig. 18.6 constitute formal mathematical models.

In the following sections, three game-theoretic systems of systems methodological applications are used to illustrate the modeling of interactions among decision makers at the strategic level where information tends to be unstructured, more qualitative, and soft. In the first application, Section 18.4.2, The Graph Model for Conflict Resolution (Kilgour et al., 1987; Fang et al., 1993) is used to analytically study the Aral Sea confrontation arising among countries competing over a stressed water resources system of system. The Graph Model for Conflict Resolution is an expansion of the theoretical scope and practicality of both metagame analysis (Howard, 1971) and conflict analysis (Fraser and Hipel, 1984), where each decision maker in a conflict has a directed graph that records each unilateral move available to the decision maker as the conflict evolves from one state to another. Usually, ordinal preference information is used, although a graph model can handle intransitive preferences. In Section 18.4.3, the general multiple participant-multiple objective system of systems fair resources allocation, a procedure based on cooperative game theory, is described using the application of the South Saskatchewan River Basin in Canada for the allocation of scarce resources among competing stakeholders based on fairness and rights principles of monotonicity, priority, and equitability. Finally, multiple criteria decision analysis (MCDA) is applied in Section 18.4.4 to the problem of managing water levels in the Great Lakes of North America between the states and provinces that share this important and very large-scale complex system of systems. MCDA consists of a set of principles and tools that are employed to assist stakeholders and users solve tactical and strategic decision problems by finding, selecting, sorting, or ranking a set of actions, which are evaluated based on a number of criteria. (See textbooks by authors such as Hobbs and Meier (2000) and Belton and Stewart (2002) for a good description of a range of MCDA techniques, and Chen et al. (2006a,b) for an explanation of classifying alternatives.) In MCDA, alternative solutions are evaluated according to multitude environmental, economic, social, and political criteria, which are usually conflicting. It is, therefore, uncommon to find a single optimal solution that satisfies all criteria; rather the outcome of applying the procedure represents tradeoffs and a compromise solution that reconciles all the different criteria. An important input to MCDA, as is in the other approaches, is the decision maker's preferences expressed over the criteria.

18.4 SYSTEMS OF SYSTEMS METHODOLOGICAL APPLICATIONS

18.4.1 Introduction

As outlined in the previous section, a rich variety of systems of systems engineering decision tools are available for addressing complex environmental as well as other types of systems of systems problems arising in the real world. Of particular concern is the global warming crisis, which is now escalating on a worldwide basis. In fact, if the global warming problem described in Section 18.2 remains unchecked, many regional ecosystems of systems could collapse or be seriously damaged around the world. For example, the shrinking of glaciers in the Rocky Mountains of Western Canada brought about by global warming means decreasing flows in the Saskatchewan River and its tributaries that transport water from west to east across the Canadian prairie provinces of Alberta and Saskatchewan to Lake Winnipeg in the Province of Manitoba, because a large part of these flows are sustained by melt-water from glaciers, especially during the dry summer months. Over time, Lake Winnipeg could substantially contract in size and thereby become the Canadian version of the Aral Sea debacle in central Asia, which is addressed in Section 18.4.2. Fortunately, if appropriate political action is taken by all governments of the world, at least the magnitude of this type of disaster can be reduced. Hipel and Obeidi (2005) carry out a generic conflict study of the ongoing conflict taking place between the forces of unchecked global trade and those supporting sustainable trade that is environmentally, socially and economically responsible. They explain why a strategic settlement is possible, but, as emphasized by the IPCC (2007), Stern (2007), and others, collective political action must be taken now.

The purpose of this section is to explain how systems of systems decision technologies can be gainfully employed to tackle the multiple participant-multiple objective decision making aspects of three large-scale environmental systems of systems, each of which is under increasing stress due to global warming. In the next subsection, The Graph Model for Conflict Resolution (Fang et al., 1993) is employed for strategically studying a dispute among upstream and downstream nations along the Syr Darya river located in the Aral Sea Basin, over how water can be better shared in a very dry region currently undergoing immense environmental pressures. In Section 18.4.3, a comprehensive optimization model structured upon ideas from economics, cooperative game theory and hydrology is employed for rigorously investigating the equitable allocation of scarce water resources in the South Saskatchewan River Basin situated within the western Canadian province of Alberta. In the final section, multiple criteria decision analysis techniques are utilized to assess and compare alternative solutions to manage fluctuating water levels in the five Great Lakes in North America, which straddle the border between Canada and the United States.

18.4.2 Strategic Study of the Aral Sea Confrontation

The infamous large-scale devastation of the Aral Sea, along with much of the surrounding river basin in which the sea is cradled, constitutes the greatest and most widely known ecological disaster of the twentieth century. This human-induced tragedy demonstrates at a regional level how civilization can seriously damage a natural system of systems by purposefully and, many would argue, unethically, putting perceived economic gain in the short-term ahead of long-term environmental stewardship and the principle of sustainability. As explained by Diamond (2005), many ancient civilizations had collapsed throughout history, largely as a result of environmental damage, exacerbated by ineptness in group decision-making mechanisms that led to failure to respond to these environmental problems. However, the demise of the Aral Sea is starkly striking because it occurred over a few short decades, it is well-documented, and many of the people who participated in this "war on the environment" are still alive today.

As shown in Fig. 18.7, the Aral Sea Basin is located in the heart of the Eurasian continent, extending over the territories of the five Central Asian Republics, consisting of Kazakhstan, Turkmenistan, Kyrgyzstan, Tajikistan, and Uzbekistan. Parts of the basin lie in Afghanistan, China, and Iran. The two main rivers flowing into the Aral Sea are the Amu Darya, which starts in the mountains of Afghanistan and Tajikistan and flows through Turkmenistan and Uzbekistan to the Aral Sea, and Syr Darya, which originates in Kyrgyzstan and flows through Tajikistan, Uzbekistan, and Kazakhstan to the Aral Sea.

The main problem with the Aral Sea is that it has dramatically shrunk in size since 1960 because of large diversions of water from the Amu Darya and Syr Darya rivers for irrigation purposes. In fact, since 1960 the surface area of the Aral Sea has decreased by

FIGURE 18.7 The Aral Sea Basin

about 74% and its volume by 90% (Micklin, 2007). Along with its size decrease, the lake has become significantly saltier and has been heavily polluted by raw sewage runoff as well as industrial and military pollutants. The huge plains left by the receding sea are covered with salt and toxic chemicals, which are dispersed by the wind as toxic dust to the surrounding areas. The fishing industry in the Aral Sea has been obliterated and farming in the region closer to the sea has been destroyed by salt deposited on the land. A large body of information regarding the environmental, societal, and economic aspects of the Aral Sea Basin is available, including contributions by Ellis (1990), Seiko (1998), Vinogradov and Langford (2001), Antipova et al. (2002), McKinney (2003), and Micklin (2000, 2007).

In 1918, the Soviets made the decision to divert water from Amu Darya and Syr Darya rivers to irrigate the desert to grow cotton, rice, melons, and cereals. Cotton, or "white gold," became a major export for the Soviet Union and at the present time is still a key export product for Uzbekistan. The construction of irrigation canals, which leaked badly, began in earnest in the 1930s. Between 1960 and 1980 both the volume of water taken from the rivers and the production of cotton doubled. The surface areas of the Aral Sea in 1960 and 2006 were 67,499 and 17,382 km^2, respectively (Micklin, 2007).

In an attempt to revive the Aral Sea, in 1988 the Soviet Union decreed that cotton growing was to be decreased. However, since the disintegration of the Soviet Union in 1991, central control of the region ended and the Aral Sea crisis has been in the hands of the five independent central Asian nations shown in Fig. 18.7. Independent actions by these countries rapidly increased water usage and this ongoing "tragedy of the commons" intensified. Moreover, these countries have failed to construct a regional approach to sharing water and energy, as was done under the Soviet system. Hence, conflict over a myriad of issues abounded within and among nations of this region. At an even higher level, these countries are part of the overall global conflict over trade versus the environment, including global warming, which is taking place worldwide (Hipel and Obeidi, 2005). Therefore, conflict resolution tools are needed for thoroughly investigating environmental system of systems disputes to lead negotiations in a positive direction to arrive at "win-win" resolutions in which all parties gain according to their particular interests.

Consider now an ongoing dispute that is taking place in the Aral Sea Basin, to outline how conflict resolution tools from systems engineering can be readily applied in practice. At the overall basin level, the three downstream countries comprised of Kazakhstan, Turkmenistan, and Uzbekistan are major consumers of water for their expanding agricultural sectors, while the economically weaker upstream countries consisting of Kyrgyzstan and Tajikistan wish to use more water for electricity generation and farming. In 1992, the five states established the Interstate Coordinating Water Commission (ICWC) to manage water resources in the Aral Sea Basin. Nandalal and Hipel (2007) employed The Graph Model for Conflict Resolution (GMCR) (Fang et al., 1993) to systematically study a conflict occurring in the northern subsystem of the Aral Sea Basin among countries along the Syr Darya. Specifically, the upstream country, Kyrgyzstan, trades water in the Syr Darya River to the downstream nations consisting of Uzbekistan and Kazakhstan for energy in the form of gas, coal, or oil (see Fig. 18.7). Because energy deliveries have been unreliable,

Kyrgyzstan responded by releasing more water through its hydropower dam in winter, which resulted in downstream flooding and less water for summer irrigation in the downstream countries. Kyrgyzstan's demand to be paid for water has been resisted by the downstream nations. After independence, Uzbekistan, for example, found itself dependent on food imports as its self-sufficiency in food had been compromised by the predominance of cotton production in the republic. For Uzbekistan to purchase food in world markets, the government needed hard currency and obtaining this money required the continuing maintenance of a cotton monoculture. Shortages of food caused socioeconomic strain in the region, but the country's continuing demand for large amounts of water in order to grow cotton created international political conflict. The report of the International Crisis Group (2002) furnishes many examples of conflict over sharing water that have arisen among the countries along the Syr Darya river. Unfortunately, each country in the region interprets ongoing disputes as a zero-sum game in which one country's gain in, say, water volume is another nation's loss.

Wolf (1998) views conflict over water as an opportunity to foster cooperation and mutual gain among the affected nations. Conflict tools provide a formal mechanism to ascertain if and how this can be strategically accomplished. As is argued by Hipel and Fang (2005), game-theoretic methods have a key role to play in studying both technological and societal systems of systems since conflict and associated value systems constitute inherent components of environmental (natural-world), societal (real-life), intelligent (artificial-life), and integrated (mixed-life) systems of systems.

Because of the ubiquitous nature of conflict, research on conflict resolution has taken place in a wide range of disciplines. As explained in an overview paper by Hipel (2002) and in articles contained within the theme on Conflict Resolution in the Encyclopedia of Life Support Systems (EOLSS), a rich range of psychological, sociological, operational research, game theory, systems engineering (Hipel et al., 2007; Sage and Rouse, 2008) and other kinds of models have been developed for systematically studying conflict and its resolution. With regards to formal systems engineering methods, shown in the left branch in Fig. 18.6, the Graph Model for Conflict Resolution (Fang et al., 1993) constitutes an improvement and extension of conflict analysis (Fraser and Hipel, 1984), which in turn is an enhancement of metagame analysis (Howard, 1971) and also has advantages over drama theory (Howard, 1999). In particular, the Graph Model can take into account irreversible moves in which movement is allowed only in one direction. When, for example, a body of water, such as the Aral Sea, is severely polluted, the damage incurred cannot be immediately reversed. This flexible approach can also handle common moves in which two or more decision makers can cause a conflict to change from one state to the same final state. In addition, the Graph Model can take into consideration a wide range of different types of human behavior under conflict when stable states for a given decision maker as well as equilibria or compromise resolutions are determined.

The decision support system GMCR II (Hipel et al., 1997, 2001; Fang et al., 2003a,b) permits the graph model methodology to be conveniently employed in practical applications. Figure 18.8 shows the overall design of the GMCR II, which has been developed within a Microsoft Windows® environment. Extensions to the graph model methodology (Kilgour et al., 2005) include its capability to model coalitions (Kilgour

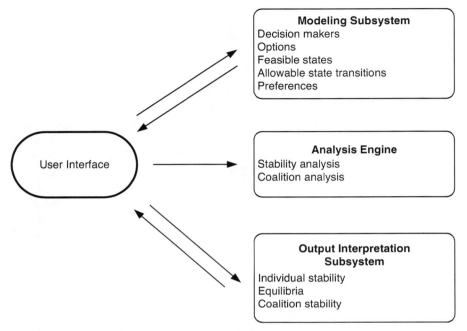

FIGURE 18.8   Overall design of GMCR II

et al., 2001; Inohara and Hipel, 2008a,b), uncertain preferences (Li et al., 2004a), strength of preference (Hamouda et al., 2004, 2006), emotions (Obeidi et al., 2005, 2006), attitudes (Inohara et al., 2007), policies (Zeng et al., 2007), large conflicts (Xu et al., 2007), and the evolution of a conflict to a final outcome (Li et al., 2004b, 2005, 2006).

As shown in the top-right box in Fig. 18.8, the decision makers and options must first be determined at the modeling stage. Figure 18.9 displays the conflict model for the Syr Darya over sharing water as developed by Nandalal and Hipel (2007). As can be seen, this conflict model consists of the three main decision makers and a total of eight options. Subsequent to determining the feasible states, allowable state transitions, and relative preferences of each decision maker, one can calculate the stable states and equilibria according to a wide variety of human behavior under conflict using the analysis engine embedded in GMCR II depicted in Fig. 18.8. Figure 18.10 portrays how the Syr Darya conflict evolved from a status quo-state on the left via a transition state to the final equilibrium on the right. Each of the three feasible states in Fig. 18.10 consists of a column of Y and N, where "Y" indicates "yes," the option opposite Y is selected by the decision maker controlling it, and "N" means "no," the option is not taken.

In Fig. 18.10, the arrows indicate the location and direction of option changes during the evolution of the conflict. At the status quo, upstream Kyrgyzstan ignores the agreement and uses more water in the winter and downstream countries apply political or military pressure not to release excess water in the winter. Nonetheless, attempts by the upstream country to follow the agreement would encourage

Decision makers		Options
Downstream Two downstream countries	Uzbekistan and Kazakhstan (needs water for irrigation)	**1. Less fuel**: supply less fuel than specified. **2. Pressure**: exert political and military pressure. **3. Follow agreement**: follow the present agreement.
Upstream One upstream country	Kyrgyzstan (requires water for generating power)	**4. Ignore agreement**: generate more hydroenergy in winter than agreed upon. **5. Follow agreement**: do not release too much water for hydropower generation during winter and release sufficient water for irrigation during summer. **6. Revise agreement**: revise agreement to meet winter hydropower demand and allow for summer irrigation.
ICWC	Interstate Commission for Water Coordination	**7. Enforce agreement**: enforce the present agreement. **8. Revise agreement**: revise agreement to try to meet new demands of both upstream and downstream countries.

FIGURE 18.9   Decision makers and their options in the Syr Darya water sharing dispute

downstream countries also to adhere to the agreement as shown in the joint transition from the status quo to the transition state. Therefore, by exhibiting a spirit of cooperation, downstream and upstream can jointly move the conflict to the transition state, which is more preferred than the status quo by both decision makers. Moreover, the transition state is not a strong equilibrium as explained in the detailed stability analysis executed by Nandalal and Hipel (2007). The final strong equilibrium, shown in the right column in which both the upstream and the downstream countries agree to a revised agreement, could be an acceptable solution for all of the parties. Accordingly, an important insight that can be drawn from this strategic analysis is that both upstream

	Status Quo	Transition State	Final Outcome
Downstream			
1. Less fuel	N	N	N
2. Pressure	Y ⟹	N	N
3. Follow agreement	N ⟹	Y	Y
Upstream			
4. Ignore agreement	Y ⟹	N	N
5. Follow agreement	N ⟹	Y	Y
6. Revise agreement	N	N	N
ICWC			
7. Enforce agreement	N	N	N
8. Revise agreement	N	N ⟹	Y

FIGURE 18.10   Evolution of the Aral Sea conflict from the status quo to the final outcome

Kyrgyzstan and downstream Uzbekistan and Kazakhstan wish to neither ignore the agreement nor exert political or military pressure over the sharing of water in the Syr Darya River as their most preferred strategy toward obtaining water.

18.4.3   Fair Resources Allocation in the South Saskatchewan River Basin in Canada

How scarce resources should be allocated fairly among competing stakeholders in a complex system of systems is a common problem. For example, various stakeholders in a river basin or watershed have different uses for water and there are differences in economic benefits that can be generated by different uses. Therefore, fair water resources allocation in a river basin is an example of a system of systems problem for which informed decisions should be made. The entire river basin is a system and each stakeholder can be considered as a system, but it is also a part of a larger system, connected by the main river and its tributaries.

The South Saskatchewan River Basin (SSRB) in southern Alberta, Canada, consists of four subbasins: the Red Deer, Bow and Oldman River subbasins, and the portion of the South Saskatchewan River subbasin included within Alberta, as shown in Fig. 18.11. The SSRB within Alberta drains about 120,000 km^2 of land, has a primarily semiarid climate, and had a population size of 1.5 million as of 2004 (Dyson et al., 2004). Population growth and economic expansion put significant pressure on the region's water resources. The Provincial Government of Alberta approved the SSRB Water Management Plan in August 2006, which stipulates that new water license applications for the Bow, Oldman, and South Saskatchewan subbasins will no longer be accepted by Alberta Environment (2006). New water allocations must be obtained through water allocation transfers.

To achieve the objective of equitable and efficient water allocation among competing stakeholders at the basin level, a general mathematical programming-based methodology, called the Cooperative Water Allocation Model (CWAM), has been developed (Wang et al., 2003, 2007a,b, 2008). Concepts from cooperative game theory, economics, and hydrology are integrated within the CWAM framework. The CWAM methodology consists of two main steps: initial water rights allocation to stakeholders based on legal water rights systems or agreements, and subsequent water and net benefits reallocation to achieve efficient use of water and equitable redistribution of net benefits, as summarized in Fig. 18.12. Both water quantity and quality are considered by the CWAM. For a basin wide water allocation study, the CWAM involves formulating and solving a large number of large-scale nonlinear mathematical programming problems.

The node-link river basin network system of the SSRB is shown in Fig. 18.13. Six typical case scenarios are studied in terms of the combination of water demands, hydrologic conditions, and initial water rights allocation methods. To account for Alberta's existing prior water rights system, the priority-based maximal multi-period network flow (PMMNF) approach is used to allocate the initial water rights in Cases A, B, and C. In Cases D, E, and F, initial water rights are assigned by utilizing the lexicographic minimax water shortage ratios (LMWSR) method to explore

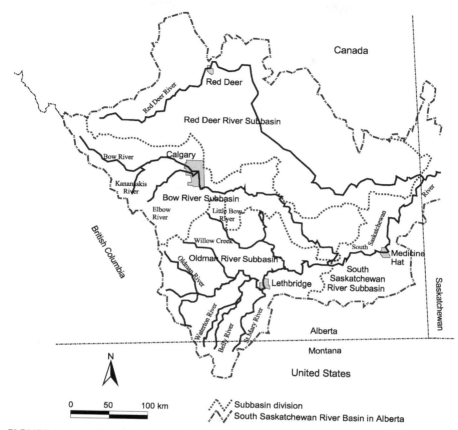

FIGURE 18.11   The South Saskatchewan River Basin within the Canadian Province of Alberta

allocations under an assumed public regime. Case A (1995 wet and PMMNF) and Case D (1995 wet and LMWSR) correspond to the actual situations of water inflows and demands in 1995. Case B (2021 normal and PMMNF) and Case E (2021 normal and LMWSR) explore initial water right allocations under the long-term mean (1912–2001) monthly water inflows and the forecasted water demands in 2021. Case C (2021 drought and PMMNF) and Case F (2021 drought and LMWSR) consider the hydrologic conditions of an assumed drought year and the forecasted water demands in 2021. Case A is used for calibrating model parameters as well.

For each given time period, the PMMNF and LMWSR methods can allocate initial water rights to each of all demands, compute the release of each of all reservoirs, and calculate the outflow from the SSRB to the downstream province of Saskatchewan. Based on the water allocation to each demand node in a given time period, water supply/demand satisfaction ratio for this node can be calculated. The most apparent difference between the PMMNF and LMWSR methods is that the satisfaction ratios of

Network model of the basin SoS

Node-link river basin network, where a node denotes a physical component of interest and a link represents a natural or man-made water conduit.

Initial water rights allocation

- Priority-based maximal multiperiod network flow (PMMNF) programming, which can be applied under prior, riparian, and public water allocation regimes.
- Lexicographic minimax water shortage ratios (LMWSR) programming, which can be used under a public water rights regime.

Equitable water and net benefits reallocation

- Irrigation water planning model (IWPM), which is utilized for deriving benefit functions of irrigation water for all time periods.
- Hydrologic-economic river basin model (HERBM), which is used for performing coalition analysis, for finding optimal water allocation schemes and net benefits of various coalitions of stakeholders.
- Cooperative reallocation game of the net benefit of a given coalition, which utilizes cooperative game theoretic approaches to carry out equitable allocation of the net benefit of a given coalition. Core-based (nucleolus, weak nucleolus, proportional nucleolus, and normalized nucleolus) and noncore-based (Shapley value) concepts are employed.

FIGURE 18.12   Cooperative water allocation model (CWAM)

all demand nodes obtained by the LMWSR approach are more evenly distributed than those by PMMNF. This is because PMMNF assigns water to nodes strictly according to a priority order while the LMWSR method ensures that every node receives its share of the water during shortage periods in accordance with equivalent weights of all nodes.

Rights-based water allocations usually do not make efficient use of water for the entire river basin because different stakeholders can produce different economic benefits for a given amount of water. Therefore, water can be reassigned among stakeholders to achieve the objective of basin-wide economic optimization or optimization involving any group of stakeholders. To estimate the net benefits of water uses at demand nodes, the monthly net benefit functions of municipal and industrial demand nodes, hydropower stations, agriculture water uses, stream flow requirements, and reservoirs are derived within the CWAM framework. An integrated hydrologic-economic river basin model (HERBM) is developed to find the maximum net benefit of water uses involving a group of stakeholders, called a coalition of stakeholders. Cooperative game theoretic approaches are used to carry out equitable allocation of the net benefit of a given coalition. Water transfers based on initial water

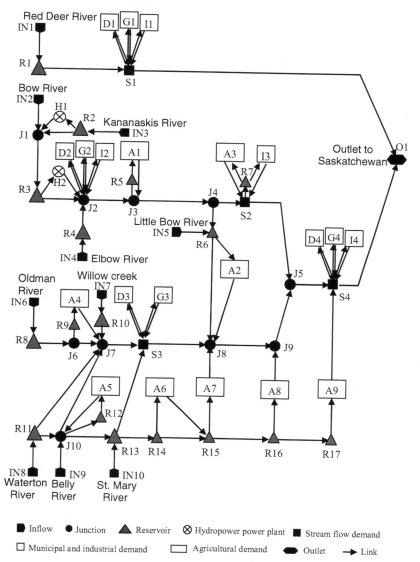

FIGURE 18.13 Network system of the South Saskatchewan River Basin in Southern Alberta (Wang et al., 2007b)

rights achieve the objective of economically efficient use of water under a given coalition.

For the SSRB case study, comparing the initial rights and basin-wide optimal scenarios under both Cases C and F, shows that only four stakeholders (hydropower plants on the Bow River, the city of Calgary, irrigation regions in Bow River subbasin, and irrigation regions in the Oldman River subbasin) have significant changes of inflows and net benefits. A coalition analysis of these four stakeholders can produce value of participation in this coalition for each stakeholder, which is the additional gain over the optimal benefit that can be generated based on his or her own initial water rights. Values of participation for stakeholders can be used as a basis for negotiation among the stakeholders to achieve the objective of basin-wide fair water resources allocation.

18.4.4   Managing Water Levels in the Great Lakes of North America

The Laurentian Great Lakes, shown in Fig. 18.14, constitute an excellent example of an important natural system of systems. Five large lakes lying on or near the Canada–United States border, together with connecting water bodies, form the world's largest repository of fresh water. Each lake can be understood as a system, but it is also a node of a larger system, with a strait, a small lake, and four connecting rivers as links.

The five large lakes, often called "inland seas," are Lake Superior (the world's largest lake by surface area), Lake Michigan, Lake Huron, Lake Erie, and Lake Ontario. In total, the Great Lakes system contains about 20% of the world's fresh water, enough to cover the contiguous 48 states of the United States to a uniform depth of 2.9 m. Their shoreline, shared by the province of Ontario and the states of Minnesota, Wisconsin, Michigan, Illinois, Indiana, Ohio, Pennsylvania, and New York, measures 17,549 km, about 88% of the length of the ocean coastline of the

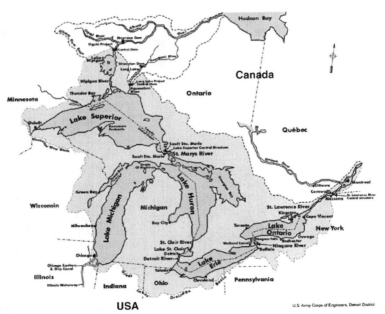

FIGURE 18.14   The Laurentian Great Lakes shared between Canada and the United States. *Source*: http://www.great-lakes.net/

contiguous states of the United States (Wikipedia, 2007). Large cities located on the Great Lakes include Chicago, Detroit, and Toronto.

The Great Lakes system supplies fresh water to about 40 million people living in the region. The Saint Lawrence Seaway, a system of canals that makes the entire system accessible to ocean-going vessels, facilitates shipping to, within, and from the Great Lakes. The economic value of shipping is, however, limited by the fact that most of the Great Lakes freeze over in winter. Moreover, the Seaway does not accommodate the wider container ships that have become more common since it opened in 1959. The ecological importance of the Great Lakes is enhanced by differences among the lakes, reflecting climatological variation as well as anthropogenic effects. The shoreline, and the lakes and rivers themselves, supply important recreational opportunities. Commercial and recreational fisheries throughout the system remain important, but are proving vulnerable to invading species.

Many system of systems techniques were applicable to a comprehensive Great Lakes water level management study that began in the 1980s. In 1985 and 1986, water in all of the Great Lakes except Lake Ontario was at its highest level of the century, a consequence of almost 20 years of above-average precipitation and below-average evaporation. Because of these high water levels, storms caused severe flooding, shoreline erosion, and damage to lakeshore properties. In response to widespread public concern, the governments of Canada and the United States asked the International Joint Commission (IJC) to study how water levels in the Great Lakes-St. Lawrence system could be better managed.

The International Joint Commission is an independent binational organization established and funded by the United States and Canada, under the terms of the International Boundary Waters Treaty of 1909 and subsequent agreements. Its purpose is to prevent or resolve disputes about the use and quality of boundary waters, and to advise the two countries on water resources issues. The IJC administers continuing programs related to its general mandate, sponsors conferences and public discussions, and investigates issues concerning joint water bodies – but only when requested to do so by both national governments. Its recommendations are not binding.

In the 1986 reference, the IJC was asked to comment on the issue of fluctuating water levels in the Great Lakes-St. Lawrence System. The term "fluctuating" reflected that both high and low water levels can cause environmental and economic damage. In fact, water levels fell to the normal range during 1987 and in 2007 they were very low in all of the Great Lakes.

The IJC created the Levels Reference Study Board, assigning primary responsibilities to the U.S. Army Corps of Engineers and Environment Canada. Federal agencies on both sides of the border, state and provincial governments, and citizens' groups also contributed to the study.

The IJC issued a preliminary report in the late 1986, and created a task force to study technical aspects of measures to reduce water levels. The task force reported in 1988 with recommendations, including that better coordination of activities be instituted between the two countries. But there were few concrete actions taken; one reason the report was not persuasive was its failure to account for social and political linkages, such as the effects on lower lakes of measures applied to the upper lakes. In 1990, the

IJC again established the Levels Reference Study Board, with a mandate to conduct a broader study including not just measures to regulate water levels, but also actions to protect shorelands, and regulations and incentives governing their use (Levels Reference Study Board, 1991).

The Levels Reference Study Board conducted a multiple criteria decision analysis study of a wide range of actions designed to regulate water levels in the Great Lakes system, and to mitigate the effects of fluctuations. It established Working Committees, with balanced membership, to conduct technical and scientific investigations, and an 18-member Citizens Advisory Committee, which was represented on every Working Committee. (One of the authors, K. W. Hipel, was a member of Working Committee 4, on "Principles, Measures, Evaluation, Integration, and Implementation," and also served as a facilitator.)

Four main criteria for the assessment of actions, and nine subcriteria by which impacts were to be measured, were developed during a series of meetings of committees and stakeholders in 1990–1991. The criteria and subcriteria are shown in Fig. 18.15. For example, under subcriterion EC1, Economic Benefits and Costs, the implementation costs of an action were balanced against economic benefits including prevention of further damage, increases in revenue, and avoidance of future costs, which were determined for the commercial shipping industry, shoreline property owners, the commercial fishery, the recreational boating industry, and hydropower utilities. Evaluations on the second criterion, Environmental Impacts, were based in large part on their impacts upon wetlands, which were identified as accurate indicators of the basin's aquatic health that were relatively easy to measure.

The Board set out to recommend a set of actions to regulate water levels in the Great Lakes system, and to mitigate the effects of fluctuations. Initially over 120 possible alternatives or actions were proposed. One issue that increased the number of actions was whether the outflow from every lake should be regulated; at the time, only Lake Superior and Lake Ontario had structures for this purpose. Such regulation would be relatively easy to construct at the outlet of Lake Erie, so many actions were considered for two lakes (Superior and Ontario only), three lakes (Superior, Erie, and Ontario), and all five Great Lakes. Evaluation of all 120 alternatives would have been a huge undertaking; for example, hydrological analysis, including stage–damage curves and detailed studies of example sites, would be required to evaluate the two, three, and five lake actions according to economic and environmental criteria.

	Main Criteria	Subcriteria
I	Economic impacts	EC1. Benefit cost analysis
		EC2. Other economic and social impacts
II	Environmental impacts	EN1. Ecological productivity
		EN2. Environmental purity
III	Distribution of impacts	D1. Distribution among affected interests
		D2. Distribution among affected regions
IV	Feasibility	F1. Technical feasibility
		F2. Operational feasibility
		F3. Legal and public policy feasibility

FIGURE 18.15   Criteria for the IJC lake levels study

Therefore, a series of committee and stakeholder meetings was convened to reduce the list of possible alternative actions by removing irrelevant or ineffective alternatives and by combining complementary ones. Afterward, 32 alternatives remained—17 aimed at regulating water levels, 8 at regulating land use along the shoreline, 4 at providing new incentives for property owners, and 3 at providing additional protection for shorelands.

The problem of the IJC was then a more standard multiple-criteria selection problem. However, even though only one alternative might have been appropriate for a specific location in the Great Lakes, a combination of alternatives was necessary in the context of the entire system. Standard multicriteria evaluation techniques were among the tools that IJC used to solve these combinations of alternatives problem (Yin et al., 1999). The IJC made its recommendations in its final report, delivered on March 31, 1993 (Levels Reference Study Board, 1993).

A later study (Rajabi et al., 2001) showed that the choice procedure carried out by the IJC could have been handled using multiple-criteria screening techniques— procedures for identifying the potentially optimal alternatives in a large set. This problem is difficult when a combination of alternatives is to be selected, since in this case a dominated alternative can be potentially optimal (Kilgour et al., 2004). The difficulty is exacerbated when the actions are interdependent, as in the Great Lakes case. Nonetheless, procedures that use partial information about criterion weights are available, and give a good approximation to the subset of actions recommended by the IJC.

In summary, the Great Lakes provide a clear and instructive example of a large-scale and complex system of systems. The Great Lakes Levels Reference Study conducted by the IJC in the early 1990s demonstrates that Multiple Criteria Decision Analysis tools can provide important information about system of systems decision problems, and can be a major contributor to their solution.

18.5 INSIGHTS AND DISCUSSION POINTS

As portrayed in Fig. 18.1, the concept of a system of systems, described in Section 18.1, furnishes an insightful paradigm to depict important relationships within and among societal and natural systems of systems. As can be seen, by-products from societal systems of systems can have negative effects upon both natural and societal systems of systems. For example, Fig. 18.4 depicts how energy systems of systems from the societal realm can influence atmospheric systems of systems. In fact, as explained in Section 18.2.2, the release of greenhouse gases from societal systems of systems is creating the ongoing global warming that could have devastating consequences on both societal and natural systems of systems. Global warming may, in reality, be the worst potential disaster that society has faced since the Cuban missile crisis of October 1962, when the world was extremely close to being destroyed by a full-scale nuclear war between the United States of America and the Union of the Soviet Socialist Republic (USSR). Fortunately, the system of systems decision methodologies and tools discussed in Section 18.3 and elsewhere could assist in solving or mitigating large-scale systems of systems problems, like climate change. The application of systems of systems tools to large-scale environmental problems are illustrated for the cases of conflict over water utilization in the Aral Sea Basin, fair allocation of water among water users in the South Saskatchewan River Basin in Canada, and fluctuating water levels in the Great Lakes systems of systems in Sections 18.4.2, 18.4.3, and 18.4.4, respectively.

In the future, one does not know what negative emergent properties may arise among the societal and natural systems of systems given in Fig. 18.1.

- Can you think of a possible unforeseen consequence, like global warming and climate change, that could emerge unexpectedly in the future?
- What types of systems of systems tools, governance bodies, and environmental policies do you feel should be in place to control unforeseen future events that may spring forth and cause widespread desolation and chaos?
- What types of new decision tools in systems management do you think need to be designed and developed for tackling current and future large-scale systems of systems problems?
- How do you think you can personally contribute to these and other great challenges of our times as well as those of future generations, so that wise decisions can be made at this most critical point in human history?

REFERENCES

Ackoff, R.L., 1973, Science in the systems age: beyond IE, OR and MS, *Operations Research*, 21(3): 661–671.

Ackoff, R.L., Emery, F.E., 1972, *On Purposeful Systems*, Aldine-Atherton, Chicago.

Alberta Environment, 2006, Approved Water Management Plan for the South Saskatchewan River Basin (Alberta), Edmonton, Alberta, Canada.

Antipova, E., Zyrynov, A., McKinney, D., Savitsky, A., 2002, Optimization of Syr Darya water and energy uses, *Water International*, 27(4): 504–516.

Bar-Yam, Y., 2005, *Making Things Work: Solving Complex Problems in a Complex World*, Knowledge Press, Cambridge, MA.

Belton, V., Stewart, T.J., 2002, *Multiple Criteria Decision Analysis: An Integrated Approach*, Kluwer, Boston, MA.

Blanchard, B.S., Fabrycky, W.J., 1990, *Systems Engineering and Analysis*, 2nd edition, Prentice-Hall, Englewood Cliffs, NJ.

Braha, D., Minai, A.A., Bar-Yam, Y. (Eds.), 2006, *Complex Engineered Systems: Science Meets Technology*, Springer Verlag, Cambridge, MA.

Calvano, C.N., John, P., 2004, Systems engineering in an age of complexity, *Systems Engineering*, 7(1): 25–34.

Carlock, P.G., Fenton, R.E., 2001, System of systems (SoS) enterprise systems for information-intensive organizations, *Systems Engineering*, 4(4): 242–261.

Casti, J.L., 1997, *Would-be Worlds: How Simulation is Changing the Frontiers of Science*, Wiley, New York.

Chen, Y., Hipel, K.W., Kilgour, D.M., 2006a, Multiple criteria sorting using case-based distance models with application in water resources management, *IEEE Transactions on Systems, Man and Cybernetics, Part A*, 37(5): 680–691.

Chen, Y., Kilgour, D.M., Hipel, K.W., 2006b, Multiple criteria classification with an application in water resources planning, *Computer and Operations Research*, 33(11): 3301–3323.

Diamond, J., 2005, *Collapse: How Societies Choose to Fail or Succeed*, Viking Books, New York.

Dyson, I., Paterson, B., Fitzpatrick, D., 2004, The state of Southern Alberta's water resources, *Confronting Water Scarcity: Challenges and Choices Conference*, July 13–16, Lethbridge, Alberta, Canada.

Ellis, W.S., 1990, A Soviet sea lies dying, *National Geographic*, 177(2): 73–93.

Fang, L., Hipel, K.W., Kilgour, D.M., 1993, *Interactive Decision Making: The Graph Model for Conflict Resolution*, Wiley, New York.

Fang, L., Hipel, K.W., Kilgour, D.M., Peng, X., 2003a, A decision support system for interactive decision making, part 1: model formulation, *IEEE Transactions on Systems, Man and Cybernetics, Part C, Applications and Reviews*, 33(1): 42–55.

Fang, L., Hipel, K.W., Kilgour, D.M., Peng, X., 2003b, A decision support system for interactive decision making, part 2: analysis and output interpretation, *IEEE Transactions on Systems, Man, and Cybernetics, Part C, Applications and Reviews* 33(1): 56–66.

Flannery, T., 2005, *The Weather Makers*, Harper Collins, Toronto.

Fraser, N.M., Hipel, K.W., 1984, *Conflict Analysis: Models and Resolutions*, North Holland, New York.

Gharajedaghi, J., 2006, *Systems Thinking: Managing Chaos and Complexity: A Platform for Designing Business Architecture*, Butterworth-Heinemann, Burlington, MA.

Gore, A., 2006, *An Inconvenient Truth: The Planetary Emergency of Global Warming and What We Can Do About It*, Rodale, New York.

Grunwald, M., 2007, The threatening storm-hurricane Katrina: two years later, *Time Magazine*, Special Report, 17(7): 18–29.

Gunderson, L., 1999, Resilience, flexibility and adaptive management—antidotes for spurious certitude, *Conservation Ecology*, 3(1): 7. Website: http://www.consecol.org/vol3/iss1/art7/. Accessed March 7, 2007.

Gunderson, L., Holling, C.S., Light, S.S., 1995, *Barriers and Bridges to Renewal of Ecosystems and Institutions*, Columbia University Press, New York.

Hamouda, L., Kilgour, D.M., Hipel, K.W., 2004, Strength of preference in the graph model for conflict resolution, *Group Decision and Negotiation*, 13: 449–462.

Hamouda, L., Kilgour, D.M., Hipel, K.W., 2006, Strength of preference in graph models for multiple decision-maker conflicts, *Applied Mathematics and Computation*, 179: 314–327.

Hipel, K.W. (Ed.), 1992, Multiple objective decision making in water resources, Monograph Series No. 18, American Water Resources Association, Middleburg, VA.

Hipel, K.W., 2002, Conflict resolution, Theme Overview Paper, in Conflict Resolution, *Encyclopedia of Life Support Systems (EOLSS)*, EOLSS Publishers, Oxford, UK. Website: http://www.eolss.net.

Hipel, K.W., Ben-Haim, Y., 1999, Decision making in an uncertain world: information-gap modeling in water resources management, *IEEE Transactions on Systems, Man, and Cybernetics—Part C, Applications and Reviews*, 29(4): 506–517.

Hipel, K.W., Fang, L., 2005, Multiple participant decision making in societal and technological systems, In: Arai, T., Yamamoto, S., Makino, K., editors, *Systems and Human Science— For Safety, Security, and Dependability*, Selected Papers of the 1st International Symposium, SSR2003, Osaka, Japan, Chapter 1, 3–31, Published by Elsevier, Amsterdam, The Netherlands.

Hipel, K.W., Jamshidi, M.M., Tien, J.M., White, C.C., III, 2007, The future of systems, man and cybernetics: application domains and research methods, *IEEE Transactions on Systems, Man, and Cybernetics, Part C, Applications and Reviews*, 37(5): 726–743.

Hipel, K.W., Kilgour, D.M., Fang, L., Peng, X., 1997, The decision support system GMCR II in environmental conflict management, *Applied Mathematics and Computation*, 83(2/3): 117–152.

Hipel, K.W., Kilgour, D.M., Fang, L., Peng, X., 2001, Strategic decision support for the services industry, *IEEE Transactions on Engineering Management*, 48(3): 358–369.

Hipel, K.W., Kilgour, D.M., Rajabi, S., Chen, Y., 2008, Operations research and refinement of courses of action, In: Sage, A.P., Rouse, W.B., (Eds.), *Handbook of Systems Engineering and Management*, 2nd edition, Wiley, New York.

Hipel, K.W., McLeod, A.I., 1994, *Time Series Modelling of Water Resources and Environmental Systems*, Elsevier Scientific Publishing Company, Amsterdam, New York.

Hipel, K.W., Obeidi, A., 2005, Trade versus the environment: Strategic settlement from a systems engineering prespective, *Systems Engineering*, 8(3): 211–233.

Ho, F., Kamel, M., 1998, Learning coordination strategies for cooperative multiagent systems, *Machine Learning*, 33: 155–177.

Hobbs, B.F., Meier, P., 2000, *Energy Decisions and the Environment: A Guide to the Use of Multicriteria Methods*, Kluwer, Boston, MA.

Howard, N., 1971, *Paradoxes of Rationality: Theory of Metagames and Political Behaviour*, MIT Press, Cambridge, MA.

Howard, N., 1999, *Confrontation Analysis: How to Win Operations Other Than War*, CCRP Publications, Pentagon, Washington, DC.

Inohara, T., Hipel, K.W., 2008a, Coalition analysis in the graph model for conflict resolution, *Systems Engineering*, 11(4).

Inohara, T., Hipel, K.W., 2008b, Interrelationships among noncooperative and coalition stability concepts, *Journal of Systems Science and Systems Engineering*, 17(1): 1–29.

Inohara, T., Hipel, K.W., Walker, S., 2007, Conflict analysis approaches for investigating attitudes and misperceptions in the war of 1812, *Journal of Systems Science and Systems Engineering* 16(2): 181–201.

International Crisis Group, 2002, *Central Asia: Water and Conflict*, Asia Report 34, Brussels.

IPCC, Intergovernmental Panel on Climate Change, 2007, Climate change 2007: climate change impacts, adaptation and vulnerability, summary for policymakers, Working Group II Contribution to the Intergovernmental Panel on Climate Change. Fourth Assessment Report. Website: http://www.ipcc.ch. Accessed February 15, 2007.

Jaccard, M., 2005, *Sustainable Fossil Fuels: The Unusual Suspect in the Quest for Clean and Enduring Energy*, Cambridge University Press, United Kingdom.

Keating, C., Rogers, R., Unal, R., Dryer, D., Sousa-Poza, A., Safford, R., Peterson, W., Rabadi, G., 2003, Systems of systems engineering, *Engineering Management Journal*, 15(3): 36–45.

Keeney, R.L., 1992, *Value Focused Thinking: A Path to Creative Decision Making*, Harvard University Press, Cambridge, MA.

Kilgour, D.M., Hipel, K.W., 2005, The graph model for conflict resolution: past, present, and future, *Group Decision and Negotiation*, 14(6): 441–460.

Kilgour, D.M., Hipel, K.W., Fang, L., 1987, The graph model for conflicts, *Automatica*, 23(1): 41–55.

Kilgour, D.M., Hipel, K.W., Fang, L., Peng, X., 2001, Coalition analysis in group decision support, *Group Decision and Negotiation*, 10(2): 159–175.

Kilgour, D.M., Rajabi, S., Hipel, K.W., Chen, Y., 2004, Screening alternatives in multiple criteria subset selection, *INFOR*, 42(1): 42–60.

Knabb, R.D., Rhome, J.R., Brown, D.P., 2006, Tropical cyclone report: hurricane Katrina, August 23–30, 2005, *National Hurricane Center*. Website: http://www.nhc.noaa.gov/pdf/TCR-AL122005_Katrina.pdf. Accessed September 5, 2007.

Levels Reference Study Board, 1991, Levels reference study: Great Lakes—St. Lawrence River Basin, International Joint Commission, Washington, DC, and Ottawa, Canada.

Levels Reference Study Board, 1993, Principles, measure, evaluation, integration, and implementation, *Working Committee 4 Final Report*, International Joint Commission, Washington, DC, and Ottawa, Canada.

Li, K.W., Hipel, K.W., Kilgour, D.M., Fang, L., 2004a, Preference uncertainty in the graph model for conflict resolution, *IEEE Transactions on Systems, Man, and Cybernetics, Part A*, 34(4): 507–520.

Li, K.W., Kilgour, D.M., Hipel, K.W., 2004b, Status quo analysis of the flathead river conflict, *Water Resources Research*, 40(5), W05S03, doi:10.1029/2003WR002596 (9 pages).

Li, K.W., Kilgour, D.M., Hipel, K.W., 2005, Status quo analysis in the graph model for conflict resolution, *Journal of the Operational Research Society*, 56(6): 699–707.

Li, K.W., Hipel, K.W., Kilgour, D.M., Noakes, D.J., 2006, Integrating uncertain preferences into status quo analysis with application to an environmental conflict, *Group Decision and Negotiation*, 14(6): 461–479.

Luskasik, S.J., 1998, Systems, system of systems, and the education of engineers, *Artificial Intelligence for Engineering Design, Analysis, and Manufacturing* 12(1): 55–60.

Maier, M., 1999, Architecting principles for system-of-systems, *Systems Engineering*, 1(4): 267–284.

McKinney, D.C., 2003, *Cooperative Management of Transboundary Water Resources in Central Asia*, Website: http://www.ce.utexas.edu/prof/mckinney/papers/aral/CentralAsia-Water-McKinney.pdf. Accessed September 21, 2007.

Micklin, P., 2000, *Managing water in Central Asia*, Russia and Eurasia Programme, The Royal Institute of International Affairs, London.

Micklin, P., 2007, The Aral Sea disaster, *Annual Review of Earth and Planetary Sciences*, 35: 47–72. Website: http://arjournals.annualreviews.org/doi/pdf/10.1146/annurev.earth.35.031306. 140120.

Miller, J.H., Page, S.E., 2007, *Complex Adaptive Systems: An Introduction to Computational Models of Social Life*, Princeton University Press, Princeton, New Jersey.

Nandalal, K.W.D., Hipel, K.W., 2007, Strategic decision support for resolving conflict over water sharing among countries along the Syr Darya River in the Aral Sea Basin, *Journal of Water Resources Planning and Management*, 133(4): 289–299.

NAERC, North American Electric Reliability Corporation, 2004, Final report on the August 14, 2003 blackout in the United States and Canada: causes and recommendations, *U.S.-Canada Power System Outage Task Force*, Princeton, New Jersey and Washington, DC.

Newbern, D., Nolte, J., 1999, Engineering of complex systems: understanding the art side, *Systems Engineering*, 2(3): 181–186.

Obeidi, A., Hipel, K.W., Kilgour, D.M., 2005, The role of emotions in envisioning outcomes in conflict analysis, *Group Decision and Negotiation*, 14(6): 481–500.

Obeidi, A., Hipel, K.W., Kilgour, D.M., 2006, Turbulence in Miramichi Bay: The Burnt Church conflict over native fishing rights, *Journal of the American Water Resources, Association,* 42 (12): 1629–1645.

Quinn, J.B., Baruch, J.J., Paquette, P.C., 1987, Technology in services, *Scientific American*, 257 (6): 50–58.

Rajabi, S., Hipel, K.W., Kilgour, D.M., 2001, Multiple criteria screening of a large water policy subset selection problem, *Journal of the American Water Resources Association,* 37(3): 533–546.

Rosenschein, J.S., Zlotkin, G., 1994, *Rules of Encounter: Designing Conventions for Automated Negotiation among Computer*, MIT Press, Cambridge, MA.

Rouse, W.B., 2003, Engineering complex systems: implications for research in systems engineering, *IEEE Transactions on Systems, Man, and Cybernetics—Part C: Applications and Reviews*, 33(3): 154–156.

Rouse, W.B., 2007, Complex engineered, organizational and natural systems, *Systems Engineering*, 10(4): 260–271.

Sage, A.P., 1992, *Systems Engineering*, Wiley, New York.

Sage, A.P., 1999, Simulation and model driven experimentation in systems engineering, *Systems Engineering*, 2(2): 57–61.

Sage, A.P., Biemer, S.M., 2007, Processes for system family architecting, design, and integration, *IEEE Systems Journal*, 1(1): 5–16.

Sage, A.P., Cuppan, C.D., 2001, On the systems engineering and management of systems of systems and federations of systems, *Information Knowledge Systems Management,* 2(4): 325–345.

Sage, A.P., Rouse, W.B., (Eds.), 2008, *Handbook of Systems Engineering and Management,* 2nd edition, Wiley, New York.

Seiko, T.S., 1998, Geographical and socio-economic dimensions of the Aral Sea Crisis and their impact on the potential for community action, *Journal of Arid Environments*, 39(2): 225–238.

Senge, A., 1994, *The Fifth Discipline*, Doubleday, NY.

Stern, N., 2007, *Economics of Climate Change: the Stern Review*, Cambridge University Press, New York.

Syme, G.J., Nancarrow, B.E., McCreddin, J.A., 1999, Defining the components of fairness in the allocation of water to environmental and human uses, *Journal of Environmental Management*, 57(1): 51–70.

Van Heerden, I., Bryan, N., 2006, *The Storm: What Went Wrong and Why During Hurricane Katrina—The Inside Story from One New Orleans Scientist*, Penguin Group, NY.

Vinogradov, S., Langford, V. P. E., 2001, Managing transboundary water resources in the Aral Sea Basin: in search of a solution, *International Journal of Global Environmental, Issues* 1 (3/4): 345–362.

Von Neumann, J., Morgenstern, O., 1953, *Theory of Games and Economic Behavior*, 3rd edition, Princeton University Press, Princeton, NJ.

Walters, C., 1997, Challenges in adaptive management of riparian and coastal ecosystems *Conservation Ecology*, 1(2): 1, Available online at http://www.consecol.org/vol1/iss2/art1/. Accessed February 21, 2007.

Wang, L., Fang, L., Hipel, K.W., 2003, Water resources allocation: a cooperative game theoretic approach, *Journal of Environmental Informatics*, 2(2): 11–22.

Wang, L., Fang, L., Hipel, K.W., 2007a, Mathematical programming approaches for modeling water rights allocation, *Journal of Water Resources Planning and Management*, 133(1): 50–59.

Wang, L., Fang, L., Hipel, K.W., 2007b, On achieving fairness in the allocation of scarce resources: measurable principles and multiple objective optimization approaches, *IEEE Systems Journal*, 1(1): 17–28.

Wang, L., Fang, L., Hipel, K.W., 2008, Basin-wide cooperative water resources allocation, *European Journal of Operations Research*, 190(3): 798–817.

Wikipedia, 2007, Great Lakes article, http://en.wikipedia.org/wiki/Great_Lakes. Accessed August 15, 2007.

Wolf, A.T., 1998, Conflict and cooperation along international waterways, *Water Policy*, 1(2): 251–265.

Xu, H., Kilgour, D.M., Hipel, K.W., 2007, Matrix representation of solution concepts in graph models for two decision-makers with preference uncertainty, *Dynamics of Continuous, Discrete and Impulsive Systems, Supplement on Advances in Neural Networks – Theory and Applications*, 14, 703–707.

Yin, Y., Huang, G., Hipel, K.W., 1999, Fuzzy relation analysis for multicriteria water resources management, *Journal of Water Resources Planning and Management*, 25(1): 41–47.

Young, H.P., 1994, *Equity in Theory and Practice*, Princeton University Press, Princeton, NJ.

Zeng, D.-Z., Fang, L., Hipel, K.W., Kilgour, D.M., 2007, Policy equilibrium and generalized metarationalities for multiple decision-maker conflicts, *IEEE Transactions on Systems, Man, and Cybernetics, Part A, Systems and Humans*, 37(4): 456–463.

Chapter 19

Robotic Swarms as System of Systems

FERAT SAHIN

Rochester Institute of Technology, Rochester, NY, USA

19.1 INTRODUCTION

In this chapter, we study and evaluate a robotic swarm approach in the context of the system of systems (SoS) concepts. Swarms and/or robotic swarms show strong system of systems characteristics such as interoperability, integration, and adaptive communications. On the basis of these common characteristics, robotic swarms can be evaluated and studied as systems of systems. This chapter presents a design and evaluation of a robotic swarm in the context of system of system. In addition, the chapter presents system of systems characteristics of robotic swarms. Next, we evaluate the characteristics of both swarms and system of systems.

19.1.1 Swarm Intelligence

Large groups of small insects constitute swarms in nature. In these swarms, members perform a simple task but their actions produce complex behaviors as a whole (Hinchey et al., 2007). This emergent behavior is also visible in higher order animals such as ant colonies, bird flocks, and packs of wolves. These groups present swarm behaviors in many ways. Many areas in computer science and engineering have explored the complex problem-solving capability of swarms (Hinchey et al., 2007).

Swarm intelligence term was first introduced by Beni, Hackwood, and Wang (Beni, 1998; Beni and Wang, 1989; Hackwood and Beni, 1992) in the context of cellular robotics. Intelligent swarms have members that can present independent intelligence. Intelligent swarm members can be homogenous or heterogeneous based on their environmental interactions (Hinchey et al., 2007). Bonabeau, Dorigo, and Theraulaz

System of Systems Engineering: Innovations for the 21st Century, Edited by Mo Jamshidi
Copyright © 2009 John Wiley & Sons, Inc., Publication

zadc provided a more general definition of swarm intelligence, which is "any attempt to design algorithms or distributed problem-solving devices inspired by the collective behavior of social insects and other animal societies" (Kennedy and Eberhart, 1995). The behavior of the social insects is determined by the use of pheromones. For instance, ants are able to forage in unknown environments by laying and following pheromone trails. Ant colony-based algorithms are generally based on short-range recruitment (SRR) and long-range recruitment (LRR) behaviors of the ants (Kumar and Sahin, 2002; Kumar and Sahin, 2003a; Kumar and Sahin, 2003b). Dorigo and Di Caro (1999) provided a common framework, ant colony optimization, for the algorithms related to ant colonies (Dorigo and Di Caro, 1999).

On the other hand, swarm intelligence techniques, different from intelligent swarm, are population-based stochastic methods and mainly used in combinatorial optimization problems (Kennedy and Eberhart, 2001; Yavuz et al., 2006; Sahin and Devasia, 2007; Sahin et al., 2007a). In swarm intelligence methods, the collective behavior is created by the members' local interactions with their environment to come up with functional global patterns (Hinchey et al., 2007). Particle swarm optimization (PSO) is the main example of such swarm intelligence techniques (Kennedy and Eberhart, 2001).

The PSO algorithm involves casting a population of co-operative agents, called particle, randomly in the multidimensional search space. Each particle has an associated fitness value, which is evaluated by the fitness function to be optimized, and a velocity that directs its motion. Each particle can keep track of its solution that resulted in the best fitness as well as the solutions of the best performing agents in its neighborhood. The trajectory of each particle is dynamically governed by its own and its companions' historical behavior. Kennedy and Eberhart (2001) view this adjustment as conceptually similar to the crossover operation utilized by genetic algorithms. Such an adjustment maximizes the probability that the particles are moving toward a region of space that will result in a better fitness. At each step of the optimization, the particle is allowed to update its position by evaluating its own fitness and the fitness of the neighboring particle. The PSO algorithm is terminated when the specified maximum number of generations is reached or when the best particle position of the entire population cannot be improved further after a sufficiently large number of generations. A simple pseudocode describing the functioning of the optimizer taken from Taşgetiren and Liang, (2003) is shown below.

```
Initialize parameters
Initialize population
Evaluate
Do {
    Find particlebest
    Find globalbest
    Update velocity
    Update position
    Evaluate
} While (Termination)
```

In recent years, ant colony optimization (Sim and Sun, 2003; Dorigo et al., 2006) and particle swarm optimization (Kennedy and Eberhart, 2001; Yavuz et al., 2006; Sahin and Devasia, 2007; Sahin et al., 2007a) are mostly studied and applied to various problems. Swarm intelligence has been applied to important problems and domains such as telecommunications (Di Caro and Dorigo, 1998; Kassabalidis et al., 2001; Günes et al., 2002; Montresor et al., 2002), business (Bonabeau and Meyer, 2001), robotics (Beckers et al., 1994; Hayes et al., 2001; Kumar and Sahin, 2002; Kumar and Sahin, 2003a; Kumar and Sahin, 2003b,; Leung et al., 2003; Chapman, 2004; Chapman and Sahin, 2004), and optimization (Dorigo et al., 1996; Dorigo and Stützle, 2004). This chapter focuses on swarm intelligence applied to robotics.

19.1.2 Swarm Robotics and Robotic Swarms

Swarm robotics is defined by Hinchey et al. (2007) as the application of swarm intelligence techniques to the analysis of activities in which swarm members are physical robotic devices that can change their environments by intelligent decision making based on various inputs. These are mobile robots with various locomotion capabilities such as legged or wheeled ground robots (Chapman and Sahin, 2004; Sahin, 2004; Azarnoush et al., 2006; Groß et al., 2006) underwater robots (Gracias et al., 2003; Yu et al., 2007) and flying robots (Hart and Craig-Hart, 2004). In addition, there are efforts in designing microrobots and MEMS-based swarm robots (Donald et al., 2006; Kim et al., 2007).

Control strategies developed for swarming robots are generally inspired from biological systems (Pack and Mullins, 2003; Hart and Craig-Hart, 2004; Stewart and Russell, 2004; Fritsch et al., 2007; Meng et al., 2007) as well as hierarchical (Kloetzer and Belta, 2007) and cooperative control schemes (Clark and Fierro, 2005; Bishop, 2006). In almost all control schemes, the swarm members have a common communication medium and very similar hardware and software components. In addition, modular robots are mainly employed in swarm robotics applications because they are easy to assemble, and by definition they share the same design architecture that makes them highly compatible (Sahin, 2004; Groß et al., 2006).

Modular robotics field is a rapidly progressing research area (Rus et al., 2002; Yim et al., 2002; Sahin, 2004; Groß et al., 2006). Modular robotic systems are inherently robust and flexible. These properties are becoming increasingly important in real-world robotics applications. Recently, special attention has been paid to *self-reconfigurable* robots (Groß et al., 2006); that is, modular robots whose components can autonomously organize into different connected configurations. In most of the current modular robotics implementations, modular robots are initially manually assembled and, once assembled, they are incapable of assimilating additional modules without external direction. In this chapter, we present a modular robot design with hot swappable modules, called GroundScouts (Sahin, 2004).

Since the concept of swarming robots exhibits similar characteristics to system of systems characteristics, it would be best to define and evaluate the characteristics of system of systems theory.

19.2 SYSTEM OF SYSTEMS

There has been a growing recognition that significant changes are necessary in governments and industries, especially in the aerospace and defense areas. Recently, major aerospace and defense manufacturers, including (but not limited to) Boeing, Lockheed-Martin, Northrop-Grumman, Raytheon, and BAE Systems, include some version of "large-scale systems integration" as an integral component of their business strategies. In some companies, there are even business units dedicated to systems integration activities (Crossley, 2004).

Recently, there has been a strong interest in SoS concepts and strategies. The performance optimization among group of heterogeneous systems to achieve a common task is becoming the focus of a diverse range of applications including military, security, aerospace, and disaster management (Lopez, 2006; Wojcik and Hoffman, 2006). There is an increasing interest in generating synergy between these independent systems to achieve the most desired overall system performance (Azarnoush et al., 2006). In the literature, several researchers addressed the issue of coordination and interoperability in a system of systems (Crossley, 2004; DiMario, 2006; Abel and Sukkarieh, 2006).

The concept of SoS is essential to more effectively implement and analyze large, complex, independent, and *heterogeneous* systems working (or made to work) cooperatively (Azarnoush et al., 2006). The main thrust behind the desire to view the systems as an SoS is to obtain higher capabilities and performance than that would be possible with a traditional system view. The SoS concept presents a high-level viewpoint and explains the interactions among the independent systems. However, the SoS concept is still at its developing stages (Abbott, 2006; Meilich, 2006). The literature has revealed that much of the recent work introduces new concepts toward an SoS approach. However, very few researchers have attempted application to real-world scenarios (Abel and Sukkarieh, 2006).

System of systems are supersystems comprised of other elements that themselves are independent complex operational systems interact among themselves to achieve a common goal. Each element of an SoS achieves well-substantiated goals even if it is detached from the rest of the SoS. SoS exhibit behavior, including emergent behavior, not achievable by the component systems acting independently (Pearlman, 2006). SoS are considered as metasystems that are diverse in their components' technologies, context, operation, geography, and conceptual framework (Keating, 2006). For example, a Boeing 607 airplane, as an element of an SoS is not SoS, but an airport is a SoS or a rover on Mars is not a SoS, but a *robotic colony* (or a *robotic swarm*) exploring the red planet is an SoS. Associated with SoS, there are numerous problems and open-ended issues that need a great deal of fundamental advances in theory and verifications. In fact, there is not even a universal definition among system engineering community (Jamshidi, 2008).

Based on the literature survey on system of systems, there are several definitions (Manthorpe, 1996; Kotov, 1997; Luskasik, 1998; Pei, 2000; Carlock and Fenton, 2001; Sage and Cuppan, 2001; Jamshidi, 2008). Detailed literature survey and discussions on these definitions are given in (Sahin et al., 2007b). All the definitions of SoS have their own merits, depending on their application. Our favorite definition is "Systems of systems are large-scale concurrent and distributed systems that are comprised of complex systems." The primary focus of this definition is *information systems*, which emphasizes the *interoperability* and *integration* properties of an SoS (Sahin et al., 2007c).

The *interoperability* in complex systems (i.e., multiagent systems) is very important as the agents operate autonomously and interoperate with other agents (or nonagent entities) to accomplish better actions. Interoperability requires successful communication among the agents (systems). Thus, the systems should carry out their tasks *autonomously* as well as communicate with other systems in the SoS to take actions for the overall goodness of the SoS, not just for themselves (Sahin et al., 2007c).

The *Integration* implies that each system can communicate and interact (control) with the SoS components regardless of its hardware and software characteristics. This means that systems need to have the ability to communicate with the SoS or a part of the SoS without compatibility issues such as operating systems, communication hardware, communication protocol, and so on. For this reason, a SoS needs a common language that the SoS components can speak. Without having a common language, the SoS components cannot be fully functional, and the SoS cannot be adaptive in the sense that new components cannot be integrated to the SoS without a major effort. Integration should also consider meaningful control aspects of the SoS (Sahin et al., 2007c). For example, a system within an SoS should understand commands and/or control signals from other SoS components.

In real-world systems, the problem is addressed in a higher level where the systems send and receive data from other systems in the SoS and make a decision that leads the SoS to its global goals. Let us take the military surveillance example where different units of the Army collect data through sensors to locate a threat or to determine the identity of a target. In this type of situations, army command center receives data from these heterogeneous sensor systems such as AWACS, ground RADARS, and submarines. These systems are the parts of the SoS that make the decision, say the command and control station. However, they may be developed with different technologies. This will create a huge barrier in data aggregation and data fusion using the data received from these systems because they would not be able to interact successfully without hardware and/or software compatibility (Sahin et al., 2007c). In addition, the data coming from these systems are not unified, which will add to the barrier in data aggregation and cooperation among the SoS components.

One solution to the problem is to modify the communication medium among the SoS components. Two possible ways of accomplishing this task are as follows:

Create a Software Model of Each System: In this approach, each component in the SoS talks to a software module embedded in itself. The software module collects data from the system and generates outputs and sends to the other SoS components through the software model. If these software modules are written with a common architecture and a common language, then the SoS components can communicate effectively regardless of their internal hardware and/or software architectures (Sahin et al., 2007c).

Create a Common Language to Describe Data: In this approach, each system can express its data in this common language so that other SoS components can parse the data successfully.

The overhead that needs to be generated to have software models of each system on an SoS could be enormous and must be redone for new member of the SoS. In addition, this requires the complete knowledge of the state-space model of each SoS components, which is often not possible.

Consequently, recent research on SoS concepts focus on the study of coming up with a common language to describe the data of each system in an SoS to create an effective communication medium in the SoS. There are some research activities on using a language based on Extensible Markup Language (XML) in order to wrap data coming from different sources in a common way (Sahin et al., 2007b; Sahin et al., 2007c). In these research activities, the XML is used to describe components of the SoS and their data in a unifying way. If XML-based data architecture is used in an SoS, the only requirement for the SoS components is to understand/parse XML file received from the components of the SoS (Sahin et al., 2007b). In addition, Chapter 6 presents an example simulation of an SoS using XML-based language and discrete event systems simulation tools.

In a robotic swarm, similar approaches to integration are utilized to achieve effective communication among the swarm components. The next section explores how we can use SoS concepts on robotic swarms.

19.3   SYSTEM OF SYSTEM APPROACH TO ROBOTIC SWARMS

As stated in the introduction, the robotic swarms have common characteristics with system of systems such as interoperability and integration. Interoperability and integration require an effective adaptive communication scheme among the systems (swarm members). Adaptive nature of the communication is critical since it allows the SoS components to be able to communicate regardless of the system dynamics. The following sections explore the interoperability and integration characteristics of the systems of systems and their reflections to robotic swarms.

19.3.1   Interoperability

In a swarm, members of the swarm can survive independently although their individual performance is generally dismal compare to the performance of the swarm overall. On the other hand, in a system of systems, the systems are often independently operable and their individual performance may be same as the performance they would have in a system of systems. This suggests that interoperability is stronger in an SoS than in a robotic swarm. In other words, interoperability of a system of systems is more general compare to the limited interoperability of a swarm member. Thus, to study robotic swarms as system of systems, the interoperability of the swarm should be achieved, but it should not be an integral part of the design. In a swarm, if the members leave the swarm, they can survive by themselves, but they immediately start looking for ways of joining back to the swarm. In order for a swarm members to find their way back into the swarm, they need to be able to survive and operate independently. Robotic swarm design presented in this chapter has modular microrobots with various sensors and they can autonomously operate in the environment. Thus, they definitely satisfy the interoperability characteristics of an SoS. The architecture of the robot design for interoperability and modularity are presented in Section 19.3.

19.3.2   Integration

Integration is extremely important for both SoS and robotic swarms since both need to expand and shrink with ease during the operation. In an SoS, it is critical to have the ability to bring a system into the SoS without any major problem such as redesign, restart, and/or reprogram the system. Integration in an SoS has two main components. The first one is to have adaptive communication schemes that allow the inclusion of a new system member. The second one is to have a common language that any new system can send its data to other systems in the SoS without any difficulty or major effort. As presented in Section 19.2, the common language such as XML is a critical necessity for an SoS. The architecture for such as language can be incorporated with a discrete event simulation tool so that the SoS concepts can be simulated before they are applied to real problems. Detailed explanations of SoS simulation and XML-based language are presented in Chapter 6.

Consequently, the integration characteristics of an SoS require effective and adaptive communication among the systems. When the systems in an SoS are heterogeneous in hardware and software, the XML-like common language becomes very essential. However, since all the members in a swarm have either identical or very similar hardware and software components, necessity of a common language is not so critical. Thus, in a robotic swarm, the design need not have an XML-like common language. Instead, a common protocol or a common format will suffice. Thus, the design efforts are not focused on the XML-like common language in the robotic swarm presented in this chapter.

As stated earlier, another component of the integration characteristics is to have adaptive communication schemes. This characteristic is more important for a swarm robot since the members of a swarm move in and out of the swarm very often. Thus, the communication scheme needs to have the ability to adapt itself for the population changes in the swarm. To achieve this goal, processes for accepting new members to the swarm and removing a member from a swarm need to be implemented as part of the communication scheme. The details of the communication scheme employed in the robotic swarm are presented in this chapter.

19.4   IMPLEMENTATION OF SYSTEM OF ROBOTS: GROUNDSCOUTS

There has been a significant amount of research in reconfigurable and modular robotics in recent years (Murata et al., 1999; Yoshida et al., 1999; Lee and Sanderson, 2000; Yim et al., 2002; Sahin, 2004; Bishop, 2006; Groß et al., 2006). Most of the research has been on multiple identical modules that construct a single robot. The modular design presented in this chapter has a vertical modularity in which the modules are not identical to one another (Sahin, 2004). It slices a robot into functional abstract modules such as locomotion, control, sensors, communication, and actuation. Any mobile robot can be constructed by combining the above abstract modules for a specific application. A submodule is a piece of hardware, that accomplishes the functionality of an abstract module, that is, a wireless communication submodule for a communication module. A similar modular robotics approach is explored in Fryer and McKee (1998). These authors focus on sensor-oriented modular robotics systems (MRS) where modules contain a mobile base, a camera mount, and a camera (Fryer and McKee, 1998). In our architecture (Sahin, 2004), the same robot can be constructed by combining a locomotion module, a sensory module, and an actuator module. In Fryer and McKee, (1998), the software modularity has not been intensively explored. Our MRS model (Sahin, 2004), shown in Fig. 19.1, has both hardware and software modularity.

In the model, each abstract module can be implemented by the corresponding hardware. For example, a sensor module may be an ultrasound sensor board, a proximity sensor board, or perhaps both. The submodules are designed such that they have a unique signature and standard pin connections. The submodules can be added at any level, the position does not effect its operation. Since modules can be combined in any order, an application-specific robot can be quickly constructed. For example, if a new problem domain requires legs rather than wheels, the wheeled submodule can be instantly swapped with the legged submodule. This is essential for swarm intelligence applications since agents might be equipped with complementary abilities instead of having the same abilities. For example, in ant colonies there are different kinds of ants handling distinct tasks using their special abilities (Sim and Sun, 2003; Dorigo et al., 2006).

Programming robotic applications is far from being standard. The primary reason is that each robot is composed of very special hardware designed for a specific goal or problem domain. As a result, the software components also become specific to the

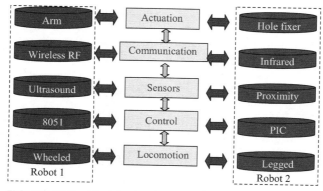

FIGURE 19.1 Modular hardware architecture (Sahin, 2004)

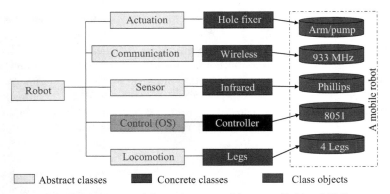

FIGURE 19.2   Modular software architecture (Sahin, 2004)

robot. Thus, systems currently available are not capable of providing modules that are easily maintained and reusable (Denneberg and Fromm, 1998). Denneberg and Fromm attempted to develop open software for autonomous mobile robots (Denneberg and Fromm, 1998). Their approach, open software concept for autonomous robots (OSCAR), is based on a layered model with four software levels: command layer, execution layer, image layer, and hardware layer. Each layer acts as an interface between the upper layer and the lower layer. Therefore, the upper layer does not require knowledge of the lower layer (Denneberg and Fromm, 1998).

Our MSR approach employs software modularity that matches the hardware modularity mentioned above. In OSCAR, the hardware layer deals with only the hardware components, whereas in our approach each hardware module corresponds to a software module as well. Modular software development methodologies are directly related to the object-oriented programming paradigm. Our MSR approach consists of abstract classes and their concrete subclasses. The abstract classes are *robot, locomotion, control, communication, sensor,* and *actuators*. The concrete classes are derived from these abstract classes. For example, a *wireless* layer may be instantiated from the *communication* class. The *wireless* object represents the robot's communication module. The concrete classes inherit functionality from the abstract classes, in addition to their hardware-specific functionality. Even though the modules contain their own software modules, the central controller (or operating system) identifies newly inserted modules and instantiates the corresponding objects and interfaces to the hardware module. These objects are going to be the tasks dynamically loaded and executed in the operating system. Fig. 19.2 shows the class structure for the software architecture.

19.5   HARDWARE–MODULARITY

On the basis of the approach described in the previous section, modular microrobots are designed by taking advantage of a layered design approach called *GroundScouts*. Even though there are five levels in the hardware and software architectures, the

FIGURE 19.3 Robot base with four wheels and legged (Sahin, 2004)

implementations (submodules) of the levels (modules) can be more than one. In addition, each level may also involve multiple closely related functionalities. The following sections presents hardware (sub)modules in the GroundScouts: *locomotion, control, sensor, communication,* and *actuation.* More detailed explanations of the modules can be found in Sahin (2004).

19.5.1 Locomotion

Locomotion module has a mechanical base and locomotion module hardware (submodule). The base of the robot consists of an aluminum frame, two DC motors, some gearing, four wheels and associated ball bearings, and the batteries. The base is designed by CAD tools and machined with high-precision CNC machines. Figure 19.3 (left) shows the base with wheels, gears, and motors. A legged version of the base is also designed as an alternative locomotion to be used in different applications. The battery selected for the robot is an AA form factor NiMH rechargeable cell. Four of these cells connected in series are used in the system. Two are placed in the front of the robot and two are in back.

The locomotion, the first electronic level shown in Fig. 19.4, is the most critical level in the operation of the robot. It contains the circuitry for an H-Bridge, a charger circuitry, and a power system that consists of a DC–DC converter and some passive components. The power system provides +5 V for the entire robot, and will accept

FIGURE 19.4 Locomotion submodule top (left) and bottom (right) view (Sahin, 2004)

FIGURE 19.5   Microcontroller board top (left) and bottom (right) view (Sahin, 2004)

from 1.5 to 15 V on its input. This provides plenty of flexibility if a different battery system is put in place. Finally, this layer contains three proximity sensors, one centered at the front and two facing 45° off center on each side of the front sensor. These sensors allow the detection of objects in close proximity to the robot. This level attaches to the base with four screws and houses connectors for the battery system and the motors.

19.5.2   Control

Housed in this layer, seen in Fig. 19.5, is the main controller, an 8051-based microcontroller running at 11.0592 MHz. A 64 k × 8 flash memory IC, a 32k × 8 SRAM IC, and an 8-bit latch are the rest of the components on this level. The flash memory is in a dip package and is attached via an IC socket, so it can be removed to be reprogrammed. With improved memory architecture, GroundScouts are able to run a modular embedded operating system explained later. The microcontroller currently in place is a Philips 80C552-5, which has no internal ROM and vectors directly to the external flash for its program. As for I/O, it has five 8-bit registers, two of which are used to access external program and data memory. One of these registers has an 8-bit analog-to-digital converter. In addition to the five registers, the microcontroller has two PWM lines and some program and data memory access lines. Because of the microcontroller's built-in address space, the only external address decoding required was a single FET to enable/disable the SRAM and the flash ROM.

19.5.3   Sensor

There are two sensor submodules in GroundScouts. The infrared submodule, shown in Fig. 19.6, has five infrared sensor pairs. Each pair consists of an emitter and a detector. The IR emitters of all five pairs of sensors are enabled simultaneously via a logic-level FET and are turned on by the microcontroller. The detectors require that the emitters be driven at 38 kHz modulated on 833 Hz, which is done by the microcontroller. The detectors have built-in filters and provide the distance from the closest object to the microcontroller. The IR layer uses a total of six microcontroller I/O lines: one to drive the emitters and five to read each of the detectors.

FIGURE 19.6   Infrared submodule (Sahin, 2004)

The ultrasonic sensor submodule, shown in Fig. 19.7, is designed for three ultrasonic receiver and transmitter pairs. These sensors allow the robot to determine its distance to an object in the range of 2–300 cm. In order for the object to be detected, it must have at least one section that is normal to the sensors as well as be of a material that will reflect a majority of the ultrasonic energy. The three sensors can be in one of two possible configurations: (1) one front and centered, and two on either side at 45° from center or (2) one front and centered and two facing rearward at 45° from center .

19.5.4   Communication

The communication submodule, shown in Fig. 19.8, consists of a PIC microcontroller that controls a LINX wireless transceiver. The modulation scheme for the transceiver is frequency shift keying (FSK) centered at 913 MHz with a 20 MHz bandwidth. A PIC microcontroller is used to form packets and to do address decoding. The PIC communicates serially with the main controller (80552). When multiple users are

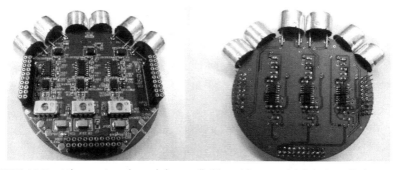

FIGURE 19.7   Ultrasonic submodule top (left) and bottom (right) view (Sahin, 2004)

FIGURE 19.8   Communication submodule for GroundScouts (Sahin, 2004)

present, a medium access control (MAC) protocol must be applied in order for data to be successfully transmitted. Time division multiple access (TDMA) is implemented as main protocol because it are easier to design and fairly well-known.

Since there are many individuals in a robotic swarm and they may not see/communicate with each other at any given time, a better protocol needs to be designed for robotic swarms. Conventional TDMA is reliable but has some inefficiencies that exist when the maximum amount of users are not present. Unused time slots could be used to increase the transfer rate of the users who are present. An adaptive TDMA (ATDMA) protocol was developed to utilize these unused time slots (Chapman, 2004; Sahin, 2004). As users get in and out of the network, the number of time slots will change accordingly, creating a TDMA network with no unused time slots. The cost of this protocol is one time slot per frame. This time slot is designated for users to enter the network. The data link layer incorporates error detection with the use of a checksum, packet numbering to ensure reliable transmissions, and a transmit/acknowledge procedure to allow the erroneous packets to be retransmitted. The ATDMA protocol has been successfully tested with two swarm intelligence-based applications (Chapman and Sahin, 2004; Opp and Sahin, 2004).

19.5.5 Actuation

A GPS submodule, shown in Fig. 19.9, is designed as the actuation level for the GroundScouts. A CCD camera and a gripper are also considered as possible actuator submodules. The chosen GPS receiver was the Lassen SQ GPS receiver. It operates on the standard L1 carrier frequency of 1575.42 MHz and has the capability to automatically get GPS satellites and track up to eight of them. It can also compute location, speed, heading, and time.

Data are sent to and from the GPS receiver via serial communications. The communication protocol is dependent on the standard chosen, either Trimble Standard Interface Protocol (TSIP) or National Marine Electronics Association (NMEA). TSIP was chosen for this application, having an input and output baud rate of 9600, 8 data bits, odd parity, 1 stop bit, and no flow control. TSIP automatically sends information

FIGURE 19.9 GPS submodule top (left) and bottom (right) view (Sahin, 2004)

on GPS time, position, velocity, receiver health/status, and satellites in view. GPS time and receiver health/status are sent every 5 s, and all others are sent every 1 s. Position and velocity measurements are capable of being sent as singles or doubles depending on the desired precision. In this application, single precision was chosen to save space and calculation time. GPS information is received and stored by a PIC16F73 using its onboard UART at the aforementioned settings. I2C is used to communicate between the main microcontroller (80552) and the PIC when GPS information is desired by the 80552. When an I2C message is received by the PIC, it finishes the current GPS message in its entirety and then services I2C to guarantee complete up-to-date information delivery.

This completes the modular hardware building blocks of GroundScouts. Pictures of a completed GroundScout are shown in Fig. 19.10. The next section presents an implementation of the modular software architecture.

FIGURE 19.10 A GroundScout with all submodules (Sahin, 2004)

19.6 SOFTWARE MODULARITY

Controlling many cooperative robots is not an easy task. The software needed is complex and must allow multitasking. To mitigate the complexity, many developers have begun using real-time operating systems (RTOSs). In addition to providing the ability to precisely synchronize multiple events, RTOSs give the application programmer predefined system services and varying degrees of hardware abstraction, both of which are aimed at making software development easier and more organized (Sahin, 2004). Thus, in addition to the software architecture described previously, a micro real-time operating system is developed with the GroundScouts' software architecture. This was critical especially for dynamic task uploading, which could send executable software components through the communication medium.

19.6.1 Operating System

Based on the software architecture described previously, a significantly small-size basic operating system (OS) is developed with dynamic task loading over a serial port (via wireless transceiver). Each robotic submodule has an associated task (driver). Each task corresponds to an object of the corresponding concrete class defined earlier. These drivers are separated from the OS in order to reduce the OS size. This separation makes the dynamic task loading and execution a crucial component of the operating systems.

The current operating system is a real-time operating system capable of accommodating the modular nature of the robots (Kauler, 1999; Santo, 2001). The hardware platform is developed around the Philips 80C552 microcontroller, a derivative of the 8051 microcontroller originally designed by Intel. This 8-bit microcontroller may support up to 64k of external program memory and an additional 64k of data memory. The primary properties of the operating system are

1. An operating system that is generic and promotes portability.
2. Programmed in C except for processor-specific assembly.
3. Provides modular source code.
4. A multitasking operating system that supports dynamic loading and task execution.

19.6.2 Dynamic Task Uploading (DLE)

The challenge when implementing DLE using a RTOS stems from the fact that there are fundamental operating system limitations. A RTOS such as the µC/OS-II is preemptive, meaning that the highest priority task waiting for execution will be executed unless it is suspended, delayed, or waiting for an event. The µC/OS-II operating System, written by Labrosse (1999), was specifically designed for embedded systems. The µC/OS-II runs on most 8, 16, 32, and even 64 bit microprocessors including the 8051. The easiest solution is to use µC/OS-II to create a shell task as part of the OS-application image into which the dynamically loaded task can be loaded at a

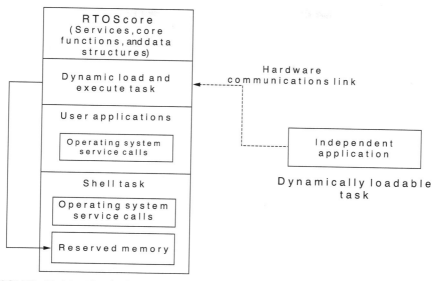

FIGURE 19.11   The high-level view of μC/OS-II with dynamic load and execute functionality (Sahin, 2004)

later time. This shell task remains dormant, that is not executed, until a dynamic task is loaded into the shell. A generic shell task contains a reserved memory location that holds the dynamically loaded task program code and the necessary operating system services code. If an independent task is loaded into the reserved memory location and the shell task is dynamically activated and executed by the CPU (Sahin, 2004). Figure 19.11 presents the dynamic load and execute feature.

The human supervisor via the software interface and wireless communication link sends a task. The operating system receives the task and forward it to a reserved memory space. Finally, the task is executed when needed. The human supervisor can also be replaced by a master robot that would have a library of tasks (drivers) to be dynamically uploaded. In our robotic swarm experiments, we did not use the dynamic task uploading since swarm intelligence techniques we explored had identical swarm members.

19.7   COMMUNICATION: ADAPTIVE AND ROBUST

To fully understand the development of the adaptive protocol, the different layers will be defined as well as how they interact with one another. A picture of these layers is shown in Figure 19.12.

The data link layer links the data together with the addition of different information, depending on the protocol. The new packet then enters the medium access control layer, where the protocol controls the data being placed on the physical layer in order to

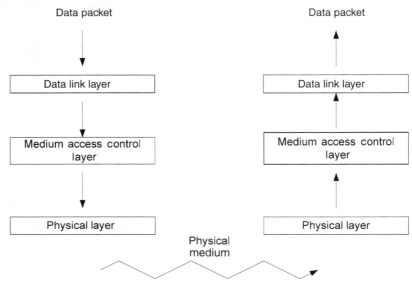

FIGURE 19.12   Layered network structure (Chapman, 2004)

prevent collisions that could occur on the physical layer. The physical layer is composed of the actual physical medium that the signal is going to travel across, and in this case the air. The specific layers are outlined and developed below.

19.7.1   Physical Layer

The communication module on the GroundScouts is comprised of a PIC microcontroller that controls a LINX transceiver that uses frequency shift keying (FSK) to transmit the data. FSK varies the frequency corresponding to whether or not the data is a 0 or a 1. The biggest advantage of this technique is that the reception of the signal is not as susceptible to signal power as other techniques. The data can still be successfully received down to a certain power threshold.

19.7.2   Medium Access Control Layer

Since there will be multiple users on the network, a medium access protocol (MAC) must be applied to ensure that data are successfully transmitted through the physical medium. Several MAC protocols were analyzed to see how feasible their application was to the hardware.

Frequency division multiple access (FDMA) separates the users in frequency by placing the messages on different carriers such that their bandwidths do not overlap. The difficulty with this protocol is that the receiver must know what carrier the signal that is being transmitted is on or else the data will not be received. This protocol is not feasible because the carrier frequency is set in hardware by the transceiver and cannot be altered.

Code division multiple access (CDMA) incorporates frequency separation as well as time separation, creating a signal that is only coherent to the receiver that has the valid "code" word. The signal is transmitted in a spread form that spreads the power spectrum in the frequency band such that its resulting signal power is below the noise floor. The receiver then uses a correlator that is based on the code word to acquire and decode the spread signal. The spectrum cannot be spread with the current hardware (Chapman, 2004). A correlator is difficult to implement. It can be done with either a DSP or an FPGA. Also, an A to D converter must be present. The current hardware does not have an A to D converter.

The last MAC protocol that was analyzed is time division multiple access (TDMA). This protocol separates the users in time, giving each user a time slot to transmit while all of the other users are listening to the channel. This was found very feasible to implement using the current hardware and characteristics of swarming robots. One of the microcontroller's timers must be used to keep track of the time so that each node knows when to transmit. Thus, this protocol was selected for the implementation.

The basis of TDMA is to divide the users by giving each node (or link) a "slot" to transmit data. The time it takes for all users on the network to have a chance to transmit is referred to as a "frame."

19.7.2.1 *Slot Allocation Strategies*   Two distinct types of slot allocation strate-
gies emerge from the topology of TDMA networks. They are node allocation and link allocation (Stevens and Ammar, 1990). Node allocation is based on slots being allocated to nodes such that each node has an opportunity to transmit to all other nodes. It has its advantages and disadvantages. One advantage is that only one transmit queue is necessary since the node will simply transmit the next ready packet. One of the disadvantages is that only one node can transmit data in each time slot.

The second approach is the link allocation that allocates time slots to link communications. The advantage is that more than one packet can be sent at once. The disadvantage is that multiple queues are necessary since each link needs its own queue. This is shown in Fig. 19.13.

Note that the topology in the Fig. 19.13 is such that a node can only hear the nodes next to it. The numbers represent which slot the corresponding transmission took place in. Experiments were done in Stevens and Ammar (1990) that compared the two allocation strategies to one another with different network densities and topologies. All the results showed that node allocation strategies outperformed link allocation strategies in all of the tests, even when broadcast packets were sent. These results illustrate the need for a node allocation strategy to be developed.

FIGURE 19.13 Node allocation (**a**) and link allocation (**b**) (Chapman, 2004)

19.7.2.2 Adaptive TDMA Development Under bursty traffic conditions, TDMA-based protocols suffer from low performance since a large number of slots remain idle (Papadimitriou and Pomportsis, 1999). This observation raises the question of whether or not a simple adaptive strategy can be developed that can utilize unused slots to increase the performance of the nodes that need more bandwidth. This has motivated researchers to study this topic. The following will outline a few distinct protocols that were researched that attempted to create the ideal adaptive network.

Kanzaki et al. (2003) developed a dynamic protocol for ad hoc networks that introduced a few different techniques for slot allocation. The protocol is based on the prior work of Young (1996) in the *Unifying Slot Assignment Protocol* (USAP) and *USAP-Multiple Access* (USAP-MA). The protocol begins with the first slot being designated for new nodes to transmit control packets for requesting slot assignments. There are four different control packets associated with the protocol. They are request (REQ), information (INF), suggestion (SUG), and reply (REP) packets. When a REQ packet is received, the network switches into a control mode where all of the other nodes send INF packets. The new node will accordingly set its frame length. It then needs to get a time slot to transmit in. This is done in one of the three ways. It either takes an unused slot. If there are no unused slots, it asks a node that has multiple slots to release one. If there are no nodes with multiple slots, it will double the frame length, effectively splitting the time slot with one of the nodes. These are illustrated in Fig. 19.14.

Conflicting slots are treated in a similar fashion where if they both have single slots, the frame is doubled and they share the slot in opposite frames. If one has multiple slots, it gives the conflicting slot to the other node. This strategy proved to be affective but seemed fairly complicated to implement.

Burr et al. (1994) created an adaptive strategy that was based on interference in the channel. The forward error correction (FEC) coding technique used was changed according to the amount of error correction needed to successfully transmit the packet. The trade-off is bandwidth. When low interference is detected by real-time channel evaluation (RTCE), the FEC code is lowered, which will increase the transmission rate, therefore, exploiting the channel characteristics. When the bandwidth is changed, slots will either become vacant or more slots will be needed. The protocol will

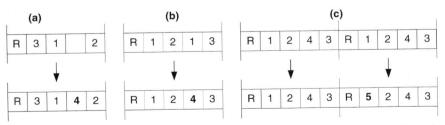

FIGURE 19.14 Slot allocation, node taking unused slot (a), node taking multiple slot (b), node doubling frame to split slot (c) (Chapman, 2004)

constantly be monitoring the slots and allocating them accordingly. When interference occurs, packets will be lost. Therefore, this protocol incorporates a one packet buffer that will hold a packet until it is successfully transmitted. This is a useful feature of this protocol.

Ali et al. (2002) propose a protocol that has three distinct advantages; the information that is necessary for the protocol is gathered locally, global coordination is not required, and changes to the state of a specific terminal do not need to be distributed to all the other terminals in the network. Two algorithms were developed that create these advantages. The first algorithm adjusts the slot assignments if an added link is detected. The second algorithm modifies slot assignments to increase the efficiency when it is determined that two terminals are no longer in range of one another. One constraint is that the two scenarios, a node leaving and a collision occurring, must be detectable. The algorithm uses a flag header that is appended to the beginning of each packet. This header is used to send control-type signals for new users to enter the network. The algorithms are described in detail in Ali et al. (2002). The important thing to note is that the flag field is a waste of bandwidth when not in use and that the network can only handle one change at a time. The second change will be addressed after the network stabilizes from the first change.

These protocols led to the development of a protocol that attempts to exploit all the positive features of each network to produce a more efficient and reliable adaptive network. The first slot of each frame is designated for new users to enter the network. This idea was formulated from Kanzaki et al. (2003). The complex control signals will be minimized in an attempt to create a more robust and simpler network for implementation. The one packet buffer from Burr et al. (1994) will allow packets to be transmitted without any lost packets. The constraints from Ali et al. (2002) are placed on this protocol, allowing the network to adjust to users entering and leaving the network. The overall product is an adaptive protocol that only has contention when a new user is attempting to enter. The network is at maximum efficiency at all times with the only cost being one time slot per frame designated for new users to enter the network.

19.7.2.3 *Adaptive TDMA Protocol for Swarm Robots*   The background for the TDMA protocol has been introduced as well as some key features. The algorithms for nodes entering and leaving the network will be defined.

1. At network startup, no master is present. The node will wait for a sink signal. When it times out, it will assume master responsibilities. The initial frame structure for the network is illustrated in Fig. 19.15. Assume that the new node is node A. Node A assumes the first slot as its transmit window.
2. Node B receives the sink signal and locks on to the master. The node then requests a TX window. If there is no contention, the TX window will be granted. Node B is let into the network in the slot right after the request slot, that is, node A's current slot. This simplifies the procedure. The master now does not need to know who is on the network, rather how many people are on the network. The

New node	Node A TX	Sink

FIGURE 19.15 Frame structure at startup (Chapman, 2004)

master knows that the sink signal needs to be sent directly after the master has had his TX window. The new frame structure is shown in Fig. 19.16.

3. If contention occurs in the new node slot, the two nodes will select a random number between 0 and 7, which represents how many more frames the node will wait before reattempting to request a TX window.
4. This procedure continues as more nodes enter the network.

The network senses that a node left the network when a packet was not transmitted in a slot. When a window is deemed unused, the slot is closed by having all of the nodes that are in later time slots decrement their time slot for the next frame. This will close the slot effectively. This can, however, create some contention. It is a constraint on the network that all nodes can hear one another. The horsepower was not available to give the network the functionality it would have needed to approach this sort of problem. If, however, the situation does occur, the two nodes that are colliding can sense a collision when an acknowledge packet is not received, meaning the packet was trashed. Both the nodes will withdraw their transmit windows and reapply to the network. No packets are lost during this transition, since one packet buffer is used.

19.7.3 Data Link Layer

The design of the overall system was traditional in that the layers are transparent to one another. The only interactions are forwarding packets up and down through the layers. When a packet is forwarded up from the microcontroller to the data link layer (DLL), two things are appended to the packet, a packet number and a checksum.

19.7.3.1 Packet Numbering Packet numbering is present to ensure that packets are reliably transmitted from node to node. Since a send and wait technique is used for error control, only one bit is necessary for the packet number. An example will illustrate this statement.

New node	Node B TX	Node A TX	Sink

FIGURE 19.16 Frame structure after second node has joined (Chapman, 2004)

If node 1 wants to transmit a packet to node 2, it sends a packet number of 0. Node 2 is expecting packet 0 and replies with an acknowledge packet (ACK) since it received packet 0. The next packet that node 1 sends to node 2 will have a packet number of 1. Node 2 will be expecting a packet number of 1.

Let us say that the same transmission occurs as above except that the acknowledge being sent back is lost. Node 2 thinks that it got the packet, but node 1 does not know that ACK was lost, so node 1 retransmits the packet. If there is no packet numbering, node 2 would think that it is the next packet even though it is not. Since the packet number is still a 0, node 2 knows that it is the same packet as before being retransmitted, so it discards the packet. The network will sustain successful packet transmission through the addition of a 1-bit packet number.

This process involves the addition of two look-up tables (LUT), each containing a bit for each possible user on the network. The first LUT stores the packet number for the next transmitted packets to each user on the network. The second LUT contains the expected packet number for the next received packet from each user. It is important to note that a hardware reset will cause the robots LUTs to reinitialize to all 0s. This means that if the master has already sent a packet to a slave, resetting the slave will mess up the packet numbering, and the following packet will be discarded by the reset slave. The next packet will arrive fine though. This is important to note when debugging.

19.7.3.2 Checksum (Error Detection)

19.7.3.2 Checksum (Error Detection)   A checksum gives the receiver the ability to sense whether or not a packet contains errors. The checksum is simply the sum of all the bytes in the packet. The receiver simply receives the packet, adds the bytes, and then sends either an ACK or a not ACK (NACK) to the transmitter, signaling whether or not there were errors in the packet. The transmitter will then retransmit if it was not an ACK.

19.7.4 Implementation Results

The detailed descriptions of the implementations of the communication protocol, the 8051 driver, and the results can be found Chapman (2004). The system was implemented for both regular TDMA and ATDMA. The TDMA system has the following specifications.

- *Static Master*: Different source code for the master and the slave
- *Error Control*: Packet Numbering and Checksum
- *Fixed Frame Length*: Fixed amount of robots on the network (numbers 1–15).

The TDMA was first implemented and was found useful for the swarm intelligence. The robots could not hear one another when placed directly next to each other because the signal was too strong, causing the receiver to become saturated. This destroyed the ability for ATDMA to reliably transmit data. The TDMA proved to be a good solution to the problem and worked quite well. The ATDMA had the following specifications.

- Dynamic master
- Ability to create subnetworks
- Error control—packet numbering and checksum
- Dynamic frame length—could adapt to a maximum of 255 users on the network
- Minimization of unused slots—maximizes data transfer rates.

When the nodes were far away from one another (>5 feet), ATDMA worked excellent. However, in a normal robotic swarm operation, the robots will be detected before they are too close to each other. Thus, in the tests, this did not cause any major failure. The only disadvantage of the ATDMA system is that robots need to send a NULL packet when they have no data to transmit. This is a large waste of power. Comparatively speaking though, the communication module on the robot does not use very much power, so this waste of power is considered negligible.

19.8 APPLICATION: MINE DETECTION WITH ANT COLONY-BASED SWARM INTELLIGENCE

The ant algorithm was first developed by Colomi et al. (1991). They performed an experiment that illustrated how ants lay *pheromone* trails as they move, creating a high probabilistic path for other ants to follow. As more ants follow the path, the scent becomes stronger. This is a form of an *autocatalytic* behavior (Colomi et al., 1991), where the more the ants follow the path, the higher the probability that other ants will follow the path. The result is a group of agents having low-level interactions that create a cooperative network.

To test the results of the observation, Dorigo applied the algorithm to the traveling salesman (optimisation) problem (TSP) (Colomi et al., 1991). Dorigo has published multiple papers since with small improvements to the TSP that illustrate better results (Dorigo and Gambardella, 1997; Dorigo et al., 2006). This raised the curiosity of other researchers in the application of this technique to other sorts of optimization problems (Sim and Sun, 2003).

In the world of communications, researchers have been analyzing different means of solving the ever-growing network routing problem. Different researchers have shown great success in applying this technique to map out various optimal routes for networks (Günes et al., 2002; Wang and Xie, 2000). Hoshyar et al. (2000) have even been working on finding different interleaving patterns for turbo codes using the ant colony based algorithm.

Other studies have been done in using the ant algorithm in data mining (Parpinelli et al., 2002) and scheduling (Tsai et al., 2002) with good results. These developments led to the research of applying the algorithm to the mine detection optimization problem.

Simulation studies of the application of the ant colony algorithm to mine detection has been done by Multi Agent Bio-Robotics Lab group at Rochester Institute of

Technology (Kumar and Sahin, 2002; Kumar and Sahin, 2003a; Kumar and Sahin, 2003b). The results showed that the algorithm was well suited for the application. Then, we have implemented the ACO-based mine detection algorithm on GroundScouts and tested their performance. Next, we give some details about the ACO algorithm implemented on GroundScouts and the results on their performance.

19.8.1 Behaviors of Ants in Ant Colonies

The behavior of the ants can be classified into three different states: *foraging, scent following, and waiting* (Kumar and Sahin, 2003a). A diagram of the states is shown below in Fig. 19.17. Next, we will evaluate each state for the mine detection problem.

19.8.1.1 Foraging   Foraging is defined as the ants' random search for mines. The action is characterized by random movements throughout the search space. The addition of cognitive maps to the state can increase the efficiency of the foraging state by giving the ant a memory mechanism so it does not continue to study the same location multiple times (Kumar and Sahin, 2003a).

19.8.1.2 Scent Following   This state is characterized by an ant following a scent (pheromone) left by another ant. For simulation purposes, the scent was represented as an exponential function that grew stronger as an ant got closer to the source. This provided a means for the ants to have a path to follow (Kumar and Sahin, 2003a).

19.8.1.3 Waiting   This state is also referred to as the recruitment state. This involves the ants releasing the pheromone to attract other ants to come and assist. In an ant colony, there are two different types of recruitment, short-range recruitment (SRR) and long-range recruitment (LRR) (Kumar and Sahin, 2003a). SRR occurs when an ant is at the food and is releasing the pheromone to attract ants that are close by. LRR

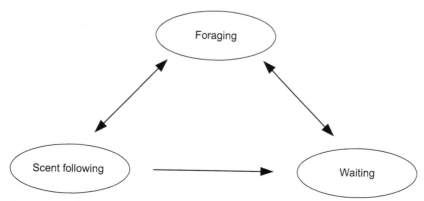

FIGURE 19.17   State diagram of the different ant behaviors (Chapman, 2004)

occurs when an ant goes back to the nest to leave a trail for other ants to follow to the food, therefore, recruiting them from far away.

19.8.2   The Mine Detection Problem

The mine detection problem is an optimization problem that is comprised of a number of mines randomly dispersed throughout a space. The goal for the algorithm is to find and disarm all of the mines in as little time as possible using minimal effort from each agent (Kumar and Sahin, 2003a,b). The ant colony algorithm is an ideal algorithm for mine detection since it utilizes the foraging capability of the ants, which allows them to search a large space in a short time. It then uses the recruiting capabilities to attract more ants to collectively disarm the mine.

19.8.3   State Transition Events

Three events could cause transitions from state to state: finding a mine, finding a scent, and waiting time-out.

19.8.3.1   *Finding a Mine*   Regardless of what state the ant is currently in, if a mine is found, it will instantly switch into the waiting state since the goal of the algorithm is to find and disarm all of the mines.

19.8.3.2   *Finding a Scent*   If an ant is currently in the foraging state, it will switch to the scent following state if a scent is found. If it is in the waiting state, it will not follow the scent since it is already at a mine.

19.8.3.3   *Waiting Timeout*   This occurs when an ant has been waiting for a long time for other ants to come and assist in disarming the mine. The ant will leave the mine and begin foraging again after certain amount of time, called threshold. This occurrence is a safeguard to prevent freezing. Freezing happens when all of the ants have found mines, but they are all waiting for other ants to come to assist. This would create an infinite waiting loop where the algorithm will cease to function if there is no limit to the amount of waiting an ant needs to do (Kumar and Sahin, 2003a).

The simulation results from Kumar et al. (2003b) illustrated that by varying the number of ants, the size of the mine field, and the number of mines, the ants could indeed disarm all of the mines. The results indicated that a real-time application was necessary to test the difficulty in implementing the algorithm.

19.8.4   Implementation of Mines

The mines consist of a communication module and an infrared beacon that constantly transmits an infrared signal modulated at 38 KHz in all directions (Chapman and Sahin, 2004). A picture of the beacon is shown in Fig. 19.18. The infrared signal can be sensed by the robot within 7 feet radius through the infrared sensor layer. In the

(a)   (b)

FIGURE 19.18 Outside view (a) and inside view (b) of the mine (Chapman and Sahin, 2004)

foraging stage, the robot is constantly searching for the signal. As soon as the signal is found, the robot knows that it is close to the mine and tries to move toward the mine.

Directionality is found by viewing the five infrared sensors in the infrared sensor layer of the robot. Early attempts were made to find the sensor with the best signal and assume that the mine is in that direction. This is proved to be difficult since the sensors are somewhat omnidirectional, creating a number of sensors having a good signal, making it difficult to really pinpoint the exact direction of the mine. It was concluded that finding the direction could be simplified by looking for the two sensors that have the worst signal. The robot could then move in the opposite direction, which would be directly toward the mine. The mine is disarmed using the GroundScouts communication module. A communication board was placed on the top of the mine as shown in Fig. 19.18a. When enough robots are surrounding the mine to disarm it, a message is sent by the command center to the communication board telling it to disarm the mine so that the robot can start foraging to find other mines in the field. The PIC on the communication board will then toggle a pin that will turn the mine off. The robots will then shift back into the foraging state since signal from the mine will not be available after the mine is disarmed.

19.8.5 Implementation of the Scent

The scent is implemented using the communication board. Packets are transmitted that represent the scent for the robots to follow. According to the algorithm, the scent can only travel a certain distance (Chapman, 2004). This poses a problem, How can the robots be programmed to ignore messages that are outside their listening, or smelling, radius?

The solution to this problem is to create an internal coordinate system on the robots so they know where they are in reference to one another. The robots maintain an X and Y coordinate as well as a direction. All the robots are initialized to start at the same position pointing in the same direction. When the robots send a packet to other robots, the coordinates are attached. Then, each robot calculates distances to other robots using the other robots' coordinates. There is a limit to the distance a robot can sense another robot. If the distance is larger than this limit, the robots ignore the messages from those robots. This is effectively emulating vaporizing pheromone in an ant colony.

There are three different messages that are sent from robot to robot. The first message signals other robots that a mine was found. This mimics the scent. The second message signals that a robot is leaving a mine. This happens when a robot times out after the waiting threshold has been reached. This is also used to allow robots to know exactly how many other robots are surrounding the mine. The third message signals that a robot has disarmed the mine. This lets the other robots that are waiting around the mine know that the mine was disarmed.

19.8.6 Main Program

The main program functions as a state machine with three possible states: foraging, scent following, and waiting. A flow chart for the state machine is shown in Fig. 19.19.

The algorithm in Fig. 19.19 describes the *mine found* algorithm. The *scent found* event occurs when a packet is received saying that the robot is at a mine and currently

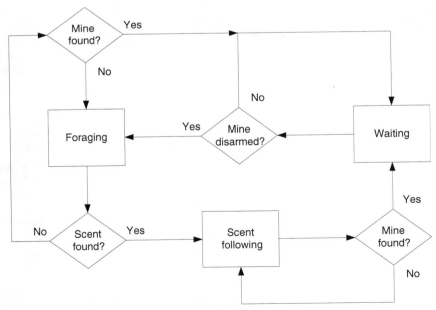

FIGURE 19.19 State machine for the overall program (Chapman, 2004)

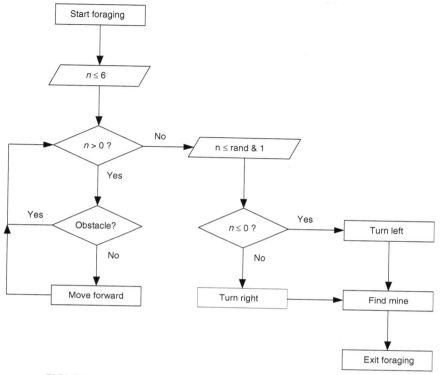

FIGURE 19.20   Flow chart for the foraging state (Chapman, 2004)

in the foraging state. The robot will then switch immediately to the scent following state, where it will go to the location of the mine. The *mine disarmed* event occurs when the front infrared sensor no longer picks up a signal, meaning that the mine is off. The individual states are described below.

19.8.6.1 Foraging   A flow chart for the foraging state is shown in Fig. 19.20.

Every movement of the robot is set to be 6 in. Setting $n = 6$ makes the robot move forward 3 feet, assuming there are no obstacles ahead. When this is finished, the robot will choose a random number. If the number is even, then the robot will turn left. The robot will turn right otherwise. The find mine algorithm is then called to see if the robot has found a mine. If it is not, it returns to the foraging state.

19.8.6.2 Scent Following   A flow chart of the scent following algorithm is shown below in Fig. 19.21.

The algorithm appears to be complex. However, all it does is compare the Y coordinate corresponding to the location of the mine to the current Y coordinate and decides the direction that the robot has to travel in to get closer to the mine. It does the

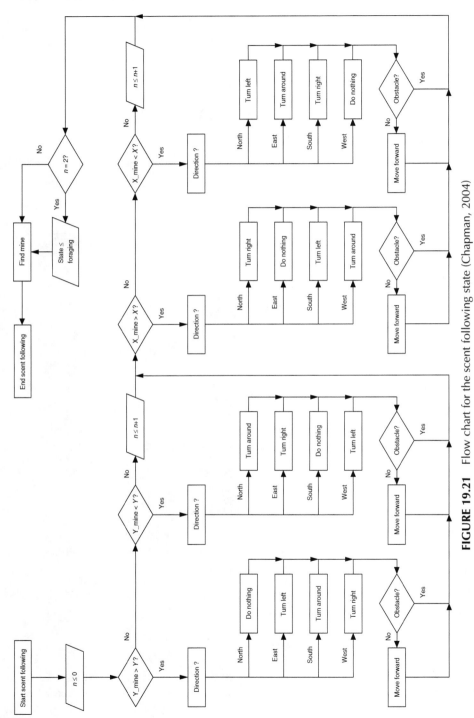

FIGURE 19.21 Flow chart for the scent following state (Chapman, 2004)

same for the X coordinates. At the end, the find mine algorithm is run. If $n = 2$, then the robot is at the location of the mine and did not find the mine. The robot will then go back into the foraging state. This is a safeguard against the situation where the internal coordinates of the robots become off center, causing the mine to be in a different place than the robot thinks it is.

19.8.6.3 *Waiting*

The waiting state periodically transmits a packet that informs the other robots that it has found a mine and where the mine is. The flow chart is shown in Fig. 19.22.

The front IR *if* statement of the chart is checking to see if the mine has been disarmed. The following part of the algorithm is to see if the robot timed out. If the robot did indeed time out, then it will turn around and travel 15 feet away. This ensures that the robot does not instantly find the same mine again.

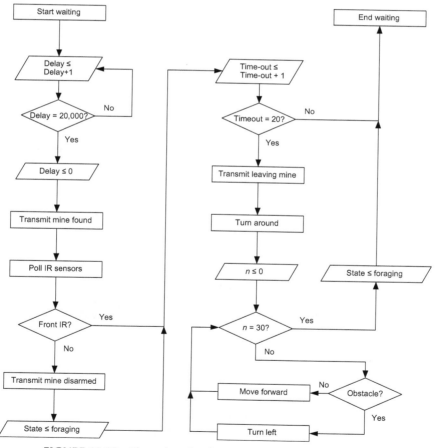

FIGURE 19.22 Flow chart for the waiting state (Chapman, 2004)

FIGURE 19.23   Starting point of the experimental setup (Chapman, 2004)

19.8.7   Experimental Results

The experiments were performed in a gymnasium so that the robots had plenty of room to work with. A picture of the starting point of the experiments is shown in Fig. 19.23.

The robots were turned on one at a time and allowed to move about 3 feet before the next robot was turned on. At the start of the algorithm, the robots are foraging, as shown in Fig. 19.24.

Figure 19.24 clearly shows the robots randomly searching for mines, foraging state. The robots near the top of the figure are beginning to find the first mine. The algorithm used to find the mines using the infrared sensors brings the robots toward the mine. The robot will then move until the back infrared sensors have no signal and the ultrasonic sensors are picking up an object that is within 6 in. Figure 19.25 shows a picture of a robot at a mine.

FIGURE 19.24   Robots in the foraging state (Chapman, 2004)

FIGURE 19.25 Robot at the mine (Chapman, 2004)

As soon as the robot reaches the mine, it will begin sending out the recruitment signal to other robots. A few problems were encountered with this. Since the implementation of the internal coordinate system neglects slippage, over time the robots' internal coordinates will begin to become off-center. As a robot at a mine sends out the recruitment signal other robots that are within the physical distance may not hear this signal since they are outside of *listening* range according to the coordinate system. Another problem is that sometimes a robot would hear the signal but would be going to the wrong location, since this is where the robot thinks the mine is.

As soon as four robots are surrounding the mine, the mine can be turned off. The robots decide that a mine is turned off by checking their front infrared sensor. If no signal is detected, then the robots conclude that the mine has been disarmed, and they then instantly switch into the foraging stage, which incorporates obstacle avoidance. This is shown in Fig. 19.26.

FIGURE 19.26 Robots having disarmed the mine and leaving (Chapman, 2004)

Some problems were encountered doing this. First of all, the robots have difficulty seeing each other using the ultrasonic sensors since their physical structure have a limited amount of surface area to bounce off. They are now back in the foraging state where they are searching for more mines. The implementation of the algorithm proved to be successful, but there were some problems that were faced during this implementation. They are outlined as follows.

- The internal coordinate system became very inaccurate as time progressed. This caused scent following problems.
- Robots that were right next to each other could not hear one another because the transmit power was full, causing the received signal to be saturated, so it was impossible for a robot to know how many other robots were around the mine.
- The algorithm that brought the robots next to the mine using the infrared sensors worked but not as efficiently as desired. The robots would sometimes go directly next to the mine, miss it, and have to turn around and try again.

As stated before, the implementation of a swarm intelligence technique, ant colony optimization, for a real problem was accomplished successfully. The swarming robots are designed with the characteristics of a swarm and/or a system of systems. Mainly, their modular architecture and adaptive TDMA for robust communication among the swarm members are inspired from these characteristics such as integration. The GroundScouts are also used to test another swarm intelligence technique, artificial immune systems, for the same mine detection problem (Opp and Sahin, 2004). This also proved that GroundScouts modular design and their communication scheme are very suitable for swarm intelligence applications.

19.9   CONCLUDING INSIGHTS

In this chapter, we have presented a design of a robot for swarm intelligence applications, called GroundScouts, using some of the concepts from system of systems. As mentioned before, a swarm is a system of systems as each member can operate independently and each of them can interact to perform a task together. These satisfy the integration and interoperability characteristics of a system of systems. With these characteristics in mind, we have presented how the hardware and software architectures can be designed and tested on a real-world problem, mine detection.

As mentioned in the introduction, the system of systems field is new, and there are some confusion about the terminology and theory. We have evaluated existing definitions and their characteristics. Then, we presented a parallel between robotic swarms and system of systems concepts. As the concepts of system of systems become clearer and a better theory of system of systems exist, one can design better swarm robots with system of systems characteristics. In fact, robotic swarms could be a testing ground for the system of system concepts and theory. With enough flexibility in hardware and software architecture, robotic swarms can even have members who have different hardware components and subsystems in order to evaluate the SoS concepts where systems are not identical.

ACKNOWLEDGMENT

The author would like to express his appreciation to Eric Chapman, a recent graduate of Department of Electrical Engineering Department at Rochester Institute of Technology. He has contributed to the resultant chapter.

REFERENCES

Abbott, R., 2006, Open at the top; open at the bottom; and continually (but slowly) evolving, *Proceedings of IEEE International Conference on System of Systems Engineering*, Los Angeles.

Abel, A., Sukkarieh, S., 2006, The coordination of multiple autonomous systems using information theoretic political science voting models, *Proceedings of IEEE International Conference on System of Systems Engineering*, Los Angeles.

Ali, F.N., Appani, P.K., Hammond, J.L., Mehta, V.V., Noneaker, D.L., Russell, H.B., 2002, Distributed and adaptive TDMA algorithms for multiple-hop mobile networks, *IEEE Military Communications Conference*, Vol. 1, pp. 546–551.

Azarnoush, H., Horan, B., Sridhar, P., Madni, A.M., Jamshidi, M., 2006, Towards optimization of a real-world robotic-sensor system of systems, *Proceedings of World Automation Congress (WAC) 2006*, Budapest, Hungary.

Beni, G., 1998, The concept of cellular robotic systems, *Proceedings of the IEEE Intelligent Control Symposium, Arlington, VA*, pp. 57–62.

Beni, G., Wang, J., 1989, Swarm intelligence in cellular robotic systems, *Proceedings of the NATO Advanced Workshop on Robots and Biological Systems, Il Ciocco, Tuscany, Italy.*

Beckers, R., Holland, O.E., Deneubourg, J.L., 1994, From local actions to global tasks: stigmergy and collective robotics, *Proceedings of Artificial Life IV, Cambridge, MA, MIT Press*, pp. 181–189.

Bishop, B.E., 2006, Dynamics-based control of robotic swarms, *Proceedings of the 2006 IEEE International Conference on Robotics and Automation*, pp. 2763–2768.

Bonabeau, E., Meyer, C., 2001, Swarm intelligence, a whole new way to think about business, *Harvard Business Review*, 79(5): 106–114.

Burr, A.G., Tozer, T.C., Baines, S.J., 1994, Capacity of an adaptive TDMA cellular system: comparison with conventional access schemes, *IEEE International Symposium on Personal, Indoor, and Mobile Radio Communications*, Vol. 1, pp. 242–246.

Carlock, P.G., Fenton, R.E., 2001, System of systems (SoS) enterprise systems for information-intensive organizations, *Systems Engineering*, 4(4): 242–261.

Chapman, E., 2004, GroundScouts: application of swarm intelligence to the mine detection problem, *Master Research Paper*, Rochester Institute of Technology.

Chapman, E., Sahin, F., 2004, Application of Swarm Intelligence to the Mine Detection Problem, *Proceedings of 2004 IEEE International Conference on Systems, Man & Cybernetics*, Vol. 6, pp. 5429–5434.

Clark, J., Fierro, R., 2005, Cooperative hybrid control of robotic sensors for perimeter detection and tracking, *Proceedings of 2005 American Control Conference*, pp. 3500–3505.

Colomi, A., Dorigo, M., Maniezzo, V., 1991, Distributed optimization by ant colonies, *European Conference on Artificial Life*, pp. 134–142.

Crossley, W.A., 2004, system of systems: an Introduction of Purdue University Schools of Engineering's Signature Area, *Engineering Systems Symposium*, Tang Center - Wong Auditorium, MIT.

Denneberg, V., Fromm, P., 1998, OSCAR: an open software concept for autonomous robots, *Proceedings of the 24th Annual Conference of the IEEE Industrial Electronics Society*, pp. 1192–1197.

Di Caro, G., Dorigo, M., 1998, AntNet: distributed stigmergetic control for communications networks, *Journal of Artificial Intelligence Research*, 9: 317–365.

DiMario, M.J., 2006, System of systems interoperability types and characteristics in joint command and control, *Proceedings of IEEE International Conference on System of Systems Engineering*, Los Angeles.

Donald, B.R., Levey, C.G., McGray, C.D., Paprotny, I., Rus, D., 2006, An untethered, electrostatic, globally controllable MEMS micro-robot, *Journal of Microelectromechanical Systems*, 15(1): 1–15.

Dorigo, M., Di Caro, G., 1999, Ant colony optimization: a new meta-heuristic, *Proceedings of the 1999 Congress on Evolutionary Computation*, Vol. 2, pp. 1470–1477.

Dorigo, M., Gambardella, L.M., 1997, Ant colony system: a cooperative learning approach to the traveling salesman problem, *IEEE Transactions on Evolutionary Computation*, Vol. 1, pp. 53–66.

Dorigo, M., Stützle, T., 2004, *Ant Colony Optimization*, MIT Press, Cambridge, MA.

Dorigo, M., Maniezzo, V., Colorni, A., 1996, Ant system: optimization by a colony of cooperating agents, *IEEE Transactions System, Man, Cybernetics: Part B*, 26(1): 29–41.

Dorigo, M., Birattari, M., Stutzle, T., 2006, Ant colony optimization: artificial ants as a computational intelligence technique, *IEEE Computational Intelligence Magazine*, pp. 28–39.

Fryer, J.A., McKee, G.T., 1998, Resource modeling and combination in modular robotics systems, *Proceedings of the 1998 IEEE International Conference on Robotics and Automation*, pp. 3167–3172.

Fritsch, D., Wegener, K., Schraft, R.D., 2007, Control of a robotic swarm for the elimination of marine oil pollutions, *Proceedings of the 2007 IEEE Swarm Intelligence Symposium (SIS 2007)*.

Gracias, N.R., Van der Zwaan, S., Bernardino, A., Santos-Victor, J., 2003, Mosaic-Based Navigation for Autonomous Underwater Vehicles, *IEEE Journal of Oceanic Engineering*, 28(4): 609–624.

Groß, R., Bonani, M., Mondada, F., Dorigo, M., 2006, Autonomous self-assembly in swarm-bots, *IEEE Transactions on Robotics*, 22(6): 1115–1130.

Günes, M., Sorges, U., Bouazizi, I., 2002, ARA—the ant colony based routing algorithm for MANETs, *Proceedings of the IEEE ICPPW*, pp. 79–85.

Hackwood, S., Beni, G., 1992, Self-organization of sensors for swarm intelligence, *Proceedings of the IEEE International Conference on Robotics and Automation*, Vol. 1, pp. 819–829.

Hart, D.M., Craig-Hart, P.A., 2004, Reducing swarming theory to practice for UAV Control, *Proceedings of 2004 IEEE Aerospace Conference Proceedings*, pp. 3050–3063.

Hayes, A.T., Martinoli, A., Goodman, R.M., 2001, Swarm robotic odor localization, *Proceedings of IEEE/RSJ International Conference IROS*, pp. 1073–1087.

Hinchey, M.G., Sterritt, R., Rouff, C., 2007, Swarms and Swarm Intelligence, *Computer Society Magazine*, pp. 111–113.

Hoshyar, R., Jamali, S.H., Locus, C., 2000, Ant colony algorithm for finding good interleaving pattern in turbo codes, *IEE Proceedings on Communications*, Vol. 147, pp. 257–262.

Jamshidi, M., 2008, Introduction to system of systems, in: Jamshidi, M. (Ed.) *System of Systems—Principles and Applications*, Chapter 1, Taylor Francis CRC, Boca Raton, FL.

Keating, C., 2006, Critical Challenges in the Maturation of the System of Systems Engineering Paradigm, *Keynote speech, Proceedings of IEEE International Conference on System of Systems Engineering*, Los Angeles.

Kanzaki, A., Uernukai, T., Hara, T., Nishio, S., 2003, Dynamic TDMA slot assignment in ad hoc networks, *International Conference on Advanced Information Networking and Applications*, pp. 330–335.

Kassabalidis, I., El-Sharkawi, M.A., Marks, R.J., Asabshahi, P., Gray, A.A., 2001, Swarm intelligence for routing in communication networks, *Proceedings of IEEE Globecom, San Antonio, TX*, pp. 3613–3617.

Kauler, B., 1999, *Flow Design for Embedded Systems*, R&D Books, Lawrence Kansas.

Kennedy, J., Eberhart, R., 1995, Particle swarm optimization, *Proceedings of the IEEE International Conference on Neural Networks*, Vol. 4, pp. 1942–1948.

Kennedy, J., Eberhart, R.C., 2001, *Swarm Intelligence*. Morgan Kaufmann Publishers, San Francisco.

Kim, S., Knoll, T., Scholz, O., 2007, Feasibility of inductive communication between millimeter-sized wireless robots, *IEEE Transactions on Robotics*, 23(3): 605–609.

Kloetzer, M., Belta, C., 2007, Temporal logic planning and control of robotic swarms by hierarchical abstractions, *IEEE Transactions on Robotics*, 23(2): 320–330.

Kotov, V., 1997, Systems of Systems as Communicating Structures, Hewlett Packard Computer Systems Laboratory Paper HPL-97-124, pp. 1–15.

Kumar, V., Sahin, F., 2002, A swarm intelligence based approach to the mine detection problem, *Proceedings of the IEEE International Conference on Systems, Man and Cybernetics*, Vol. 3.

Kumar, V., Sahin, F., 2003a, Foraging in ant colonies applied to the mine detection problem, *Proceedings of IEEE International Workshop on Soft Computing in Industrial Applications*, pp. 61–66.

Kumar, V., Sahin, F., 2003b, Cognitive maps in swarm robots for the mine detection problem, *Proceedings of IEEE International Conference on Systems, Man and Cybernetics*, Vol. 4, pp. 3364–3369.

Labrosse, J.J., 1999, *MicroC/OS-II: The Real Time Kernel*, CMP Books, Lawrence Kansas.

Lee, W.H., Sanderson, A.C., 2000, Dynamics and distributed control of modular robotic systems, *Proceedings of the 26th Annual Conference of the IEEE Industrial Electronics Society*, pp. 2479–2484.

Leung, H., Kothari, R., Minai, A., 2003, Phase transition in a swarm algorithm for self-organizing construction, *The American Physical Society: Physical Review E*, 68(4): 046 111. 1–046 111. 5.

Lopez, D., 2006, Lessons learned from the front lines of the aerospace, *Proceedings of IEEE International Conference on System of Systems Engineering*, Los Angeles.

Luskasik, S.J., 1998, Systems, systems of systems, and the education of engineers, *Artificial Intelligence for Engineering Design, Analysis, and Manufacturing*, 12(1): 11–60.

Manthorpe, W.H., 1996, The emerging joint system of systems: a systems engineering challenge and opportunity for APL, *John Hopkins APL Technical Digest*, 17(3): 305–310.

Meilich, A., 2006, System of systems (SoS) engineering & architecture challenges in a net centric environment, *Proceedings of IEEE International Conference on System of Systems Engineering*, Los Angeles.

Meng, Y., Kazeem, O., Muller, J.C., 2007, A hybrid ACO/PSO control algorithm for distributed swarm robots, *Proceedings of the 2007 IEEE Swarm Intelligence Symposium (SIS 2007)*.

Montresor, A., Meling, H., Babaoglu, Ö., 2002, Messor: Load-balancing through a Swarm of Autonomous Agents, Department of Computer Science, University of Bologna, Bologna, Italy, Tech. Rep. UBLCS-2002-11.

Murata, S., Yoshida, E., Tomita, K., Kurokawa, H., Kamimura, A., Kokaji, S., 1999, Hardware design of modular robotic system, *Proceedings of the 1999 IEEE/RSJ International Conference on Intelligent Robots and Systems, Korea*, pp. 2210–2217.

Opp, W.J., Sahin, F., 2004, An artificial immune system approach to mobile sensor networks and mine detection, *Proceedings of 2004 IEEE International Conference on Systems, Man & Cybernetics*, Vol. 1, pp. 947–952.

Pack, J.P., Mullins, B.E., 2003, Toward finding a universal search algorithm for swarm robots, *Proceedings of the 2003 IEEE/RSL International Conference on Intelligent Robots and Systems*, pp. 1945–1950.

Papadimitriou, G.I., Pomportsis, A.S., 1999, Self-Adaptive TDMA Protocols for WDM Star Networks: a Learning Automata Based Approach, *IEEE Photonics Technology Letters*, Vol. 11, pp. 1322–1324.

Parpinelli, R.S., Lopes, H.S., Freitas, A.A., 2002, Data mining with an ant colony optimization algorithm, *IEEE Transactions on Evolutionary Computation*, Vol. 6, pp. 321–332.

Pearlman, J., 2006, Creation of a system of systems on a global scale: the evolution of GEOSS, *Keynote speech, Proceedings of IEEE International Conference on System of Systems Engineering*, Los Angeles.

Pei, R.S., 2000, Systems of systems integration (SoSI)—a smart way of acquiring Army C4I2WS systems, *Proceedings of the Summer Computer Simulation Conference*, pp. 134–139.

Rus, D., Butler, Z., Kotay, K., Vona, M., 2002, Self-reconfiguring Robots, *Communications, ACM*, 45(3): 39–45.

Sage, A.P., Cuppan, C.D., 2001, On the systems engineering and management of systems of systems and federations of systems, *Information, Knowledge, Systems Management*, 2(4): 325–334.

Sahin, F., 2004, GroundScouts: architecture for a modular micro robotic platform for swarm intelligence and cooperative robotics, *Proceedings of 2004 IEEE International Conference on Systems, Man & Cybernetics*, Vol. 1, pp. 929–934.

Sahin, F., Devasia, A., 2007, Distributed particle swarm optimization for structural bayesian network learning, *Swarm Intelligence: Focus on Ant and Particle Swarm Optimization*, Advanced Robotics Systems International.

Sahin, F., Yavuz, C.M., Arnavut, Z., Uluyol, O., 2007a, Fault diagnosis for airplane engines using bayesian networks and distributed particle swarm optimisation, *Parallel Computing*, 33(2): 124–143.

Sahin, F., Jamshidi, M., Sridhar, P., 2007b, A discrete event XML based simulation framework for system of systems architectures, *Proceedings of IEEE International Conference on System of Systems Engineering*, San Antonio, Texas.

Sahin, F., Sridhar, P., Horan, B., Raghavan, V., Jamshidi, M., 2007c, System of systems approach to threat detection and integration of heterogeneous independently operable systems, *Proceedings of IEEE International Conference on Systems, Man and Cybernetics*, Montreal, Canada.

Santo, B., 2001, Embedded Battle Royale, *IEEE Spectrum*, pp. 36–41.

Sim, K.M., Sun, W.H., 2003, Ant Colony Optimization for Routing and Load-Balancing: Survey and New Directions, *IEEE Transactions on System, Man, and Cybernetics – Part A*, 33(5): pp. 560–572.

Stevens, D.S., Ammar, M.H., 1990, Evaluation of slot allocation strategies for TDMA protocols in packet radio networks, *IEEE Military Communications Conference*, Vol. 2, pp. 835–839.

Stewart, R.L., Russell, R.A., 2004, Building a loose wall structure with a robotic swarm using a spatio-temporal varying template, *Proceedings of 2004 IEEE/RSL International Conference on Intelligent Robots and Systems*, pp. 712–716.

Taşgetiren, M.F., Liang, Y.-C., 2003, A binary particle swarm optimization algorithm for lot sizing problem, *Journal of Economic and Social Research*, 5(2): 1–20.

Tsai, C.F., Tsai, C.W., Tseng, C.C., 2002, A new approach to solving large traveling salesman problems, *Proceedings of the International Joint Conference on Neural Networks*, Vol. 2, pp. 1540–1545.

Wang, Y., Xie, J., 2000, Ant colony optimization for multicast routing, *IEEE Asia-Pacific Conference on Circuits and Systems*, pp. 54–57.

Wojcik, L.A., Hoffman, K.C., 2006, Systems of systems engineering in the enterprise context: a unifying framework for dynamics, *Proceedings of IEEE International Conference on System of Systems Engineering*, Los Angeles.

Yavuz, C.M., Sahin, F., Arnavut, Z., Uluyol, O., 2006, Generating and exploiting bayesian networks for fault diagnosis in airplane engines, *Proceedings of IEEE International Conference on Granular Computing, GrC 2006*, pp. 250–255.

Yim, M., Zhang, Y., Duff, D., 2002, Modular Robots, *IEEE Spectrum*, 39(2): 30–34.

Yoshida, E., Kokaji, S., Murata, S., Kurokawa, H., Tomita, K., 1999, Miniaturized self-reconfigurable system using shape memory alloy, *Proceedings of the 1999 IEEE/RSJ International Conference on Intelligent Robots and Systems*, Korea, pp. 1579–1585.

Young, C.D., 1996, USAP: a unifying dynamic distributed multichannel TDMA slot assignment protocol, *IEEE Military Communications Conference*, Vol. 1.

Yu, J., Wang, L., Tan, M., 2007, Geometric optimization of relative link lengths for biomimetic robotic fish, *IEEE Transactions on Robotics*, 23(2): 382–386.

Chapter **20**

Understanding Transportation as a System of Systems Problem

DANIEL A. DELAURENTIS

Purdue University, West Lafayette, IN, USA

20.1 INTRODUCTION

20.1.1 Synopsis of Challenge

The national transportation system (NTS) is a collection of networks composed of heterogeneous systems within which the air transportation system (ATS) and its national airspace system (NAS) is one sector. Research on each sector of the NTS is generally conducted independently, occasionally missing important interactions between sectors. This isolated treatment is also often manifest within a sector, with infrequent consideration of the full-scope dimensions (e.g., multimodal impacts and policy, societal, and business enterprise influences) and levels of aggregation (e.g., layered dynamics within a scope category). Under these limitations, modifying the transportation system may not necessarily have the intended effect or impact. A systematic method for modeling these interactions is essential to the formation of a more complete model and understanding of the NTS and/or ATS. In this chapter, it is proposed that a system of systems (SoSs) approach fulfills this objective and, if adopted, could ultimately lead to better outcomes from high-consequence decisions in technological, socio-economic, operational, and political policy-making context. Such decision support is especially vital as decision makers in both the public and private sector, in venues such as the Joint Planning and Development Office (JPDO) (Arbuckle et al., 2007), face the challenge of transforming the current ATS architecture to a desired future state in the midst of increasing system complexity, demand, as well as deep uncertainty.

System of Systems Engineering: Innovations for the 21st Century, Edited by Mo Jamshidi
Copyright © 2009 John Wiley & Sons, Inc., Publication

Absent a framework and modeling approach for the system of systems, effective system analysis for decision support in transportation quickly becomes unmanageable, with multiple, heterogeneous, distributed systems involved and couched in networks at multiple levels. The situation is exacerbated by the fact that most organizations (and their processes) remain in the "stovepipe" mode; their people and tools are configured to study within narrow bins with little or no analysis across those bins. Current frames of reference, thought processes, analysis, and design methods are not complete for these SoS problems, exemplified in this chapter by seeking to understand air transportation and its possible evolution.

Thus, the development of system of systems engineering (SoSE) capabilities is a lynchpin to enabling decision makers to discern whether related infrastructure, policy, and/or technology considerations together are good, bad (or indifferent) over time (DeLaurentis and Callaway, 2004). And this need is urgent, for such problems frequently involve decisions that commit large amounts of money, for which ultimate failure or success carries heavy consequences that impact large segments of the public over several generations. The aforementioned JPDO, facing the daunting challenge of transforming the current architecture to a desired future state, exemplifies this situation. The costs associated with a chosen path may still be high, but the intent is to maximize the probability that these costs result in good outcomes rather than bad.

A small (but growing) number of research groups have begun formal study of this problem class from distinct points of origin, thus bringing a rich set of perspectives, formulations, and preferred analysis techniques. However, it is unlikely that any single set of algorithms from traditional academic domains can solve the SoS problem. Thus, in addition to new methods, *integration interfaces* are needed at multiple levels to understand the problem in full context and ensure that interconnections between related entities from different domains can be identified and exploited. Unlike in individual system optimization, a tenet of both SoSE methods and solutions must be the recognition that the organization of systems/algorithms is just as important as the nature of the systems/algorithms to be organized. Further, this philosophical connection between traits of the problem and traits of the solution approach can be exploited when planning implementation via object-oriented approaches.

Other chapters in this volume describe important developments in the nascent SoSE domain. Without repeating these, but to provide context, the following section outlines how several views and approaches on SoS as well as a background in design theory have motivated and informed the views on air transportation developed in this chapter.

20.1.2 Drawing on Diverse Domains

A symposium on system of systems was held in the summer of 2004 seeking to identify areas of common understanding as well as the most pressing research questions. The participants constituted a diverse group of individuals from government, academia, and industry struggling to understand SoS. The gathering's conversation focused on the status of the thinking and current practices (ad hoc or otherwise) associated with SoS. A report of the salient outcomes from the symposium is available

(Popper et al., 2004). Three findings are summarized here. First, the SoS problem is a new problem class (or, at least, a significantly extended form of existing decision-making problems), though not yet well formed into a definable field of study. Second, existing fields of study provide incomplete models and partial solutions but are important for the intellectual foundation for SoS study. Finally, progress toward dealing with SoS problems depends upon addressing the information gap between the data produced by engineers/analysts and the insights best utilized by decision makers. This gap can result in much effort being expended for little benefit (to decision maker and society both). In the summary report, all participants agreed that formal and expanded research in this area as in urgent need, and the existence of the volume for which this chapter appears, is a sign of progress toward that charge.

The second finding, detailing the need for knowledge synthesis from many domains, is germane to this chapter's objective. Until recently, the application domain in which "system of systems" most often appears is that of military systems development and acquisition. It is now part and parcel of major programs, appearing throughout key documents such as the Operational Requirements Document (ORD) and forcing a reexamination of especially systems engineering. Emerging from the network centric warfare concept (Alberts et al., 1999), system of systems is part of the strategy by which independent but related platforms are to be procured, fielded, and manipulated according to warfighter needs in any particular circumstance. Thus, the concept of connectivity between heterogeneous systems comes to the fore in all stages of systems design and development, including very early phases where joint warfighter requirements are being discerned at the same time as technologies are being explored (Schrage et al., 2002). However, even the strongest advocates of the SoS perspective would admit that the methods are not yet complete in this domain; further, it remains unclear whether they transmute appropriately to other applications, especially those in the commercial sector. This latter observation motivates investigation for the air transportation domain.

As a result of defense system related transformations, perhaps the most active group discussing SoS is the systems engineering community. Several contributions in the recent literature have come from this community where there is a growing recognition that systems engineering processes are not complete for the new challenges posed by SoS (Keating et al., 2003). Specifically, Sage presented a working collection of traits for SoS that points to a "new federalism" as a construct for dealing with SoS (Sage and Cuppan, 2001). In this setting, the federalism offers a means of bringing clarity to the degree of operational independence systems may possess in an SoS. Earlier, the work of Maier 1998 was important in delineating the salient traits of SoS that will be explored later in this chapter.

Operations research (OR) is another research community for which SoS has kinship. Mathematical modeling and results from OR in several of its thrusts (transportation logistics, network design, etc.) have relevance for SoSE. Even before the primary developments in OR, the idea of general system theory (later often known simply as system theory) which traces back to Ludwig Bertanflly's *general systems theory* certainly also rings true to the SoS challenge (Von Bertalanffy, 2003). The subject matter of this theory is the formulation of principles that are valid for any "system" whatever the nature of components. Similarities of observed phenomena across fields of study lend credence to the idea that commonalities can be instructive and useful. The prime thrust is a common set of principles applicable to any system across scientific fields. The focus on interconnections within systems, in general, is an important aspect for SoS, and especially important for the "unintended consequences" behavior that will be mentioned later. Modern proponents of this line of thinking include the very well-known work of Ackoff (1979), who has proposed that the real roots of OR are/should be in systems theory.

Understanding the implications of interconnections in an SoS leads to another community of interest: modern network theory. This community examines basic structure and behavior in networks that appear in the seemingly vastly different domains—biology, computer science, and sociology, etc. The work of Barabási (2002) is an example of a journey from standard mathematical models for the structure of networks to the fascinating realization that these models largely failed in the face of data from real-world networks (e.g., disease transmission, the World Wide Web, transportation). Already two recent studies have been published to begin to map these concepts more rigorously to air transportation applications (Guimera et al., 2005; DeLaurentis and Han, 2006).

The final community of interest to be discussed in relation to SoS research for air transportation is that of aerospace design. Aerospace design has most often focused on singular product (system) design. Design method innovations have come primarily in how the system is characterized (i.e., treatment of the constituent disciplines and design variable vector, *(x)* and what are the objectives *(y)* used to compare alternative solutions and/or search for the best (i.e., optimization). This evolution is summarized in very broad themes in Fig. 20.1. The design-for-performance philosophy seeks the singular solution that maximized performance needs of the customer. The arrival of emphasis on affordability (performance per unit cost) changed matters. For example, the advent of concurrent engineering (adding especially a process perspective) and then multidisciplinary design optimization, MDO, (adding especially formal treatment of the interacting disciplines) gave rise to new methods that led to better solutions through exploitation of knowledge of disciplinary interactions.

Philosophy	Method	Synopsis
Design for performance	Deterministic optimization	$\max y(\mathbf{x})$
Design for affordability	Concurrent engineering and multidisciplinary design optimization (MDO)	$\max y(\mathbf{x}_{aero}, \mathbf{x}_{struct}, \mathbf{x}_{coupling})$
	Parametric design (metamodels)	$y = b_o + \sum_{i=1}^{k} b_i x_i + \sum_{i=1}^{k} b_{ii} x_i^2 + \sum_{i=1}^{k-1} \sum_{j=i+1}^{k} b_{ij} x_i x$
	Robust design	$\max P(y(\mathbf{x}) > y_{nom})$

FIGURE 20.1 Synopsis of key developments in aerospace design formulations

Subsequently, advancement in parametric methods applied to aerospace design improved matters further by allowing higher fidelity contributing analyses to be employed and more clearly exposing these interaction sensitivities in the resultant multidimensional design space. Finally, realization of the importance of uncertainty led to the robust design family of approaches. Attributions for all these innovations are far too numerous to cite here and their details are not germane to this chapter, but accessing good survey articles would be recommended for further exploration of these topics.

Implicit in these methods is some sense of self-contained purpose for the vehicle as specified, for example, by the design mission. However, one might ask, Would such optimal air vehicle designs necessarily improve air transportation service metrics? Understanding the system of systems for which air vehicles are one part is the starting point for finding an answer.

SoSE methods must explore transdomain interactions and cooperation among independent systems. A challenge will arise in those cases where an SoS solution demands *suboptimality* in some dimension of the vehicle objective function space. Additionally, as robust design sought to find configurations that performed well in the face of modeled uncertainty, SoS solutions must be found that are robust (i.e., asymptotically good over time) in the face of deep uncertainty and an ensemble of plausible futures (Lempert et al., 2003). But, as seen shortly, there are SoS traits that require advancements beyond the insights and tools available today.

20.1.3 Chapter Overview

The purpose of this chapter is then threefold. First, the distinguishing traits of the system of systems problem class (as currently understood) are highlighted. Second, a framework for understanding, characterizing, and measuring behavior for SoS problems is presented. Finally, these concepts are mapped to the application domain of air transportation and prospects for new concepts and implications for modeling and simulation are addressed. While detailed presentation of analysis and simulation results lies outside the scope of this chapter, the most important of these results are referenced and can be accessed for further investigation. A mission statement for the on-going research could be summarized as follows: generation of system of systems formulations/tools/processes to understand and bound complex problems and create an ability to determine sensitivities and guide policies/ decisions/ visions.

20.2 GENERAL PROBLEM CHARACTERIZATION

20.2.1 Distinguishing Traits of an SoS

Rather than focusing on rigid definitions, a basic conception of an SoS problem is summarized via distinctive traits, as shown in Fig. 20.2. The first four are referred to as the Maier's Criteria (Maier, 1998). The first three primarily describe the problem

Trait	Description	Comment
Operational and managerial independence	Constituent systems are useful in their own right and do operate independent of other systems (i.e., with unique intent provided by the owner/operator)	- Intent for interoperability and global status of system coupling variables are not easily discerned - At the SoS level, *no one is completely in charge*
Geographic distribution	Constituent systems are not physically colocated, but they can communicate	- Spatial dynamics important for SoS behavior - Communication rules/capabilities are critical
Evolutionary behavior	The SoS is never completely, finally formed, constantly, changes and has a "porous" problem boundary, that is, a living system	- Behavior and constitution is time varying - Presence of multiple timescales, possibly spanning a generation in some applications
Emergent behavior	Properties appear in the SoS that are not apparent (or predicted) from examination of the constituent systems	- Drives nonmonotonic character of evolution - Not readily addressed in current design methods - Unpredictable, but manageable
Networks*	Networks define the connectivity between independent systems in the SoS through rules of interaction	- Need methods for network topology design - Adaptability/scalability of topologies are crucial to SoS performance and are source of emergence
Heterogeneity*	Constituent systems are of significantly different nature, with different elementary dynamics that operate on different timescales	- Primarily a challenge in modeling complexity - Many methods focus on single system design - Variety of system analysis techniques required
Transdomain*	(Proposition) Effective study of SoS requires unifying knowledge across fields of study: engineering ∪ economy ∪ policy ∪ operations	- Dramatic advancement in problem definition, lexicon, and domain interfaces are needed (the spirit of MDO, but across domains instead of disciplines)

FIGURE 20.2 Distinguishing traits of system of systems and implications for modeling *These traits identified as important by the author above and beyond the traits listed in Maier (1998) and sources cited therein

boundaries and mechanics of the interacting elements while the latter two describe overall behavior. Maier contends that if a majority of these criteria are met, one should treat the problem as an SoS. In particular, emergent behavior—the manifestation of behavior that develops out of complex interactions of component systems that are not present for the systems in isolation—presents a particular challenge. Emergent behavior is unpredictable, is often nonintuitive, and can be manifested in a positive manner (e.g., a new capability arises) or negative manner (e.g., a new failure is created). The well-known phrase "unintended consequences" well encapsulates for many the emergence concept. A primary challenge in the study of SoS with greater effectiveness is to understand the mechanism of emergent behavior, develop cues to detect it, and create a methodology for managing it intelligently. The latter three traits, introduced by the author, are derived from these original ones and have direct implications on modeling. Also included for each trait are comments on the difficulties that arise in modeling for executing design-oriented analysis.

20.2.2   Characterizing an SoS

The traits summarized above are useful in sketching the dimensions of the SoS problem and collection of the challenges is important because they will have to be overcome eventually. Both the traits and the challenges point toward key structures whose relevance will be clear when examining simulation approaches. However, a prerequisite ingredient is an effective language to allow communication among diverse parties over this transdomain landscape.

A lexicon has been formed to provide structure with which the numerous participating entities can be understood (DeLaurentis and Callaway, 2004). By definition, systems in an SoS have connectivity, often set in a hierarchical manner. The lexicon enumerates these interacting levels of various components with Greek letters. Starting from α, the levels proceed upward, leading to γ, δ, or higher depending on the complexity of the problem and the depth of analysis desired. Further, a β-level constituent represents a network of α-level entities; a γ-level constituent represents a network of β-level entities, and so on.

In addition to these hierarchical labels representing levels of complexity, a set of scope dimensions are defined. These dimensions highlight the transdomain aspects of SoS, since not all entities within the levels are similar in their basic characters. Thus, to properly distinguish constituent systems, they are categorized primarily into resources, operations, policy, and economics (ROPE). Each of these dimensions independently comprises the previously described levels, thereby completing the SoS lexicon. The categorization of the levels lends clarity in dealing with the different facets of the problem while maintaining the lucidity provided by the levels. The relationship between the categories and the levels can be conceived as a pyramid, indicating that the number of systems decreases at higher levels. This pyramid representation is shown in Fig. 20.3.

This lexicon intuits an important point about SoS problems: the behavior of the SoS is driven by the structure and organization interactions between levels, not exclusively by the characteristics of the α-level entities. This also resonates with the importance ascribed to the emergence property that, by definition, can only be perceived at the β-level or above. In other words, from a design perspective, a relevant question might be: How does the preferred or observed behavior at the upper levels (e.g., γ-level) affect the possibilities for alternatives at the lower levels (α and β)? Not coincidentally, these upper levels are also the levels where the most consequential decisions arise.

While a systematic representation of scope and structure is crucial, the ability to characterize SoS problems in the analysis domain is the next required step. Further, the SoS

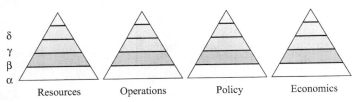

FIGURE 20.3   ROPE lexicon: exposing hierarchic networks and scope dimensions

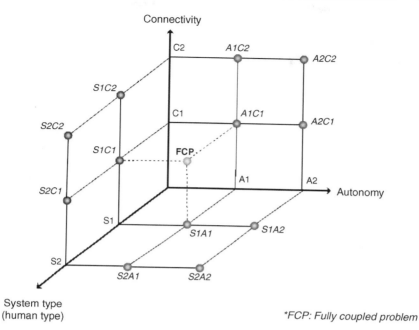

FIGURE 20.4 Three key dimensions for system of systems

of interest must be defined in a way that exposes appropriate level(s) and categorical scope dimensions. Toward these objectives, a taxonomy has been proposed (DeLaurentis and Crossley, 2005) consisting of a three-dimensional space characterized by system type, autonomy, and connectivity, illustrated in Fig. 20.4. Analysis methods must be appropriate for the *system types (S-axis)* that constitute the SoS. Some SoSs consist predominantly of technological systems—independently operable mechanical (hardware) or computational (software) artifacts. Technological systems have no purposeful intent; that is, these resources must be operated by, programmed by, or activated by a human. Other SoSs consist predominantly of humans and human enterprise systems—a person or a collection of people with a definitive set of values. The second SoS dimension is the *degree of control (A-axis)* over the entities by an authority or the autonomy granted to the entities. This relates to Maier's discussion of operational independence and managerial independence of systems within an SoS. Emphasizing the importance of control/autonomy, others refer to a collection of systems with operational, but limited managerial, independence as a "system of systems" and a collection of systems with little central authority as a "federation of systems" (Sage and Cuppan, 2001). Finally, systems involved in a system of systems are interrelated and *connected (C-axis)* with other (but likely not all) systems in the SoS. These interrelationships and communication links form a network. A key focus for design methods research in an SoS context lies in analysis and exploitation of interdependencies (i.e., network topology) in addition to the attributes of systems in isolation. These dimensions serve as a taxonomy to guide the formulation and analysis of the SoS design problem. A particular SoS can therefore be considered as a "point" or "region" in the three-dimensional space formed by the aforementioned dimensions as axes. Based on its spatial location and other indicators of problem structure, the approach and methodology necessary for analysis and design can be more intelligently selected.

20.2.3 Evolutionary and Emergent Behaviors

In light of the unique traits and characteristics of an SoS, modeling and measuring performance in an SoS context will require new approaches. To approach this in more depth and looking ahead to application for air transportation, a current approach in aerospace systems design is contrasted with SoS needs. In particular, the implications that arise from the evolutionary nature (calling for a dynamic response measure) and emergence (introducing nonmonotonic behavior) are discussed.

The robust design method for aerospace systems was introduced only briefly in Fig. 20.1. Though there are a variety of formulations reported in the literature, the popular one combines sampling-based and simulation-based perspectives. Typically, a Monte Carlo sampling technique operates on expected variability in selected parameters in the simulation model expressed via probability distribution functions. The simulation model (or the surrogates models, or metamodels) that describes the system remain fixed. This method produces outcome probability distributions that estimate likely performance (Mavris et al., 1999). This and the related approaches (Chen and Yuan, 1997) are important for moving beyond the deterministic design paradigm. Yet, the result remains only a point estimate focused on a single system, fixed in time and tied to the uncertainty distributions chosen. The latter observation illustrates why these are primarily *risk-based approaches*, rather than uncertainty-based, since the variability of uncertain parameters is assumed known.

In contrast, for an SoS problem, both the unknown variations in external factors and the internal structure of the of the SoS may change with time. The evolutionary trait of SoS can be traced to two important points—most SoS applications of interest already exist and their *internal structure* will change over time. The transportation system is a clear example. The NTS/ATS operates today and will do so even after it is transformed; however, at future points in time its constituent systems and networks will be different. Legacy systems will mix with new systems and operational paradigms. Evolutionary behavior implies that the genetics of what exists today must be understood and considered when analyzing future realizations.

All dynamic systems exhibit evolutionary behavior in some sense. Studies in control theory have developed many methods to deal with them. There are unique aspects of SoS problems that complicate the measurement of robustness over time. The following example illustrates the insight, captured in Fig. 20.5, and highlights that both heterogeneity and networks traits are at play. Initially, three heterogeneous systems at the α-level of an SoS are presumed to be present at time t_1, configured in a particular β-level network, β_i. Other β-level networks may certainly exist (not shown), themselves organized at the γ-level. As time evolves, the constituent systems, their connectivity to others, and their self-interested objectives may change. Over a sufficient period, a significant portion of the initial set of α-units present at t_1 could

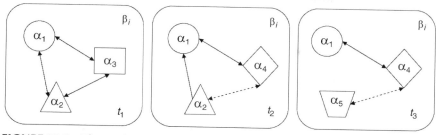

FIGURE 20.5   The evolving nature of the SoS: shifting constituents and connectivity

have been replaced. Yet, from the design perspective, we still speak of the same entity, the "so-and-so system of systems", as the subject of our design. As the evolution depicted, Fig. 20.5 shows two new α-level systems appear and by the third time epoch the connectivity of the SoS is significantly changed. New configurations lead to altered interactions (emergence) and this emergence in turn affects the course of the future make-up of the β-level (evolution).

The concept of emergence has now been mentioned twice, first as an SoS trait and second under the recognition that its appearance occurs (and is detected) only at the upper levels of the SoS hierarchy. Indeed, mergence implies that the "whole is greater than the sum of the parts." However, acknowledging the emergence may occur, while interesting, does not immediately assist in forming mental (and, later, simulation) models of SoS problems and using these models to make better design decisions. It would of value, as pointed out by Holland (1998), to be able to recognize recurring patterns of emergence and thus seek to identify underlying mechanisms at play. To do this, understanding the impact of emergence on the time-dependent behavior that results from the evolutionary behavior is required. Further, the proper analysis methods for ensuring the emergent properties can be examined must be determined; on this issue, it is proposed that simulation is the optimum method for this task. This latter topic will be covered later in this chapter.

20.3   AIR TRANSPORTATION AS A SYSTEM OF SYSTEMS

20.3.1   Introduction—Transportation Is an SoS Problem

Modeling and evaluating the behavior of the NTS to track the cascade of interrelated technological, infrastructure-related and societal perturbations over significant time horizons is overwhelmingly difficult task! Mapping of generic SoS traits to transportation (Fig. 20.6) clearly indicates, however, that the NTS may best be conceived as a system of systems—a diverse collection that evolves over time, organized at multiple levels, to achieve a range of (likely) conflicting objectives, never quite behaving as planned.

20.3.2   Framework

Casting the development of future transportation concepts as an SoS problem means that the implications of the associated SoS traits must be faced. Strategies are under

SoS trait	Transportation mapping
Operational and managerial independence	Numerous transportation systems are operated independent of other systems, though at higher levels some cooperation is evident, for example, a commercial aircraft operates on a schedule dictated solely by its company, but must coordinate while in flight with other aircraft (traffic)
Geographic distribution	The nature of transportation systems makes obvious the presence of this trait, though the increased sharing of *information* between distributed systems is of particular interest for the future NTS
Evolutionary behavior	Measures of effectiveness for transportation are dynamic in nature, due primarily to delayed response to major inputs as well as inherent feedback mechanisms; static performance measures are not often used (though they are often used for evaluating new systems within the NTS)
Emergent behavior	Some behavior is only evident when two or more effects are combined (e.g., bad weather in isolated places during peak evening activity can produce system-wide impacts)
Heterogeneity	Transportation systems are diverse, including categories such as vehicles (aircraft, trains, autos, etc.), infrastructure, service providers, users, and regulatory bodies
Networks	Transportation network topologies exist for both vehicles and the business enterprises that operate them; each topology can range from pure hub-and-spoke to fully distributed, with hybrid (and possible adaptive) ones also present
Transdomain	The variety of engineered systems and human organizations (stakeholders) in the transportation SoS indicate the numerous domains that must be spanned for a holistic study approach to be fully adopted

FIGURE 20.6   Mapping of SoS traits to the NTS

development to deal with these challenges and some portions have already been applied to transportation. In the remainder of this chapter, a broad outline of a framework for analysis and decision making—an SoSE analysis framework—is presented along with reference to articles that contained more detailed results. Undoubtedly, very much remains to be done in this endeavor.

The graphic in Fig. 20.7 gives an overarching view of the major steps in the proto-method for SoS problems as currently conceived. The major activity in the *Definition Phase* is one of understanding—characterize the *status quo* of the SoS (especially legacy systems, barriers to enhanced performance, and operational contexts) and establish categories and levels that will later be required to detect evolutionary and emergent properties. In the *Abstraction Phase*, the main actors, effectors, disturbances, and networks are drawn out and placed in context corresponding to their real

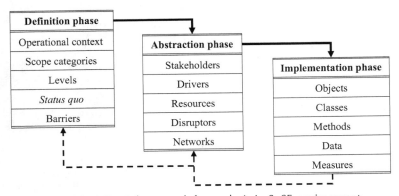

FIGURE 20.7   A framework for analysis in SoSE environment

interrelation. The focus of this phase is in grappling with the overwhelming complexity by abstracting the main entities; it is *not* intended to construct a detailed hierarchical decomposition. Finally, the *Implementation Phase* instantiates all or part of the abstraction within a modeling and simulation environment. It is in this final phase that specific hypotheses about the SoS can be proposed and tested.

20.3.3 Definition Phase

The objectives of this phase are to sketch the mental map of the SoS as it exists (embodied in categories, levels, and forces at play) and identify barriers to more preferred behavior. Execution of a portion of this phase for transportation is summarized in Fig. 20.8, illustrating how the NTS could be considered in an SoS frame of reference. This use of the lexicon introduced in Fig. 20.6 provides value at two levels: first, the breadth of the problem and subsequent imperative to move beyond (across) domain stovepipes is evident, and second, the categorizations help effectively guide the *Abstraction Phase*. The variety of decision makers involved in transportation can be identified, engaged, and included in the discussion. Human values and the organizations in which they participate influence the SoS through application of self-interests and perspectives (via operational and managerial independence). Dynamics from this influence take place at multiple levels and under multiple timescales. For example, humans can operate and implement policies for α-level systems. In air transportation, these roles include pilots, maintenance crew, inspectors, etc. Further, actions at these lower levels of hierarchy tend to evolve on a short timescale; they typically are near-term actions of a practical variety in response to the situation at hand and are operational in nature. Through subsequent modeling, the probabilities for solutions at the γ- or δ-levels can be formed by aggregating the α- and β-level entities. It is also important to note that the number of entities increase tremendously, likely by orders of magnitude, as one moves from the higher levels down to the α-level.

| | Level | System of systems dimensions | | | |
		Resources	Operations	Economics	Policy
Base level	α	Vehicles and infrastructure (e.g., aircraft, runway, terminal area)	Operating a resource (e.g., aircraft, control tower)	Economics of building/operating/buying/selling/leasing a single resource	Policies relating to single-resource use (e.g., flight procedures)
↑ Network of networks ↓	β	Collections of resources for a common function (e.g., airport)	Operating resource networks for common function (e.g., airline)	Economics of operating/buying/selling/leasing resource networks	Policies relating to multiple vehicle use (e.g., local airport noise policies)
	γ	Resources in a transport sector (e.g., air transportation system)	Operating collection of resource networks (e.g., commercial air operations)	Economics of a business sector (e.g., airline industry)	Policies relating to sectors using multiple vehicles (e.g., FAA certification)
	δ	National transportation system	Operators of total national transportation system	Overall transportation forecasts/market	National transportation policies
	ε	Global transportation system	Global operators in the world transportation system	WTO, global marketplace	Global transportation system policies

FIGURE 20.8 Transportation SoS- lexicon matrix

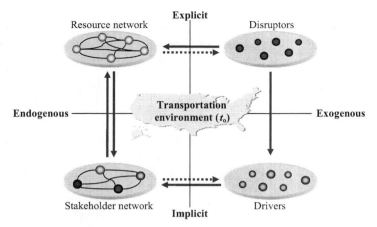

FIGURE 20.9   Entity-centric abstraction model

20.3.4   Abstraction Phase

An abstraction for transportation that builds from the levels and categories has been carefully developed and reported (Lewe et al., 2006). A brief summary is given here to highlight the role of this phase in the SoS proto-method. Two pairs of entity descriptors emerge from the abstraction process: explicit–implicit and endogenous–exogenous. Unlike under the reductionism mindset, the role of the descriptors is not to facilitate break down of the entities into separate pieces. Instead, it is only to organize them by articulating their inherent natures. Four entity categories are generated based on the descriptors: resources, stakeholders, drivers, and disruptors. All these entities are interwebbed by networks that define the linkages among them. This is summarized in Fig. 20.9.

Vehicles and infrastructure are examples of resources that consumers physically experience (explicit) while traveling or sending shipments. Further, they are under partial or full control of the imagined architect, thus endogenous. But there are "other than physical" entities that desire to exert forces on the architecture for their own interests. This type of endogenous entity is called stakeholder, and in most circumstances their behaviors and decisions are not manifested in an explicit manner to the consumers (implicit). The stakeholders reside in both private and public sectors, ranging from the actual consumers of transportation services to the providers of those services.

Each stakeholder holds objectives they wish to pursue in the transportation environment. For example, an individual values doorstep-destination (DD) speed, cost per mph, mobility flexibility, etc. However, from a societal perspective, there may be desires to minimize total energy expended, maximize the robustness of the system to disturbance, etc. Often, members of these two groups move in tandem (e.g., robustness of system and mobility freedom), but other times they may not (e.g., possible increase in energy per mile traveled in an on-demand system).

While the stakeholders and resources are considered endogenous building blocks, the transportation environment contains exogenous entities that are outside of the architect's controllable domain. This entity class has been traditionally treated as given assumptions, circumstances, and constraints about the transportation environment (e.g., population, weather). Driver entities are largely concerned with economic, societal, and psychological circumstances that influence the stakeholder network by implicit means. On the other hand, disruptor entities explicitly affect the resource network and/or a portion of the driver entities by reducing the efficiency of the resource network, disabling particular nodes or links of the network.

The network concept has appeared both in the *Definition Phase* and the *Abstraction Phase*. As alluded to in Fig. 20.2, networks ("graphs" in the mathematics community) can be viewed at a simple level as constructs that define the connectivity (links) between entities (nodes). The transportation abstraction in Fig. 20.9 identifies networks existing in both the resource and the stakeholder domains. For example, air transportation resource networks arise when nodes (airports) are linked by aircraft either in a scheduled or unscheduled (on-demand) manner. The connectivity provided by the commercial airline industry is determined by predefined schedules and is presently by far the predominant means of air transportation. By studying the statistical properties of these networks, measures of performance, robustness, and vulnerability can be discovered. For example, the graphic in Fig. 20.10 displays the topology that represents connectivity in the 2004 U.S. domestic scheduled air service

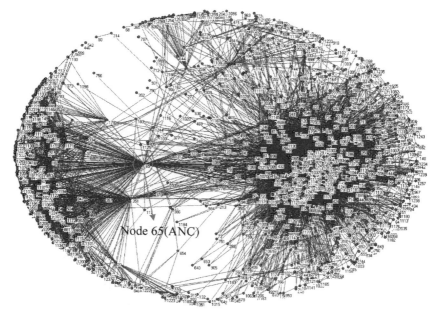

FIGURE 20.10 Visualization of 2004 U.S. air capacity network topology. [*Note*: The visualization is not representative of actual geography; nodes are configured to illustrate betweenness centrality of node 65 (ANC)]

534 UNDERSTANDING TRANSPORTATION AS A SYSTEM OF SYSTEMS PROBLEM

network (where a link is defined as at least one flight service route executed annually). While there are numerous measures that can be used to indicate the importance of nodes (DeLaurentis and Han, 2006), one that is visually apparent in Fig. 20.10 is that of betweenness centrality. Nodes with high values of this trait link otherwise separated parts of a network, which further indicates such an entity would be crucial to operation of an SoS. In Fig. 20.10 this highlighted node is Anchorage, Alaska airport which serves as a local transfer airport to transport passengers among the many locales in Alaska served by commercial air service. According to transport data for 2004, 685 airports are located in Alaska comprising about 25% of total airports in the BTS database and resulting in two distinguishable clusters.

Presently, charter, fractional ownership, and personally owned options exist, but they remain at least one order of magnitude higher in cost than the commercial scheduled service. Demand, which theoretically drives schedules, is an important component of the real networks. Thus, in contrast to the present network, unscheduled (on-demand) models are also being increasingly studied. On-demand implies that the systems organize themselves to serve a particular demand of a particular customer at a particular time. In this case, ad hoc networks would be formed to respond to demand, requiring a flexible set of vehicles and traffic management rules to do the job. A very insightful encapsulation of these possible networks as well as the various layers within the resource networks has been presented by Holmes (2004).

But how these networks actually arise, how they evolve, and how robust they are to disruption are topics of current research. Further, such networks can only arise if there is an economic imperative to do so. For example, Crossley et al. (2004) studied the problem of a simplified airline as a system of systems, employing a combination of MDO techniques with integer programming to determine how a new aircraft design can be optimized for the greater good of the existing airline operation. In that study, comparative studies of SoS performance versus alternative network topologies and vehicle performance levels were generated. The measures of merit were profitability of a service provider attempting to operate this system in a competitive commercial environment. Overall, the entire abstraction phase, including entities and networks, is important for framing the relevant hypotheses about the SoS problem at hand and thus the most appropriate means for implementation.

20.3.5 Implementation Phase

Prospects for a purely analytical solution to most system of systems problems are dim at best. Thus, the research direction of most interest for the *Implementation Phase* involves instantiating the aforementioned conceptual abstraction for transportation via simulation. Several elements of such a simulation for this endeavor have already been created and exercised by the author and collaborators (DeLaurentis et al., 2006). However, a more comprehensive simulation is required, one in which the higher levels of organization (the layered networks) can be considered when forming the concept space for future transportation architectures. The simulation must properly charac-terize the salient SoS traits, including the emergent properties that can arise when individual systems are empowered to adapt and form tailored networks. Further, it is

System view	Methods	Transportation modeling implication
Systems of hierarchical mapping	Systems engineering (SE), decomposition methods	The proven record of SE for individual system development demands its consideration for SoS problems, especially techniques for systems analysis; however, the shifting nature of transportation systems and networks may limit use of this view in its traditional form
Systems of uncertain state equations	System dynamics, control theory	Robustness methods when facing uncertainty are well established in this view and needed for exploring future transportation SoS; however, the inability to characterize the "plant dynamics" and to handle variables that switch between controllable and uncontrollable may severely limit direct application of algorithms
Systems of nonlinear models	Chaos theory, networks	The notion that simple mechanisms can be the source of complex behavior in several applications has proved intriguing; identifying those mechanisms in transportation may be difficult, although the network theory experience may be a start
Systems of autonomous agents	Agent-based modeling (ABM)	ABM most closely matches the SoS traits for transportation outlined in Fig. 20.6 and already has some applications reported

FIGURE 20.11   System Views, Methods, and Transportation SoS Modeling Implications

intended to use the extended simulation to draw out recurring patterns within alternate configurations that seem to perform well over an ensemble of plausible scenarios (i.e., they are scalable). Two major topics are outlined with regards to modeling and simulation: alternative system views/methods and schemes for effective implementation.

20.3.5.1   System Views and Methods

A recent article by Rouse (2003) postulated that among the implications brought by complex systems,[1] an important one is to consider the variety of possible "system views," especially since such views are tightly coupled with subsequent modeling and simulation approaches. The four views proposed are systems of hierarchical mappings, systems of uncertain state equations, systems of discontinuous nonlinear models, and systems of autonomous agents. Some related methods for each view as well as comments on the applicability to transportation SoS studies are offered in Fig. 20.11. While each view (and method) is distinct, the magnitude of most SoS problems may require an integration of several or all of these. The choice of which in particular might be needed should be guided by use of a taxonomy, for example, the initial one is described in Section 20.2.

Two of the methods that are perhaps most distinct from those traditionally used in air transportation research are system dynamics and agent-based modeling (ABM). They are highlighted here because they both address the implications of SoS problems that are poorly addressed by existing aerospace design approaches: evolutionary and emergent behavior.

---

[1]In this chapter, we overlook the differences, both semantic and real, between "complex systems" and "system-of-systems".

The modern field of system dynamics was pioneered in the early 1960s, and has since matured into an active field often used in system studies and policy analysis (Forrester, 1969; Sterman, 2000). Through the use of causal loop diagrams and stock and flow dynamic models that capture key feedback mechanisms, system dynamic approaches attempt to determine policies and functions that result in optimal outcomes over time. In recent years, researchers have focused specifically on business dynamics, emphasizing the identification and improvement of behavior patterns. This field could be particularly applicable to analysis of future transportation service provider business models as they evolve along with the technology and regulatory aspects (Kang et al., 2003). The type of time responses notionally shown in Fig. 20.5 above is not uncommon in the system dynamics approach. Potential drawbacks, however, are also apparent, such as the inability to explore emergent behavior as it is presently defined.

ABM is an area growing in diversity of formulation and application. The premise of ABM is that the global behavior of a complex system is derived from the low-level interactions among its constituent elements, whose behavior is encoded based on a variety of factors. Autonomous agents representing these elements are individually defined, as is the environment within which they interact. The collective action of the agents produces an "emergent behavior" that often produces real-world insight, even if the model itself is relatively simple in form. The major strength of ABM comes from the fact that it is a simple and versatile method that is well suited for studies of complex nonlinear systems. Whereas agent-based models can be made arbitrarily realistic by capturing processes or mechanisms that drive individual components, they can also be made quite abstract in an attempt to understand the essence of the problem (Hood, 1998). Agent-based simulations can reveal both qualitative and quantitative properties of the real system, so ABM/S can be deemed as *computational laboratories* (Dibble, 2001) to perform experiments to test nearly any kind of imaginable hypotheses.

This modeling approach holds significant potential to capture the complex inter-actions existing among the NTS stakeholders from airlines and air traffic control, to passengers and policy makers. Capturing these interactions at various levels of organization in the SoS has been undertaken by some researchers in the SoS community in an effort to improve the understanding of SoS dynamics. For example, in the air transportation domain, traveler preference models (an α-level behavior) have been modeled by Lewe et al. (2006) and Baik and Trani (2005). Further, models of β-level entities have been developed, such as MITRE's JetWise model of airlines (Niedringhaus, 2004), and higher level dynamics in DeLaurentis et al. (2006). The mathematical representation of agent rules is often quite simple, but the resultant system-wide behavior is often more complicated, unexpected, and thus instructive (Bonabeau, 2003). ABM for systems analysis emanated largely from the study of complex adaptive systems (Waldrop, 1992), which represents a problem class with many of the same dynamic behaviors of SoS problems (e.g., emergence). For air transportation, Donohue (2003) has specifically called for a CAS approach. The aforementioned recent advancements in analyzing complex networks also provide useful tools for addressing the global connectivity of a given SoS problem without reliance on low-level interactions (Barabási and Albert, 1999). However, a need remains for an increased ability to understand and replicate the complex human factors involved in a human-technical system of systems.

A significant challenge to the acceptance of ABM simulation results (and, more broadly, results from any of the modeling approaches) is validation. In the sciences, validation comes from the comparison of theory with experiment. In an SoS context, these two are intertwined: the theory and the experiment both are embedded in the simulation for the "true" experimental apparatus is the transportation system itself. Thus, the equivalents of the theoretician and experimental communities in SoS applications will likely continue the struggle to find agreeable means for validating the soundness of the methods, not the preciseness of any prediction.

20.3.5.2 *Object-Oriented Implementation Schemes*

The second topic of the *Implementation Phase* briefly discussed here concerns the implementation scheme associated with instantiating the collection of networks and entity models into a simulation architecture that is effective, flexible, comprehensive, and verifiable. The chosen system views, associated methods, and data from the α-, β-, γ-, δ-levels must be combined to test a myriad of possible hypotheses, especially at the upper levels of organization. For this task, and object-oriented (OO) approach appears well suited. As the name implies, the OO approach is based on the concept of object. Objects are constructs that are responsible for themselves. When a specific problem is at hand, objects are specific methods that can be called by other objects. When a particular computer program that represents the problem is run, objects exercise their methods with their data, all self-contained. Further advances in modeling languages/frameworks from the systems engineering realm (e.g., SysML) will take on increasing importance as the need for more systematic SoS problem representations are recognized, especially as the development process proceeds toward building/fielding.

Perhaps this is no surprise, as the OO approach is now standard across a wide swath of domains. However, there remain "corners" where this shift is still in its infancy (e.g., aircraft design). Further, there appears to be an intriguingly tight connection in philosophy between the system of systems perspective and the OO paradigm, one that is just beginning to be explored (see Fig. 20.12). The central role of abstraction in

SoS trait	Object-oriented construct/trait
Operational and managerial independence	*Objects* are imbued with responsibility for their own independent behavior (the implementation form of an "entity," which is a *class* for individual systems in a SoS)
Distribution in networks	*Objects* are not just resources, but can also represent distributed behavior of other objects
Emergent behavior	*Polymorphism*: the same calls to different derivations of the same class produce different results; supports the notion that the same entities appear differently in separate views
Heterogeneity*	Separate *abstract class* represents each family of conceptually related systems; *concrete class* defines the methods and data types for different systems
Transdomain*	The *extensibility* claim-to-fame for OO approaches indicates hope for wide applicability across domains through provision of interfaces

FIGURE 20.12 Philosophical relation between SoS traits and OO principles

properly building a mental model of the problem is at the heart of this connection. More specifically, though the close connection between the concepts of object, entity (introduced in the *Abstraction Phase*), and agent (introduced earlier in the present *Implementation Phase*) are abundantly clear. The hope is that the use of the generalized and abstract objects with certain attributes and methods that modify those attributes can generate the holistic views required for system of systems. Further, OO constructs appear to fulfill the need for rapid exchange of model entities and networks within a transportation system analysis environment. What remains is to identify the obstacles that do exist in OO implementations of transportation SoS simulations. The answers remain to be uncovered.

20.4   SUMMARY

The purpose of this chapter was to articulate how transportation, in particular, air transportation, can be understood as a system of systems problem. Such a perspective is needed as a basis to develop effective analysis and design approaches that account for the significant complexity in this domain and enable transition to a superior future state. An initial survey of the traits of system of systems and the mapping of these traits to transportation indicate that indeed there is urgent need for new methods. Toward this end, a framework for analysis in system of systems engineering was presented, including discussions of its three major phases: Definition, Abstraction, and Implementation. Within each, steps are taken to ensure that the most challenging behaviors found in systems-of-systems are properly treated: evolutionary and emergent behavior. This grander web exhibits new traits not observed in the collection of systems viewed in isolation. It is these hard-to-observe traits that are the real goal of design in this context. This emergent behavior, and the already present complication of evolutionary behavior that proceeds over several timescales (some even generational), summarizes perhaps the greatest challenge posed by the SoS problem class.

While the ultimate fruits of a system of systems field of study are still a ways off, even conceiving of a larger frame of reference provides increased understanding for our seemingly complex pursuit of better aerospace vehicles and air transportation architectures. In this larger frame, the ill-advised practice of individual system (e.g., aircraft, air traffic rule, infrastructure, etc.) optimization may become clearer as the role of other systems and their influence becomes evident. By acknowledging the presence of the multiple levels of organization, we can work out some of the vexing challenges facing us in developing tomorrow's transportation solutions.

20.5   FUTURE CHALLENGES

This chapter concluded with the realization that the success of future aerospace vehicles and airspace management concepts will depend on how well they harmonize with the dynamics of the system of systems of which they have membership. And the success of the future systems of systems for air transportation will depend on the extent to which design methods can properly pose and solve problems in this context.

Within this state of affairs, there are several key questions that provide a starting point for researchers to become engaged investigating SoSE as an approach for transportation. We offer them here to the reader:

1. Thinking of your own experience in using transportation, can you think of additional entities, interactions, and dynamics beyond those described in Figs. 20.6, 20.8 and 20.9?
2. While this chapter focused upon understanding the nature of an SoS and ideas for representation and modeling, the ultimate objective is to develop capability for design and decision making as part of SoSE. What kind of objective functions do you envision for this task, both generically and for air transportation?

As exemplified in the air transportation example, an SoS is composed of participating systems who often have little control, but potentially significant influence, on other participants. In what ways can the concept of "influence" be employed in fielding more effective SoSs?

REFERENCES

Ackoff, R., 1979, The future of operations research is past, *Journal of Operations Research Society*, 30(2): 93–104.

Alberts, D.S., Garstka, J.J., Stein, F., 1999, *Network Centric Warfare- Developing and Leveraging Information Superiority*, 2nd edition, CCRP Publishing, Washington, DC, Available at www.dodccrp.org.

Arbuckle, D., Rhodes, D., Andrews, M., Roberts, D., Hallowell, S., Baker, D., Burleson, C., Howell, J., Anderegg, A., 2007, U.S. vision for 2025 air transportation, *Journal of Air Traffic Control*, 1st quarter issue (Jan-March), 2007.

Baik, H., Trani, A.A., 2005, A Transportation systems analysis model (TSAM) to study the impact of the small aircraft transportation system (SATS), *23rd International Conference of the System Dynamics Society*, July 17–21, 2005, Boston, MA.

Barabási, A., 2002, *Linked- The New Science of Networks*, Perseus Book Group, Cambridge, MA.

Barabási, A.-L., Albert, R., 1999, Emergence of scaling in random network, *Science*, 286: 509–512

Bonabeau, E., 2003, Agent-based Modeling: Methods and Techniques for Simulating Human Systems, *Proceedings of the National Academy of Science, USA*, 99(3): 7280–7287. Accessed February 6, 2006, from http://www.pnas.org/cgi/reprint/99/suppl3/7280.

Chen, W., Yuan, C., 1997, A Probabilistic-Based Design Model for Achieving Flexibility in Design, *Proceedings of DETC'97, ASME Design Engineering Technical Conference*, September 14–17, 1997, Sacramento, CA.

Crossley, W.A., Muharrem, M., Nusawardhana, 2004, Variable resource allocation using multidisciplinary optimization: initial investigations for system of systems, *Proceedings of 10th AIAA/ISSMO Symposium on Multidisciplinary Analysis and Optimization*, August 30 – September 1, 2004, Albany, NY. AIAA-2004-4605.

DeLaurentis, D.A., Callaway, R.K., 2004, A system-of-systems perspective for future public policy, *Review of Policy Research*, 21(6): 2004.

DeLaurentis, D., Crossley, W., 2005, A taxonomy-based perspective for systems of systems design methods, *Proceedings of IEEE System, Man, & Cybernetics Conference*, Oct. 10–12, 2005, Hawaii, Paper 0-7803-9298-1/05.

DeLaurentis, D., Han, E., 2006, A Network Theory-Based Approach for Modeling System-of-Systems, *11th AIAA/ISSMO Multidisciplinary Analysis and Optimization Conference*, September 6–7, 2006, Portsmouth, Virginia. AIAA-2006-6989.

DeLaurentis, D., Han, E.-P., Kotegawa, T., 2006, Establishment of a network-based simulation of future air transportation concepts. *6th AIAA Aviation Technology, Integration and Operations Conference (ATIO)*, September 25–27, 2006. Wichita, Kansas, AIAA 2006-7719.

DeLaurentis, D.A., Kang, T., Lim, S., 2004, Solution space modeling and characterization for conceptual air vehicles, *AIAA Journal of Aircraft*, 41(1): 73–84.

Dibble, C.H., 2001, Theory in a Complex World: GeoGraph Computational Laboratories, Ph.D. dissertation, University of California, Santa Barbara, 2001.

Donohue, G., 2003, Air transportation is a complex adaptive system: not an aircraft design, *AIAA International Air and Space Symposium and Exposition: The Next 100 Years*, July 14–17, 2003. Dayton, Ohio. AIAA Paper 2003–2668.

Forrester, J.W., 1969, *Urban Dynamics*, The M.I.T. Press, Cambridge, MA.

Guimera, R., Mossa, S., Turtschi, A., Amaral, L.A.N., 2005, The worldwide air transportation network: anomalous centrality, community structure, and cities global roles, *Proceedings of the National Academy of Science, USA*, 102: 7794–7799.

Holland, J., 1998, *Emergence*, Perseus Books, Cambridge, MA, p. 45.

Holmes, B., 2004, Transformation in air transportation systems for the 21st century, *Proceedings of the 24th Congress of the International Council on the Aeronautical Sciences (ICAS)*, August 29–September 3, 2004, Yokohama, Japan. Plenary Paper.

Hood, L., 1998, Agent based modeling, [online article], URL: http://www.brs.gov.au.social_sciences/kyoto/hood2.html [accessed 18 Apr. 2002].

Kang, T., Lim, S., DeLaurentis, D., Mavris, D., A system dynamics model of the development cycle for future mobility vehicles, *Proceedings of the 21st International Conference of the System Dynamics Society*, July 20–24, 2003, New York, NY.

Keating, C., Rogers, R. Unal, R., Dryer, D., Sousa-Poza, A., Safford, R., Peterson, W. Rabadi, G., 2003, System of Systems Engineering, *Engineering Management Journal*, 15,(3): 36–45.

Lempert, R.J., Popper, S.W., Bankes, S.C., 2003, *Shaping the next one hundred years: New methods for quantitative, long-term policy analysis*, RAND Corporation, Santa Monica, CA.

Lewe, J., DeLaurentis, D., Mavris, D., Schrage, D., 2006, Modeling abstraction and hypothesis of a transportation architecture, *Journal of Air Transportation*, 11: 3.

Maier, M.W., 1998, Architecting principles for system-of-systems, *Systems Engineering*, 1(4): 267–284.

Mavris, D.N., Bandte, O., DeLaurentis, D.A., 1999, Robust design simulation: a probabilistic approach to multidisciplinary design, *AIAA Journal of Aircraft*, 36(1): 298–307.

Niedringhaus, W.P., 2004, The Jet:Wise model of national air space system evolution, *Simulation*, 80(1): 45–58.

Popper, S., Bankes, S., Callaway, R., DeLaurentis, D., 2004, System-of-Systems Symposium: Report on a Summer Conversation, July 21–22, 2004; online at http://www.potomacinstitute.org/academiccen/sos.htm.

Rouse, W., 2003, Engineering complex systems: implications for research in systems engineering, *IEEE Transactions on Systems, Man, and Cybernetics- Part C Applications and Reviews*, 33(2).

Sage, A.P., Cuppan, C.D., 2001, On the systems engineering and management of systems of systems and federations of systems, *Information, Knowledge, Systems Management*, 2(4): 325–345.

Schrage, D.P., DeLaurentis, D.A., Taggart, K., 2002, IPPD concept development process for future combat system, *Proceedings of the 9th AIAA/ISSMO Symposium on Multidisciplinary Analysis and Optimization*, September 4–6, 2002, Atlanta, GA. AIAA-2002-5619.

Sterman, J.D., 2000, *Business Dynamics*, Mcgraw-Hill, Irwin.

Von Bertalanffy, L., 2003, *General System Theory: Foundations, Developments, Applications*, revised edition, George Braziller, New York 2003 (first published in 1969).

Waldrop, M.M., 1992, *Complexity: The Emerging Science at the Edge of Order and Chaos*, 1st edition, Simon & Schuster Inc., NY.

Chapter **21**

Health Care System of Systems

NILMINI WICKRAMASINGHE,[1] **SURESH CHALASANI,**[2]
RAJENDRA V. BOPPANA,[3] **and ASAD M. MADNI**[4]

[1]Illinois Institute of Technology, Chicago, IL, USA
[2]University of Wisconsin-Parkside, Kenosha, WI, USA
[3]University of Texas, San Antonio, TX, USA
[4]Crocker Capital, San Francisco, CA, USA

21.1 INTRODUCTION

Health care systems are inherently complex in nature and address the needs of several stakeholders in health care management. Key stakeholders in health care industry include patients, physicians, nurses, hospitals, health care organizations, pharmacies, government regulatory & licensing agencies, and government funding agencies; for the web of health care players see Fig. 21.1. Interests of key stakeholders translate into myriad business rules that lead to heterogeneity in health care delivery and management.

A "patient-centric" approach to health care delivery requires that at all times the patients' needs and interests are the focus. Patient-centric health care, though simple in concept, is difficult to implement in practice due to the complex nature of independent systems that govern health care in the United States. After Kotov (1997), a health care system of systems (HSoS) can be defined as a collection of independent, large-scale complex, distributed systems. HSoS exhibit several key characteristics of a general system of systems as suggested by Sage and Cuppan (2001). First, they exhibit operational and managerial independence. For example, the hospitals and organizations such as HMOs, though they work together, work independently of each other. Similarly, government funding agencies such as Medicare/Medicaid and physicians, hospitals work independently of each other. Second, they are geographically distributed. It is possible that a patient who resides in Los Angeles needs to be treated in New York or even overseas. Third, health care systems exhibit evolutionary

System of Systems Engineering: Innovations for the 21st Century, Edited by Mo Jamshidi
Copyright © 2009 John Wiley & Sons, Inc., Publication

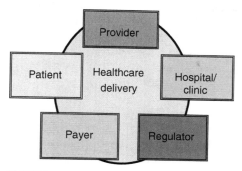

FIGURE 21.1 Web of players in healthcare

development in the sense that they change continuously in response to government regulation, advances in medical technology, and new threats such as bioterrorism and the need to cope with large-scale catastrophic events.

In this chapter, we first discuss health care systems and how they fit the definition of system of systems. We discuss different components of the health care system of systems and their characteristics. We also present a preliminary model that indicates the scheduling constraints of HSoS.

The remaining part of this chapter is organized as follows. Section 21.2 presents a brief literature survey on the system of systems and the health care system of systems. Section 21.3 discusses health care system of systems in more detail. Section 21.4 presents a network-centric model for HSoS. Section 21.5 presents the World Health Information Grid (WHIG) concept as an enabler for network-centric health care. Section 21.6 concludes this chapter.

21.2 LITERATURE SURVEY

System of systems has become a widely accepted term for describing complex systems in recent years. There is no universal definition of system of systems. Jamshidi (2005) provides an excellent summary of the different definitions of the system of systems. Here we repeat a few definitions of the SoS. Sage and Cuppan (2001) require SoS to exhibit a majority of the following five characteristics: operational & managerial independence, geographic distribution, emergent behavior, and evolutionary development. Kotov (1997) defines SoS as "large-scale concurrent and distributed systems that are comprised of complex systems." Carlock and Fenton (2001) discuss Enterprise Systems of Systems Engineering as being "focused on coupling traditional systems engineering activities with enterprise activities of strategic planning and investment analysis." Luskasik (1998) describes SoSE as "the integration of systems into systems of systems that ultimately contribute to evolution of the social infrastructure." A significant amount of research has been done in health care systems. For a compilation of research articles on health care management, see Wickramasinghe and Geisler (2006).

Madni (2006) designed smart reconfigurable wireless sensor networks for SoS applications. The main thrust of this chapter is in applying the definition of SoS to health care systems and examines the properties of SoS that are valid for health care systems. This chapter also discusses the network-centric aspects of HSoS.

21.3 HEALTH CARE SYSTEM OF SYSTEMS

For the purposes of this chapter, we assume the health care system of systems to be comprised of the systems shown in Fig. 21.1. Managed care companies, hospitals, clinics, physicians, and governmental agencies and programs such as Medicare and Medicaid work together to serve patients. Figure 21.2 simplifies the concept of the patient-centric approach. Below we discuss each one of the systems depicted in Fig. 21.2.

1. *Managed Care Companies:* Managed care companies have become popular in the last decade primarily because of their efforts to contain medical costs. PPOs (Preferred provider organizations) and HMOs (Health Maintenance Organizations) are examples of managed care companies. Managed care companies employ a variety of techniques to contain costs including gate-keeping (requiring mandatory authorization for hospitalization), capitation (payment of a fixed amount per member per month), generic drug substitution for brand name drugs (Getzen and Allen, 2007). A variety of information systems exist within managed care companies to enroll patients, maintain their claim records, audit claims, and track physician services.

2. *Hospitals:* Hospitals are financed by a variety of sources including payments from managed care companies, Medicare and Medicaid programs. Diagnosis related groups (DRGs) payments reimburse the hospitals based on the diagnosis and treatment. Hospital care in the United States is characterized

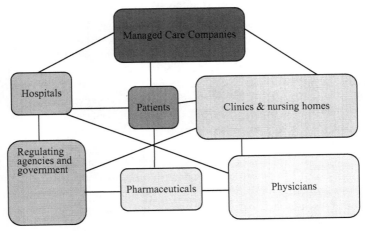

FIGURE 21.2   Different systems that are part of HSoS

by technologies for diagnosis, treatment to electronic maintenance of patient health records.

3. *Physicians:* Groups of physicians often practice together serving patient needs. Physician groups work with managed care companies via insurance contracts, utilize hospitals for inpatient care and work with governmental agencies such as Veteran's Administration hospitals, Medicare and Medicaid programs, and so on. Physicians utilize a number of technologies and information systems to diagnose and monitor patient's health status.

4. *Clinics:* Clinics are utilized by physician groups for treating outpatients. Clinical information systems are utilized for managing patient health records to scheduling patient visits and billing managed care organizations for patient visits.

5. *Governmental Agencies:* Governmental agencies provide oversight of physicians, clinics, hospitals, and nursing homes. Special units exist for dealing with veteran's health and for administering programs such as Medicare and Medicaid.

Based on the above discussion, we next describe how health care systems fit into the different definitions of system of systems.

21.3.1 Sage and Cuppan's Definition of HSoS

Health care systems exhibit the following properties.

1. *Operational and Managerial Independence:* It is clear each individual system—from managed care systems to governmental systems—exists on its own independent of the other systems. Together they serve the patient needs, while operating independently. In addition, each has its managerial independence with the management structure ranging from governmental management to private sector management.

2. *Geographical Distribution:* Health care systems are inherently distributed with no central organizational structure. Clinical systems, hospital systems, regulatory systems are dispersed geographically.

3. *Evolutionary Development:* Health care systems undergo evolutionary development. For example, pharmaceutical companies bring new drugs to the market after years of research, while hospital technologies are evolving to incorporate sensor-based monitoring of inpatients.

4. *Emergent Behavior:* An example of emergent behavior is to track the quality of care using the data from the individual systems such as the clinical systems and the hospital systems. For example, insurance companies such as the United Health certify physicians on whether they meet the quality standards based on national averages on quality care obtained from information on physician treatment of patients across the nation.

Based on the above discussion, Sage and Cuppan's definition applies to HSoS.

21.3.2 Application of Other Definitions of SoS to Health Care Systems

Kotov (1997) defines SoS as "large scale concurrent and distributed systems that are comprised of complex systems." Each individual system in Fig. 21.1 is a complex system that works concurrently of the other systems. Hence, Kotov's definition is applicable to HSoS. Luskasik's definition discusses systems of systems ultimately contributing to the evolution of the social infrastructure. This is the case for health care systems since each individual system is intricately tied to the patient population and the social infrastructure.

21.4 NETWORK-CENTRIC HEALTH CARE

The previous sections presented health care from a system of systems perspective. In this section, we present an architectural view that enables the health care system of systems. Network-centric health care is a critical component in realizing HSoS. This section discusses network-centric health care and the next section presents the WHIG as an example of network-centric health care.

As outlined by von Lubitz and Wickramasinghe (2006a, 2006b, 2006c) and von Lubitz, Yanovsky and Wickramasinghe (2006) network-centric health care operations are conducted at the confluence of three critical domains including informational, physical, and cognitive (refer to Table 21.1). In essence, these domains serve to cumulatively capture and then process all critical information and data so that effective and efficient value-driven health care operations may ensue.

The key challenges regarding e-health use include (1) cost effectiveness, that is, less costly than traditional health care delivery, (2) functionality and ease of use, that is, they should enable and facilitate many uses for physicians and other health care participants by combining various types and forms of data as well as be easy to use, and (3) they must be secure. One of the most significant legislative regulations in the United States is the Health Insurance Portability and Accountability Act (HIPAA, 2001).

TABLE 21.1 The Three Domains in Network-Centric Health Care

Domain	Description
Physical	Encompasses the structure of the entire environment health care operations intend to influence directly or indirectly, for example, elimination of disease, fiscal operations, political environment, patient and personnel education, and so on.
Information	Contains all elements required for generation, storage, manipulation, dissemination/sharing of information, and its transformation and dissemination/sharing as knowledge in all its forms. It is here that all aspects of command and control are communicated and all sensory inputs gathered.
Cognitive	Relates to all human factors that affect operations, such as education, training, experience, political inclinations, personal engagement (motivation), "open-mindedness," or even intuition of individuals involved in the relevant activities.

Given the nature of health care and the sensitivity of health care data and information, it is incumbent on governments not only to mandate regulations that will facilitate the exchange of health care documents between the various health care stakeholders but also to provide protection of privacy and the rights of patients. Moreover, irrespective of the type of health care system, that is, whether 100% government driven, 100% private or a combination thereof, it is clear that some governmental role is required to facilitate successful e-health initiatives.

A network-centric perspective to health care delivery also serves to underscore the inextricable connection and intertwining of e-health and e-government that, to date, has rarely been researched let alone acknowledged. Furthermore, for such an approach to become adopted successfully it requires governments to develop policies and protocols that will in turn facilitate its usability. We identify four key areas that will have an important impact on the development of the necessary policies and protocols as IT education, morbidity, cultural/social dimensions, and world economic standing as elaborated upon below. It is interesting to note that these areas also impact the development of numerous e-government initiatives.

21.5 WORLD HEALTH INFORMATION GRID (WHIG)

The backbone of network-centric health care operations is the WHIG, a set of interconnected web enabled information networks with intelligence capabilities that support the seamless transferring of all necessary data and information to the point of care so that the physician (or decision maker) is always making decision based on the best possible data, information, and knowledge (Fig. 21.3).

Specifically, the nodes on the grid (circles at various grid intersections in Fig. 21.3) represent the portal entry to the grid and various systems that capture store and then transmit data and information as depicted in Fig. 21.3. The power of this interconnected system is that at all times current data and information are accessible to all decision makers no matter where they are located; hence, decision makers are supported with relevant information and germane knowledge anywhere anytime, unlike in most health care systems that are platform-centric in design and hence information chaos results that leads to volumes of irrelevant and more often erroneous data that in turn lead to suboptimal or poor quality decision making.

WHIG provides the technology backbone for network-centric health care delivery. However, for WHIG to function as intended, various protocols and procedures must be developed at a global level. Without such standardization even the simplest of functions such as the exchange of documents and other procurement information, connectivity and e-commerce enabled benefits become problematic while the critical goal of decreasing information asymmetry becomes unattainable. Unfortunately, standardization is woefully lacking in too many areas of health care, let alone e-health. We indicate below two key areas that must be addressed.

1. *Information Communication Technology (ICT) Architecture/Infrastructure:* The generic architecture for most e-health initiatives is similar to that required

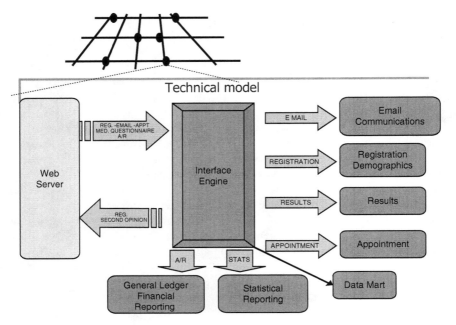

FIGURE 21.3 Schematic of World Health Information Grid

by WHIG. However, this infrastructure, which is made up of diverse technologies—phone lines, fiber trunks and submarine cables, T1, T3, and OC-xx, ISDN, DSL as well as satellites, earth stations, and teleports, must be available globally.

A sound technical infrastructure is an essential ingredient to the undertaking of e-health. Such infrastructures should also include telecommunications, electricity, access to computers, number of Internet hosts, number of ISP's (Internet Service Providers), and available bandwidth and broadband access. To offer a good multimedia content and thus provide a rich e-health experience, one would require a high bandwidth. ICT considerations are undoubtedly one of the most fundamental infrastructure requirements.

In addition, networks and telecommunications are a vital piece of the infrastructure for Internet access. One of the pioneering countries in establishing a complete and robust e-health infrastructure is Singapore, which is in the process of connecting 98% of homes and offices with a broadband cable network (Panagariya, 2000; APEC, 2001).

2. *Security and Trust:* Consumers are also concerned about a number of dimensions of trust: for example, trust in the security of value passed during electronic transactions with organizations that are "virtual" in a disconcertingly ineffable way, and trust in the privacy of personal data arising from electronic transactions (Ghosh and Swaminatha, 2001; Fjetlande, 2002; Roquilly, 2002).

In addition, ethical concerns are inevitably related to health care operations. The notion of the governmental bodies having ready access to health care information of the citizens is among the major concerns in the United States; similar reservations are also voiced in Europe. The possibility of security breaches, similar to those that recently affected millions of credit card customers in the United States, demands a very stringent layer of protective layers that will assure prevention of commercial misuse of health care data.

Figure 21.3 describes a model of network-centric health care operations that can support the different systems that comprise the HSoS. Hence network-centric health care mandates the conceptualization of health care as system of systems.

In a recent book, Porter and Tiesberg (2006) note that the problem with the U.S. health care system is that it is engaged in zero-sum competition. This means that the web of players compete with each other. Porter recommends that health care be redefined to encourage, as in other industries, positive sum competition. In doing so, it will be possible for all actors in the web of health care to benefit and most importantly for health care to be value-driven and patient-centric. However, Porter falls short of describing a model that can enable this to take effect. Health care is a knowledge-rich environment where data and information are critical for diagnosis and then the consequent prescription of the appropriate health care treatment. In today's information age clearly ICTs play an integral role in enabling the correct data and information to reach the decision maker when and where it is required. However, given the complex nature of health care operations the careful design of ICTs in health care is crucial. The proposed system of systems model coupled with the ideas of network-centric health care appears to hold the key.

21.6 CONCLUDING REMARKS

As both the United States and Europe move forward on their respective agendas to incorporate e-health and electronic medical records or computerized patient records, the concept of health care systems of systems becomes more important if we are to fully realize the benefit of ICT use in health care delivery. In the context of network-centric health care operations it becomes a strategic necessity. We therefore, close, with a strong call for more research in this area as it relates to health care.

REFERENCES

APEC, 2001. The new economy and APEC, Report to the Economic Committee of APEC, http://www.diw.de/deutsch/produkte/veranstaltungen/docs/apec-report.html. A Report Prepared for Asia Pacific Foundation 2002.

Carlock, P.G., Fenton, R.E., 2001, System of systems (SoS) enterprise systems for information-intensive organizations, *Systems Engineering*, 4(4): 242–261.

Carlock, P.G., Decker, S.C., Fenton, R.E., 1999, Agency level systems engineering for systems of systems, *Systems and Information Technology Review Journal*, 99–110.

Fjetland, M., 2002, Global commerce and the privacy Clash, *Information Management Journal*, 36(1): 54–58.

Getzen, T.E., Allen, B.H., 2007, *Healthcare Economics*, John Wiley.

Ghosh, A.K., Swaminatha, T.M., 2001, Software security and privacy risks in mobile e-commerce, *Communications of the ACM*, 44(2): 51–58.

Health Insurance Portability and Accountability Act (HIPPA) 2001, Privacy Compliance Executive Summary, May 2001, Protegrity Inc.

Jamshidi, M., 2005, System-of-systems engineering - a definition, *IEEE SMC 2005*, Big Island, Hawaii, URL: http://ieeesmc2005.unm.edu/SoSE_Defn.htm.

Keating, C., Rogers, R., Unal, R., Dryer, D., Sousa-Poza, A., Safford, R., Peterson, W., Rabadi, G., 2003, System of systems engineering, *Engineering Management Journal*, 15 (3):36–45.

Kotov, V., 1997, Systems of systems as communicating structures, *Hewlett Packard Computer Systems Laboratory* Paper HPL-97-124, pp. 1–15.

Luskasik, S.J., 1998, Systems, systems of systems, and the education of engineers, *Artificial Intelligence for Engineering Design, Analysis, and Manufacturing*, 12(1): 55–60.

Madni, A.M., Smart configurable wireless sensors and actuators for system of systems (SoS) applications. Keynote Address, *Proceedings of the IEEE SMC International Conference on System of Sytems Engineering*, April 2006, Los Angeles.

Panagariya, A., 2000, e-commerce, WTO and developing countries, *The World Economy*, 23 (8): 959–978.

Porter, M., Tiesberg, E., 2006, *Redefining Health Care*, Harvard Business School Press, Boston.

Roquilly, C., 2002, Closed distribution networks and e-commerce: antitrust issues, *International Review of Law, Computers & Technology*, 16(1): 81–93.

Sage, A.P., Cuppan, C.D., 2001, On the systems engineering and management of systems of systems and federations of systems, *Information, Knowledge, Systems Management*, 2(4): 325–345. *Proceedings of the IEEE International Symposium on System of Systems*, April 2006, Los Angeles, CA.

von Lubitz, D., Wickramasinghe, N., 2006a, Healthcare and technology: the doctrine of networkcentric healthcare, *International Journal of Electronic Healthcare(IJEH)*, 4: 322–344.

von Lubitz, D., Wickramasinghe, N., 2006b, Networkcentric healthcare: applying the tools, techniques and strategies of knowledge management to create superior healthcare operations, *International Journal of Electronic Healthcare (IJEH)*, 4: 415–428.

von Lubitz, D., Wickramasinghe, N., 2006c, Key challenges and policy implications for governments and regulators in a networkcentric healthcare environment, *International Journal of Electronic Government*, 3(2): 204–224.

von Lubitz, D., Yanovsky, G., Wickramasinghe, N., 2006, Networkcentric healthcare operations: the telecommunications structure, *International Journal of Networking and Virtual Organizations*, 3(1): 60–85.

Wickramasinghe, N., Geisler, E., (Eds.), 2006, *Proceedings of the 5th International Conference on the Management of Healthcare and Medical technology*, August 2006, Chicago, IL.

Chapter **22**

System of Systems Engineering of GEOSS[*]

RYOSUKE SHIBASAKI[1] and JAY S. PEARLMAN[2]

[1]University of Tokyo, Japan
[2]Boeing Company, Seattle, WA, USA

22.1 GEOSS: ITS BACKGROUND AND OBJECTIVES

"Understanding the Earth system—its weather, climate, oceans, atmosphere, water, land, geodynamics, natural resources, ecosystems, and natural and human-induced hazards—is crucial to enhancing human health, safety and welfare, alleviating human suffering including poverty, protecting the global environment, reducing disaster losses, and achieving sustainable development. Observations of the Earth system and the information derived from these observations provide critical inputs for advancing this understanding.

The GEO (Group on Earth Observations), a voluntary partnership of governments and international organizations, was established at the Third Earth Observation Summit in February 2005 to coordinate efforts to build a Global Earth Observation System of System (GEOSS). As of November 2007, GEO's members include 72 governments and the European Commission. In addition, 46 intergovernmental, international, and regional organizations with a mandate in Earth observation or related issues have been recognized as participating organizations.

The 10-Year Implementation Plan Reference Document of GEOSS states the importance of the Earth observation and the challenges to enhance human and societal welfare. This implementation plan, for the period 2005–2015, provides a basis for GEO to construct GEOSS. The plan defines a vision statement for GEOSS, its purpose

[*]This chapter is based on personal experiences of the authors in GEOSS and in developing SoS applications and does not represent the official opinions of GEO.

System of Systems Engineering: Innovations for the 21st Century, Edited by Mo Jamshidi
Copyright © 2009 John Wiley & Sons, Inc., Publication

FIGURE 22.1   Concept of GEOSS

and scope, and the expected benefits. Prior to its formal establishment, the ad hoc GEO (established at the First Earth Observation Summit in July 2003) met as a planning body to develop the GEOSS 10-Year Implementation Plan.

The purpose of GEOSS, as illustrated in Fig. 22.1, is to achieve comprehensive, coordinated, and sustained observations of the Earth system to meet the need for timely, quality long-term global information as a basis for sound decision making, initially in nine societal benefit areas. These nine societal benefit areas are as follows:

1. Reducing loss of life and property from natural and human-induced disasters;
2. understanding environmental factors affecting human health and well-being;
3. improving management of energy resources;
4. understanding, assessing, predicting, mitigating, and adapting to climate variability and change;
5. improving water resource management through better understanding of the water cycle;
6. improving weather information, forecasting, and warning;
7. improving the management and protection of terrestrial, coastal, and marine ecosystems;
8. supporting sustainable agriculture and combating desertification;
9. understanding, monitoring, and conserving biodiversity.

GEOSS builds on and adds value to existing Earth observation systems by coordinating their efforts, addressing critical gaps, supporting their interoperability, sharing information, reaching a common understanding of user requirements, and improving delivery of information to users.

GEOSS is a step toward addressing the challenges articulated by United Nations Millennium Declaration and the 2002 World Summit on Sustainable Development (WSSD), including the achievement of the Millennium Development Goals. GEOSS will also further the implementation of international environmental treaty obligations.

22.2   ORGANIZATIONAL STRUCTURE FOR BUILDING GEOSS

The way SoSE (system of systems engineering) is applied to building GEOSS is determined by the governance of GEO. Organizational structure of GEO is summarized in this section. Four committees, with the support of GEO secretariat, are responsible for leading the building process of GEOSS as shown in Fig. 22.2. These committees report to the GEO Plenary, which is the governing body of GEO. In addition, an executive committee of GEO provides operational and policy guidance, interpreting the decisions and intent of the GEO Plenary. In developing the architecture, the committees play a role in both development and operation approaches for the SoS (System of Systems). Each committee has a specific charter of "Terms of Reference," which guides its mission.

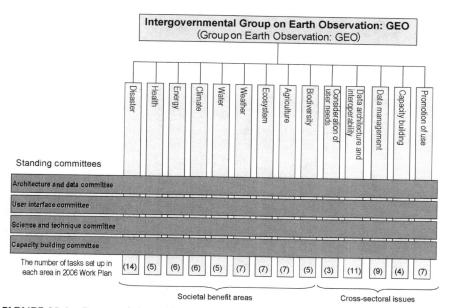

FIGURE 22.2   Responsibility of the four committees for building GEOSS (GEO, 2005)

The missions of the four committees are defined as follows (GEO home page, http://www.earthobservations.org/index.html):

1. *Architecture and Data Committee (ADC):* The Architecture and Data Committee supports GEO in all architecture and data management aspects of the design, coordination, and implementation of the GEOSS for comprehensive, coordinated, and sustained Earth observations. The objectives are to

   a. enable GEO, based upon user requirements and building on existing systems and initiatives, to define the components of GEOSS, to converge or harmonize observation methods, and to promote the use of standards and references, intercalibration, and data assimilation;

   b. enable GEO to define and update interoperability arrangements to which GEO members and participating organizations agree to adhere, including technical specifications for collecting, processing, storing, and disseminating shared data, metadata, and products;

   c. enable GEO to facilitate data management, information management, and common services, and help to promote data sharing principles in support of the GEO Plenary for the full and open sharing and exchange of data and information, recognizing relevant international instruments and national policies and legislation.

2. *Capacity Building Committee (CBC):* The Capacity Building Committee supports GEO in strengthening the capability of all countries, in particular developing countries, to use Earth observation data and products in a sustainable manner and to contribute observations and systems to GEOSS. The GEO capacity building strategy follows the WSSD concept of a global partnership between those whose capacity needs developing and those who are able to assist in the process, recognizing that activities have intertwined social, environmental, and economic impacts. The objectives are to

   a. facilitate Earth observation capacity building activities among GEO members, in concert with GEO participating organizations;

   b. build global capacity to access, retrieve, analyze, include into appropriate models, and interpret relevant data from global data systems;

   c. build global capacity to integrate Earth observation data and information with data and information from other sources, improving understanding of problems in order to identify sustainable solutions;

   d. develop a coordinated capacity building strategy among GEO members and participating organizations based on the principles articulated in the GEOSS 10-Year Implementation Plan Reference Document;

   e. recommend strategies for resource mobilization.

3. *Science and Technology Committee (STC):* The Science and Technology Committee engages the scientific and technological communities in the development, implementation, and use of a sustained GEOSS to ensure that

GEO has access to sound scientific and technological advice. The objectives are to

a. enable GEO to make decisions on best available and sound scientific and technological advice, through the solicitation of input from a broad, trans-disciplinary scientific and technological community;

b. ensure scientific and technological integrity and soundness of GEO annual work plans;

c. monitor and review output and deliverables of GEO annual work plans;

d. collaborate with GEO members and participating organizations, and through transparent processes, to identify individual experts and groups to participate in GEO working groups;

e. facilitate linkages and partnerships with major relevant international research programs as well as organizations willing to contribute to GEO activities.

4. *User Interface Committee (UIC):* The User Interface Committee engages users in the nine societal benefit areas (SBAs) in the development, implementation, and use of a sustained GEOSS that provides the data and information required by user groups on national, regional, and global scales. The User Interface Committee has a specific goal to address cross-cutting issues by coordinating user communities of practice, ensuring continuity and avoiding duplication. The objectives are to

a. enable GEO to address the needs and concerns of a broad range of user communities in developing and developed countries, across issues and transdisciplinary needs, with a particular focus on fostering new or less organized communities.

b. enable GEO, in the implementation of GEOSS, to engage a continuum of users, from producers to the final beneficiaries of the data and information;

c. Facilitate linkages and partnerships between established communities of practice and new groups or organizations interested in collaborating.

User requirements are extracted and summarized by UIC, while ADC helps GEO lay the groundwork for the system of systems, improving interoperability of existing and future data and observation systems. CBC supports GEO in improving human and organization capacities of data acquisition, processing, and utilization for better informed decision making. STC represents scientific user community and provides advice on the data/information utilization of GEOSS from scientific viewpoint.

22.3 SYSTEM OF SYSTEMS ENGINEERING AND ITS APPLICATION TO GEOSS

22.3.1 SoSE Approach

At a glance, an observer may imagine that GEOSS is a typical system of systems. GEO aims at building GEOSS on existing Earth observation systems by adding value through synergy and not building a system level capability from scratch. For this reason, GEOSS is being developed using the SoSE approach.

A clear definition of a "system of systems" is needed to understand to what degree GEOSS can employ a conventional system of systems approaches or where the approach needs to be adapted to special conditions that are driven by GEOSS governance principles. "A system of systems is a *supersystem* comprised of elements that are themselves complex, independent systems that interact to achieve a common goal." In the context of this definition, attributes of a system of systems are that the component systems achieve well-substantiated purposes in their own right even if detached from the overall system. In fact, the component systems are managed in large part for their own purposes rather than the purposes of the system of systems. To then justify the creation of a system of systems, the SoS must exhibit behavior, including emergent behavior, not achievable by the component systems acting independently. Thus, the SoS is offering significantly new capabilities that justify the "overhead" associated with the SoS (Butterfield, *IEEE Systems Journal* special issue).

Examining these attributes of a system of systems, there are a number of distinct features that are observed. For example, the component systems can operate independently to produce products or services both relevant to the system of systems and satisfying their own specific customer objectives. The prioritization of these systems remains with the originator, as well as the resource allocations for both development and operations. How then is synergy achieved? Typically, this is done through interoperability arrangements that do not require tight coupling or strong integrations, but still allow for synergies in product development or distribution. Since participating systems are run for their own purposes, it is expected that there will be times when they are unavailable to support the objectives of the system of systems. This, in turn, leads to the attribute that a system of systems must retain its inherent operational character even as systems join or disengage from the system of systems. The ability to maintain functionality in this environment is termed "adaptability for dynamic participation."

Dynamic participation is enhanced through recent developments in hardware and software capability and communication protocols to handle large volumes of information. Consider networked computers and the Internet, where various diverse computing systems interact for data creation, processing, and display. Computer languages have been developed to facilitate interoperability (using techniques such as HTML), to seamlessly share sensor and display information, coordinate information flow, or provide subsystem-like functionality for systems that are part of a greater "system of systems." Clearly, as information technology has extended operational capability to include the utilization of disparate resources by sharing remotely located components, there is a greater need to understand how previously disparate, independent systems can efficiently and effectively interoperate.

To take advantage of the processes and technologies available for advanced system of systems such as GEOSS, an engineering approach is desired that provides systematic means to evaluate the design and implementation of the system of systems. To this end, the SoSE Process used should be architecture-centric, model-based, and user-driven (Pearlman and Butterfield, 2006). Architecture-centric means that the SoS architecture model serves as the primary artifact for conceptualizing, constructing,

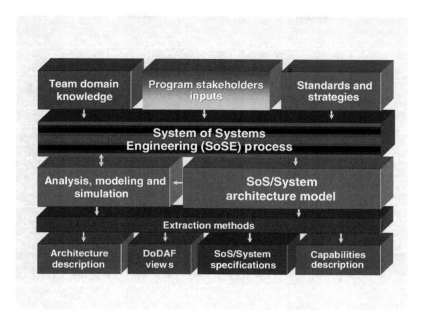

FIGURE 22.3 A block diagram of the SoS provides a visualization of the process flow

managing, and evolving the SoS under development (see Fig. 22.3). The architecture model facilitates dialogue between stakeholders by having a common notation and an intuitive depiction of the system structure and behavior. This approach provides a continuous model stream from the SoS goal level to the architecture subsystem base level in a notation that supports all of the associated engineering design disciplines. Here, the need cannot be underestimated for creating the means for clear communication between the user supplying high-level core requirements and the system architect and engineers who are designing and implementing the system of systems. This could be as simple as agreeing to a clear lexicon for stimulating dialogue, but is more likely the creation of a model that can view different aspects of the system of systems. These aspects or "architectural views" define various facets of the SoS in ways that illuminate the major characteristics of the SoS. The use of such a model also significantly improves the traceability required across a product's lifecycle by clearly defining functional requirements and system elements. This is essential if the SoS is built through a phased or evolutionary development. The architecture model results in a system architecture definition at all levels, the associated specification definition, and the needed design comprehension and realization.

In a conventional system development, the emphasis is on stakeholder (customer) input flowing into detailed requirements and the system design. With the importance of interoperability in the SoS and the likely inclusion of legacy or existing systems operating for their own purposes, and their own life cycle, there are additional input factors such as key standards and the participants' knowledge and domain capabilities.

Such factor into the constraints that act upon the SoS during formation and in operation.

These inputs are seen in the upper level of Fig. 22.3. They are analyzed and integrated through the SoS engineering process that defines the attributes and develops the construct for the SoS. The hypothetical performance of this SoS construct should be tested before further development. This can be done most effectively by creation of a system of systems model and then running simulations of its operations in a number of projected environments or use cases. This is not a single flow through, but is a highly iterative process where the outcomes of the simulations are fed back into the SoS engineering process and the construct is updated. It is only after the testing generates acceptable levels performance that detailed characteristics of the SoS are extracted from the architectural model.

Extraction of detailed SoS attributes and their implementation can be done through a process that is defined in a number of architectural approaches such as the DoD Architectural Framework (DoDAF) or the Federal Enterprise Architecture. The key is to provide the developer with a set of descriptions including processes, physical constructs, and operational relations. For GEOSS, a set of descriptions was developed by the Architecture Implementation Pilot (AIP) team (Percivall, 2007a,b) (Fig. 22.4). For the AIP, a particular construct was chosen, the RM-ODP or ISO/IEC10746, Information technology—Open distributed processing—Reference model. The RM-ODP standards are used in multiple geospatial and Earth observation architectures, for example, the ISO 19100 series of geographic information standards, and the OpenGIS Reference Model.

An example, the engineering view, is provided in Fig. 22.5 (Percivall, 2007c). This engineering viewpoint is focused on the GEO Web Portal, the GEOSS Clearinghouse, and interfacing with information and data components and services.

From a practical approach and as discussed above, GEOSS interoperability arrangements are based on the view of complex systems as assemblies of components that interoperate primarily by passing structured messages over network communication

Viewpoint Name	Description of RM-ODP Viewpoint as used Herein
Enterprise	Articulates a "business model" that should be understandable by all stakeholders; focuses on purpose, scope, and policies.
Information	Focuses on the semantics of the information and information processing performed.
Computational	Service-oriented viewpoint that enables distribution through functional decomposition of the system into objects that interact at interfaces.
Engineering	Identification of component types to support distributed interaction between the components.
Technology	Identification of component instances as physical deployed technology solutions, including network descriptions.

FIGURE 22.4   Architectural viewpoints of the GEOSS information system

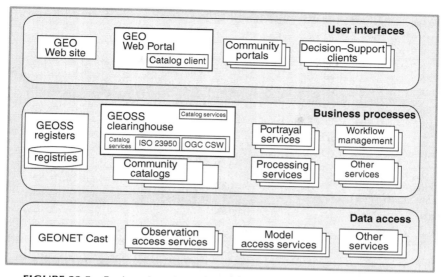

FIGURE 22.5   Engineering viewpoint of the GEOSS architecture description

services. By expressing interface interoperability specifications as standard service definitions, GEOSS system interfaces assure verifiable and scalable interoperability, whether among components within a complex system or among discrete systems (GEO, 2005). Thus, a foundation of the architecture is both the use of structured messages and the availability of descriptive data (metadata) for each of the components and related services of the SoS. These are used in the information system shown in Fig. 22.6.

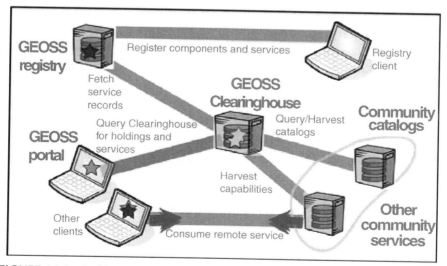

FIGURE 22.6   Architecture components support users in discovery of and access to diverse information sets

22.3.2 Challenges of GEOSS from SoSE Perspective

Considering the SoSE approach described in Section 22.3.1for developing GEOSS, the question to be addressed is whether GEOSS is similar to "conventional system of systems" or whether there are enough differences that the approach needs to be adapted in some significant way. In fact, for a number of reasons, the GEOSS development process is quite different from examples of SoSE applications reported in Butterfield et al. (2007) and Martin (2007), which refer to the cases of Boeing and NASA. Private firms and governmental organizations such as Boeing and NASA have a hierarchical and/or network structure of specialized divisions to efficiently achieve missions of the organizations. This hierarchical structure has the exclusive power of designing the kinds of system divisions that need to be created and how they are interlinked.

To build a system of existing systems within such organizations, centralized governance structure can assure the implementation of the design of the SoS. What an SoSE architect does, in this case, is to present an appropriate design of an SoS and a roadmap for the implementation. The challenge of SoSE, in this traditional context, is to show how to realize efficiently, with minimum risk, an SoS that contributes to achieving the organizational mission through integrating large-scale and complex systems.

In contrast, while GEO is also an organization consisting of members that share the common objectives, it is a voluntary partnership of national governments and international nonprofit organizations. GEO cannot force the participation of members to develop GEOSS.

To understand the differences, it is valuable to compare two cases, one of an SoS built by an organization with a single large customer focal point, typically under contract, and the other of an organization that creates the SoS from voluntary contributions (GEOSS).

Attributes of the two cases are shown in Fig. 22.7 (Pearlman and Butterfield, 2006). The differences are more than just the contractual versus voluntary nature of the SoS. GEOSS has very large cultural and technical capability differences that impact the way the SoS needs to be constructed so that it effectively serves the customer base, that is, the scientists, industry members, and government managers that create and use

Traditional SoS Environment	GEOSS Environment
• Uniformity of objectives	• Disparate motivations
• Single corporate direction	• Competing agendas
• Cultural uniformity	• Diverse backgrounds
• Common technology	• Significantly varying technology levels
• Data standards and quality established	• Multiple views of data standards and quality

FIGURE 22.7 Conventional SoS and GEOSS differ in the underlying constraints that drive their development and operation

information for societal benefits. An interesting example of the issues is in standards. One might ask, "Aren't standards globally standard?" Many instances show that different cultures independently derive standards, as they do languages. There are many words in different languages to address the same object or thought. So there are multiple definitions of how to measure physical attributes of the Earth, such as sea level. In some cases the conversion between alternative techniques has not been fully considered. This leads to the challenges in GEOSS of creating translatable taxonomies, standards, and measurement techniques to attain a global observation and information set. A major thrust of GEO is that differences in systems cannot be a barrier to tasks that must span multiple systems. Yet the system of systems must be "constructed" without imposing significant new constraints on the existing or legacy systems. Thus, the critical question in formulating the SoS is "what few things must be the same so that everything else can be different."

GEOSS has additional constraints that evolved from its basic principles of development and operation. For example, in addition to the cultural and standards diversities mentioned above, GEO cannot provide financial support and human resources to develop GEOSS, because the mission of GEO is limited to the coordination of activities to facilitate the development such as organizing meetings and sponsoring outreach. Development of an SoS usually requires additional cost for modification of interfaces and enhancement of systems for an expected increase in computational load and so forth. Moreover, to provide equal opportunities to join in the development of and then GEOSS, GEO needs to use nonproprietary and open interface standards as much as practical.

Consequently, to develop GEOSS, consensus has to be built among governments and participating organizations on the targets and the process of design and implementation, that may lead to the endorsement of actions on a voluntary basis.

Under such circumstances, the objectives of applying SoSE to GEOSS are to

1. maximize the participation of governments and the other organizations;
2. encourage voluntary activities such as the contribution of data and systems, promotion of system integration to achieve GEOSS targets;
3. coordinate the activities efficiently to build usable and reliable GEOSS;
4. create and maintain a clear understanding of the social impacts and benefits of synergy derived from operation of the SoS.

These provide a new challenge of SoSE as follows:

1. How to accelerate consensus building among diversified participants not only on abstract objectives of SoS development, but also on concrete design and deployment process of SoS?
2. How to encourage and organize coordinated activities on a voluntary basis toward an SoS development?
3. How to make sure that voluntary activities lead to the consistent and robust SoS development that can successfully fulfill user requirements of nine SBA's?

4. How to assure that the wide range of technical and domain knowledge can be supported by the SoS without reducing the products and infrastructure to the lowest common level?

In the future, an SoS will expand beyond the boundary of individual organizations to provide a variety of common and advanced services to society. ITS (intelligent transport system) would be a good example, because it requires tight collaboration among car manufacturers, part manufacturers, wireless communication carriers, road administrative organizations, polices, and so forth. In the case of disaster warning and mitigation, a number of public organizations (not limited to disaster prevention agencies) must coordinate to provide emergency services such as rescue and replacement of critical infrastructure. How to efficiently develop an SoS at societal scale beyond the boundary of individual organizations is a very important research challenge that provides large benefit to society.

One of the most successful and popular examples of a system of systems is the WWW (World Wide Web). A number of systems are now interconnected through Web technologies. The Web itself, however, evolved only after a limited number of basic interface standards such as HTML, HTTP, and associated technologies were established. It may be referred to as a successful evolution model in analyzing what kinds of standards and technologies would be well accepted by a wide range of developers and users. However, the WWW is different from GEOSS in the sense that GEOSS has to be developed based on existing systems that already run on established standards that may not be completely interoperable or accepted by different communities.

"The Cathedral and the Bazaar" (Raymond, 2006) illustrates the contrast between the new and traditional challenges of SoSE. The author compares building a cathedral to top-down and centralized development of systems, while distributed and collaborative system development based on many voluntary works (such as Linux) is called "Bazaar" model. His major finding for the success of the Bazaar model is the fact that a complete working prototype system and not just parts of a system, though it may be small and has only limited functionalities, has to be presented early to attract and stimulate the interests and voluntary participation of software engineers. Once enough volunteers are mobilized, the initial prototype system can be improved and sometimes completely replaced with the new one, to be a higher quality and reliable system.

22.3.3 Participants for Building GEOSS

Pursuing the Bazaar model of evolutionary development, who are the critical initial participants that need to be mobilized to build an initial operating capability for GEOSS.

Participants in the building GEOSS are categorized as follows:

1. Data providers
2. System providers

3. Users in SBA (beneficiaries of GEOSS, i.e., decision makers in SBA)
4. SoS builders (building an SoS using contributed data and systems to support decision making of SBA users).

A more detailed description of each of these is important to understand the directions of GEOSS development.

1. Data providers include large observing and data collection organizations such as WMO, NOAA, ESA, or JAXA. They can also be smaller organizations or even individuals who make measurements and make them available.
2. System providers are organizations that provide data, derived information, data processing, and management software. Commercial companies, however, cannot join GEO. Instead, international organizations or governments may represent their constituents that provide wide range of vendors, including commercial firms and research institutes. Major space agencies such as NASA and ESA may also belong to both (1) and (2) because they provide not only satellite data, but also software products and services for data dissemination and utilization.

Data providers (1) and system providers (2) basically have their own data and software products that could be contributed to GEOSS. Their incentives to participate in GEOSS include the expectation of obtaining more users of their products and strengthening of visibility through the contribution to solving global environmental issues and improving human and societal welfare in nine societal benefit areas. However, decision support services that users (category 3 above) would like to receive for application in the various SBAs may be complex and may require a lot of resources. If users are satisfied with the simple overlay of several geographical information system (GIS) layers in a restricted geographical region, they may not participate in GEO, because the technologies and data products are already available. SoS builders (4) are expected to provide solutions to SBA users by integrating data and systems contributed to GEO to provide either unique synergistic products or the means for interoperability to create such products.

For practical reasons of both implementation and achievement of early impacts, specific tasks were established at the formation of GEO in 2005. This was updated and extended to a three-year plan (2007–2009) to increase continuity when addressing multiyear efforts. Participants in GEO usually work on tasks (individual projects) defined in this work plan (GEO, 2007b). These tasks report to the Plenary through the four GEO committees. The committees provide the environment for efficient and well-coordinated task progress, checking scientific validity and effectiveness and supporting capacity building and outreach activities.

Values of GEOSS are realized through helping SBA users make better informed decisions. To tackle complex issues of SBA users, it is necessary to propose solutions by integrating data and systems. Coordination among (1) data providers, (2) system providers and (4) SoS builders is of critical importance. While the infrastructure approach is maturing as described above, the details of the architecture are still evolving.

22.4   SoSE ACTIVITIES FOR GEOSS

As described in Section 22.3.1, the key to success of GEOSS depends on how to

1. maximize the participation of governments and international participating organizations;
2. encourage voluntary activities such as the contribution of data and systems, promoting system of systems integration to achieve GEOSS targets;
3. efficiently coordinate the activities to build a usable and reliable GEOSS.

To meet these objectives, three types of complementary approaches should be considered: visualizing both the available resources and the benefits from committing these and additional resources; providing incentives for membership through reduced net cost or equivalent benefits; and promoting coordination and evolution of formulation through work plan tasks that lead to understandable and unique efficiencies.

22.4.1   Visualizing Available System and Technical Resources

The first approach is to visualize available technical resources for the building of GEOSS and a future blueprint of GEOSS that presents a direction for both the development and the structure of the outcomes. They are categorized as follows:

1. Blueprint or future prospects of GEOSS;
2. User requirements;
3. Data systems, services, and vocabulary used for description of metadata;
4. Standards and interoperability arrangements;
5. Best practices and quality assurance;
6. Case studies, lessons to be learnt, and human resources.

22.4.1.1   Blueprint or Future Prospects of GEOSS   As the GEOSS 10-Year Implementation Plan states, "GEOSS builds on and adds value to existing Earth observation systems by coordinating their efforts and addressing critical gaps," but no detailed plan or design is presented. Instead, 10-Year Implementation Plan Reference Document (2005) provides a conceptual model (Fig. 22.8).

Figure 22.8 illustrates areas that are included and others that are outside the GEOSS framework. GEOSS is focused on providing information for decision makers of society and thus incorporates both the collection of observations and the translation of these observations into information through models and analyses. It does not address specific decision support tools that take the information and develop options for decision makers. This may seem to be a strange and artificial boundary, but follows the logic that decisions, while using global information as the underlying information set, incorporate local/regional issues and constraints.

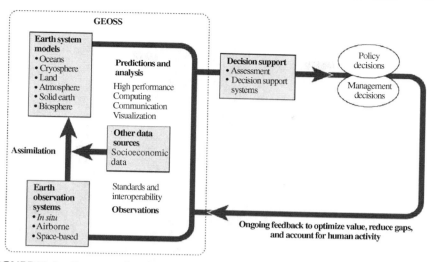

FIGURE 22.8 Conceptual architecture of GEOSS. "This diagram demonstrates the end-to-end nature of data provision, the feedback loop from user requirements and the role of GEOSS in this process. The primary focus of GEOSS is on the left side of the diagram." (GEO, 2005)

22.4.1.2 User Requirements

User requirements are important resources because they provide the basis of system design. In the practice of ordinary system engineering and SoSE, user requirements are not regarded as "resources" for system development and integration because clearly stated user requirements are premises for the plan and design. GEOSS users, as discussed earlier, are very diversified and geographically dispersed in different socioeconomic and cultural contexts. Also, in this environment, some factors may be "invisible" in the early engineering phase, even though they may play important roles in alleviating human sufferings in local environment. This creates the necessity that activities of exploring and identifying user communities and of extracting/organizing their requirements need to be clearly defined in the process of SoSE for GEOSS. User requirements have to be registered and visualized for GEO members or participating organizations so they may be balanced and incorporated in the planning of the SoS. A registry of user requirements is now being designed under UIC initiative with the support from ADC.

22.4.1.3 Data Systems, Services, and Vocabulary Used for Description of Metadata and Standards

In addressing the GEOSS concept, the engineer tends to focus on data and the systems for acquisition, storage, and distribution. This focus is too narrow, because the products are information sets that are derived from models and analyses. Thus, the systems engineer must consider interoperability and standardization of model outputs. In the context of GEOSS, this extension can be challenging because the system is bringing together disparate communities such as biodiversity and climate that have limited common descriptions and standards.

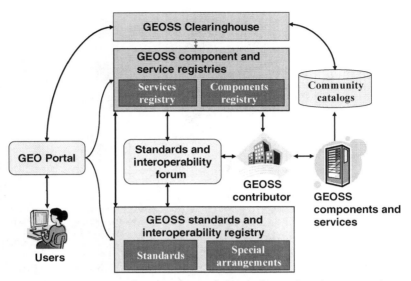

FIGURE 22.9   Initial operating capabilities of GEO (Pearlman, 2007)

Thus, data, models, systems, and services are building blocks of the SoS. Since the focus is on bringing together existing and new systems, information on available and well-accepted standards is indispensable in interlinking systems and in disseminating and sharing data and information. GEO established component/service registries and standard registries for this purpose (http://geossregistries.info/). Figure 22.9 shows the registries in GEOSS information system structure. The component registry allows contributors to register metadata on components (system and data) together with those on interfaces to access systems and data. Sixty-six components and seventy-seven service interfaces are already registered (as of December 2007). Each of these components may offer one or more services, generally with automated interfaces. Thus a companion registry, the service registry, enables users to understand the full range of services available through GEOSS. The engineer might question the reason for maintaining a component registry. By clearly defining the components through entry into the components registry, both the heritage of the services can be traces, and the component supplier is committing to follow the interoperability agreements of GEOSS.

22.4.1.4 Standards and Interoperability Arrangements   The interoperability agreements fall into two classes, those which are formally defined standards and those employing special arrangements that are commonly recognized, but not (yet) formalized through international standards organizations. There are cases of special interoperability arrangements in commonly accepted arrangements such as the Adobe "pdf," which are in wide use. The special interoperability arrangements are needed because, as noted earlier, GEOSS is a voluntary system and the SOS operators cannot impose significant new demands on existing contributed components. Standards and special arrangements, registries for both GEOSS recognized standards and GEOSS interoperability arrangements have already been established. The primary focus initially was on standards used by the contributed observing components. This has been expanded and it is anticipated that interoperability of models will be added as a focus for the standards efforts of GEOSS. The registration process was initiated in the June 2007 and further improvement of the registries will continue based on feedbacks from users.

Data vocabulary such as data item names and underlying definitions provide a semantic basis to support the consistent and easy-to-understand description of metadata, a key to GEOSS interoperability. This is important both for machine readability and for working in a cross-cultural, cross-discipline environment. So far, the standardization on format and encoding rules, that is, syntactic aspects of data and interfaces have been a primary concern in discussions about interoperability. These are typically done within a technical disciple. However, when users need to discover and integrate data and services provided across different disciplines, semantic interoperability will be one of the key issues. A registry of data vocabulary has not yet been established though some terminologies are already registered as components in the component registry. Development of the registry of data semantics or vocabulary is being discussed by taking into account the balance between the complexity of descriptions and the value of detailed description of meanings.

22.4.1.5 *Best Practices and Data Quality*

A method to significantly improve interoperability is to delineate best practices that are followed in the collection and analyses of data. In some instances, this would correlate closely with the work on semantics discussed in the previous section. An example would be the definition of sea level, which is not globally uniform. A corollary would be to define a best practice in the measurement techniques and perhaps instrumentation for such sea level measurements. While it is recognized that best practices occur in all aspects of GEOSS, they are difficult to refine in detail due to the variety of techniques and the availability of technology. Since it is unlikely that a best practice is absolute in nature, it requires consensus development through peer review to converge to recommended best practices. To accelerate the creation of a compendium of best practices while maintaining the peer review process, GEOSS has instituted a Web-based wiki (http://wiki.ieee-earth.org). While this requires some editorial oversight, the open forum nature of the wiki allows broad community participation.

Community-accepted best practices are one step in providing the nontechnical user with some assurance that the information available is reliable and of consistent "high quality." The challenge of quality assurance is normally addressed through peer review in the scientific community. As GEOSS supports a broad range of technical and nontechnical users, traditional methods may need to be adapted. Adding the concept of "certification" or assurance in a system of systems is challenging. Doing this in a voluntary organization where there can be significant cost implications for assurance implementations will require innovation. A GEOSS task in quality assurance led by CEOS and IEEE is formulating options for consideration by the GEOSS community.

22.4.1.6   Case Studies, Lessons to be Learned, and Human Resources   Information on SoS design and implementation for individual SBA applications, ranging from the extraction and definition of user requirements to the delivery of observation-based information, is very important in promoting and helping the SoS design and applications in associated fields. Such information could be considered as best practices, lessons learned, and case studies associated with information applications. There has been substantial development of case studies, those used for assessing and validating interoperability for cross-discipline developments (Khalsa, 2007a) and those which look in depth at a single discipline (AIP use cases; Percivall, 2007b).

22.4.2   Providing Incentives and Reducing Cost

Once the registries are established to visualize available system and technical resources for GEO members, the critical steps are populating the registries and encouraging their usage as well as the use of other resources of the SoS. More specifically, for data and system providers, the incentives and cost of contributing components are of importance. SBA users may be concerned about the cost of finding information, the appropriate best practices and contributing user requirement information. This includes also the issue of stability and sustainability of the data streams. The issue here is not only the monetary expenses, which should be minimal, but also the personnel time to use the system. To reduce the cost of registration, metadata items for registration should be minimized. Present GEO component, service, and standards registries require only a minimum set of metadata, basically with a free text description, which enables users to find relevant information.

To contribute to reducing the cost of building an SoS, more detailed and structured description of components should be encouraged. Also standards used in the contributed components should be preferentially nonproprietary and open ones, such as ISO and IEEE standards. Only from technical viewpoints of reducing system integration cost will GEO request the generation of new standards and request GEO members to these where appropriate. When GEO identifies gaps between the available standards and the demand for better interoperability, it will encourage existing standardization organizations to work with GEO and members/participating organizations. When contributed components useful to GEOSS are not consistent with internationally accepted interface standards, GEO may consult with the component providers to alleviate the interoperability issues. The Standards and Interoperability Forum (SIF), a group of voluntary experts under the supervision of ADC, was formed to help providers register standards/interoperability arrangements and to recommend the creation of new standards to standardization bodies. The terms of reference provide more details of the SIF charter and operations (Khalsa, 2007b).

In addition, ADC developed the Strategic and Tactical Guidance documents for GEO members (GEO/ADC, 2007a,b). The Strategic Guidance document presents the benefits of contributing components to GEOSS, while Tactical Guidance document provides instruction on how to register and use components.

To the further promotion of component registration and uses for SoS building, incentives have to be visualized, in addition to the cost reduction. Counting the number of contributed components by organizations and information on who downloaded which components could benefit component providers. Registration of practices in the Best Practice wiki can more directly lead to the benefits to data/system providers and SoS builders as well, but improving and facilitating interoperability early in the data and information creation. The incentives of contributors are basically determined by how visible GEO activities are from the world audience. For example, 91 organizations offered participation and supported development in the Architecture Implementation Pilot to present demonstrations of GIS-based data integration during 2006 and 2007 at the Ministerial Summit at Cape Town in November 2007.

Contribution of GEOSS to society, however, has to be evaluated by how GEOSS can contribute to better decisions of SBA users. Outcome indicators of GEOSS need to be developed. This aspect will be touched upon in the next section.

22.4.3 Promoting Coordination and Evolution of Tasks in the Work Plan

In the work plan of GEO, tasks are defined to achieve and promote the development and application of GEOSS. These tasks were originally proposed by GEO members and driven by voluntary participation. While initial operating capabilities such as registries of components and standards are established to visualize and distribute the available resources, supporting task activities in pursuing the targets and encouraging GEO members to propose new necessary tasks in a timely manner are crucially important for the success of building GEOSS. In addition, to ensure that task teams can help each other, and to maximize the synergy effects of task achievements, encouraging and coordinating better collaboration among tasks is also needed.

For the task coordination, GEO needs to share an "overview map" showing

1. how well the goals of GEOSS for each SBA are achieved by GEO activities, and how well individual tasks contribute to the achievement of overall goals;
2. where tasks are positioned and how they are associated with each other in the flow of Earth environmental data from observation to SBA applications, and how contributed components and standards are used by the tasks; and
3. which areas need to be covered by new tasks and which tasks need to be strengthened to fill the gap between present capabilities and user requirements?

Visualizing how each task is linked with the achievement of the overall goals of GEOSS is quite effective, as shown by the application of architecture model to Earth observation systems of NASA (Martin, 2007 (Fig. 22.10)).

GEO presently shares only the work plan (GEO, 2007b) and the progress of tasks is monitored by the individual committees. No maps are yet generated to show how individual tasks are associated with each other to achieve the overall goals. No performance indicators are yet developed, though draft ideas were presented at the Plenary at Cape Town in November 2007 (GEO, 2007a (Fig. 22.11)).

As with all system of systems, metrics are an essential part of the construct. Metrics need to be applied at multiple levels from overall SoS performance in each of the societal benefit areas to metrics for more detail activities such as registry effectiveness.

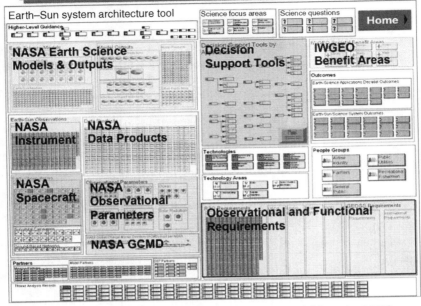

FIGURE 22.10 NASA Enterprise Architecture Model (Martin, 2007)

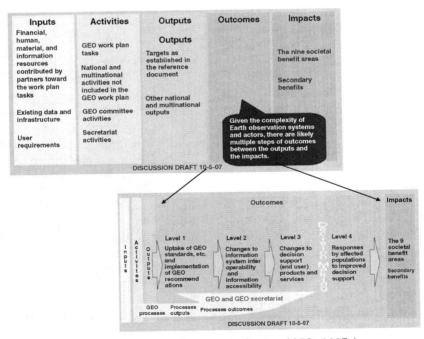

FIGURE 22.11 Performance indicators (GEO, 2007a)

GEOSS metrics are under development from both a technical performance viewpoint and an impact (cost–benefit analysis) perspective. An example of potential metric applications, Fig. 22.11 applies five-stage indicators from "input" to "impact" to GEOSS building activities. Though no concrete indicators are yet defined, with the indicators, we can evaluate the achievements of individual tasks for stages of "input" to "outcome" and how well the overall goals for each SBA are achieved by GEO activities. By sharing the map among GEO members, each task team and the associated GEO members can initiate new activities to strengthen collaborative links with the other tasks and can propose new tasks from the strategic viewpoints of filling gaps and focusing efforts.

22.5 CONCLUDING REMARKS AND FUTURE PROSPECTS

GEOSS is a significant international endeavor to strengthen links from Earth observation to SBA applications through the integration of individual observation systems with the use of advanced information technologies. From the viewpoint of SoSE, building GEOSS is also an important challenge of how to enable heterogeneous systems developed by different organizations in different contexts to behave like a unified system of systems in order to realize such large societal benefits that could have never been provided by individual systems operating by themselves. The voluntary nature of GEOSS and its breadth as a global collaboration make this SoS development unique. GEO has finished building an initial operating capacity that includes the basic structural components such as registries, portal interfaces, and data discovery capabilities. Using the success of the "Bazaar" model approach, GEOSS is establishing the building blocks in the registries that will accelerate spontaneous activities toward the more complete and comprehensive development of the SoS. In the next steps during 2008, GEOSS will broaden the use of these capabilities through individual task achievements focusing on applications that provide substantial benefits to society.

REFERENCES

Butterfield et al., 2007, Systems engineering for a global system-of-systems, *IEEE GEOSS Workshop on Implementing a System of Systems,* April 15, 2007, Hawaii.

GEO, 2005, *10-Year Implementation Plan Reference Document.* Available at http://www.earthobservations.org/about_geo.shtml.

GEO, 2007a, *GEOSS Outcome Performance Indicators*, Document 26, submitted to GEO-IV for information.

GEO 2007b, *GEO 2007–2009 Work Plan Toward Convergence*, as accepted as a Living Document at GEO-III. Available at http://www.earthobservations.org/docs/GEO-III/Plenarydocs/11-2007-2009_Work_Plan.v3.pdf.

GEO (Group on Earth Observation) home page. Available at http://www.earthobservations.org/index.html, January 2008.

GEO/ADC, 2007a, Tactical Guidance for Current and Potential Contributors to GEOSS, Document 24, submitted to GEO-IV for information.

GEO/ADC, 2007b Strategic Guidance for Current and Potential Contributors to GEOSS, Document 25, submitted to GEO-IV for information.

Khalsa, S.R.J., 2007a, Report on the Interoperability Process Pilot Project, in: Pearlman, F. (ed.), *Proceedings of the User and the GEOSS Architecture Workshop, GEOSS Interoperability and Applications to Biodiversity,* July 2007, Barcelona. Available at http://www.ieee-earth.org.

Khalsa, S.R.J., 2007b, Terms of reference of the Standards and Interoperability Forum. Available at http://earthobserbvations.org.

Martin, J.N., 2007, Value assessment of GEOSS using an architectural model, *IEEE GEOSS Workshop on Implementing a System of Systems,* April 15, 2007. Available at http://www.ieee-earth.org/.

Pearlman, J., 2007, *Report of the GEO Architecture and Data Committee (ADC) for the GEO Plenary,* November 28–29, 2007 Cape Town. Available at http://www.ieee-earth.org.

Pearlman, J., Butterfield, M., 2006, Creation of a system of systems on a global scale: the evolution of GEOSS, keynote address, *IEEE International Conference on Systems of System Engineering,* April 24 2006, Los Angeles.

Percivall, G., 2007a, GEO Task AR-07-02 Architecture Implementation Pilot, *GEO Architecture and Data Committee Meeting,* September 12–13, 2007, Washington, D.C.

Percivall, G., 2007b, *AIP Call for Participation,* Annexur B, p. 29, April 13, 2007.

Percivall, G. 2007c, *Architecture Implementation Pilot Report,* Section 5.3, November 20, 2008.

Raymond, E.S., 2006, The Cathedral and the Bazaar. Available at http://www.catb.org/~esr/writings/cathedral-bazaar/.

Author Index

Ackoff, R.L., 445, 446, 459, 523
Aerts, D., 175, 181
Agusdinata, D.B., 264
Alberts, D.S., 151, 537
Ammar, M.H., 499
Anderson, D.G., 199
Anderson, P.M., 396
Arnold, S., 48
Ashby, R., 203
Azani, C., xiii, 21

Baik, H., 536
Barabási, A., 523, 537
Bar-Yam, Y., 200
Beer, S., 180, 184, 191
Ben-Jacob, E., 23
Berg, D., 293, 295, 297
Beroggi, B.E.G., 305
Bhasin, K., xiii, 10, 359, 361, 363
Biemer, S.M., 46, 48, 446
Boardman, J., xiii, 6, 191, 193, 200, 205, 207, 213–214
Bonabeau, E., 91
Boppana, R.V., xiii, 542

Brown, S.L., 205
Bryan, N., 443

Cahn, M.F., 297
Callaway, R.K., 521, 526
Cameron, K.S., 194
Cannon, W.B., 27
Carlock, P.G., 200, 543
Carson, E., 176
Cazangi, R., 95
Chalasani, S., xiii, 14, 542
Chapman, E., 499
Checkland, P., 174, 175
Chen, P., 79, 83
Chen, W., 528
Clothier, J., 79
Cloutier, R. J., xiii, 6, 150
Correa, Y., 91
Crossley, W.A., 527, 534
Cuppan, C.D., 44, 46, 59, 527, 542, 543

Dagli, C., xiii, 4, 77, 83, 86, 96
Dahmann, J.S., xiii, 7, 17, 103, 219
Davenport, T.H., 310

System of Systems Engineering: Innovations for the 21st Century, Edited by Mo Jamshidi
Copyright © 2009 John Wiley & Sons, Inc., Publication

Subject Index